FUNDAMENTALS OF COLLEGE MATHEMATICS

Gary Clendenen
University of Texas, Tyler

Julie Kern
Palo Alto College

HarperCollins*CollegePublishers*

Sponsoring Editor: Karin E. Wagner
Developmental Editor: Maxine Effenson Chuck
Project Editor: Cathy Wacaser
Design Administrator: Jess Schaal
Text Design: Miller & Seper
Production Administrator: Randee Wire
Cover Design: Lesiak/Crampton Design Inc
Cover Illustration/Photo: 92 Lightscapes/The Stock Market
Compositor: CRWaldman Graphic Communications
Printer and Binder: R.R. Donnelley & Sons Company
Cover Printer: R.R. Donnelley & Sons Company

Fundamentals of College Mathematics

Library of Congress Cataloging-in-Publication Data

Clendenen, Gary.
Fundamentals of college mathematics / Gary Clendenen, Julie Kern.
—Student ed.
p. cm.
Includes index.
ISBN 0-06-501068-X (Student Edition)—ISBN 0–06–502179–7
(Annotated Instructor's Edition)
1. Mathematics. I. Kern, Julie. II. Title.
QA39.2.C585 1994
513′.14—dc20 93-30178
CIP

3 4 5 6 7 8 9 10-DOW-00999897

CONTENTS

PREFACE

Fundamentals of College Mathematics is designed for a one-semester or one-quarter course in arithmetic. There is no prerequisite other than some familiarity with numbers. This text was designed to help students become better problem solvers and to help them integrate mathematics into their daily experiences.

Approach

Mathematics is a tool that everyone can use. Far too often, students come away from a math class never having seen how math threads its way into the very fabric of modern life. We have taken great pains to blend reading, writing, mathematics, and critical thinking. The numerous word problems tend to be a bit longer and somewhat more realistic than those used in other texts. Applications are taken from consumer math, business, geography, history, medicine, and biology among other areas.

Every chapter ends with a longer application word problem that guides a student towards solutions. Journal questions at the end of every section require students to think and to write. The group application word problems in Appendix A force students to work in groups and compile their results in a written report. If time permits, it would be a natural extension to ask students to orally present their results.

End-of-Chapter Elements

Each chapter ends with four special features to help students synthesize the material.

Chapter Reviews at the end of each chapter allow students to find the definition of major terms quickly and to see the application of a particular concept.

Application Word Problems are longer applications that guide the student along a path of logic needed to solve more complicated problems. Each chapter ends with one such multipart application. The topics of these applications vary from real life, to history, to other more advanced concepts.

Chapter Review Problems include problems for students to use in reviewing the major concepts of the chapter. For a sample of these review problems, see pages 71–72.

Chapter Tests provide a good study tool for the student upon completion of the material. A sample chapter test appears on pages 75–76.

Exercises

At the end of each section we include exercises so students have plenty of practice and can test their understanding of what they have just learned. Sometimes a few problems in a set will require skills that were presented earlier in the test. The student must then think carefully to determine which skill should be used; these exercises also help students to review previously learned skills and to integrate these skills. The text contains more than 5,500 graded exercises. To see a sample exercise section, turn to page 47.

Applications are incorporated into most of the exercise sets to help students further develop their reading and thinking skills. See page 54 for an example of applications in an exercise set.

Journal Exercises appear at the end of each exercise section. Nearly 250 in all, the exercises require that students think about the concepts learned in the section and put their thoughts in writing. The student should keep a separate mathematics journal and answer these Journal Exercises in it. The first Journal Exercise follows this preface.

Cumulative Tests after every other chapter include exercises that cover major topics from two or three related chapters. For a look at a sample cumulative test, turn to page 135.

Final Examination This test at the end of the text allows students to see if they have mastered the concepts in the text before taking any final exam for their course grade.

Group Application Word Problems Located in Appendix A, these relatively unstructured problems can be assigned to a group of students. They require a group to work together and write a report about their findings.

Features of the Text

Our text offers several features that make it distinctive from other texts on the market. These features include:

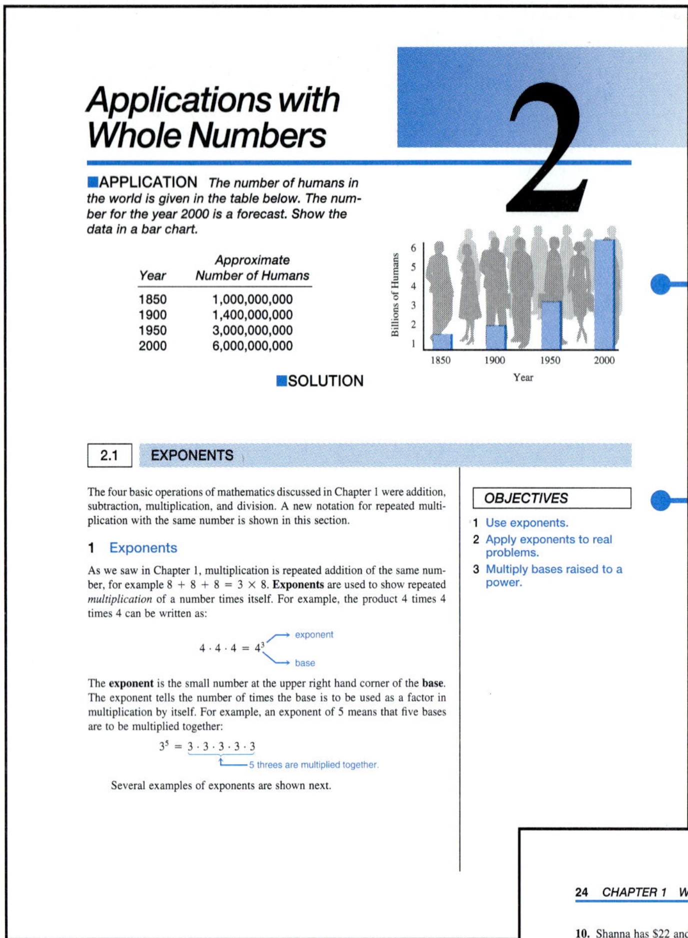

Applications with Whole Numbers

2

■APPLICATION *The number of humans in the world is given in the table below. The number for the year 2000 is a forecast. Show the data in a bar chart.*

Year	*Approximate Number of Humans*
1850	1,000,000,000
1900	1,400,000,000
1950	3,000,000,000
2000	6,000,000,000

■SOLUTION

2.1 EXPONENTS

The four basic operations of mathematics discussed in Chapter 1 were addition, subtraction, multiplication, and division. A new notation for repeated multiplication with the same number is shown in this section.

1 Exponents

As we saw in Chapter 1, multiplication is repeated addition of the same number, for example $8 + 8 + 8 = 3 \times 8$. **Exponents** are used to show repeated *multiplication* of a number times itself. For example, the product 4 times 4 times 4 can be written as:

$$4 \cdot 4 \cdot 4 = 4^3$$

(exponent → 3; base → 4)

The **exponent** is the small number at the upper right hand corner of the **base**. The exponent tells the number of times the base is to be used as a factor in multiplication by itself. For example, an exponent of 5 means that five bases are to be multiplied together:

$$3^5 = 3 \cdot 3 \cdot 3 \cdot 3 \cdot 3$$

5 threes are multiplied together.

Several examples of exponents are shown next.

OBJECTIVES

1 Use exponents.
2 Apply exponents to real problems.
3 Multiply bases raised to a power.

Chapter-opening Pages

Each chapter begins with a real-life application of a concept that is discussed in the chapter.

Learning Objectives

Each section is organized around objectives that are listed at the beginning of each section. The objectives are used as subheadings within the section.

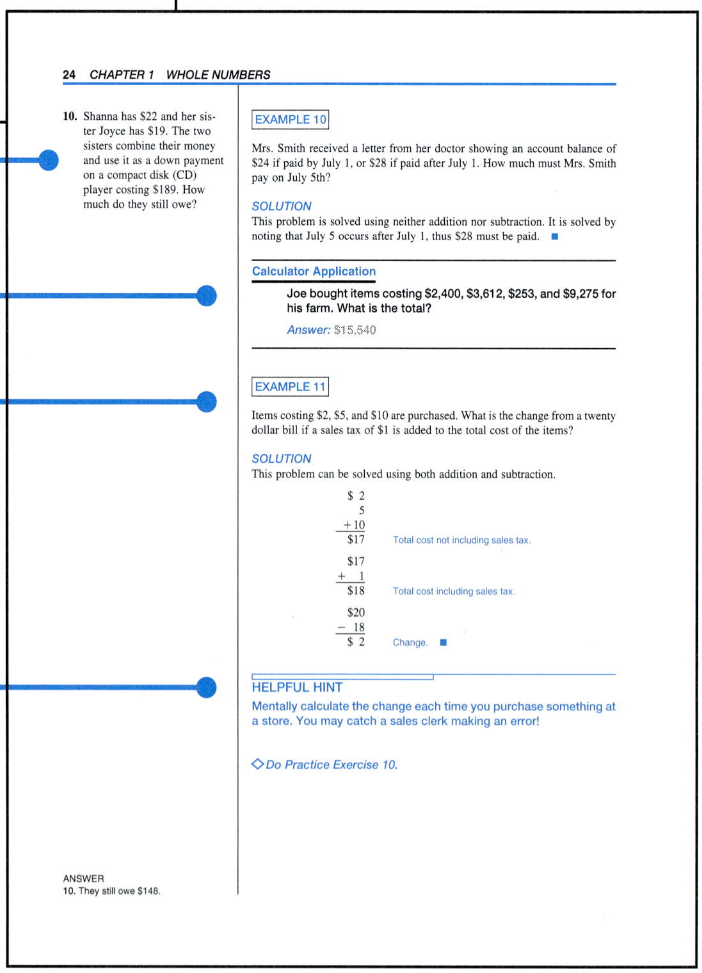

24 *CHAPTER 1 WHOLE NUMBERS*

10. Shanna has $22 and her sister Joyce has $19. The two sisters combine their money and use it as a down payment on a compact disk (CD) player costing $189. How much do they still owe?

EXAMPLE 10

Mrs. Smith received a letter from her doctor showing an account balance of $24 if paid by July 1, or $28 if paid after July 1. How much must Mrs. Smith pay on July 5th?

SOLUTION

This problem is solved using neither addition nor subtraction. It is solved by noting that July 5 occurs after July 1, thus $28 must be paid. ■

Calculator Application

Joe bought items costing $2,400, $3,612, $253, and $9,275 for his farm. What is the total?

Answer: $15,540

EXAMPLE 11

Items costing $2, $5, and $10 are purchased. What is the change from a twenty dollar bill if a sales tax of $1 is added to the total cost of the items?

SOLUTION

This problem can be solved using both addition and subtraction.

$$\begin{array}{r} \$\ 2 \\ 5 \\ +10 \\ \hline \$17 \end{array} \quad \text{Total cost not including sales tax.}$$

$$\begin{array}{r} \$17 \\ +\ \ 1 \\ \hline \$18 \end{array} \quad \text{Total cost including sales tax.}$$

$$\begin{array}{r} \$20 \\ -\ 18 \\ \hline \$\ 2 \end{array} \quad \text{Change. ■}$$

HELPFUL HINT

Mentally calculate the change each time you purchase something at a store. You may catch a sales clerk making an error!

◇*Do Practice Exercise 10.*

ANSWER
10. They still owe $148.

Practice Exercise with Answers

occur frequently within each section so students can put what they have just learned into practice.

Calculator Applications

show the use of calculators for solving more complex problems.

Examples

are numerous in each section and are detailed to show a justification for each step.

Helpful Hints

are offered to students in boxes for emphasis.

Application Word Problems

are longer applications that guide the student along a path of logic needed to solve more complicated problems. Each chapter ends with one such multipart application. The topics of these applications vary from real life, to history, to other more advanced concepts.

APPLICATION WORD PROBLEM

NAPOLEON BONAPARTE'S INVASION OF RUSSIA

Napoleon Bonaparte was born on August 15, 1769. His life was one of many battles. In 1805, he fought the Austrians at Austerlitz. In 1807, he fought the Russians at Friedland. In 1809, he fought the Austrians again, but this time at Wagram. One of Napoleon's greatest tragedies was his 1812 invasion of Russia. He began the invasion in February of 1812 with a "grand army" of 422,000 men. His army faced many battles, forced marches, and starvation.

1. How old was Napoleon when he invaded Russia?

Near Moscow in a city called Borodino, the Russians fought and lost a major battle to the French. One of his soldiers describes the battlefield:

> Here (center of the battlefield) the corpses lay piled higher ever higher. . . . The ditches were filled to the brim with bodies. . . . Men and horses had been gashed and maimed in every conceivable way. . . .

Source: *With Napoleon in Russia 1812* by Henrich Vossler. Published by Folio Society of London, 1969.

The Russians let Napoleon and his remaining army of 100,000 enter Moscow without resistance. Napoleon waited in Moscow for their surrender, but they did not. His army began the retreat back to France. The retreat was marked by starvation, more battles, and the horrible cold of the Russian winter. Only 10,000 soldiers of the grand army survived to return to France.

2. Complete the following graph using the above information.

The number of soldiers alive versus location of the invading army is shown in the graph below. The number of soldiers alive is shown by the width of the band. The darker band shows the retreat of the grand army. This graph was taken from one drawn in 1869 by French engineer, Charles Joseph Mitard.

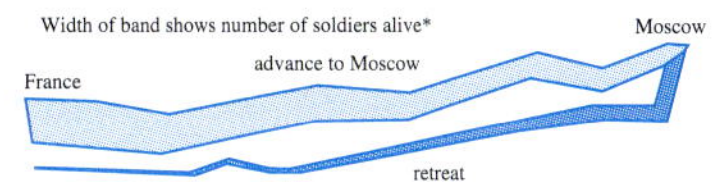

* Army began from France with 422,000 soldiers. 100,000 soldiers remained at Moscow. Only 10,000 soldiers returned alive to France.

127

CHAPTER 2 REVIEW

SECTION	TOPIC OR TERM	EXAMPLE OR DEFINITION
2.1 Exponents	Exponent	An exponent is the number of times a value is to be multiplied times itself: $4^2 = 4 \cdot 4$.
	Squared	Squared means that a value is to be raised to the second power: three squared $= 3^2$.
	Cubed	Cubed means that a value is to be raised to the third power: one cubed $= 1^3$.
2.2 Order of Operations	Order of Operations	The order of operations are the rules for the order in which math operations in an expression are calculated, for example: $3 \times 5 - 1$; $15 - 1$ Multiply first.; 14 Result.
2.3 Tests for Divisibility	Divisible	One number is divisible by a second if the second number divides into the first without remainder.
2.4 Prime Numbers and Prime Factorization	Prime Number	A prime number is a number whose only factors are itself and 1, such as 17.
	Composite Number	A composite number is a number with factors other than itself and 1, such as 10 whose factors include 2 and 5.
	Prime Factorization	Prime factorization is the unique set of primes that when multiplied together produce the original number. The prime factorization of 55 is 5 times 11.
2.5 Least Common Multiples	Least Common Multiple	The LCM of two numbers is the smallest number that both divide. The LCM of 8 and 12 is: $8 = 2 \cdot 2 \cdot 2$; $12 = 2 \cdot 2 \cdot 3$; LCM $= 2 \cdot 2 \cdot 2 \cdot 3 = 24$
	Greatest Common Divisor	The GCD of two numbers is the largest number that is a factor of both. The GCD of 8 and 20 is: Divisors of 8 are: 1, 2, 4, and 8. Divisors of 20 are: 1, 2, 4, 5, 10, and 20. The GCD is 4.

125

Chapter Reviews

at the end of each chapter allow students to find the definition of major terms quickly and to see the application of a particular concept.

Content Highlights

Chapter 1 covers critical thinking, including inductive and deductive logic (a TASP requirement), and introduces variables from an intuitive perspective at the end of the chapter. Tables, charts, and graphs are introduced in Chapter 2 since they are important concepts in our data-rich society. Chapter 3 presents fractions, and Chapter 4 presents mixed numbers, ratios, and proportions.

Decimal notation is described in Chapter 5. Chapters 2, 3, 4, 5, and 9 include sections on the order of operations. Chapter 6 introduces percents and devotes three sections to real-life applications. Chapter 7 covers the U.S. and metric measurement systems and conversion between the two systems.

Chapter 8 covers basic concepts of geometry, including surface areas and calculating the perimeter and/or circumference, area, and volume of common geometric shapes. Signed numbers are covered in Chapter 9, with one section devoted exclusively to an intuitive understanding of positive and negative numbers. Chapter 10 introduces algebra, covering variables, terms, and expressions, as well as solving first-degree equations and related word problems. Basic concepts of statistics are introduced in Chapter 11.

Supplement Package to Accompany the Text

For the Instructor

The **Annotated Instructor's Edition** includes answers to all of the problems in the text in place on the page. The only exception is that answers to Journal Exercises are not given, since they will vary.

The **Instructor's Manual: Solutions and Tests** contains worked-out solutions for all of the even-numbered exercises in the text excluding the Journal Exercises. It also has print versions of tests. All answers are supplied in a convenient format.

The **HarperCollins Test Generator/Editor for Mathematics with QuizMaster** is available in IBM and Macintosh versions and is fully networkable. The Test Generator enables instructors to select questions by objective, section, or chapter or to use a ready-made test for each chapter. The editor enables instructors to edit any preexisting data or to create their own questions. The software is algorithm-driven, allowing the instructor to regenerate constants while maintaining problem type, for a nearly unlimited number of available test or quiz items in multiple-choice or open-response formats. The system features printed graphics and accurate mathematics symbols. **QuizMaster** enables instructors to create tests and quizzes using the Test Generator/Editor and save them to disk so that students can take the test or quiz on a stand-alone computer or network. QuizMaster then grades the test or quiz and allows the instructor to create reports on individual students or classes.

A **videotape series** is available to accompany *Fundamentals of College Mathematics.*

For the Student

A **Student's Solution Manual** includes complete, worked-out solutions to the odd-numbered problems in the text except for Journal Exercises. To order, use ISBN 0-06-501964-4.

An **Interactive Tutorial Software with Management System** is available in IBM and Macintosh versions and is fully networkable. As with the Test Generator/Editor, this software is algorithm-driven and automatically regenerates constants so that a student will not see the numbers repeat in a problem type if he or she revisits any particular section. The tutorial is self-paced and provides unlimited opportunities to review lessons and to practice problem solving. When students give a wrong answer, they can request to see the problem worked out. The program is menu-driven for easy use, and on-screen help can be obtained at any time with a single keystroke. Students' scores are automatically recorded and can be printed for a permanent record. The optional Management System lets instructors record student scores on disk and print diagnostic reports for individual students or classes. For student home purchase use ISBN 0-06-502188-6 for IBM and use ISBN 0-06-502189-4 for Macintosh.

ACKNOWLEDGMENTS

We are deeply indebted to the following reviewers for their many valuable comments and suggestions:

Sharon Abramson, Nassau Community College

Emily Bell, Indiana Vocational Tech—Northwest

Dora Cantu, Laredo Junior College

Robert Carlson, Bunker Hill Community College

James Condor, Manatee Community College

Ann Corbeil, Massasoit Community College

Lucy Dechéne, Fitchburg State College

Beth Grovenstein, Weatherford College

Constance C. Holden, University of Maine—Bangor

Barbara Jordan, Barton County Community College

Patricia McCann, Franklin University

James Marcotte, Cincinnati Technical College

Frank Post, South Seattle Community College

Samuel Richardson, Augusta College

Deborah Ritchie, Moorpark College

John Rossi, Daytona Beach Community College

Jack W. Rotman, Lansing Community College

Arlene Sego, Cuyahoga Community College—West

Gordon Shimizu, Santa Rosa Junior College

Jon D. Weerts, Triton College

Special thanks go to our editors at HarperCollins who have worked with us patiently throughout the publishing process: Maxine Effenson Chuck, Karin E. Wagner, Ginny Guerrant, Anne Kelly, and Linda Youngman. The work would not have been completed without the committed efforts of these individuals.

For helping to insure the accuracy of the answers to the problems in the text, we owe our thanks to Phyllis Arnold of Palo Alto College and Susan Boyer of University of Maryland, Baltimore County. We also wish to thank Dennis Kern, Ling Chen, and Reading Chang for typing the solutions manuals.

We especially thank our families: Vicki and Jason as well as Dennis, Michael, Marcie, and Andy.

We welcome feedback from both students and instructors. Please send your comments and suggestions to us or submit them to the publisher.

Gary Clendenen
Julie Kern

Journal Assignment #1

Dear Student,

This book is about math, but it is also about thinking in a logical step-by-step fashion. Both math and mathematical thinking are required for many jobs in society. In addition, writing is important since our ideas are communicated through writing. Even mathematicians spend a large amount of their time writing.

Your professor may ask you to keep a Journal for this class. A Journal is simply a collection of your writings on math-related topics. A Journal differs from homework assignments, which are made up of math problems. The purpose of a Journal question is to make you think and explain your thoughts in writing. A Journal can help you develop your own style of thinking.

Your first Journal assignment is to write about a prior experience with math. Perhaps you have had a math teacher you really enjoyed, or perhaps math is difficult for you. It may be that math tests are very stressful for you. Try to recall past experiences with math, both good and bad, and write a few paragraphs about them.

The professor is more interested in your ideas and thoughts about math than in your grammar. However, it is hard for anyone to read a messy paper or one with many grammatical errors. Try to make your Journal writings neat, and work at stating your ideas as clearly as possible.

Best of Luck,
Gary Clendenen
Julie Kern

HOW TO STUDY MATHEMATICS

We often overhear students say, ''You just can't study for a math test.'' Actually, there *are* study strategies you can learn to follow that will help you succeed in your math courses. Here are some suggestions:

1. Regular practice is extremely important. Math is not something you can memorize in one or two nights before a test. It is important to do all of the homework, even if it will not be graded. Otherwise, taking a test is like trying to play the piano at a concert without ever having practiced the pieces.

2. Try to meet with other people in your class. Study groups are an excellent way to do your homework. Don't hesitate to approach someone you see working on the assignment in the library or learning lab and suggest that you work together. By explaining the problems to each other, you both will understand them better.

3. Likewise, don't hesitate to go to the professor during office hours for additional explanations and examples.

4. Make use of tutoring centers and learning labs. Many colleges have them available at no charge to students.

5. At the end of each class, take a minute to write down one or two sentences summarizing the most important points of the lecture for the day.

6. At the end of each chapter, list the important facts on an index card. Summarize them in your own words, and list any steps required to solve a particular type of problem. Keep the index card handy and read it over when you have a few minutes.

It is important that you do not skip over any math concept you do not understand. Chances are, you will need to use that concept again. If you do not understand something:

a. Reread the section in the book and work the Examples and Practice Exercises,

b. ask a fellow student for help, or

c. ask the professor for help.

Finally, know that with practice you CAN do all of the problems in this book.

PRETEST

Work each problem.

1. $\begin{array}{r} 509 \\ \times\ 38 \\ \hline \end{array}$

2. $2015 \div 13$

3. $3{,}625 + 19{,}078$

4. $\begin{array}{r} 1{,}700{,}239 \\ -\quad 10{,}573 \\ \hline \end{array}$

5. $2\frac{4}{9} + 5\frac{1}{2}$

6. $3\frac{3}{7} \cdot 2\frac{1}{3}$

7. $\frac{11}{15} \div 1\frac{1}{3}$

8. $4\frac{2}{3} - 2\frac{1}{5}$

9. $3\frac{1}{4} - 1\frac{4}{5}$

10. $20.8 + 9.14 + 0.5$

11. $13.05 - 1.123$

12. 7.1×2.03

13. $1.48 \div 0.2$

14. $1\frac{1}{2} - 0.75$

15. $16 - 5\frac{1}{2}$

16. $12\frac{1}{5} - 0.2$

17. $(-8)(-7)$

18. $(+46) \div (-5)$

19. $(-7) + (-4)$

20. $12 - (-13)$

21. $-4 - 28$

22. 17^2

Evaluate each expression.

23. $27 - 9 \cdot 2$

24. $(3 - 1) \cdot 85$

25. $6^2 - 42$

26. $27 \div 3 + \frac{1}{2} \cdot \frac{1}{4}$

27. $\frac{1}{2} \div \frac{1}{3} + \frac{3}{8}$

28. $4\frac{1}{3} \cdot \left(3 - 2\frac{1}{4}\right)$

29. $(-6) + (-3)(+2)$

30. $-12 - 4 \cdot 8.5$

Work each problem.

31. Write a ratio for 13 boys to 12 girls.

32. Write a ratio for 4 car wrecks to 1,000 drivers.

33. Solve $\frac{1}{3} = \frac{x}{12}$ for x.

34. Solve $\frac{3}{y} = \frac{15}{49}$ for y.

35. Find the area and perimeter of the following figure.

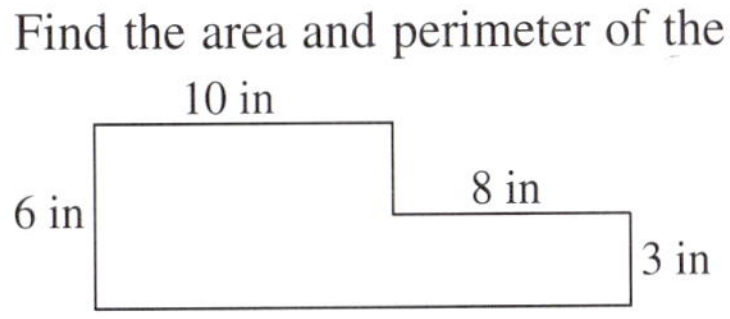

36. Find the area of the following figure.

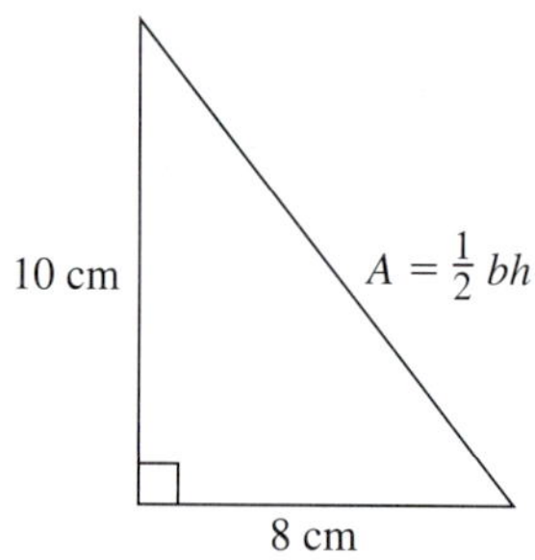

37. Find the measure of angle x in the figure.

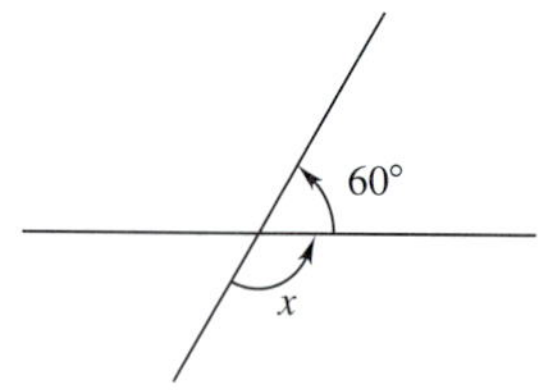

38. Find the least common multiple of 12 and 30.

39. Find the prime factorization of 220.

40. Write from smallest to largest: $2\frac{1}{4}$, 2.5, $2\frac{3}{10}$, 2.09.

41. Round 15.286 to the nearest tenth.

42. Write 45% using decimal notation and using fraction notation.

43. The monthly payment on a \$55,000 home is \$520. At the same rate, what would be the monthly payment on a \$77,000 home?

44. 65 is what percent of 250?

45. Gasoline costs were \$1.14, \$1.09, and \$1.04 per gallon at one month intervals. Find the average cost per gallon over the three readings.

46. If you answer 42 out of 70 questions correctly on an exam, what is your grade as a percent?

47. How many square feet of carpet are needed to cover the floor of a room measuring 15 feet by 16 feet? If carpet sells for \$13 per square *yard*, what would it cost to carpet the room? (*Hint:* 1 square yard = 9 square feet.)

48. A new automobile is priced at \$8,900. What would be the total cost including a sales tax of 8.5%?

49. Is $x + 9 > y - 12$ a true statement given that $x = 21$ and $y = -7$?

50. Solve $3x - 5 = 21$ for x.

Journal Assignment #1

Dear Student,

This book is about math, but it is also about thinking in a logical step-by-step fashion. Both math and mathematical thinking are required for many jobs in society. In addition, writing is important since our ideas are communicated through writing. Even mathematicians spend a large amount of their time writing.

Your professor may ask you to keep a Journal for this class. A Journal is simply a collection of your writings on math-related topics. A Journal differs from homework assignments, which are made up of math problems. The purpose of a Journal question is to make you think and explain your thoughts in writing. A Journal can help you develop your own style of thinking.

Your first Journal assignment is to write about a prior experience with math. Perhaps you have had a math teacher you really enjoyed, or perhaps math is difficult for you. It may be that math tests are very stressful for you. Try to recall past experiences with math, both good and bad, and write a few paragraphs about them.

The professor is more interested in your ideas and thoughts about math than in your grammar. However, it is hard for anyone to read a messy paper or one with many grammatical errors. Try to make your Journal writings neat, and work at stating your ideas as clearly as possible.

Best of Luck,
Gary Clendenen
Julie Kern

HOW TO STUDY MATHEMATICS

We often overhear students say, ''You just can't study for a math test.'' Actually, there *are* study strategies you can learn to follow that will help you succeed in your math courses. Here are some suggestions:

1. Regular practice is extremely important. Math is not something you can memorize in one or two nights before a test. It is important to do all of the homework, even if it will not be graded. Otherwise, taking a test is like trying to play the piano at a concert without ever having practiced the pieces.
2. Try to meet with other people in your class. Study groups are an excellent way to do your homework. Don't hesitate to approach someone you see working on the assignment in the library or learning lab and suggest that you work together. By explaining the problems to each other, you both will understand them better.
3. Likewise, don't hesitate to go to the professor during office hours for additional explanations and examples.
4. Make use of tutoring centers and learning labs. Many colleges have them available at no charge to students.
5. At the end of each class, take a minute to write down one or two sentences summarizing the most important points of the lecture for the day.
6. At the end of each chapter, list the important facts on an index card. Summarize them in your own words, and list any steps required to solve a particular type of problem. Keep the index card handy and read it over when you have a few minutes.

It is important that you do not skip over any math concept you do not understand. Chances are, you will need to use that concept again. If you do not understand something:

a. Reread the section in the book and work the Examples and Practice Exercises,

b. ask a fellow student for help, or

c. ask the professor for help.

Finally, know that with practice you CAN do all of the problems in this book.

Whole Numbers

1

APPLICATION *An adult's heart beats at the rate of 75 beats per minute. How many times will the heart beat in one year?*

SOLUTION *The answer is found using multiplication.*

$$\frac{75\text{ beats}}{1\text{ minute}} \times \frac{60\text{ minutes}}{1\text{ hour}} \times \frac{24\text{ hours}}{1\text{ day}} \times \frac{365\text{ days}}{\text{year}}$$

$$= 39{,}420{,}000 \text{ heart beats per year}$$

The adult's heart beats more than thirty-nine million times per year.

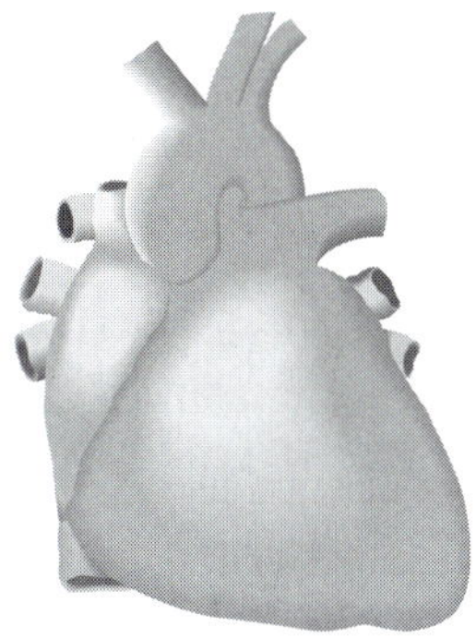

1.1 PLACE VALUES

The purpose of this chapter is to introduce numbers and basic operations with numbers. The Hindu-Arabic or decimal number system that we will be studying is used in North America and throughout much of the world. We will learn how to read and write whole numbers then to add, subtract, multiply, and divide them.

OBJECTIVES

1 Identify whole numbers.
2 Name place values.
3 Read and write large numbers.
4 Write a number using expanded notation.

1 Whole Numbers

Our number system is based on the **counting numbers**:

1, 2, 3, 4, 5, 6, 7, 8, 9, 10, 11, 12, 13, 14, 15, . . .

There is no largest counting number. No matter how large a number is, there is always another number that is larger. One million is a large number yet both two million and three million are larger.

The **whole numbers** are the counting numbers AND the number 0:

0, 1, 2, 3, 4, 5, 6, 7, 8, 9, 10, 11, 12, 13, 14, 15, . . .

The **even numbers** are those whole numbers that end in a 0, 2, 4, 6, or 8. The **odd numbers** are those whole numbers that end in a 1, 3, 5, 7, or 9.

◇ *Do Practice Exercise 1.*

1. Assume that you are able to place all of the whole numbers in a box. Assume that you reach into the box and pull out all of the counting numbers. What number(s) remain(s) in the box?

ANSWER
1. Only the number 0 remains in the box.

2 Place Values

Every **digit** in a number has a **place value**. The common place value places are shown in the table that follows.

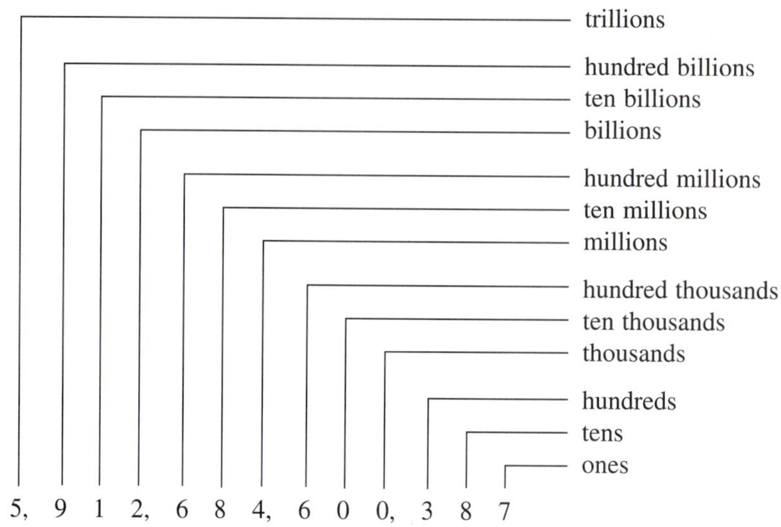

The 4 in the number 45 is in the tens place. This means that there are 4 tens in 45. The 4 is in the hundreds place in 412, meaning that there are 4 one hundreds in 412.

EXAMPLE 1

Name the place value of:

(a) the 4 in 4,160
(b) the 7 in 22,170
(c) the 0 in 102,331,114

SOLUTION

(a) thousands place
(b) tens place
(c) ten millions place ■

EXAMPLE 2

Find the place value of each digit in 421,035.

SOLUTION

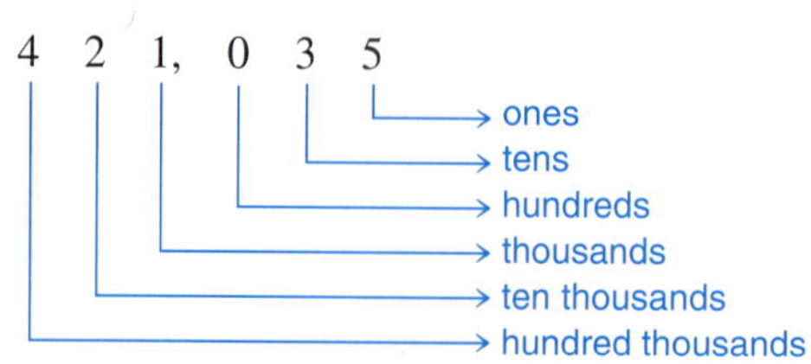

■

◇ *Do Practice Exercises 2 through 4.*

2. Find the place value of the 7 in 6,789.

3. Find the place value of the 8 in 98,000.

4. Find the place value of the 1 in 275,321.

ANSWERS
2. hundreds place
3. thousands place
4. ones place

3 Large Numbers

The first step when writing a large number is to place the commas correctly. Beginning on the right side of a whole number and moving to the left, place a comma after every three digits. For example, using 5912684600387:

Calculator Application

Some calculators place commas in the correct position for you and some do not. Check to see if yours does by keying in the largest number possible on your calculator.

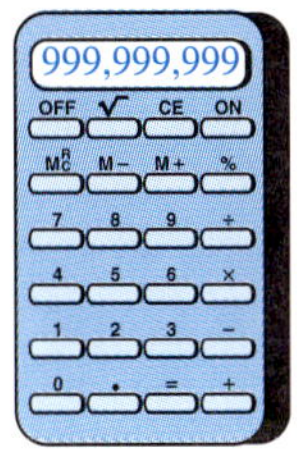

Each group of three digits within a whole number is called a **period**. The most common periods are trillions, billions, millions, thousands, and ones. The periods in the number 5,912,684,600,387 are:

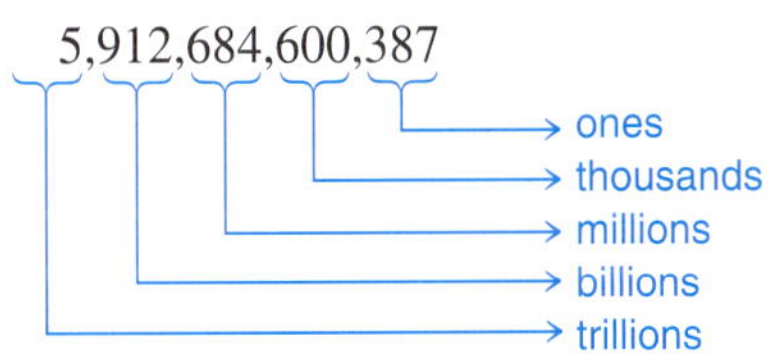

EXAMPLE 3

Name the periods in 750,021,647,000.

SOLUTION

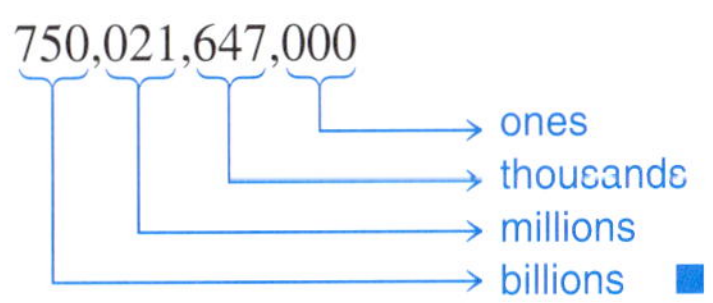

When reading the digits of a number, read the digits within each period followed by the name of that period. For example, the large number 5,912,684,600,387 is read as:

5, five trillion,
912, nine hundred twelve billion,
684, six hundred eighty-four million,
600, six hundred thousand,
387 three hundred eighty-seven

5. Write each number using words:

(a) 1,050

(b) 26,000

(c) 14,000,000

(d) 14,500,250

6. Write each number using words:

(a) 27,912

(b) 105,210,000

(c) 9,999,000

(d) 47,000,200

ANSWERS

5. (a) one thousand, fifty
(b) twenty-six thousand
(c) fourteen million
(d) fourteen million, five hundred thousand, two hundred fifty

6. (a) twenty-seven thousand, nine hundred twelve
(b) one hundred five million, two hundred ten thousand
(c) nine million, nine hundred ninety-nine thousand
(d) forty-seven million, two hundred

EXAMPLE 4

Write each number using words:

(a) 8,750
(b) 16,000
(c) 4,800,000

SOLUTION

(a) eight thousand seven hundred fifty
(b) sixteen thousand
(c) four million eight hundred thousand ■

HELPFUL HINT

Do not use the word "and" when reading or writing a whole number.

◇ *Do Practice Exercise 5.*

When reading a number, it is not necessary to mention any period that contains only zeros. The next example shows this idea.

EXAMPLE 5

Write 2,000,059,750 using words.

SOLUTION

The number has two billions, no millions, fifty-nine thousands, and seven hundred fifty ones. It is not necessary to say no millions when reading the number. It is read as two billion, fifty-nine thousand, seven hundred fifty. ■

EXAMPLE 6

Write each using words:

(a) 4,000,001
(b) 13,100,007
(c) 3,000,000,075,000

SOLUTION

(a) four million, one
(b) thirteen million, one hundred thousand, seven
(c) three trillion, seventy-five thousand ■

◇ *Do Practice Exercise 6.*

The digits of a number can be found from the written number. For example, the digits of seventy million, three hundred forty thousand, fourteen are:

seventy million ⟶ 070,
three hundred forty thousand ⟶ 340,
fourteen ⟶ 014

The digits are 70,340,014.

EXAMPLE 7

Find the digits of each:

(a) forty-five thousand, twenty
(b) twenty thousand
(c) six hundred thousand, four hundred eight

SOLUTION
(a) 45,020
(b) 20,000
(c) 600,408 ■

◇ *Do Practice Exercises 7 through 9.*

If a number has a digit in the millions period, it **must** have a period for thousands and for ones. If a number has a digit in the trillions period, it *must* have periods for billions, millions, thousands, and ones.

For example, look at the number seven trillion, six hundred thousand, seventy-five. Neither millions nor billions is mentioned. The fact of no billions and of no millions is shown by placing three zeros in both the billions and millions periods.

seven trillion ⟶ 7,
no billions ⟶ 000,
no millions ⟶ 000,
six hundred thousand ⟶ 600,
seventy-five ⟶ 075

The digits of the number are 7,000,000,600,075.

EXAMPLE 8

Write the digits for sixteen million, twenty-five.

SOLUTION

16, sixteen million
000, no thousands
025 twenty-five

The digits are 16,000,025. ■

7. Write the digits for seven thousand.

8. Write the digits for nine hundred one thousand, six hundred twenty-four.

9. Write the digits for forty-one million, seven hundred thirty thousand, four hundred twelve.

ANSWERS
7. 7,000
8. 901,624
9. 41,730,412

10. Write the digits for six hundred thousand, ten.

11. Write the digits for one million, one.

12. Write the digits for one trillion, three million.

13. Write 205,146 as an expanded sum.

14. Write four million, seventy-eight using expanded notation.

15. Write six billion, four thousand using expanded notation.

ANSWERS
10. 600,010
11. 1,000,001
12. 1,000,003,000,000
13. 200,000 + 5,000 + 100 + 40 + 6
14. 4,000,000 + 70 + 8
15. 6,000,000,000 + 4,000

EXAMPLE 9

Write the digits of six billion, eighty thousand.

SOLUTION

6, six billion
000, no millions
080, eighty thousand
000 no ones

The digits are 6,000,080,000. ■

◇ *Do Practice Exercises 10 through 12.*

4 Expanded Notation

Expanded notation means the separation of a number into the sum of its place value pieces. For example, the number 9,236 is broken apart into the sum of its place value parts:

9,000	9 thousands
200	2 hundreds
30	3 tens
+ 6	6 ones
9,236	

The expanded sum can also be written horizontally:

$$9{,}296 = 9{,}000 + 200 + 30 + 6$$

(9,000 + 200 + 30 + 6: expanded notation)

EXAMPLE 10

Write 200,840 as an expanded sum.

SOLUTION

200,000	2 hundred thousands
800	8 hundreds
+ 40	4 tens
200,840	

The expanded sum is 200,000 + 800 + 40. ■

EXAMPLE 11

Write eight billion, seventy thousand using expanded notation.

SOLUTION

8,000,070,000 = 8,000,000,000 + 70,000 ■

◇ *Do Practice Exercises 13 through 15.*

EXERCISE SET 1.1

Place commas correctly, then write each number using words.

1. 3840

2. 1205

3. 4900

4. 21900

5. 7250

6. 81000

7. 96050

8. 40008

9. 148200

10. 340000

11. 600221

12. 200200

13. 16240

14. 99000

15. 125200

16. 405030

17. 100000

18. 300009

19. 180000

20. 9050078

21. 4642000

22. 7000275

23. 3600000

24. 7035000

25. 10000000

26. 20008000

27. 6500080

28. 700090090

29. 800000050

30. 1400067033

Write the digits of each number and place commas correctly.

31. Six thousand, four hundred

32. Four thousand, ninety

33. Nine thousand, three hundred twenty-one

34. Fourteen thousand, nine hundred

35. Seventy-eight thousand

36. Two hundred ten thousand, forty

37. Seventeen thousand, three hundred

38. Forty-four thousand, nine hundred ninety-nine

39. Seventy thousand, one

40. Twelve thousand

41. Twelve thousand, twenty

42. Twenty thousand, six hundred

43. Six thousand, four

44. Fourteen thousand, ninety-five

45. Seventy thousand, four hundred fifty

46. One million, forty thousand

47. Eight million, two thousand, seven

48. Nine million, forty-nine

49. Nine million, one hundred thousand, twenty-four

50. Six billion, eight hundred seventy thousand

51. Twenty billion, nine million, five

52. Two trillion, six billion, forty million, seven thousand, four hundred sixty-five

53. Ninety-nine trillion

54. Nine hundred trillion, nine hundred

Name the place value position of the digit indicated.

55. The 4 in 470

56. The 7 in 6,721

57. The 8 in 5,080

58. The 9 in 8,900

59. The 2 in 21,465

60. The 8 in 180,222

61. The 4 in 4,055,012

62. The 5 in 1,450,120

63. The 7 in 8,070,000

64. The 4 in 24,000,000

65. The 6 in 60,070,897

66. The 7 in 50,250,000,070

Write in expanded notation.

67. 245

68. 706

69. 2,137

70. 97,327

71. 3,265,890

72. 41,200,080

73. 29,000,000

74. 6,290,300,001

Work each problem.

75. Write the largest whole number possible using five digits.

76. Light travels approximately one hundred eighty-six thousand miles in one second. Write the digits of this number.

77. The average distance from the earth to the moon is two hundred thirty-nine thousand miles. Write the digits of this number.

78. The last dinosaurs are thought to have died out 65,000,000 years ago. Write this number using words.

79. If all of the capillaries (tiny blood vessels) of a human were attached end to end, they would extend 62,000 miles. Write this number using words.

80. A company had total sales of $45,327,150 in one year. Write this number using words.

JOURNAL EXERCISES

1. What is the approximate population of the city in which you live? Approximately how many of those people do you know?

2. Explain the difference between the counting numbers and the whole numbers.

3. Explain place values. Why are they needed?

4. Is the number zero useful? Why?

5. Why are numbers necessary in our society? Name some things that would change if there were no numbers.

1.2 ROUNDING AND ORDER

OBJECTIVES

1. Round whole numbers.
2. Use a number line and order numbers based on size.

1 Rounding

A large number may have more digits of accuracy than needed. In this event, its value can be approximated using **rounding**. For example, the number 23,812,067 is closer to 24,000,000 than it is to 23,000,000. Therefore, 23,812,067 rounded to the nearest one million is 24,000,000. The following rule provides the steps to round a number.

Rule: To Round a Number to a Desired Place Value

Step 1 Find the desired place value.

Step 2 Look at the digit immediately to the right.

Step 3 If this digit is five or larger increase the digit in the desired place value by one. Otherwise do not change the digit in the desired place value.

Step 4 Replace all digits to the right of the desired place value with zeros.

In the next two examples a box is placed around the desired place value for emphasis.

EXAMPLE 1

Round 2,652 to the nearest one hundred.

SOLUTION

Step 1 Find the desired place value.

2,652 — Desired place value is one hundreds.

Step 2 Look at the number to the right.

2,652 — number to the right

Step 3 The number to the right is 5 so the digit in the box should be increased by 1.

Step 4 All digits following the one hundreds place value are replaced with zeros.

2,652 → 2,700
Increase the 6 to a 7 and change the following digits to zeros.

2,652 rounded to the nearest one hundred is 2,700 ■

Rounding 2,652 to the nearest one hundred as we did in the last example is the same as deciding whether 2,652 is closer to 2,600 or closer to 2,700.

2,652
2,600 ← → 2,700
Is 2,652 closer to 2,600 or to 2,700?

The number 2,652 is closer to 2,700 so that it rounds off to 2,700.

1. Round each to the nearest one hundred:

 (a) 387

 (b) 81,224

 (c) 621,429,714

2. Round 751,620 to the nearest thousand.

3. Round 38,494,000 to the nearest million.

4. Round seventy-two million, three hundred ninety-six thousand to the nearest ten thousand.

ANSWERS
1. (a) 400 (b) 81,200
 (c) 621,429,700
2. 752,000
3. 38,000,000
4. 72,400,000

EXAMPLE 2

Round 8,450 to the nearest one thousand.

SOLUTION

Digit to the right is a 4.

8, 4 5 0

desired place value

8,450 rounded to the nearest one thousand is 8,000

8,450 is closer to 8,000 than it is to 9,000 ■

Sometimes, you may be asked to round a number to the nearest hundred meaning one hundred, or to the nearest thousand meaning one thousand. Look at the next example.

EXAMPLE 3

Round each:

(a) 72,380 to the nearest hundred
(b) 17,999 to the nearest thousand
(c) 2,300,000 to the nearest million

SOLUTION

(a) 72,400
(b) 18,000
(c) 2,000,000 ■

◇ *Do Practice Exercise 1.*

Rounding a number results in a loss of accuracy. For example, the table below shows the number 745,116,020 rounded to different place values.

Rounded to Nearest	*Rounded Value*	
hundred	745,116,000	More accurate
ten thousand	745,120,000	
hundred thousand	745,100,000	
million	745,000,000	
hundred million	700,000,000	
billion	1,000,000,000	Less accurate

Each of the rounded values above is correct. The best one to use depends upon the amount of accuracy needed.

◇ *Do Practice Exercises 2 through 4.*

EXAMPLE 4

In 1990, the largest city in the world was Tokyo-Yokahama, Japan with an estimated population of 26,952,000. This number has probably already been rounded to the nearest one thousand. Now round the population to the nearest one million.

SOLUTION

27,000,000 people

27,000,000 is not as accurate as the original number of 26,952,000. However, it is accurate enough to compare the population of Tokyo with the estimated 15,000,000 people in New York City in 1990. ■

2 Number Line and Order of Numbers

A **number line** is a convenient way to show numbers. For example, the whole numbers 4 and 12 are shown as points on the following number line:

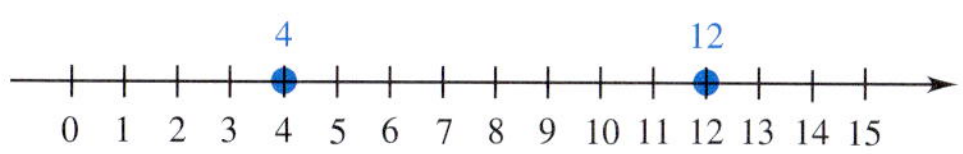

◇ *Do Practice Exercise 5.*

The numbers become smaller when moving to the left on a number line and they become larger when moving to the right:

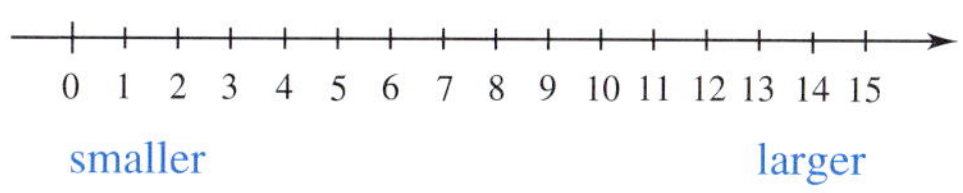

Order means that one number is either **less than** or **greater than** a different number. In mathematics, the symbol for less than is $<$ and the symbol for greater than is $>$.

$10 < 15$ 10 is LESS than 15.

$75 > 1$ 75 is GREATER than 1.

EXAMPLE 5

Draw a number line and place the points 3 and 5 on it. Is $3 > 5$ a true statement?

SOLUTION

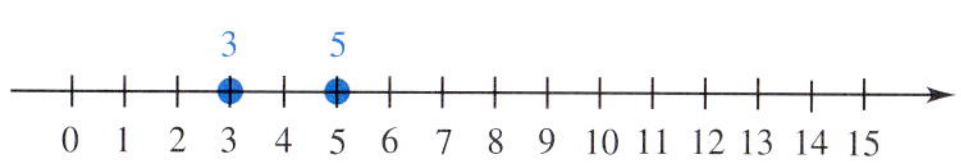

3 is *not greater* than 5 so that $3 > 5$ is false.

On the other hand, $3 < 5$ is a true statement. ■

The statement $17 > 17$ is a *false statement* since 17 is *not greater* than 17. The first number must be greater than the second to use the $>$ symbol. Also, the first number must be less than the second to use the $<$ symbol.

◇ *Do Practice Exercise 6.*

5. Draw a number line and show 5 and 8 as points.

6. State whether each is true or false:

(a) $8 < 5$

(b) $4 > 1$

(c) $9 < 15$

(d) $7 > 25$

(e) $9 < 8$

(f) $7 < 7$

ANSWERS

5.

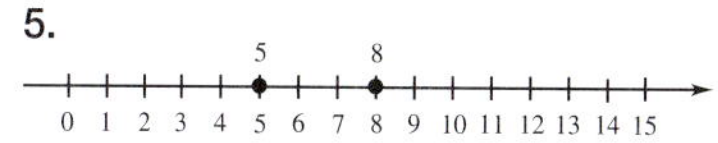

6. (a) false (b) true (c) true (d) false (e) false (f) false

name date hour

Place commas correctly, then write the number using words. (Review)

1. 5871

2. 4987

3. 25000

4. 45691

5. 32200001

6. 421008

Round each to the nearest one hundred.

7. 734

8. 899

9. 119

10. 2,275

11. 6,940

12. 95,290

13. 104,771

14. 95,100

15. 16,000

16. 250,080

17. 1,240

18. 3,400

19. 1,927

20. 6,740

21. 2,009

22. 25,800

23. 16,625

24. 75,201

25. 812,216

26. 1,914,632

Round each to the nearest ten thousand.

27. 67,200

28. 94,409

29. 107,000

30. 112,600

31. 389,007

32. 629,900

33. 400,005

34. 718,600

35. 365,400

36. 4,217,034

37. 6,299,100

38. 70,001,000

Round each to the indicated position.

		to the nearest
39.	369	one hundred
40.	3,491	one thousand
41.	92,251	ten thousand
42.	65,088	ten
43.	79,900	one hundred
44.	27,080	one hundred
45.	45,900	one thousand
46.	19,020	ten thousand
47.	276,119	ten thousand
48.	412,250	one thousand
49.	1,685,000	one hundred thousand
50.	2,004,999	one million
51.	35,971,000	one million
52.	1,906,000	ten thousand
53.	9,657,100	ten million
54.	19,274,805	one thousand

		to the nearest
55.	2,304,546,345	ten million
56.	8,675,876,345	one trillion

State which of the following are true *and which are* false.

57. $6 < 9$

58. $12 > 14$

59. $5 < 10$

60. $7 < 13$

61. $7 < 85$

62. $14 > 12$

63. $3 > 3$

64. $2 < 90$

65. $5 > 0$

66. $40 < 50$

Place either a $<$ or $>$ symbol between the two numbers as required to make the statement true.

67. 7 ? 5

68. 3 ? 13

69. 15 ? 2

70. 284 ? 35

71. 36 ? 37

72. 1 ? 0

Work each problem.

73. A new car costs sixteen thousand four hundred dollars. Round this number to the nearest thousand and to the nearest ten thousand.

74. In 1942, the United States placed 110,000 Americans of Japanese ancestry into detention (prison) camps. Round 110,000 to the nearest one hundred thousand.

75. In 1990, the number of people in prisons in the United States was 710,054. Round this number to the nearest one hundred thousand.

76. Round 389,800,100,000,000 to the nearest trillion.

77. At one firm, annual sales for 1992 were \$35,400,000 and annual sales for 1993 were \$41,500,000. Compare the two sales figures using a $<$ or $>$ symbol.

78. Alaska is the largest state in the United States with an area of 591,004 square miles. Rhode Island is the smallest state with an area of 1,212 square miles. Round the area of Alaska to the nearest hundred thousand square miles and the area of Rhode Island to the nearest one thousand square miles.

JOURNAL EXERCISES

1. A ranch is so large that the manager never knows exactly how many cattle she has. One week the manager counted 3,701 head of cattle and her ranch hands counted 3,758 head. Round both numbers to the nearest one hundred. Do they agree? If not, why not?

2. A bag contains 70 marbles. The number of red marbles in the bag, rounded to the nearest ten, is 50. The number of blue marbles in the bag, rounded to the nearest ten, is 20. How many red marbles are in the bag? List all answers that satisfy the conditions of the problem.

3. Why are large numbers necessary in our society?

4. Write a description of the process of rounding.

1 Addition

The process of finding the sum of two or more numbers is called **addition**. The symbol for addition is +. For example, two squares plus three squares is five squares:

2 squares + 3 squares = 5 squares

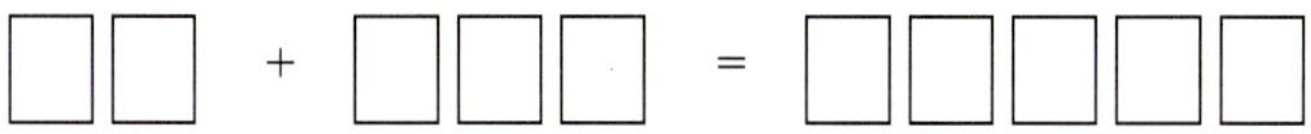

Addition problems can be written either horizontally or vertically. For example:

$$6 + 8 = 14 \quad \text{or} \quad \begin{array}{r} 6 \\ +\ 8 \\ \hline 14 \end{array} \quad \begin{array}{l} \text{addend or term} \\ \\ \text{sum or total} \end{array}$$

The numbers being added are called either **addends** or **terms** and the answer to the addition problem is the **sum** or **total**.

◇ *Do Practice Exercise 1.*

The addition of larger numbers is done by writing the numbers above each other with identical place values above one another. Addition begins by adding digits in the ones place values, then continues to the left across the number. If a column of digits adds up to ten or more, a number must be **carried** to the next column. Look at the next example.

EXAMPLE 1

Add 384 + 48.

SOLUTION

$$\begin{array}{r} 384 \\ +\ 48 \\ \hline \end{array}$$ Line up identical place values.

$$\begin{array}{r} {}^{1}\ \\ 384 \\ +\ 48 \\ \hline 2 \end{array}$$ The ones place values add to 12. Place 2 ones in the sum and carry 1 ten to the next column.

$$\begin{array}{r} {}^{1}\ \ \ \\ 384 \\ +\ 48 \\ \hline 32 \end{array}$$ The tens place values add to 13 tens. Place 3 tens in the sum and carry 1 hundred to the next column.

$$\begin{array}{r} 384 \\ +\ 48 \\ \hline 432 \end{array}$$ Result. ■

The same basic process applies when adding more than two numbers as shown in the next example.

OBJECTIVES

1 Add whole numbers.
2 Use properties of addition.
3 Subtract whole numbers.
4 Apply addition and subtraction in solving word problems.

1. Mentally add the following:

(a) 6 + 9

(b) 3 + 7

(c) 3 + 5 + 2

(d) 3 + 4

(e) 4 + 8

(f) 7 + 5 + 4

(g) 5 + 8

(h) 7 + 8

(i) 5 + 5 + 9

(j) 7 + 6

(k) 8 + 9

(l) 3 + 2 + 4 + 5

ANSWERS
1. (a) 15 (b) 10 (c) 10 (d) 7 (e) 12 (f) 16 (g) 13 (h) 15 (i) 19 (j) 13 (k) 17 (l) 14

2. Add 3,621 and 17.

3. Find the sum of seventy-four and fourteen thousand.

4. Add 2,321 plus 927 plus 18,227.

5. Which property is illustrated by each?

(a) $(3 + 41) + 1 = 3 + (41 + 1)$

(b) $71 + 0 = 71$

(c) $9 + 17 = 17 + 9$

ANSWERS
2. 3,638
3. 14,074
4. 21,475
5. (a) associative (b) identity (c) commutative

EXAMPLE 2

Find the sum of $3{,}221 + 35 + 168$.

SOLUTION

$$\begin{array}{r} \scriptstyle 1\,1 \\ 3{,}221 \\ 35 \\ +\ \ 168 \\ \hline 3{,}424 \end{array}$$ ■

◇ *Do Practice Exercises 2 through 4.*

2 Properties of Addition

There are three basic properties of mathematics that apply to addition. These properties are given next.

PROPERTIES OF ADDITION

1. **Identity property**—A number is not changed by adding 0 to it:

$$19 + 0 = 19$$

2. **Commutative property**—The order does not make any difference when adding two numbers:

$$8 + 9 = 9 + 8$$

3. **Associative property**—When adding three numbers it makes no difference which two are added first:

We can add these first or

$$(7 + 8) + 9 = 7 + (8 + 9)$$

add these first.

The commutative property of addition is the idea that terms can be added in any order without changing the answer to the addition problem. The associative property of addition is the idea that it makes no difference which two numbers are added first when adding three numbers; the answer will always be the same.

EXAMPLE 3

Which property is shown by each?

(a) $7 + 12 = 12 + 7$
(b) $18 + 0 = 18$
(c) $9 + (2 + 5) = (9 + 2) + 5$

SOLUTION

(a) Commutative property—The order in which they are added does not matter.
(b) Identity property—A number is not changed when added to 0.
(c) Associative property—It makes no difference which two are added first. ■

◇ *Do Practice Exercise 5.*

3 Subtraction

Subtraction is the process of taking something away. The symbol for subtraction is −. For example, four squares minus one square leaves three squares:

4 squares − 1 square = 3 squares

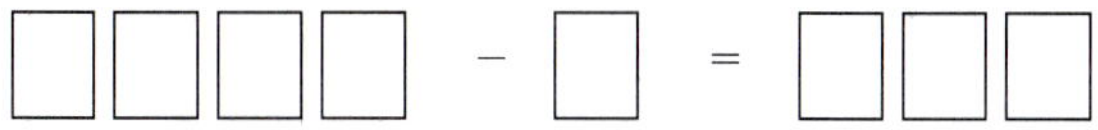

Subtraction can be written either horizontally or vertically, for example:

$12 - 7 = 5$ or

$$\begin{array}{r l} 12 & \text{minuend or term} \\ -\ 7 & \text{subtrahend or term} \\ \hline 5 & \text{difference} \end{array}$$

The first term is the **minuend** and the number being subtracted is the **subtrahend**. The result of subtraction is the **difference**.

◇ *Do Practice Exercise 6.*

The subtraction of larger numbers is done by writing the second number under the first with identical place values under one another. Subtraction then begins with the ones place values and proceeds to the left across the numbers. Look at the next example.

EXAMPLE 4

Subtract 658 − 15.

SOLUTION

$$\begin{array}{r} 658 \\ -\ 15 \\ \hline \end{array}$$ Line up place values.

$$\begin{array}{r} 658 \\ -\ 15 \\ \hline 3 \end{array}$$ The ones place values subtract to a 3. Place 3 in the difference.

$$\begin{array}{r} 658 \\ -\ 15 \\ \hline 43 \end{array}$$ The tens place values subtract to a 4. Place a 4 in the tens place of the difference.

$$\begin{array}{r} 658 \\ -\ 15 \\ \hline 643 \end{array}$$ The hundreds place values subtract to 6. Place a 6 in the hundreds place of the difference.

The result can be checked by adding 643 + 15 which equals 658. ■

◇ *Do Practice Exercise 7.*

Sometimes, we need to subtract a large digit from a small digit. In that case, it is necessary to **borrow** before subtracting. Borrowing is the idea of taking 1 from the next higher place value. Look at the next two examples.

6. Mentally subtract the following:

(a) 14 − 7

(b) 13 − 9

(c) 15 − 7

(d) 18 − 9

(e) 12 − 8

(f) 12 − 9

(g) 16 − 8

(h) 15 − 9

(i) 17 − 7

(j) 19 − 19

(k) 12 − 6

(l) 14 − 6

7. Subtract the following:

(a) 658 − 237

(b) 2,349 − 138

(c) eighty-seven thousand minus twelve thousand

ANSWERS
6. (a) 7 (b) 4 (c) 8 (d) 9 (e) 4 (f) 3 (g) 8 (h) 6 (i) 10 (j) 0 (k) 6 (l) 8
7. (a) 421 (b) 2,211 (c) 75,000

8. Subtract the following:

(a) 27 − 19

(b) 1,750 − 780

(c) 2,600 − 932

EXAMPLE 5

Subtract 230 − 14.

SOLUTION

$$\begin{array}{r} 230 \\ -\ 14 \\ \hline \end{array}$$

Line up place values.

$$\begin{array}{r} {\scriptstyle 2\,1} \\ 2\not{3}0 \\ -\ 14 \\ \hline 6 \end{array}$$

4 ones cannot be subtracted from 0 ones. It is necessary to borrow 1 ten from the 3 tens in 230. 10 ones minus 4 ones = 6.

$$\begin{array}{r} {\scriptstyle 2} \\ 2\not{3}0 \\ -\ 14 \\ \hline 16 \end{array}$$

Subtract the tens place values.

$$\begin{array}{r} 230 \\ -\ 14 \\ \hline 216 \end{array}$$

Subtract the one hundreds place value. ■

EXAMPLE 6

Subtract 2,000 − 49.

SOLUTION

9 ones cannot be subtracted from 0 ones. There is nothing in the tens or hundreds place to borrow. Therefore, borrow 1 from the thousands place as follows:

Borrow 1 thousand leaving 1 thousand and 10 hundreds.
Borrow 1 hundred leaving 9 hundreds and 10 tens.
Borrow 1 ten leaving 9 tens and 10 ones.

$$\begin{array}{r} {\scriptstyle 1\ 1} \\ \not{2},000 \\ -\ \ 49 \\ \hline \end{array} \qquad \begin{array}{r} {\scriptstyle 1\ 9\ 1} \\ \not{2},\not{0}00 \\ -\ \ 49 \\ \hline \end{array} \qquad \begin{array}{r} {\scriptstyle 1\ 9\ 9\ 1} \\ \not{2},\not{0}\not{0}0 \\ -\ \ 49 \\ \hline 1,951 \end{array}$$

Result. ■

◇ *Do Practice Exercise 8.*

4 Word Problems

A **word problem** is a problem stated with sentences. Certain words (or phrases) called **keywords** (or keyphrases) may help identify whether addition, subtraction, multiplication, or division is needed to work the problem.

ANSWERS
8. (a) 8 (b) 970 (c) 1,668

Some keywords for addition and subtraction are:

Addition	*Subtraction*
add	subtract
sum	minus
plus	take away
perimeter	difference
total	more than
more	less than
increased by	diminished by
gain of	subtract . . . from
deposit	withdrawal

EXAMPLE 7

Find the sum of eighty plus two hundred.

SOLUTION

The keyword *sum* tells us to add:

$$80 + 200 = 280 \quad \blacksquare$$

EXAMPLE 8

What is the difference between four hundred and twenty-five?

SOLUTION

The keyword *difference* tells us to subtract:

$$400 - 25 = 375 \quad \blacksquare$$

EXAMPLE 9

Subtract 12 from 49.

SOLUTION

The keyword *subtract* tells us it is a subtraction problem. The word ''from'' tells us to take 12 away from 49:

$$49 - 12 = 37 \quad \blacksquare$$

◇ *Do Practice Exercise 9.*

Each word problem is different and must be read carefully. The use of a keyword alone may not be enough to decide how to work the problem. Look at the next example.

9. Find the keyword or key phrase in each problem:

(a) Subtract sixteen from ninety.

(b) A savings account with \$400 is increased by \$150. Find the total amount in the savings account.

(c) Find the difference between ten million and one thousand.

ANSWERS

9. (a) subtract
(b) is increased by
(c) difference

10. Shanna has $22 and her sister Joyce has $19. The two sisters combine their money and use it as a down payment on a compact disk (CD) player costing $189. How much do they still owe?

EXAMPLE 10

Mrs. Smith received a letter from her doctor showing an account balance of $24 if paid by July 1, or $28 if paid after July 1. How much must Mrs. Smith pay on July 5th?

SOLUTION

This problem is solved using neither addition nor subtraction. It is solved by noting that July 5 occurs after July 1, thus $28 must be paid. ■

Calculator Application

Joe bought items costing $2,400, $3,612, $253, and $9,275 for his farm. What is the total?

Answer: $15,540

EXAMPLE 11

Items costing $2, $5, and $10 are purchased. What is the change from a twenty dollar bill if a sales tax of $1 is added to the total cost of the items?

SOLUTION

This problem can be solved using both addition and subtraction.

$$\begin{array}{r} \$\ 2 \\ 5 \\ +10 \\ \hline \$17 \end{array} \quad \text{Total cost not including sales tax.}$$

$$\begin{array}{r} \$17 \\ +\ \ 1 \\ \hline \$18 \end{array} \quad \text{Total cost including sales tax.}$$

$$\begin{array}{r} \$20 \\ -\ 18 \\ \hline \$\ 2 \end{array} \quad \text{Change.}$$ ■

HELPFUL HINT

Mentally calculate the change each time you purchase something at a store. You may catch a sales clerk making an error!

◇ *Do Practice Exercise 10.*

ANSWER

10. They still owe $148.

EXERCISE SET 1.3

Mentally work and write down the answer to each problem.

1. 6 + 12 **2.** 5 − 3 **3.** 12 − 8 **4.** 16 + 5

5. 17 − 9 **6.** 7 + 8 **7.** 20 − 15 **8.** 12 + 3

9. 14 − 7 **10.** 12 − 11 **11.** 22 − 22 **12.** 25 − 24

13. 10 − 7 **14.** 13 − 10 **15.** 8 + 12 **16.** 10 + 3 + 5

17. 15 + 2 + 3 **18.** 7 + 7 + 7 **19.** 6 + 5 + 8 **20.** 7 + 4 + 8

Work each, showing all steps.

21. 27 − 8 **22.** 204 − 19 **23.** 675 + 21

24. 47 − 41 **25.** 79 − 79 **26.** 71 + 35

27. 78 − 65 **28.** 456 + 21 **29.** 754 + 492

30. 612 − 441 **31.** 591 + 475 **32.** 6,914 − 24

33. 7,212 + 1,900 **34.** 4,501 − 4,501 **35.** 3,210 + 179

36. 4,221 + 907 **37.** 6,100 − 312 **38.** 3,700 − 499

39. 355 + 12 **40.** 1,204 + 199 **41.** 695 − 7

42. 1,244 + 75

43. 19,088 − 767

44. 723 − 489

45. 3,112 − 2,919

46. 14,421 + 1,760

47. 19,200 − 894

48. 6,000 − 41

49. 3,281 + 95

50. 2,104 + 18,001

51. 32,000 − 31,900

52. 6,240 + 78,218

53. 14,200 − 9,900

54. $\begin{array}{r} 16{,}845 \\ +\ 1{,}004 \\ \hline \end{array}$

55. $\begin{array}{r} 92{,}104 \\ -10{,}002 \\ \hline \end{array}$

56. $\begin{array}{r} 7{,}100 \\ -\quad 69 \\ \hline \end{array}$

57. $\begin{array}{r} 18{,}284 \\ -\ 1{,}300 \\ \hline \end{array}$

58. $\begin{array}{r} 19{,}400 \\ +\ 4{,}500 \\ \hline \end{array}$

59. $\begin{array}{r} 90{,}000 \\ -\quad 399 \\ \hline \end{array}$

60. $\begin{array}{r} 654{,}894 \\ 811{,}120 \\ +\ 61{,}297 \\ \hline \end{array}$

61. $\begin{array}{r} 275{,}612 \\ 857{,}253 \\ +\ 37{,}331 \\ \hline \end{array}$

62. $\begin{array}{r} 3{,}965{,}045 \\ 199{,}972 \\ +1{,}007{,}873 \\ \hline \end{array}$

63. 700 + 8,000 + 90,000

64. 7,345,772 − 452,669

65. 6,912 + 814

66. 2,000,000 − 1

Work each and write the answer using digits. (Review)

67. One thousand, eight minus four hundred, seventeen

68. Twenty thousand, one hundred plus two thousand, two

69. Four million minus ten thousand

70. Sixty-four thousand, nine hundred fifty minus fourteen thousand, six hundred seventy-four

71. Eighteen billion plus forty-four million

72. Seventeen trillion minus one

Determine whether the identity, commutative, or associative property of addition is used.

73. $6 + 7 = 7 + 6$

74. $5 + 0 = 5$

75. $12 + (9 + 7) = (12 + 9) + 7$

76. $(12 + 9) + 1 = 12 + (9 + 1)$

77. $0 + (7 + 9) = (7 + 9)$

78. $5 + 999 = 999 + 5$

79. $3 + 0 = 0 + 3$

80. $(2 + 7) + 4 = 2 + (7 + 4)$

81. $12 = 12 + 0$

82. $67 + (8 + 1) = (67 + 8) + 1$

Work each using addition and/or subtraction.

83. Find the sum of the first ten numbers of the pattern given by 1, 3, 5, 7, . . .

84. Find the sum of the first ten numbers of the pattern given by 2, 4, 6, 8, . . .

85. A supermarket has 45 cases of diet soda without caffeine, 27 cases of diet soda with caffeine, 31 cases of regular soda without caffeine, and 12 cases of regular soda with caffeine. How many cases of soda with caffeine does the store have? How many cases of diet soda does the store have?

86. Joan started school at age 6 and went to public schools for 12 years. After public school she immediately went to a community college where she just completed her first year. How old is Joan?

87. Three companies bid on a construction job. Company A bid \$257,850, Company B bid \$265,785, and Company C bid \$247,690. Which company bid the lowest? How much additional money is needed if the lowest bid is rejected and the second lowest bid is accepted?

88. A family plans to purchase a home for \$59,000 and were told to expect \$1,200 in additional closing costs. If a bank agrees to lend them \$53,500, how much cash do they need to buy the home?

89. The Humane Society handled two hundred fifty-seven animals during 1988. Seventy-four of the animals were put to sleep due to disease and twenty-one were given to a company studying the effect of pet foods on growth. An additional one hundred thirty-three animals were placed in homes. How many animals remain at the Humane Society?

JOURNAL EXERCISES

1. Explain the three basic properties of addition. Do all three properties also apply to subtraction? Why, or why not?

2. Explain the difference between numbers and mathematical operations, such as addition or subtraction.

3. Explain the idea of borrowing. When is it used?

4. Explain the idea of carrying. When is it used?

1 Multiplication

Multiplication is a shortcut to repeated addition of the same number. For example, three rows of eight squares each are shown:

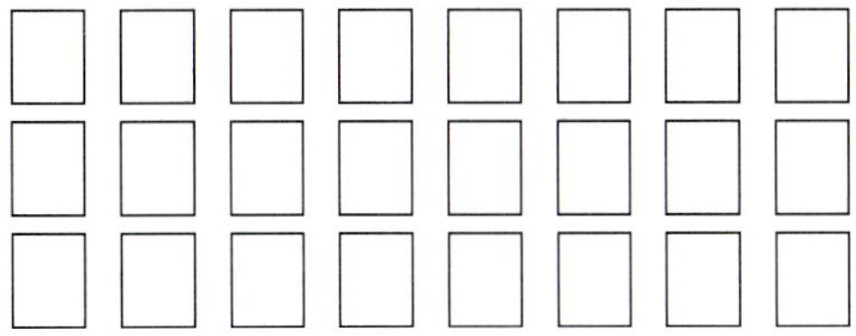

The total number of squares can be found by adding 8 + 8 + 8, or by multiplying 3 times 8.

Addition is always shown by the + symbol and subtraction is always shown by the − symbol. However, several different symbols are used for multiplication. Each of the following is three times eight:

$$3 \times 8 \qquad 3 \cdot 8 \qquad 3(8) \qquad (3)(8)$$

Multiplication problems can be written either horizontally or vertically.

$$3 \times 8 = 24 \quad \text{or} \quad \begin{array}{r} 3 \\ \times 8 \\ \hline 24 \end{array} \quad \begin{array}{l} \text{factor} \\ \text{factor} \\ \text{product} \end{array}$$

The numbers being multiplied are called **factors** and the result of the multiplication is the **product**. The keywords for multiplication include:

multiply	of	triple
product	double	total
times	twice	

The first step in learning multiplication is to *learn* the multiplication table. Knowing the multiplication table is important. You must be able to multiply in order to divide, work with decimals, and work percent problems.

Multiplication Table

×	0	1	2	3	4	5	6	7	8	9	10	11	12
0	0	0	0	0	0	0	0	0	0	0	0	0	0
1	0	1	2	3	4	5	6	7	8	9	10	11	12
2	0	2	4	6	8	10	12	14	16	18	20	22	24
3	0	3	6	9	12	15	18	21	24	27	30	33	36
4	0	4	8	12	16	20	24	28	32	36	40	44	48
5	0	5	10	15	20	25	30	35	40	45	50	55	60
6	0	6	12	18	24	30	36	42	48	54	60	66	72
7	0	7	14	21	28	35	42	49	56	63	70	77	84
8	0	8	16	24	32	40	48	56	64	72	80	88	96
9	0	9	18	27	36	45	54	63	72	81	90	99	108
10	0	10	20	30	40	50	60	70	80	90	100	110	120
11	0	11	22	33	44	55	66	77	88	99	110	121	132
12	0	12	24	36	48	60	72	84	96	108	120	132	144

OBJECTIVES

1. Multiply whole numbers.
2. Multiply by powers of ten.
3. Estimate answers.
4. Use properties of multiplication.

1. Mentally work the following:

(a) 3×4

(b) $5 \cdot 9$

(c) 4×10

(d) 7×6

(e) $(4)(7)$

(f) $6 \cdot 8$

(g) 4×8

(h) 6×3

(i) 9×7

(j) $7 \cdot 7$

(k) 9×8

(l) $(6)(6)$

ANSWERS

1. (a) 12 (b) 45 (c) 40 (d) 42 (e) 28 (f) 48 (g) 32 (h) 18 (i) 63 (j) 49 (k) 72 (l) 36

Notice that 0 times any number is equal to zero, and that 1 times any number is the number itself. For example,

$$507 \times 0 = 0 \qquad 507 \times 1 = 507$$
$$1{,}453{,}766 \times 0 = 0 \qquad 1{,}453{,}766 \times 1 = 1{,}453{,}766$$

◇ ***Do Practice Exercise 1.***

The multiplication of large numbers is based upon knowledge of the multiplication table plus a few rules. The following examples review the multiplication of larger numbers. As was the case with addition, carrying may be required. Look at the next example.

EXAMPLE 1

Multiply 43×6.

SOLUTION

$$\begin{array}{r} 43 \\ \times\ 6 \\ \hline \end{array}$$ Line up place values.

$$\begin{array}{r} {}^{1} \\ 43 \\ \times\ 6 \\ \hline 8 \end{array}$$ $6 \times 3 = 18$
Place 8 ones in the product and carry 1 ten.

$$\begin{array}{r} 43 \\ \times\ 6 \\ \hline 258 \end{array}$$ $4 \times 6 = 24$; $24 + 1 = 25$
Place 25 in the product. ■

The next example shows a more complicated multiplication problem using **partial products**. Partial products are identified in the next example.

EXAMPLE 2

Multiply 213×81.

SOLUTION

$213 \times 81 = 213 \times 8$ tens plus 213×1 one.

One partial product is $213 \times 1 = 213$.
The other partial product is 213×80, found by

Commutative property

$$213 \times 80 = 213 \times 8 \times 10 = 213 \times 10 \times 8 = 2{,}130 \times 8 = 17{,}040$$

partial product ←

The two partial products must then be added. The usual way in which multiplication is shown is:

$$\begin{array}{r} 213 \\ \times\ 81 \\ \hline 213 \\ 1704 \\ \hline 17{,}253 \end{array}$$

213 — partial product
1704 — partial product
17,253 — product

The 4 in the second partial product is placed into the tens place because we are actually multiplying 213×8 tens. ■

EXAMPLE 3

Multiply 271 × 45.

SOLUTION

$$\begin{array}{r} 271 \\ \times\ 45 \\ \hline 1355 \\ 10840 \\ \hline 12{,}195 \end{array} \qquad \begin{array}{r} 271 \\ \times\ 45 \\ \hline 1355 \\ 1084 \\ \hline 12{,}195 \end{array}$$

The 0 is frequently left off. ■

◇ *Do Practice Exercise 2.*

The next example multiplies a number with a zero in it times another number. Look at it carefully.

EXAMPLE 4

Multiply 234 times 105.

SOLUTION

$$\begin{array}{rl} 234 & \\ \times 105 & \\ \hline 1170 & \text{Multiply 5 ones times 234.} \\ 000 & \text{Multiply 0 tens times 234.} \\ 234 & \text{Multiply 1 hundred times 234 and add.} \\ \hline 24{,}570 & \text{Result.} \end{array}$$

Or, you may be familiar with the short method for working this problem:

$$\begin{array}{rl} 234 & \\ \times 105 & \\ \hline 1170 & \text{Replaces row of zeros in problem above.} \\ 2340 & \text{Note how this differs from the above.} \\ \hline 24{,}570 & \end{array}$$

An entire row of zeros was replaced by a single zero in the short method. It is important that the 0 in 2340 is lined up below the 7 in 1170. ■

Calculator Application

Multiply 678,247 times 798 by hand and with a calculator. Note the difference in the amount of time required. Which method do you think will be the more accurate as a general rule?

◇ *Do Practice Exercise 3.*

2. Multiply the following:

(a) 64 × 12

(b) 231 × 27

(c) 3,412 × 35

3. Multiply the following:

(a) 52 × 10

(b) 371 × 108

ANSWERS
2. (a) 768 (b) 6,237 (c) 119,420
3. (a) 520 (b) 40,068

4. Multiply the following:

(a) 361×100

(b) $6{,}248 \times 10{,}000$

(c) $2{,}300 \times 1{,}000$

2 Multiplication by Powers of Ten

A **power of ten** is a number with a 1 followed by one or more zeros. For example, the first few powers of ten are:

Powers of Ten		
10	=	10
100	=	10 × 10
1,000	=	10 × 10 × 10
10,000	=	10 × 10 × 10 × 10

Look at the following examples in which 394 is multiplied times different powers of ten.

$$\begin{array}{r} 394 \\ \times\ 10 \\ \hline 3{,}940 \end{array} \qquad \begin{array}{r} 394 \\ \times 100 \\ \hline 39{,}400 \end{array} \qquad \begin{array}{r} 394 \\ \times 1{,}000 \\ \hline 394{,}000 \end{array} \qquad \begin{array}{r} 394 \\ \times 10{,}000 \\ \hline 3{,}940{,}000 \end{array}$$

In each case, the product is the number 394 followed by a number of zeros. The number of zeros attached to the end of 394 is *exactly the same* as the number of zeros that were originally at the end of the power of ten:

$394 \times 10 = 3{,}940$ which is 394 with 1 zero on the end

$394 \times 100 = 39{,}400$ which is 394 with 2 zeros on the end

$394 \times 1{,}000 = 394{,}000$ which is 394 with 3 zeros on the end

In order to multiply a number by a power of ten:

1. Count the number of zeros to the right of the 1 in the power of 10.
2. Attach the same number of zeros onto the end of the number which is not the power of ten.

EXAMPLE 5

How many days are in 1,000 years? Assume that each year contains exactly 365 days.

SOLUTION

The problem requires you to multiply 365 by 1,000. There are three zeros after the 1 in 1,000. Add three zeros onto the end of 365 and place the commas appropriately.

Attach 3 zeros to the end of 365.

$365 \times 1{,}000 = 365{,}000$ days ■

◇ *Do Practice Exercise 4.*

3 Estimation

One tool used to check problems is to **estimate** or **approximate** the answer. Estimates are made by rounding before doing the math operation(s). The estimate can then be used to check the results from the exact calculation.

ANSWERS
4. (a) 36,100 (b) 62,480,000 (c) 2,300,000

EXAMPLE 6

Multiply 671 × 327, then estimate the product by rounding each number to the nearest one hundred and multiplying these values.

SOLUTION

exact		*estimate*
671	671 rounds to 700	700
×327	327 rounds to 300	×300
4697		210,000
1342		
2013		
219,417		

The estimate of 210,000 is close to 219,417. ■

◇ *Do Practice Exercise 5.*

4 Properties of Multiplication

The identity, commutative, and associative properties that apply to addition also apply to multiplication. A fourth property, the distributive property involves both multiplication and addition. These properties are shown in the following examples.

PROPERTIES OF MULTIPLICATION

1. **Identity property**—A number is not changed if multiplied by 1:

$$6 \times 1 = 6$$

2. **Commutative property**—The order in which two numbers are multiplied does not make any difference:

$$6 \times 7 = 7 \times 6$$

3. **Associative property**—When multiplying three numbers, it makes no difference which two are multiplied first:

$$8 \times (3 \times 5) = (8 \times 3) \times 5$$

4. **Distributive property**—The product of a number times a sum is the same as the sum of the products.

$$6 \times (4 + 9) = 6 \times 4 + 6 \times 9$$

The distributive property involves both addition and multiplication. It says that the number being multiplied times the sum can first be multiplied times the first term then times the second term. For example:

$$7 \cdot (3 + 8) = 7 \cdot 3 + 7 \cdot 8$$

5. Estimate by first rounding each number to the nearest one thousand:

(a) 3,621 × 2,150

(b) 9,491 + 1,200 + 3,641

(c) 111,270 − 101,800

ANSWERS
5. (a) 8,000,000 (b) 14,000 (c) 9,000

6. Which property is shown by each of the following?

(a) $7 \times 11 = 11 \times 7$

(b) $6 \times (5 \times 3) = (6 \times 5) \times 3$

(c) $4(7 + 4) = 4 \times 7 + 4 \times 4$

(d) $80 \times 1 = 80$

ANSWERS
6. (a) Commutative property
(b) Associative property
(c) Distributive property
(d) Identity property

EXAMPLE 7

Which property is shown by each of the following?

(a) $4 \times 3 = 3 \times 4$
(b) $5 \times (6 + 3) = 5 \times 6 + 5 \times 3$
(c) $3(6 + 9) = 3(6) + 3(9)$
(d) $17 \times 1 = 17$
(e) $4 \times (6 \times 3) = (4 \times 6) \times 3$

SOLUTION

(a) Commutative property—The order in which they are multiplied does not matter.
(b) Distributive property—The product equals the sum of two products.
(c) Distributive property—The product equals the sum of two products.
(d) Identity property—A number is not changed when multiplied by 1.
(e) Associative property—Any two can be multiplied first. ■

◇ *Do Practice Exercise 6.*

name date hour

Multiply the following in your head and write down the answer.

1. 6×5 **2.** 7×5 **3.** 7×7 **4.** 4×8

5. 6×7 **6.** 8×9 **7.** $8 \cdot 7$ **8.** $6 \cdot 7$

9. $4 \cdot 8$ **10.** $5 \cdot 7$ **11.** $7 \cdot 9$ **12.** $4 \cdot 3$

13. $4(9)$ **14.** $9(5)$ **15.** $(5)5$ **16.** $(7)(7)$

17. $(8)(3)$ **18.** $(9)(9)$ **19.** 9×6 **20.** $4 \cdot 4$

21. $3(9)$ **22.** $7(8)$ **23.** $8 \cdot 8$ **24.** $(5)(6)$

25. $2 \times 3 \times 5$ **26.** $4 \times 9 \times 2$

Multiply, showing all work. If you would like, check your answers with a calculator.

27. 624×7 **28.** $1{,}214 \times 8$ **29.** 999×9 **30.** $8{,}214 \times 12$ **31.** $(613)(32)$

32. 270×41 **33.** $(965)(37)$ **34.** $7{,}540 \cdot 3$ **35.** $\begin{array}{r} 750 \\ \times\ 21 \\ \hline \end{array}$ **36.** $\begin{array}{r} 697 \\ \times\ 72 \\ \hline \end{array}$

37. $\begin{array}{r} 875 \\ \times\ 75 \\ \hline \end{array}$ **38.** $\begin{array}{r} 753 \\ \times 147 \\ \hline \end{array}$ **39.** $\begin{array}{r} 648 \\ \times\ 89 \\ \hline \end{array}$ **40.** $\begin{array}{r} 789 \\ \times\ 76 \\ \hline \end{array}$ **41.** $\begin{array}{r} 251 \\ \times\ 13 \\ \hline \end{array}$

42. $\begin{array}{r} 379 \\ \times\ 47 \\ \hline \end{array}$

43. $\begin{array}{r} 832 \\ \times\ 67 \\ \hline \end{array}$

44. $\begin{array}{r} 1{,}212 \\ \times\ 44 \\ \hline \end{array}$

45. $\begin{array}{r} 3{,}871 \\ \times\ 123 \\ \hline \end{array}$

46. $\begin{array}{r} 5{,}648 \\ \times\ 230 \\ \hline \end{array}$

47. $\begin{array}{r} 372 \\ \times\ 29 \\ \hline \end{array}$

48. $\begin{array}{r} 486 \\ \times\ 45 \\ \hline \end{array}$

49. $\begin{array}{r} 798 \\ \times 178 \\ \hline \end{array}$

50. $\begin{array}{r} 2{,}912 \\ \times\ 813 \\ \hline \end{array}$

51. $\begin{array}{r} 122 \\ \times 3{,}411 \\ \hline \end{array}$

52. $\begin{array}{r} 12{,}278 \\ \times\ 481 \\ \hline \end{array}$

53. $\begin{array}{r} 96{,}785 \\ \times\ 978 \\ \hline \end{array}$

54. $\begin{array}{r} 75{,}001 \\ \times\ 8{,}775 \\ \hline \end{array}$

55. $\begin{array}{r} 2{,}065{,}221 \\ \times\ 359 \\ \hline \end{array}$

56. $\begin{array}{r} 75{,}000 \\ \times\ 100 \\ \hline \end{array}$

57. $\begin{array}{r} 815{,}100 \\ \times\ 1{,}000 \\ \hline \end{array}$

58. $\begin{array}{r} 100{,}000 \\ \times\ 10{,}000 \\ \hline \end{array}$

59. $\begin{array}{r} 2{,}344 \\ \times\ 10 \\ \hline \end{array}$

60. $\begin{array}{r} 23{,}000 \\ \times\ 100 \\ \hline \end{array}$

61. $\begin{array}{r} 34{,}001 \\ \times\ 1{,}000 \\ \hline \end{array}$

62. $\begin{array}{r} 225 \\ \times 100 \\ \hline \end{array}$

63. $\begin{array}{r} 3{,}627 \\ \times 1{,}000 \\ \hline \end{array}$

64. $\begin{array}{r} 6{,}879 \\ \times 10{,}000 \\ \hline \end{array}$

Determine whether the identity, commutative, associative, or distributive property is shown.

65. $7 \times 14 = 14 \times 7$

66. $6 \times 1 = 6$

67. $3(1 + 7) = 3 \times 1 + 3 \times 7$

68. $3(4 \times 5) = (3 \times 4)\,5$

69. $9 \cdot 7 = 7 \cdot 9$

70. $4 \times 5 + 4 \times 8 = 4(5 + 8)$

71. $70 + 0 = 70$

72. $4 + (6 + 23) = (4 + 6) + 23$

Place commas correctly, then round to the nearest thousand. (Review)

73. 76002

74. 129000

75. 800120

76. 87000060

Write the digits of each number and put commas in the correct positions. (Review)

77. Seven hundred ninety-seven

78. Sixteen thousand, four hundred forty-nine

79. Seven million, twelve

80. Eighty million, four hundred thousand

Work each problem.

81. Multiply ninety-nine times ninety-nine.

82. Find the product of seven million and eight.

83. Estimate the product of six thousand, three hundred with forty-eight. Then find the exact value.

84. What is the product of eighty-six and forty-one? Estimate the answer before calculating the exact value.

85. Multiply seventeen times forty and then subtract fifteen from the product.

86. A boy placed eighteen marbles in a blue bag, forty-nine marbles in a red bag, and twelve marbles in a brown bag. One-half of the marbles in the blue bag were moved to the brown bag. How many marbles are in the red and brown bags combined?

87. A woman wanted to estimate the number of people in a city. The number of houses in one city block was fourteen and the estimated number of city blocks in the city was eight thousand, two hundred. How many people live in the city if three people live in each house?

JOURNAL EXERCISES

1. An average dog in a small city was found to have seven fleas. How many fleas are expected to be on all of the dogs in the city? Is there enough information to answer this question? If not, what other information is needed?

2. Are calculators, electronic cash registers that scan prices at stores, and computers changing the way in which some jobs are done? Explain.

3. The distance from the earth to the moon is about two hundred thirty-nine thousand miles. What is the distance traveled by a spaceship that travels from the earth to the moon and back again? Find and write down the definition of the word *assumption.* What assumption(s) have you made?

1 Division

OBJECTIVES

1 Divide whole numbers.
2 Calculate averages.

Division is the opposite or inverse of multiplication. Division is the idea of cutting or breaking something into equal size pieces. For example, if a candy bar is divided into three equal pieces, then it is cut or broken or divided into three pieces of the same size.

The product of 6 and 7 is 42. The number 42 can be cut up into 6 groups of size 7 each, therefore 6 divides into 42 seven times. The exact same division problem can be written in three different ways as shown below:

$$42 \div 6 \qquad \frac{42}{6} \qquad 6\overline{)42}$$

Each of the first two expressions is read as forty-two *divided by* six. The last expression is read as six *divided into* forty-two. The keywords for division include:

divided by   divide   per   quotient

The result of division is called the **quotient**. The number being divided is the **dividend** and the number being divided into the dividend is the **divisor**.

$$\begin{array}{r} \text{Quotient} \\ \text{Divisor}\overline{)\ \text{Dividend}} \end{array}$$

The relationship between multiplication and division is shown by the following examples.

		Division				*Multiplication*
Dividend		*Divisor*		*Quotient*		
72	÷	8	=	9	since	$9 \cdot 8 = 72$
45	÷	9	=	5	since	$5 \cdot 9 = 45$
12	÷	4	=	3	since	$3 \cdot 4 = 12$
20	÷	10	=	2	since	$2 \cdot 10 = 20$

EXAMPLE 1

Solve each:

(a) $42 \div 7$ **(b)** $36 \div 6$ **(c)** $25 \div 5$ **(d)** $49 \div 7$

SOLUTION

(a) 6 **(b)** 6 **(c)** 5 **(d)** 7 ■

1. Mentally work the following:

(a) 63 ÷ 9

(b) 36 ÷ 6

(c) 18 ÷ 6

(d) 81 ÷ 9

(e) 24 ÷ 3

(f) 64 ÷ 8

(g) 48 ÷ 6

(h) 24 ÷ 4

(i) 56 ÷ 8

(j) 49 ÷ 7

(k) 40 ÷ 8

(l) 45 ÷ 9

ANSWERS

1. (a) 7 (b) 6 (c) 3 (d) 9 (e) 8 (f) 8 (g) 8 (h) 6 (i) 7 (j) 7 (k) 5 (l) 5

EXAMPLE 2

Jerry ran the same distance each day for eight days and ran a total of forty-eight miles. How many miles did he run each day?

SOLUTION

The answer is found by cutting the forty-eight miles up into eight equal size pieces. The problem is a division problem.

48 miles divided by 8 days = 6 miles per day ■

◇ *Do Practice Exercise 1.*

Division using the number 0 follows the same pattern. Zero divided by any number other than zero is equal to zero. The idea is shown in the next example.

EXAMPLE 3

Divide 0 by 60.

SOLUTION

$$\begin{array}{r} 0 \rightarrow \text{quotient} \\ 60\overline{)0} \\ \underline{0} \\ 0 \rightarrow \text{remainder} \end{array}$$

The number 60 divides into 0 exactly 0 times. This is true because the divisor 60 times the quotient 0 equals the dividend which is 0. ■

On the other hand the divisor of a division problem *can never* be zero, because it would make the division **undefined**. Undefined means that the problem has no answer.

EXAMPLE 4

Divide 25 by 0.

SOLUTION

There is no number such that when multiplied times 0 equals 25.

$$\begin{array}{r} ? \\ 0\overline{)25} \end{array}$$ There is no number whose product with 0 equals 25.

Thus 25 ÷ 0 is undefined, which means that it has no answer. ■

Calculator Application

Divide 25 by 0 on your calculator. What does it show?

Answer: error

HELPFUL HINT

The dividend of a division problem can be 0, but the divisor can never be 0.

Division with larger numbers requires knowledge of the multiplication table plus a few rules.

Rule: To Divide Large Numbers

Step 1 Divide.

Step 2 Multiply.

Step 3 Subtract.

Step 4 Bring down the next digit.

Step 5 Go back to Step 1.

EXAMPLE 5

Divide 105 by 5.

SOLUTION

Step 1 Divide.

$$\begin{array}{r} 2 \\ 5\overline{)105} \end{array}$$

5 divides into 10 two times; place a 2 in the quotient above the 0 in 105

Step 2 Multiply.

$$\begin{array}{r} 2 \\ 5\overline{)105} \\ \underline{10} \end{array}$$

2 times 5 is 10

Step 3 Subtract.

$$\begin{array}{r} 2 \\ 5\overline{)105} \\ \underline{10} \\ 0 \end{array}$$

10 minus 10 is 0

Step 4 Bring down next digit.

$$\begin{array}{r} 2 \\ 5\overline{)105} \\ \underline{10} \\ 05 \end{array}$$

The next digit is brought down

Step 1 Divide.

$$\begin{array}{r} 21 \\ 5\overline{)105} \\ \underline{10} \\ 05 \end{array}$$

5 divides into 5 one time; place a 1 in the quotient above the 5 in 105

Step 2 Multiply.

$$\begin{array}{r} 21 \\ 5\overline{)105} \\ \underline{10} \\ 05 \\ \underline{5} \end{array}$$

5 times 1 is 5

Step 3 Subtract.

$$\begin{array}{r} 21 \\ 5\overline{)105} \\ \underline{10} \\ 05 \\ \underline{5} \\ 0 \end{array}$$

21 → result

5 minus 5 equals 0; there are no other numbers to bring down

■

The result or quotient of a division problem can be checked using multiplication. The product of the divisor and the quotient should equal the dividend. In the last example, the divisor was 5 and the quotient was 21. The division was done correctly if $5 \times 21 = 105$, which it does.

◇ *Do Practice Exercises 2 and 3.*

A digit *must be* placed in the quotient every time a digit is brought down in a division problem. It may even be that a 0 is placed in the quotient. Look at the next example.

EXAMPLE 6

Divide 515 by 5.

SOLUTION

$$\begin{array}{r} 1 \\ 5\overline{)515} \\ \underline{5} \\ 01 \end{array}$$

Divide, multiply, and subtract.

Bring down the next digit.

$$\begin{array}{r} 10 \\ 5\overline{)515} \\ \underline{5} \\ 01 \end{array}$$

5 divides into 1 zero times; *place a 0 in the quotient above the 1 in 515*

$$\begin{array}{r} 10 \\ 5\overline{)515} \\ \underline{5} \\ 01 \\ \underline{0} \\ 15 \end{array}$$

Divide, multiply, and subtract.

Bring down the next digit.

$$\begin{array}{r} 103 \\ 5\overline{)515} \\ \underline{5} \\ 01 \\ \underline{0} \\ 15 \\ \underline{15} \\ 0 \end{array}$$

Result.

Divide, multiply, and subtract.

There are no more digits. ■

You may be more familiar with the following shortcut for the problem from the last example:

$$\begin{array}{r} 103 \\ 5\overline{)515} \\ \underline{5} \\ 015 \\ \underline{15} \\ 0 \end{array}$$

103 → result

015 ⟶ Shortcut: since the divisor 5 divides into 1 zero times, a 0 is written in the quotient and the next digit is brought down.

◇ *Do Practice Exercises 4 and 5.*

2. Divide 381 by 3.

3. Divide 2,892 by 12.

4. Divide 4,230 by 6.

5. Divide 3,255 by 31.

ANSWERS
2. 127
3. 241
4. 705
5. 105

Frequently numbers do not divide into each other **evenly**. In that event, something remains after the division. The number left over after dividing is called the **remainder**. For example,

$$\begin{array}{r} 75 \\ 3\overline{)226} \\ \underline{21} \\ 16 \\ \underline{15} \\ 1 \end{array}$$

3 divides into 226 seventy-five times, but a remainder of 1 is left over.

1 → remainder

The number 3 divides into 226 more than 75 times, but less than 76 times. The number 226 divided by 3 is 75 with a remainder of 1 which is written as:

75 r1 — Read as: Seventy-five with a remainder of 1

The answer to a division problem with a remainder can be checked by *multiplying* the divisor times the quotient and *adding* the remainder back in. For the problem above:

75	quotient
× 3	divisor
225	product
+ 1	Add remainder
226	which equals the dividend, thus the division was done correctly.

◇ *Do Practice Exercises 6 and 7.*

Calculator Application

The sun is 93,000,000 miles from the earth and light travels at 186,000 miles per second. How long does it take light to travel from the sun to the earth?

Answer: 500 seconds or a little over 8 minutes. If the sun were to vanish we would not know about it for 8 minutes.

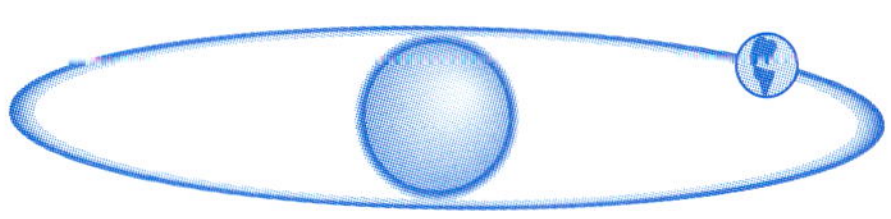

EXAMPLE 7

Four boys go together and buy a package containing fifty marbles. How many marbles are left if they split the fifty marbles evenly among themselves?

6. Divide 620 by 3.

7. Divide 6,510 by 24.

ANSWERS
6. 206 r2
7. 271 r6

8. The average of six numbers is found by dividing the sum of the six numbers by what number?

9. Find the average of 47, 54, 56, and 47.

SOLUTION

The quantity being asked for is the remainder after dividing 50 by 4.

$$\begin{array}{r} 12 \\ 4\overline{)50} \\ \underline{4} \\ 10 \\ \underline{8} \\ 2 \end{array}$$

12 ⟶ Each boy gets twelve marbles.

2 remain or are left over

Each child gets twelve marbles and there are two remaining. ■

2 Average

The **average** of two numbers is found by adding both numbers and dividing the sum by two. For example, the average of 8 and 14 is found by:

$$\frac{8 + 14}{2} = \frac{22}{2} = 11$$

The average of three numbers is found by adding the three numbers and dividing by three. The average of four numbers is found by adding them and dividing by four, and so on.

◇ *Do Practice Exercise 8.*

EXAMPLE 8

Victoria's test grades in her first college math course were 74, 82, and 75. What is her average test score?

SOLUTION

The average test score is found by adding the three test grades together and dividing the sum by three.

Add first.

$$\begin{array}{r} 74 \\ 82 \\ +\ 75 \\ \hline 231 \end{array}$$

Divide the sum by 3.

$$\begin{array}{r} 77 \\ 3\overline{)231} \\ \underline{21} \\ 21 \\ \underline{21} \\ 0 \end{array}$$

77 is the average test score ■

◇ *Do Practice Exercise 9.*

Calculator Application

Find the average test score for a group with the following grades: 72, 65, 98, 92, 85, 80, 41, 75, 78, 81, 74, and 59.

Answer: The average score is 75.

ANSWERS
8. Divide the sum by 6.
9. 51

name date hour

EXERCISE SET 1.5

Work the following in your head and write down the answer.

1. $25 \div 5$ **2.** $42 \div 7$ **3.** $56 \div 8$ **4.** $81 \div 9$

5. $72 \div 9$ **6.** $64 \div 8$ **7.** $12 \div 4$ **8.** $35 \div 7$

9. $49 \div 7$ **10.** $\frac{24}{6}$ **11.** $\frac{14}{2}$ **12.** $54 \div 9$

13. $8 \cdot 7$ **14.** $9 \cdot 6$ **15.** $7 + 14$ **16.** $17 - 9$

17. $12 \div 9$ **18.** $16 \div 2$ **19.** $\frac{48}{8}$ **20.** $24 \div 8$

21. $\frac{30}{6}$ **22.** $14 - 14$ **23.** $\frac{63}{7}$ **24.** $50 \div 2$

25. $25 - 13$ **26.** $17 \div 2$ **27.** $\frac{100}{2}$ **28.** $\frac{100}{4}$

29. $(30) \div (3)$ **30.** $(9)(6)$

Divide, showing all work. If you would like, check your answers with a calculator.

31. $60 \div 15$ **32.** $36 \div 3$ **33.** $300 \div 15$ **34.** $17 \div 4$

35. $66 \div 11$ **36.** $76 \div 3$ **37.** $360 \div 12$ **38.** $320 \div 4$

39. $651 \div 3$ **40.** $417 \div 417$ **41.** $250 \div 25$ **42.** $200 \div 12$

43. $128 \div 16$

44. $225 \div 25$

45. $512 \div 32$

46. $700 \div 35$

47. $680 \div 20$

48. $1{,}024 \div 2$

49. $1{,}024 \div 32$

50. $728 \div 8$

51. $4{,}407 \div 11$

52. $271 \div 40$

53. $2{,}704 \div 29$

54. $266 \div 45$

55. $12\overline{)480}$

56. $27\overline{)37}$

57. $75\overline{)6{,}689}$

58. $234\overline{)5{,}674}$

59. $971\overline{)2{,}980}$

60. $999\overline{)10{,}000}$

61. $76\overline{)8{,}787}$

62. $354\overline{)35{,}654}$

63. $1{,}200\overline{)35{,}900}$

64. $21\overline{)4{,}600}$

65. $750\overline{)15{,}003}$

66. $1{,}000\overline{)8{,}060{,}000}$

Work each, showing all steps. (Review)

67. 75×12

68. $380 + 14{,}600$

69. $972 - 188$

70. $65 \div 12$

71. 256×8

72. $465 - 299$

73. $6{,}000 \times 1{,}000$

74. $9{,}612 \div 452$

Work each using addition, subtraction, division, and/or multiplication as needed.

75. How many times does seventy-two divide into five hundred seventy-six?

76. What is the difference between seventeen thousand and four hundred ninety-four?

77. Divide the sum of six and forty by the number eight.

78. Find the product of sixty with the quotient of seventy divided by ten.

79. Divide four million three hundred thousand by thirty-five and write the answer using words.

80. Three calves weighed 400 pounds, 241 pounds, and 436 pounds. Find their total weight and their average weight.

81. In three different games, a high school basketball player scored 18 points, 21 points, and 27 points. What is his average number of points scored per game?

82. An individual decided to give up a two dollar per day smoking habit and save the money instead. How many days will it take to save enough to purchase a stereo costing three hundred twenty-seven dollars?

83. A pharmacist has 750 pills of a particular drug. How many containers can be filled and how many pills will be left over if 80 pills are placed into each container?

84. The earth's atmosphere is like a blanket that rests on the surface of the earth with a weight equal to a layer of water thirty-four feet thick. How many atmospheres of weight are equal to a layer of water one hundred thirty-six feet thick?

85. How much money is needed to rent an apartment for $350 per month for twenty years? Write the answer using words.

86. Terry bought thirty shares of stock at $40 per share and twenty shares of stock at $20 per share. Find the total cost of the stock and the average cost per share.

87. During 1956, Fidel Castro left Mexico and invaded Cuba where he fought the government of Batista. At different times, Castro had different numbers of troops due to deaths caused by battles and by recruitment of new troops. Find Castro's average number of troops during 1956, *assuming* the troop count at two month intervals was 81, 83, 81, 16, 27, and 120.

JOURNAL EXERCISES

1. Does the commutative property of mathematics apply to division problems? Why or why not?

2. Explain why division by zero does not make any sense.

3. Multiplication is repeated addition of a number. Division can be thought of as repeated subtraction of a number. Explain.

OBJECTIVES

1 Know the difference between inductive and deductive reasoning.
2 Solve problems using inductive reasoning.
3 Solve problems using deductive reasoning.

Both mathematics and mathematical type thinking are used to solve problems. The prior sections of this chapter have been concerned mostly with the rules and language of mathematics. Both this section and the next are about thinking.

1 Inductive versus Deductive Reasoning

It is not enough to be able to add a column of numbers or to multiply two numbers. Individuals in the workplace are expected to think through and solve problems. **Critical thinking** is the process of understanding and finding a reasonable solution to a problem.

Different types of thought processes are used in problem solving. **Inductive reasoning** is the process of first finding a pattern, then assuming that the same pattern continues. The first example shows the use of inductive reasoning in everyday life.

EXAMPLE 1

Martin and Shanna have had several dates. Shanna has noticed that Martin is always 20 to 30 minutes late when picking her up from her house. She assumes that he will be at least 20 minutes late for their date on Friday night. What type of reasoning did Shanna use?

SOLUTION

Shanna sees a pattern of Martin being late and assumes that it will continue. This is an example of inductive reasoning. ■

In contrast, **deductive reasoning** is a thought process in which accepted facts are used to draw a conclusion. The following example shows the use of deductive reasoning in everyday life.

EXAMPLE 2

Shanna is looking for her car keys, which she *always* keeps in one of three places: on top of the refrigerator, beside her bed, or in her purse. The keys are not in her purse nor are they on the refrigerator. Therefore, her keys must be beside her bed. What type of reasoning did Shanna use?

SOLUTION

The conclusion is a result of deductive reasoning which uses facts to work towards a solution. The facts in this example are:

Fact 1 The keys are in one of three places.
Fact 2 The keys are not in her purse.
Fact 3 The keys are not on the refrigerator.

Conclusion The keys must be beside her bed. ■

2 Inductive Reasoning

Humans excel at inductive reasoning or the ability to identify a pattern and repeat it. For example, a child might learn how to swing a baseball bat by watching an older player swing at a ball. A scientist might observe that the division of one bacteria into two, keeps repeating itself. The scientist watches

1. Find the next entry for each pattern:

(a) 1, 5, 9, 13, ___?___

(b) 1, 3, 9, 27, ___?___

(c)  ___?___

(d) 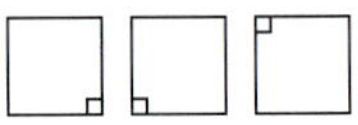 ___?___

ANSWERS

1. (a) 17 (b) 81

(c)  (d) 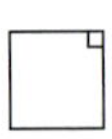

the pattern, learns from it, and perhaps publishes the results in a scientific journal. Patterns that repeat themselves are all around us, if only we can see them.

Two examples of pattern recognition are given next: one involves a list of numbers and the other involves a set of geometric figures.

EXAMPLE 3

Find the next number in the pattern:

$$4, \quad 8, \quad 16, \quad 32, \quad ?$$

SOLUTION

Each number in the pattern is equal to twice the prior number.

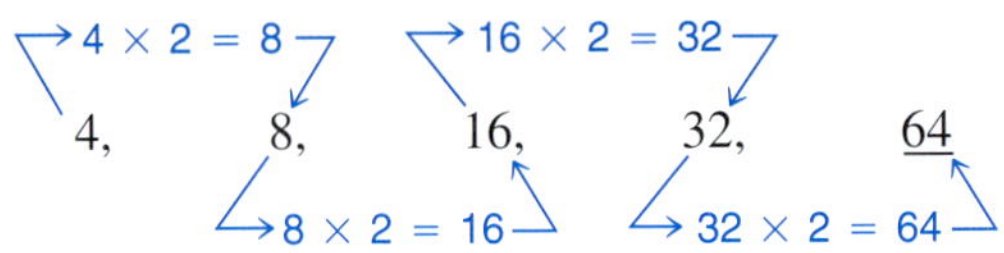

The pattern is 4, 8, 16, 32, 64, . . . , so the next number in the pattern is 64. ■

EXAMPLE 4

Find the next shape in the pattern:

SOLUTION

Every other geometric figure in the pattern is a circle and the last figure is a circle. Therefore, a circle is not the next figure in the pattern. The geometric figures remaining after the circles are removed are:

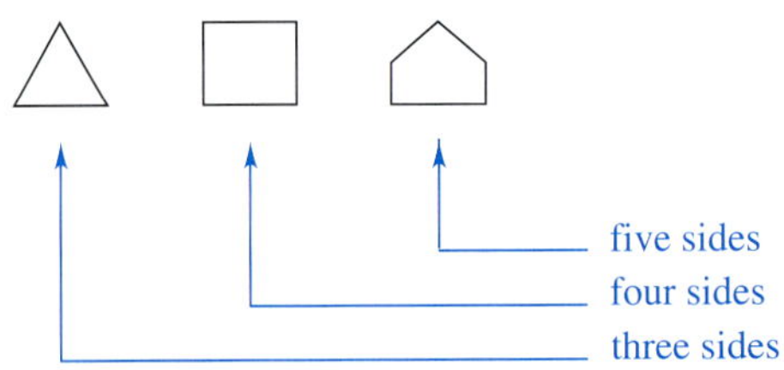

Each geometric figure has one more side than the previous figure. Thus the next geometric shape has six sides. The complete pattern including the six sided figure is:

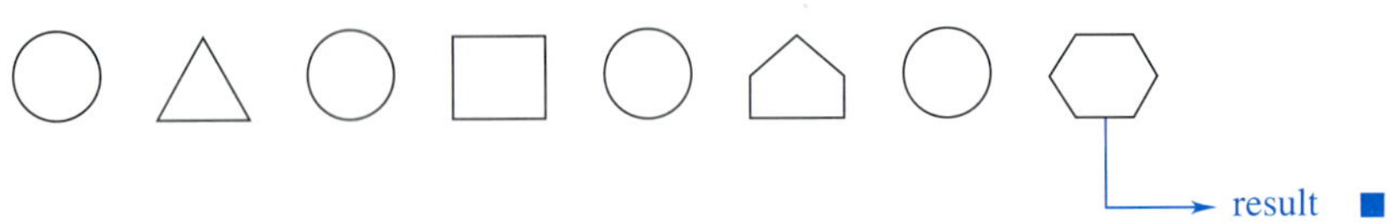

■

◇ *Do Practice Exercise 1.*

3 Deductive Reasoning

Deductive reasoning requires known facts. The facts are used to work towards a conclusion. The conclusion must follow from the facts given. The following examples apply deductive reasoning.

EXAMPLE 5

A blue bag contains 70 red marbles and a red bag contains 10 yellow marbles. One-half of the red marbles in the blue bag are moved to the red bag. How many marbles of each color end up in each bag?

SOLUTION

Fact 1 The blue bag contains 70 red marbles.
Fact 2 The red bag contains 10 yellow marbles.
Fact 3 One-half of the red marbles are moved to the red bag.

Conclusion 1 35 red marbles were moved to the red bag.
Conclusion 2 The blue bag now has 35 red marbles.
The red bag now has 35 red and 10 yellow marbles. ■

EXAMPLE 6

A math instructor gave her 30 students the choice of writing a two page report on a famous mathematician or of working 25 homework problems. Eighteen reports and 11 homework sets were turned in. How many students did not turn in either assignment?

SOLUTION

Fact 1 There are 30 students.
Fact 2 18 students wrote a report.
Fact 3 11 students worked the problems.

Conclusion $30 - 18 - 11 = 1$ student who did neither assignment. ■

◇ ***Do Practice Exercise 2.***

2. There are 130 fish in a pond of which 27 are catfish and 91 are perch. Small parasites are found on a total of 35 fish of which 17 are catfish and 15 are perch. How many fish do not have parasites and are neither catfish nor perch?

ANSWER
2. 9 fish

name date hour

EXERCISE SET 1.6

Use inductive reasoning to find the next number in each pattern.

1. 1, 3, 5, 7, 9, ____

2. 0, 2, 0, 4, 0, 6, ____

3. 1, 2, 4, 5, 7, 8, ____

4. 15, 16, 17, 18, ____

5. 2, 5, 3, 6, 4, ____

6. 1, 2, 3, 1, 2, ____

7. 5, 8, 11, 14, ____

8. 1, 2, 4, 7, 11, ____

9. 5, 20, 80, 320, ____

10. 25, 20, 15, 10, ____

11. 4, 40, 8, 44, ____

12. 15, 13, 10, 6, ____

Use inductive reasoning to find the next shape in the pattern.

13. 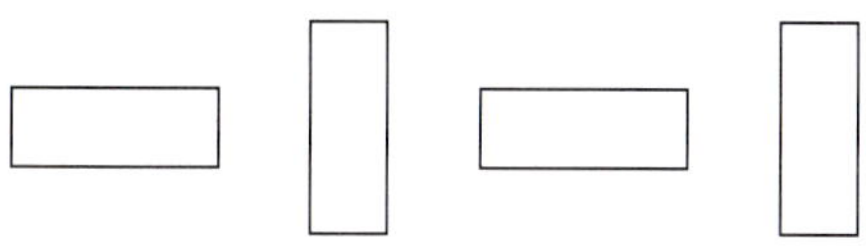 ?

14.  ?

15. 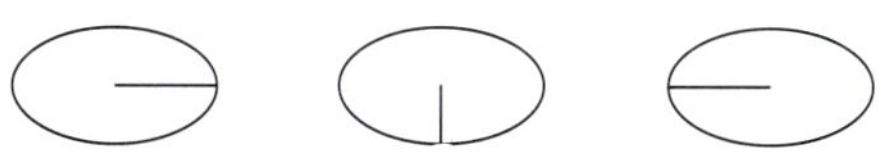 ?

16.  ?

17. 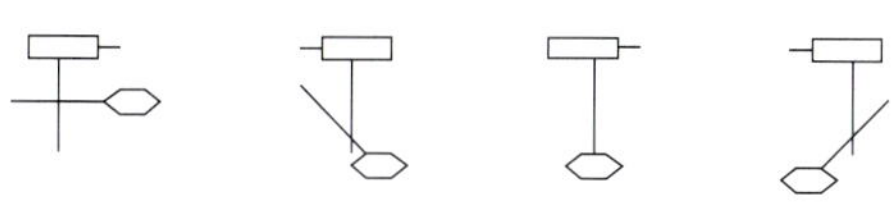 ?

18. 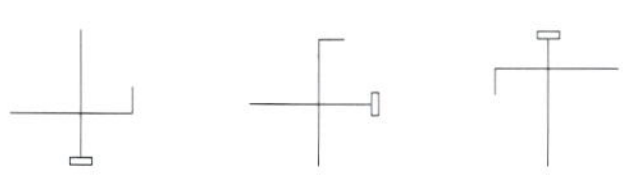 ?

Decide whether inductive or deductive reasoning is used for each conclusion.

19. Rodney has $425 and must pay the following bills: $200, $125, $85, and $50. He does not have enough money to pay his bills.

20. John gets up at 7:00 each morning. Tomorrow, John will get up at 7:00.

21. Maria has worked 40 hours per week at one firm for seven years and has always been paid on time. She worked this week and will be paid on time this week.

22. Roberta wants to buy a $200 suit, but has no money. If she saves $40 per week she can buy the suit in 5 weeks.

23. Gracie gives the following directions: Drive north on Central Avenue until you come to the third stoplight. Turn right and go two blocks. The hospital will be on your right-hand side.

24. A nurse is ordered to give 650 milligrams of a drug to a patient. She gives two 325 milligram tablets to the patient.

25. Victor was told that he must work 12 hours of overtime this week. He had already worked 45 hours where everything over 40 hours is overtime. He must work 7 more hours.

26. Tina tried to stop her car but her brakes did not work. She tried the brakes again and they still did not work. Her brakes were not going to work until repaired.

27. Jason ate 10 cookies on each of the last seven days. He ate 70 cookies.

28. In 1989, world population was approximately five billion and it was doubling every forty years. It will continue to double at the same rate for the next eighty years.

29. Jesse's pickup would not start. It is known that the problem is either the battery or the starter. Since the battery was okay, Jesse decided that his starter was bad.

30. The first piece of bread Lucy pulled from a loaf was moldy. The second and third pieces of bread were also moldy. Lucy concluded that the entire loaf was bad.

31. With great difficulty, Laurie solved the first word problem of her assignment. She also had difficulty with the second word problem. She concluded that all of the word problems would be difficult.

32. Mrs. Rosenzweig interviewed nine people for a job. Four of the nine do not have the required college degree. Two others have a college degree, but their grades are too low. Three of those with college degrees, including one with poor grades, do not meet the experience requirement. She concludes that her choice has narrowed to one person.

JOURNAL EXERCISES

1. Add 2 + 4. Did you use deductive or inductive reasoning? Explain.

2. Explain the difference between inductive and deductive reasoning. Name some every day uses of both deductive and inductive reasoning.

3. Are some everyday problems solved without using reasoning? Explain your answer.

1 Word Problems

OBJECTIVE

1 Work at solving word problems.

Word problems require thought and may require both inductive and deductive reasoning. Each word problem involves different concepts or uses words in different ways. The ability to solve word problems is learned through repeated effort. You *will become better* at solving word problems if you work at them.

The first attempt to solve a word problem may not work. Instead, it may simply lead to a "dead end" or to an answer that cannot possibly be correct. Think of the first try at solving a word problem as a learning experience that helps you understand the problem. Be patient and keep working on the problem. The helpful hints in the box may help you solve word problems.

HELPFUL HINTS FOR SOLVING WORD PROBLEMS

1. Read the problem carefully. Look words up in the dictionary if needed.
2. Look for keywords that may indicate one of the mathematical operations.
3. Look for patterns and use inductive reasoning.
4. Identify the facts and use deductive reasoning.
5. Draw sketches.
6. Work any part of the problem that you do understand.
7. If all else fails, guess at an approach or work the problem backwards.
8. Make sure your answer is reasonable.

EXAMPLE 1

Four individuals meet for the first time. Each person shakes hands *once* and *only once* with each of the others. How many handshakes occur?

SOLUTION

One approach to solving this problem is to try to mentally visualize four individuals in a room and try to count the number of handshakes. An easier approach is as follows.

Give the four individuals the names Sally, Jackie, Henry, and Tom. Then draw a picture of any one of them, say Sally, shaking hands with the other three.

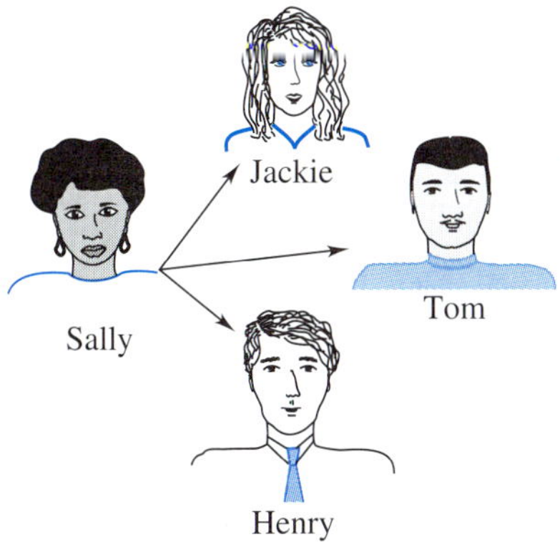

Sally shakes hands with Jackie, Tom, and Henry.

3 handshakes

1. Four children have one marble each. If each child trades his or her marble exactly once, how many trades occur?

Every handshake that includes Sally has been counted so that she can be removed from the picture. Next, Jackie is shown shaking hands with Tom and Henry.

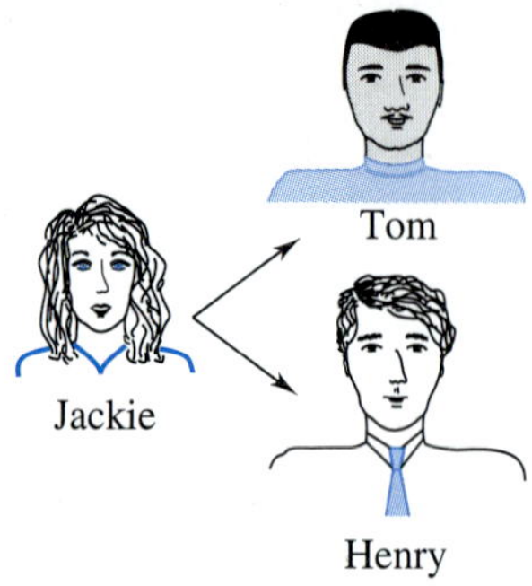

Jackie shakes hands with Tom and Henry. Jackie has already shaken hands with Sally.

2 handshakes

Every handshake that includes Jackie has been counted so that she can also be removed from the picture, leaving only Tom and Henry.

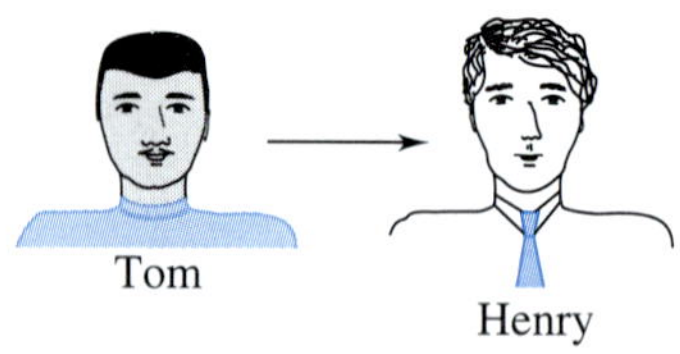

Tom and Henry shake hands.

1 handshake

A total of six different handshakes occurred. ■

Another approach to the problem in the last example will now be given. A diagram can be drawn that shows the four different individuals with the names A, B, C, and D. Each handshake is shown as a solid line.

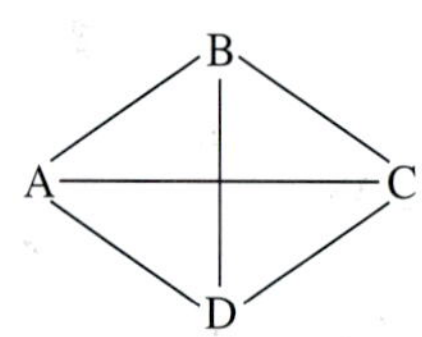

A shakes hands with B, C, and D.
B shakes hands with C, D, and A.
C shakes hands with D, A, and B.
D shakes hands with A, B, and C.
Nobody shakes hands more than once with the same person.

There are six solid lines so that six handshakes occurred.

◇ ***Do Practice Exercise 1.***

Sometimes a word problem is difficult to understand because too much information is given. Try the following suggestions.

Step 1 Read the problem carefully.

Step 2 Remove any information not needed.

Step 3 Rewrite the problem in your own words.

Step 4 Solve the problem.

Step 5 Make sure the answer is reasonable.

The next example is solved for you using the steps listed above.

ANSWER
1. Two trades occur.

EXAMPLE 2

At the end of 1985 an oil producer was selling two hundred barrels of oil per month for $28 per barrel. (Note: A barrel of oil is 42 gallons.) During the first quarter of 1986 the price of oil dropped to $12 per barrel. The producer then sold two hundred barrels per month at the lower price. By how much did the oil producer's income decrease? The oil producer was forced into bankruptcy since income at $12 per barrel was less than expenses.

SOLUTION

Step 1 Read the problem carefully.

Step 2 Remove any information not needed.

~~At the end of 1985~~ an oil producer was selling two hundred barrels of oil per month for $28 per barrel. ~~(Note: A barrel of oil is 42 gallons.) During the first quarter of 1986~~ the price of oil dropped to $12 per barrel. The producer then sold two hundred barrels per month at the lower price. By how much did the oil producer's income decrease? ~~The oil producer was forced into bankruptcy since income at $12 per barrel was less than expenses.~~

Step 3 Rewrite the problem in your own words.

An oil producer was selling two hundred barrels of oil per month at $28 per barrel. The price of oil dropped to $12 per barrel. What is the difference in income due to the price change?

Note: After rewording a problem it is important to ask if every important idea and/or number in the original problem remains. Ask yourself questions such as:

1. Is the year 1985 needed to solve the problem?
2. Do we need the information that a barrel is 42 gallons?
3. Do we need to know that the oil producer's expenses were more than income at $12 per barrel?
4. Do we need to know that the producer went bankrupt?

Step 4 Solve the problem.

Income at $28 per barrel		Income at $12 per barrel	
200	barrels	200	barrels
× 28	per barrel	× 12	per barrel
1600		400	
400		200	
$5,600	income at $28	$2,400	income at $12

$5,600	income at $28 per barrel
− 2,400	income at $12 per barrel
$3,200	**is the difference in income**

Step 5 Make sure the answer is reasonable.

Round off numbers and estimate values.

Estimates

$28 oil:	200 × 28	is approximately	200 × 30 = 6,000
$12 oil:	200 × 12	is approximately	200 × 10 = 2,000

(28 rounds off to 30; 12 rounds off to 10)

2. The following was modified from a book of nursery rhymes:

As I was walking towards St. Ives
Face to face, I met a man with seven wives
Each wife had seven sacks
Each sack had seven cats
Each cat had seven kittens
Kittens, cats, sacks, and wives
How many were going to St. Ives?

ANSWER

2. Only 1 was going to St. Ives. All of the others were coming from St. Ives.

The difference between the estimates is $4,000. The oil producer should be making approximately $4,000 less per month. The answer from the exact calculation was $3,200, which does not round off to $4,000. Perhaps a mistake was made when adding or multiplying.

$28 per barrel was *rounded up* to $30 per barrel, but $12 per barrel was *rounded down* to $10 per barrel. The rounding of these numbers in opposite directions before multiplying results in an estimate with less accuracy. Thus a mistake was not made and the answer does in fact seem reasonable. ■

◇ *Do Practice Exercise 2.*

Some problems cannot be solved with the information and/or concepts at hand. Sometimes a search for additional information is required or perhaps new concepts are needed in order to solve a problem. If this is the case, you *must* look at other resources to find the additional information or concepts needed.

EXAMPLE 3

Tod accepted a job at a salary of $23,000 per year. How much social security tax (FICA) will be withheld from his yearly salary? (Note: This problem occurs at the end of the section.)

SOLUTION

The problem cannot be solved without the social security tax rate, which is not given. Actually the tax rate is frequently changed. The tax rate for the current year can be found from the Internal Revenue Service (IRS), the library, or an accountant. *On your own, contact someone for the tax rate!* Social security tax is calculated using decimal numbers, which have not yet been covered in this book. Look at the chapter on decimals if you need help to multiply decimals, or use a calculator. ■

Sometimes a problem cannot be solved due to a lack of information. It is important to be able to recognize when additional information, concepts, or mathematical skills are needed to solve a problem.

EXAMPLE 4

An automobile trip took four hours. What was the average speed in miles per hour?

SOLUTION

This problem cannot be solved since average speed also requires distance traveled, which is not given. ■

Work each word problem.

1. In 1990, world population was appoximately five billion and doubling every forty years. What will the population be in 2030? Did you use deductive logic or inductive logic (or both) to find the answer?

2. Using information from problem 1, approximately how long will it take for the population to pass twenty billion?

3. Here is some information about two different jobs:

Job	*Annual Salary*	*Work Week (in hours)*	*Number of Weeks Worked per Year*
#1	$20,000	40	50
#2	$16,200	40	45

What is the hourly wage for each job? Which job has the highest hourly wage?

4. A blue sack contains fifteen red marbles and four blue marbles. A red sack contains twelve blue marbles and two yellow marbles. Three marbles are picked randomly from the red sack and placed in the blue sack. What is the maximum possible number of yellow marbles in the blue sack?

5. An individual has found that she must walk 20 miles in order to lose 1 pound. How far must she walk to lose 25 pounds?

6. Tod accepted a job at a salary of $23,000 per year. How much social security tax (FICA) will be withheld from his yearly salary? Use the tax rate for the current year. (*Hint:* See Example 3.)

7. An individual bought an item costing $1 and 2 items costing $3 each. What should the change be from a twenty dollar bill?

8. Five individuals walk into a room. How many handshakes occur if each person shakes hands once and only once with each of the others?

9. A company can only manufacture one product at a time. The following table gives the number of each of the items the company can manufacture, the cost to manufacture each item, and the selling price of the item. Complete the table.

Product	*Number Manufactured*	*Cost to Make Each*	*Sales Price of Each*	*Profit per Item*	*Total Profit*
1	800	$18	$22		
2	600	$22	$28		
3	1,000	$24	$28		

Which product should the company manufacture in order to have the largest possible profit?

10. Mars makes one revolution around the sun every 88 days. Earth requires 365 days to make one revolution around the sun. What is the age, in Martian years, of an individual who just turned 22 in Earth years? Assume a 90 day year for Mars and a 360 day year for Earth.

11. The total of payments and interest for a new automobile amounts to $19,780. A down payment of $2,500 is made. What will the monthly payment be if the balance is paid in 36 equal payments? If the balance is paid in 48 equal payments?

12. The depth of the Mariana Trench off of the coast of the Philippine Islands is 35,840 feet below sea level. Mount Everest in Nepal (between China and India) is 29,028 feet above sea level. What is the difference in elevation from the top of Mount Everest to the bottom of the Mariana Trench?

13. A doctor ordered enough Seconal for a daily dosage of 100 milligrams for ten consecutive days. How many capsules should the pharmacist provide if the medication comes in 20 milligram capsules?

14. Monthly expenses and income information for an apartment house are:

Item	*Monthly Expenses*	*Number of Units and Monthly Rent*
Payment	$12,500	Ten 3 bedroom units rent
Insurance	300	for $400 each
Taxes	1,200	Twenty 2 bedroom units
Utilities	1,000	rent for $350 each
Payroll	3,000	Thirty 1 bedroom units
Maintenance	2,000	rent for $310 each

Assuming the apartment house is full, find:

1. Total monthly expenses.
2. Total monthly income from rent.
3. Monthly income minus monthly expenses.

List four realistic situations or events that can reduce the rental income from an apartment complex.

JOURNAL EXERCISES

Write a short essay describing the following. Some of the questions may require you to do research at a library.

1. ABC, Inc. had sales of $600,000 in 1992 and sales of $722,000 in 1993. Estimate sales for 1994 and explain how you made the estimate.

2. Describe the life of a famous mathematician such as Newton, Gauss, Laplace, Fourier, Pascal, Euler, or Descartes.

3. Explain how both inductive and deductive reasoning are used in every day life.

4. Discuss a branch of mathematics such as arithmetic, algebra, trigonometry, geometry, or calculus.

5. Talk about how mathematics and/or mathematical type thinking can be used in your career.

6. Try to think of how computers are changing society.

The purpose of this section is to introduce the idea of a **variable**. Variables are important concepts used in algebra, economics, physics, chemistry, and biology as well as in many other areas of study.

OBJECTIVES

1 Understand the concept of variables.

2 Identify and work with terms.

3 Identify and evaluate algebraic expressions.

1 Variables

Arithmetic is one area, or branch, of mathematics. Algebra is another branch of mathematics that is similar to arithmetic in many ways. For example, whole numbers are used in algebra as well as arithmetic. The basic mathematical operations of both arithmetic and algebra are addition, subtraction, multiplication, and division. Algebra differs from arithmetic by using a letter, called a **variable**, to represent an unknown number. A variable can be any letter, such as "B," "x," "y," and so on. The idea of a variable is very important and will be introduced through the use of several examples.

EXAMPLE 1

A biologist is trying to estimate the number of fish in a lake. Use a variable for the number of fish in the lake.

SOLUTION

Let N = the total number of fish in the lake. ■

EXAMPLE 2

Tim has some marbles in his pocket but has not counted them and does not know how many are there. Use a variable for the number of marbles in Tim's pocket. Tim doubles the number of marbles in his pocket. Use the same variable to represent the new number of marbles in Tim's pocket.

SOLUTION

Let x = the number of marbles in Tim's pocket.

After doubling the number of marbles in his pocket, Tim now has $2 \cdot x$ marbles. ■

EXAMPLE 3

The owner of a cafe does not know exactly how many eggs he will use in a week. Represent the number of eggs he thinks he needs as a variable. To be safe, he orders 50 more than he thinks he needs. Write this quantity using the variable.

SOLUTION

Let y = the number of eggs he thinks he needs.

But he actually orders $y + 50$ eggs to be safe. ■

EXAMPLE 4

An individual borrows $1,000 for one year. Use a variable for the amount of interest she must pay. Use this variable to state the amount that must be repaid at the end of the year.

SOLUTION

Let I = the amount of interest at the end of one year.

She must repay $1,000 + I at the end of one year. ■

◇ *Do Practice Exercises 1 through 3.*

1. A rancher had N cattle but sold 300. Use N and 300 to state the number of cattle remaining.

2. A city has a population of P. The population then doubles. What is the new population?

3. A student has D dollars in her checking account. She adds $150 to her account. Use D and $150 to state her current checking account balance.

ANSWERS
1. $N - 300$
2. $2 \cdot P$
3. $D + 150$

4. Find the value for the variable by trial and error:

(a) $n + 3 = 12$; find the value for n.

(b) $x - 7 = 9$; find the value for x.

(c) $y + 2 = 17$; find the value for y.

A variable can be used in an equation. When it is, we may be able to figure out what the variable is equal to by using trial and error. Work the next two examples.

EXAMPLE 5

Using trial and error, find the value of x that makes the equation true:

$$x + 5 = 12$$

SOLUTION

Try 4 in place of $x \longrightarrow 4 + 5 = 9 \longrightarrow x$ is not 4.
Try 5 in place of $x \longrightarrow 5 + 5 = 10 \longrightarrow x$ is not 5.
Try 7 in place of $x \longrightarrow 7 + 5 = 12 \longrightarrow x = 7$ works.

In other words, $x = 7$. ■

EXAMPLE 6

Find the value for R which makes the equation true:

$$17 = 30 - R$$

SOLUTION

Try $R = 5 \longrightarrow 30 - 5 = 25 \longrightarrow x$ is not 5.
Try $R = 10 \longrightarrow 30 - 10 = 20 \longrightarrow x$ is not 10.
Try $R = 13 \longrightarrow 30 - 13 = 17 \longrightarrow x = 13$ works.

Here, $R = 13$. ■

◇ *Do Practice Exercise 4.*

2 Terms

As we have already seen, multiplication in arithmetic can be written without using a multiplication sign. For example:

(7)(9) seven times nine

Multiplication with variables can also be written without using a multiplication sign. For example, two times the variable x is usually written as:

$2x$ Read as two x or two times x.

No multiplication sign is written, but it is understood to be there. Thus, $5y$ is five times the variable y.

A number by itself, a variable by itself, or a number times one or more variables are examples of **terms**. Some examples of terms are given next.

Examples of Terms	*Read as*
7	seven
N	N
$3r$	three times r
$15xy$	fifteen times x times y
$12A$	twelve times A

ANSWERS
4. (a) 9 (b) 16 (c) 15

EXAMPLE 7

A bicycle rider averages ten miles per hour. Write a term for the number of miles traveled by the rider in t hours.

SOLUTION
The rider will travel $10 \cdot t = 10t$ miles in t hours. ■

EXAMPLE 8

A jet is traveling at M miles per hour. How far will it travel in 8 hours?

SOLUTION
The jet would travel $8M$ miles. ■

EXAMPLE 9

Average sales at a pharmacy are S dollars per day. What would the total sales be for a period of 7 days?

SOLUTION
The sales would be $7S$. ■

◇ *Do Practice Exercises 5 and 6.*

3 Algebraic Expressions

An **expression** occurs when two or more terms are added or subtracted. **Variable expressions** are formed by adding or subtracting terms, some of which contain variables. For example:

Variable Expression	*Read as*
$x + 10$	x plus ten
$y - 8$	y minus eight
$5z + 4$	five times z plus four

EXAMPLE 10

Tim has x marbles in his pocket. Write an expression for the number remaining after he removes two marbles from his pocket.

SOLUTION
The number of marbles remaining can be written as $x - 2$. ■

EXAMPLE 11

Write an expression for the number of marbles in Tim's pocket if he has x, triples them, then removes 5.

SOLUTION
He would then have $3x - 5$ marbles. ■

◇ *Do Practice Exercise 7.*

5. A healthy baby gains N pounds per month on average. How many pounds will it gain in 3 months?

6. A boy watches an average of H hours of television each day. How many hours will the boy watch in 30 days?

7. The speed limit in a school zone is M miles per hour. The speed limit is 10 miles per hour higher as the school zone is left behind. Write an expression for the faster speed limit.

ANSWERS
5. $3N$ pounds
6. $30H$ hours
7. $M + 10$

8. Evaluate $8Y$ given that $Y = 7$.

9. Evaluate $X + A$ given that $X = 2{,}412$ and $A = 759$.

10. Evaluate M divided by 18 given that $M = 360$.

ANSWERS
8. 56
9. 3,171
10. 20

The types of tools used in arithmetic and algebra are summarized in the following table.

	Examples of the Tools Used in	
	Arithmetic	*Algebra*
Mathematical operators	$+, -, \times, \div$	$+, -, \times, \div$
Numbers	3, 0, 243	3, 0, 243
Variables	none	x, y, z, r
Expressions	$5 + 3 \cdot 2$	$x - 12$

EXAMPLE 12

Write an expression for the sum of an unknown number plus twenty.

SOLUTION

Let $x =$ the unknown number, then the expression is

$$x + 20 \quad \blacksquare$$

A variable expression such as $x + 17$ can be **evaluated** if the value of the variable x is known. The steps to evaluate the expression are to:

1. Substitute the value for the variable.

2. Do the arithmetic.

Look at the following examples.

EXAMPLE 13

Evaluate $x + 12$ given that $x = 7$.

SOLUTION

$x + 12$	
$7 + 12$	Replace x with 7.
19	Add. ■

EXAMPLE 14

Evaluate $6xy$; $x = 3$, $y = 2$.

SOLUTION

$6xy$	
$6(3)(2)$	Replace x with 3 and y with 2.
$18(2)$	Multiply.
36	Result. ■

◇ *Do Practice Exercises 8 through 10.*

EXERCISE SET 1.8

Evaluate each expression.

1. $x + 43;\quad x = 10$

2. $21 - x;\quad x = 7$

3. $17 + y;\quad y = 5$

4. $y - 9;\quad y = 12$

5. $a + y;\quad a = 2,\quad y = 4$

6. $a + a + 1;\quad a = 12$

7. $7x;\quad x = 8$

8. $b - r;\quad b = 17,\quad r = 3$

9. $a + b + c + d;\quad a = 1,\quad b = 3,\quad c = 5,\quad d = 4$

10. $5bc;\quad b = 2,\quad c = 5$

11. $y + 6 + x;\quad y = 1,\quad x = 2$

12. $9 - r;\quad r = 5$

13. $sty;\quad s = 2,\quad t = 3,\quad y = 1$

14. $0 + t;\quad t = 1$

15. $7t;\quad t = 0$

16. $x + y;\quad x = 7,\quad y = 5$

17. $9 - y;\quad y = 4$

18. $A - B;\quad A = B = 8$

19. $A + x + y;\quad A = 1,\quad x = 2,\quad y = 9$

20. $y - 15;\quad y = 4{,}000$

21. $2{,}000 + A;\quad A = 1$

22. $4{,}000 - B;\quad B = 3{,}999$

23. $6x;\quad x = 405$

24. $9xz;\quad x = 4,\quad z = 8$

25. $3xyz;\quad x = y = 5,\quad z = 0$

26. $C - D;\quad C = 12,\quad D = 1$

27. $A + b;\quad A = 6{,}241,\quad b = 329$

28. $AB;\quad A = 7,\quad B = 9$

29. $abc;\quad a = 9,\quad b = 4,\quad c = 2$

30. $90b;\quad b = 1{,}000$

31. $20 - y;\quad y = 14$

32. $x + y;\quad x = 1,\quad y = 0$

Write an algebraic expression for the unknown.

Example: *If Tommy has y cents and spends* 25 *cents, how much is left?* Answer: $y - 25$ *cents*

June has x cents. Find an algebraic expression for the money remaining if:

33. June spends 40 cents.
34. Someone gives June 20 cents.
35. June first spends 40 cents and then receives 20 cents.

Let N be the number of marbles in a box. Find an expression for the number of marbles in the box if:

36. Fifteen marbles are removed.
37. The number of marbles are doubled.
38. The number of marbles are doubled then fifteen marbles are removed.

A herd is made of 2,800 caribou. Write an expression for the number of caribou if:

39. Wolves eat W caribou.
40. N new caribou are born.
41. Wolves eat W caribou and N caribou are born.
42. Wolves eat W caribou, an additional D caribou die, and N caribou are born.

Tom has T books and Jean has J books. Write an expression for:

43. The number of books they have together.
44. Two times the number of books Tom has.
45. The number of books Tom has after donating D books to a charity.
46. The number of books Jean has after donating K books to a charity.

Represent the unknown quantity using an algebraic expression.

Example: *Find an expression for the sum of two unknown numbers with the number twenty-seven.*
Answer: $x + y + 27$

47. Find an expression for the sum of an unkown number and the number eighteen.

48. Find an expression for the difference of two unknown numbers.

49. Write an expression for the sum of two unknown numbers.

50. Find an expression for the product of seven times an unknown number.

51. An unknown number of cattle were in a feedlot. One thousand cattle were sold. Find an expression for the number of cattle remaining in the feedlot.

JOURNAL EXERCISES

1. Explain the concept of a variable. Give several examples.

2. How are arithmetic and algebra alike?

3. Explain some of the differences between arithmetic and algebra.

4. Have you seen variables in any other courses you have taken? Which ones?

CHAPTER 1 REVIEW

SECTION	TOPIC OR TERM	EXAMPLE OR DEFINITION
1.1		
Place Values	Counting Numbers	1, 2, 3, 4, 5, 6, 7, 8, 9, 10, 11, 12, . . .
	Whole Numbers	0, 1, 2, 3, 4, 5, 6, 7, 8, 9, 10, 11, 12, . . .
	Place Value	Place value is the idea that the value of a digit depends on its position. For example, the 2 in 425 is in the tens place.
	Expanded Notation	Expanded notation is the writing of a number as the sum of its place value pieces: 3,860 = 3,000 + 800 + 60
1.2		
Rounding and Order	Rounding	2,740 rounded to the nearest one thousand is 3,000.
	Order	Order refers to the comparison of the size of two or more numbers. For example, $7 < 12$ (7 is less than 12).
1.3		
Addition and Subtraction	Sum	The sum is the answer to an addition problem.
	Identity Property	$25 + 0 = 25$
	Commutative Property	$3 + 8 = 8 + 3$
	Associative Property	$5 + (6 + 7) = (5 + 6) + 7$
	Difference	The difference is the answer to a subtraction problem.
	Keyword	A keyword is a word that indicates which math operation ($+, -, \times$, or $\div$) to use.
1.4		
Multiplication	Product	Product is the answer to a multiplication problem.
	Factor	A factor is one of two or more numbers being multiplied.
	Estimate	An estimate is an approximate answer to a problem.
1.5		
Division	Quotient	The quotient is the answer to a division problem.
	Remainder	The remainder is the number left over after dividing two numbers.
	Average	The average of four numbers is their sum divided by four. Similarly for the average of five numbers, and so on.

SECTION	TOPIC OR TERM	EXAMPLE OR DEFINITION
1.6		
Critical Thinking Using Inductive and Deductive Reasoning	Inductive Reasoning	Inductive reasoning looks for patterns and assumes they continue. For example, 5, 7, 15, 17, 25, 27, __?__ ? = 35
	Deductive Reasoning	Deductive reasoning is based on facts. For example, $5 + 8 = 13$.
1.7		
Word Problems	Word Problem	A word problem is a problem that is stated in words.
1.8		
Introduction to Variables (Optional)	Variable	A variable is a letter such as x or y or z that represents an unknown value.
	Term	$5x$, $6xy$, $20ABC$
	Expression	$x + 7$ or $3y - 12x + 5$
	Evaluate Expressions	Evaluate $x + 12$ given that $x = 4$: $x + 12$ $4 + 12$ Replace x with 4. 16 Result.

THE FAMILY BUDGET

John and Lupe Summers are newly married. John works as a cook at a restaurant and receives a salary of $1,800 per month. Lupe works as a waitress at the same restaurant and makes $400 per month in addition to tips. Her tips average two dollars for each table she waits on. The number of tables she waits on during an average week are:

Days of the Week

S	M	T	W	TH	F	S
0	0	5	15	20	22	24

1. How many tables does Lupe wait on during an average week?
2. What is Lupe's average weekly income from tips?
3. Multiply four weeks in a month times Lupe's weekly income from tips. Add this figure to her monthly salary to find her average monthly income.
4. What is the combined monthly income from both Lupe and John Summers?
5. If taxes amount to $388 per month, what is the family's take-home income after tax?

   The Summers have the following monthly bills:

Rent	$450	Car payments	$260
Utilities	45	Car maintenance	150
Telephone	40	Insurance	125
Groceries	380	Medical	80
Retirement plan	100	Miscellaneous	200

6. What are their total monthly bills not including taxes?
7. How much of their income remains after all expenses are paid? This is the amount of money they can save in a month.
8. How many months must they save in order to purchase a new dining room set costing $800? How much would they have left over?

CHAPTER 1 REVIEW PROBLEMS

Place commas in the correct places and write each number using words.

1. 6243

2. 14512

3. 920046

4. 302721

5. 5123450

6. 32400521

Write the digits of each number.

7. Seventeen thousand, four hundred seven

8. Eighty-four million, ten thousand, twelve

9. One hundred sixty-eight thousand, eight hundred

10. Nine billion, seven thousand, six hundred twenty

11. Four hundred thousand, nine hundred fifty-six

12. Six trillion, four thousand

Round each number to the indicated place value.

	to the nearest
13. 16,007	one thousand
14. 47,291	one hundred
15. 971,999	one million
16. 764,321,077	ten thousand
17. 16,741,725,000	one hundred million
18. 6,099,000,000	one billion

Work each problem.

19. $7 \cdot 8$

20. $28 \div 4$

21. $6 + 15$

22. $(6)(7)$

23. $63 \div 9$

24. $19 - 14$

25. $\begin{array}{r} 16 \\ \times 12 \\ \hline \end{array}$

26. $\begin{array}{r} 450 \\ \times\ 37 \\ \hline \end{array}$

27. $\begin{array}{r} 718 \\ \times\ 26 \\ \hline \end{array}$

28. $375 \div 5$

29. $1{,}924 \div 19$

30. $265 \div 25$

31. $\begin{array}{r} 1{,}694 \\ 327 \\ +\ 918 \\ \hline \end{array}$

32. $\begin{array}{r} 3{,}000{,}000 \\ -\quad 7{,}921 \\ \hline \end{array}$

33. $\begin{array}{r} 6{,}452 \\ \times\ 127 \\ \hline \end{array}$

34. $1{,}600 \div 40$

35. $79 \cdot 43$

36. $785 \cdot 1{,}000$

37. $684 + 1{,}285$

38. $7{,}940 - 329$

Place either a $<$ or $>$ symbol between the two numbers as needed to make the statement true.

39. 3 ? 7

40. 4 ? 1

41. 12 ? 16

42. 75 ? 120

Evaluate.

43. $x - 12;\quad x = 27$

44. $y + r;\quad y = 7,\quad r = 12$

45. $19 - s;\quad s = 7$

46. $x + y + t;\quad x = 2,\quad y = 9,\quad t = 3$

47. $3ab;\quad a = 1,\quad b = 3$

48. $78 - y;\quad y = 17$

Determine which mathematical property applies.

Example: $3 + t = t + 3$ Answer: *Commutative property of addition*

49. $7 \cdot 1 = 7$

50. $9 + 12 = 12 + 9$

51. $6(3 + 4) = 6 \times 3 + 6 \times 4$

52. $3 \times 7 = 7 \times 3$

Work each problem.

53. What is the difference between twenty-six and the sum of two and eight?

54. Write an expression for the sum of 8 plus an unknown number.

55. Let N be the number of dollars a gambler begins with. Write an expression for the amount remaining after the gambler loses seventy-five dollars.

56. A song is played at an average rate of two musical notes per each second. If the song lasts three minutes, how many notes does it contain?

57. Jeff's test scores were 65, 50, 75, and 82. Find his average score.

58. An individual wants to lose fifteen pounds. How many miles must the individual walk if it takes twenty-two miles to lose one pound?

59. Sally, who is 42, has brothers aged 24, 31, and 39. Find the average age of the four.

60. A local supermarket averages sales of one million, twenty-three thousand, fifty dollars per month. How many customers go to the supermarket in a month if the average customer spends twenty-five dollars?

CHAPTER 1 TEST

Work each.

1. $\begin{array}{r} 675 \\ \times\ 67 \\ \hline \end{array}$

2. $\begin{array}{r} 1{,}701 \\ -\ 128 \\ \hline \end{array}$

3. $974 \div 4$

4. $1{,}025 \div 12$

5. $(6)(8)(2)$

6. $\begin{array}{r} 7{,}284 \\ +\ 368 \\ \hline \end{array}$

7. $365 \cdot 12 \cdot 4$

8. $12\overline{)4{,}870}$

9. $37 \cdot 48$

10. $6{,}227 + 4{,}185$

11. $9{,}300 - 98$

12. $3{,}215 \div 30$

Which mathematical property is illustrated?

13. $4 \times (3 \times 2) = (4 \times 3) \times 2$

14. $8(7 + 5) = 8 \times 7 + 8 \times 5$

15. $16 \times 0 = 0$

16. $9 \times 21 = 21 \times 9$

Evaluate.

17. $x + 12;\quad x = 20$

18. $1{,}000 - y;\quad y = 39$

19. $x + y + z;\quad x = 17,\quad y = 4,\quad z = 2$

20. $3y;\quad y = 7$

Round each to the nearest one thousand.

21. 3,486

22. 57,912

Work each.

23. Round 7,421,300 to the nearest one hundred thousand.

24. Write 37,000,053 using words.

25. Is $16 > 24$ a true statement? Explain.

26. Monthly sales for three months were \$24,000, \$18,600, and \$19,500. Find the average sales per month.

27. A large German company had the following number of employees:

Year	*Number Employed*
1991	17,150
1992	19,597
1993	12,440

What was the average number employed during 1991–1993? Round the answer to the nearest ten thousand.

28. The land area of the seven continents is about 50,000,000 square miles and world population is near 5,000,000,000. Calculate the average number of people per square mile of land area.

29. A runner jogs the eight miles from city A to city B and walks back to city A every day of the week for ten days in a row. How many miles does he jog?

30. Four people can row a boat at 4 miles per hour in calm water. The boat is placed in a river flowing at F miles per hour. Using the variable F, find a variable expression for the speed of the boat when rowed downstream. (*Hint:* Will the boat be going faster than or slower than F miles per hour when rowed downstream?)

Applications with Whole Numbers

2

APPLICATION *The number of humans in the world is given in the table below. The number for the year 2000 is a forecast. Show the data in a bar chart.*

Year	*Approximate Number of Humans*
1850	1,000,000,000
1900	1,400,000,000
1950	3,000,000,000
2000	6,000,000,000

SOLUTION

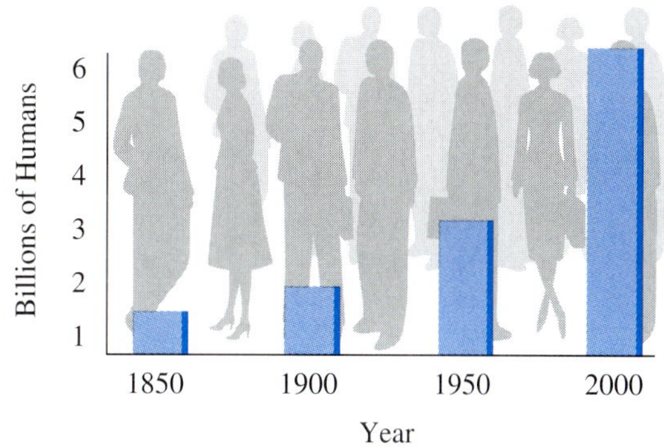

2.1 EXPONENTS

The four basic operations of mathematics discussed in Chapter 1 were addition, subtraction, multiplication, and division. A new notation for repeated multiplication with the same number is shown in this section.

OBJECTIVES

1. Use exponents.
2. Apply exponents to real problems.
3. Multiply bases raised to a power.

1 Exponents

As we saw in Chapter 1, multiplication is repeated addition of the same number, for example $8 + 8 + 8 = 3 \times 8$. **Exponents** are used to show repeated *multiplication* of a number times itself. For example, the product 4 times 4 times 4 can be written as:

$$4 \cdot 4 \cdot 4 = 4^3$$

(3 → exponent; 4 → base)

The **exponent** is the small number at the upper right hand corner of the **base**. The exponent tells the number of times the base is to be used as a factor in multiplication by itself. For example, an exponent of 5 means that five bases are to be multiplied together:

$$3^5 = \underbrace{3 \cdot 3 \cdot 3 \cdot 3 \cdot 3}$$

← 5 threes are multiplied together.

Several examples of exponents are shown next.

1. Write each as the product of bases:

(a) 8^3

(b) 5^2

(c) 10^4

2. Evaluate each:

(a) 6^2

(b) 5^3

(c) 4^3

3. Nine is three raised to what power?

ANSWERS
1. (a) $8 \times 8 \times 8$ (b) 5×5 (c) $10 \times 10 \times 10 \times 10$
2. (a) 36 (b) 125 (c) 64
3. 2nd power

Example	*Read as*	*Meaning*	*Value*
5^2	Five to the second power	$5 \cdot 5$	25
5^3	Five to the third power	$5 \cdot 5 \cdot 5$	125
5^4	Five to the fourth power	$5 \cdot 5 \cdot 5 \cdot 5$	625
12^2	Twelve to the second power	$12 \cdot 12$	144
3^4	Three to the fourth power	$3 \cdot 3 \cdot 3 \cdot 3$	81
200^2	Two hundred to the second power	$200 \cdot 200$	40,000

◇ *Do Practice Exercise 1.*

A number raised to the second power is said to be **squared** and a number raised to the third power is said to be **cubed**.

EXAMPLE 1

Evaluate six cubed and five squared.

SOLUTION

$$6^3 = 6 \cdot 6 \cdot 6 = 36 \cdot 6 = 216$$
$$5^2 = 5 \cdot 5 = 25$$ ■

EXAMPLE 2

Eight is two raised to what power?

SOLUTION
The problem is to find the exponent such that

$$2^? = 8$$

$2 \times 2 \times 2 = 8$, therefore $2^3 = 8$. The required exponent is 3. ■

◇ *Do Practice Exercises 2 and 3.*

Special cases occur when a base is raised to a power of 0 or 1. Any number raised to the first power is the number itself. Any number other than zero raised to the zero power is defined to be equal to one. For example:

$$7^0 = 1 \qquad \text{and} \qquad 15^0 = 1$$

Zero raised to the zeroth power has no meaning and is said to be **undefined**. Undefined means that it does not have any meaning in mathematics.

Exponents of One	*Exponents of Zero*
$5^1 = 5$	$7^0 = 1$
$17^1 = 17$	$12^0 = 1$
$0^1 = 0$	$195^0 = 1$
$1{,}000^1 = 1{,}000$	0^0 is undefined.

EXAMPLE 3

Evaluate 100^0 and 75^1.

SOLUTION

$$100^0 = 1 \qquad 75^1 = 75$$ ■

Whenever there is no exponent, the exponent is understood to be the number 1. For example, $7 = 7^1$, $12 = 12^1$, and so on.

EXAMPLE 4

Write 1,200 using the exponent 1.

SOLUTION

$$1{,}200 = 1{,}200^1$$ ■

◇ *Do Practice Exercise 4.*

2 An Application of Exponents

One application of exponents is in the study of population growth. The next example looks at recent trends in the population of Africa. We assume that the same trend in growth rate continues.

EXAMPLE 5

The current population of Africa is about six hundred million people, and it is estimated to be doubling every twenty-five years. If this trend continues, what will the population of Africa be in one hundred years?

SOLUTION

The population of Africa is 600,000,000. In 25 years the population will double to 1,200,000,000. In another 25 years the population will again double to 2,400,000,000. The table shows the pattern.

Year	*Population (doubles each 25 years)*	*Calculated as*
current	600,000,000	600,000,000
in 25 years	1,200,000,000	$600{,}000{,}000 \times 2$
in 50 years	2,400,000,000	$600{,}000{,}000 \times 2 \times 2$
in 75 years	4,800,000,000	$600{,}000{,}000 \times 2 \times 2 \times 2$
in 100 years	9,600,000,000	$600{,}000{,}000 \times 2 \times 2 \times 2 \times 2$

At this rate, the population of Africa will double 4 times in 100 years.

$$600{,}000{,}000 \times 2 \times 2 \times 2 \times 2 = 600{,}000{,}000 \times 2^4 = 9{,}600{,}000{,}000$$

If the population continues to grow at the same rate, Africa will have twice as many people in one hundred years as the entire world has today. ■

3 Multiplying with Exponents

There is a shortcut to multiplying a base raised to a power times the exact same base raised to another power—the exponents can be added. The idea behind this is shown below:

three tens

The exponents can be added.

$$10^2 \cdot 10^3 = \underbrace{10 \cdot 10}_{\text{two tens}} \cdot \overbrace{10 \cdot 10 \cdot 10}^{\text{three tens}} = 10^{2+3} = 10^5$$

two tens

4. Evaluate each:

(a) 25^1

(b) 20^0

(c) 0^0

(d) 1^1

ANSWERS
4. (a) 25 (b) 1 (c) undefined (d) 1

5. Write the product of 5^2 times 5^4 as 5 to some power.

6. Simplify $3^4 3^3$.

7. Simplify 200×200^3.

8. Write $4 \times 4 \times 2 \times 2 \times 2$ as the product of bases raised to the correct powers.

Ten squared is 10 times 10, and ten cubed is 10 times 10 times 10. Therefore, ten squared times ten cubed is:

$$(10 \times 10) \times (10 \times 10 \times 10) \text{ which is } 10^5$$

It is important to remember that this rule only applies *if the bases are identical.* For example,

$$8^2 3^2 = 8 \times 8 \times 3 \times 3$$

cannot be written as 8 to some power nor as 3 to some power.

EXAMPLE 6

Write $8^2 \times 8^4$ as eight raised to a power.

SOLUTION

$8^2 \cdot 8^4 = 8^{2+4} = 8^6$ which equals $8 \cdot 8 \cdot 8 \cdot 8 \cdot 8 \cdot 8$ or 262,144

→ Notice that the bases are identical. ■

The word **simplify** is used to tell you to write some quantity in a simpler way. The next example shows one use of the word.

EXAMPLE 7

Simplify $12^3 12^4$.

SOLUTION

$12^3 12^4$	Notice that the bases are identical.
12^{3+4}	Add exponents.
12^7	Result. ■

Sometimes no simplification is possible as shown in the next example. In that case, the answer is said to already be simplified.

EXAMPLE 8

Simplify $3^2 4^3$.

SOLUTION

$3^2 4^3$ cannot be simplified since the bases are different. ■

◇ *Do Practice Exercises 5 through 7.*

EXAMPLE 9

Write $3 \cdot 3 \cdot 3 \cdot 5 \cdot 5$ as the product of bases raised to the appropriate power.

SOLUTION

$$3 \cdot 3 \cdot 3 = 3^3 \quad \text{and} \quad 5 \cdot 5 = 5^2$$

$$\text{Thus} \quad 3 \cdot 3 \cdot 3 \cdot 5 \cdot 5 = 3^3 5^2 = 3^3 5^2$$ ■

◇ *Do Practice Exercise 8.*

ANSWERS
5. 5^6
6. 3^7
7. 200^4
8. $4^2 2^3$

Work each, showing all steps. (Review)

1. 65×78 **2.** $2,681 + 427$ **3.** $965 - 787$

4. $186 \div 21$ **5.** 912×420 **6.** $650 + 2,175 + 98$

Write each number using words. (Review)

7. 3,400 **8.** 18,001

9. 9,000,047 **10.** 1,000,500,000

Evaluate.

11. 5^2 **12.** 7^3 **13.** 2^3 **14.** 4^1 **15.** 9^2

16. 10^3 **17.** 1^4 **18.** 5^0 **19.** 8^2 **20.** 100^2

21. 10^4 **22.** 1^0 **23.** 2^8 **24.** 5^4 **25.** 0^0

26. 4^2 **27.** 3^4 **28.** 12^2 **29.** 20^3 **30.** 5^3

31. 75^1 **32.** 11^2 **33.** 7^4 **34.** 27^0 **35.** 4^3

36. 1^8 **37.** 3^0 **38.** 8^0 **39.** 9^1 **40.** 8^4

Write each as a base to a power or as a product of different bases to appropriate powers.

41. $2 \cdot 2$ **42.** $3 \cdot 3 \cdot 3$ **43.** $7 \cdot 7$ **44.** $2 \cdot 2 \cdot 2$

45. $1 \cdot 1 \cdot 1 \cdot 1$ **46.** $5 \cdot 5 \cdot 5$ **47.** $11 \cdot 11$ **48.** $7 \cdot 7 \cdot 7 \cdot 7$

49. $4 \cdot 4 \cdot 4$ **50.** $13 \cdot 13 \cdot 13$ **51.** $5 \cdot 5 \cdot 5 \cdot 5$ **52.** $2 \cdot 2 \cdot 5 \cdot 5$

53. $3 \cdot 3 \cdot 4 \cdot 4 \cdot 4$ **54.** $2 \cdot 2 \cdot 2 \cdot 5 \cdot 5$ **55.** $3 \cdot 3 \cdot 8 \cdot 8 \cdot 8$ **56.** $2 \cdot 2 \cdot 9 \cdot 9 \cdot 9$

57. $18 \cdot 18 \cdot 18$

58. $20 \cdot 20 \cdot 3 \cdot 3$

59. $5 \cdot 5 \cdot 7 \cdot 7 \cdot 7$

60. $3 \cdot 3 \cdot 19 \cdot 19$

Simplify each by writing as a base to a power.

61. $2^5 2^7$

62. $7^2 7^3$

63. $9^0 9^2$

64. $8^3 8^4$

65. $2^3 2^5$

66. $7^1 7^2$

67. $4^5 4^8$

68. $12^2 12^4$

69. $207^2 207^5$

70. $1^2 1^3$

71. $100^2 100^2$

72. $4^2 4^0$

Work each problem.

73. Which is larger, three to the third power or five to the second power?

74. Which is smaller, seven to the third power or twelve squared?

75. What is twenty-four thousand reduced by seventeen thousand, four hundred?

76. Is $6 > 54$ a true statement?

77. Can $4^3 5^2$ be simplified?

78. Round two million, eight hundred fifty thousand to the nearest one million, then multiply the result times three squared.

79. What is the average of these three numbers: seven squared, three squared, and two raised to the fifth power?

80. A colony is estimated to contain ten thousand bacteria and is estimated to be doubling in number every six hours. Approximately how many bacteria will there be in twenty-four hours?

81. A group of rabbits is doubling in population every year. What would the population be in ten years if they begin with five rabbits? In twenty years?

JOURNAL EXERCISES

1. Explain the differences between the mathematical operations of addition, multiplication, and raising numbers to exponents.

2. Write about the potential effects on African society if population growth continues at the same rate for 200 years from now (see Example 5). Include possible effects on natural resources such as water, oil, forests, and wildlife.

1 Arithmetic Expressions

The expression $(15 - 3 \times 4)$ evaluates to different numbers depending upon whether subtraction or multiplication is calculated first:

	subtract first		multiply first	
	$15 - 3 \times 4$		$15 - 3 \times 4$	
???	12×4	???	$15 - 12$	???
	48	⟵ Which is correct ⟶	3	

Both answers cannot be correct. We will come back to this problem in a moment. First, a set of rules is needed to define the order in which the mathematical operations in an expression are calculated. These rules are called the **Order of Operations**. Every expression should be evaluated according to these rules.

Rule: To Evaluate an Expression

Step 1 Evaluate the contents of the parentheses and above and below the fraction bar.

Step 2 Evaluate exponents.

Step 3 Evaluate multiplication and division in the order they occur when moving from left to right across the expression.

Step 4 Evaluate addition and subtraction in the order they occur when moving from left to right across the expression.

The rules above state that multiplication in an expression is calculated before addition in the same expression. Therefore,

$$15 - 3 \times 4 = 15 - 12 = 3$$

Notice that subtraction is the last operation done in this particular problem.

The next example includes several different operators in one expression. In this example, you will see that the parentheses are dropped only after its contents have been simplified to a single number.

EXAMPLE 1

Evaluate $17 - 1 \cdot (6 \div 2)$.

SOLUTION

$17 - 1 \cdot (6 \div 2)$	
$17 - 1 \cdot (3)$	contents of parentheses
$17 - 1 \cdot 3$	no longer a need for parentheses
$17 - 3$	Multiplication before subtraction.
14	Result.

Again, notice that subtraction is the *last* operation for this example. ■

◇ *Do Practice Exercises 1 and 2.*

OBJECTIVE

1 Evaluate arithmetic expressions.

1. Evaluate $25 - 3 \cdot 2$.

2. Evaluate $3 \cdot (8 - 4)$.

ANSWERS
1. 19
2. 12

3. Evaluate $27 \cdot 2 \div 3 + 12$.

4. Evaluate $(3 + 7) \div 2$.

ANSWERS
3. 30
4. 5

The third rule of the order of operations does not mean that multiplication is calculated before division. It states that multiplication and division are calculated as they occur in moving from the left to the right across the expression. Look at the next example.

EXAMPLE 2

Evaluate $6 \div 2 \cdot 4 \div 3 \cdot 8$.

SOLUTION

$6 \div 2 \cdot 4 \div 3 \cdot 8$	
$3 \cdot 4 \div 3 \cdot 8$	$\div$ and $\cdot$ from left to right
$12 \div 3 \cdot 8$	Multiply.
$4 \cdot 8$	Divide.
32	Result. ■

EXAMPLE 3

Evaluate $8 \cdot 7 - 4$.

SOLUTION

$8 \cdot 7 - 4$	
$56 - 4$	Multiply first.
52	Result. ■

◇ *Do Practice Exercises 3 and 4.*

Please read the following Helpful Hint carefully. Using these techniques will cut down on errors and help prepare you for future math classes.

HELPFUL HINT

Calculate only one mathematical operation in moving from one row to the next. The remainder of each row should be identical to the row above. This technique is used for each example in this section.

If you use the recommended technique, you will notice that:

1. The most complicated row of each problem is the first row.
2. Each subsequent row is a little simpler.
3. The answer is at the bottom.

Furthermore, the overall shape of the problem-solving process is that of a triangle as shown in the next example.

EXAMPLE 4

Evaluate $(3 + 7 \cdot 5) - 8$.

SOLUTION

$(3 + 7 \cdot 5) - 8$

$(3 + 35) - 8$ Multiplication inside parentheses.

$38 - 8$ Addition inside parentheses.

30 Result. ■

EXAMPLE 5

Evaluate $(5 + 4)^2 \div 3$.

SOLUTION

$(5 + 4)^2 \div 3$

$9^2 \div 3$ contents of parentheses

$81 \div 3$ exponents

27 Result. ■

◇ *Do Practice Exercises 5 and 6.*

EXAMPLE 6

Evaluate $(8 - 2)^2 \div (36 \div 18)$.

SOLUTION

$(8 - 2)^2 \div (36 \div 18)$

$6^2 \div 2$ contents of parentheses

$36 \div 2$ Evaluate exponents.

18 Result. ■

EXAMPLE 7

Evaluate $27 \div 3^2 + (3 - 1)^2 \cdot 5$.

SOLUTION

$27 \div 3^2 + (3 - 1)^2 \cdot 5$

$27 \div 3^2 + 2^2 \cdot 5$ contents of parentheses

$27 \div 9 + 4 \cdot 5$ Evaluate exponents.

$3 + 4 \cdot 5$ Divide.

$3 + 20$ Multiply.

23 Result. ■

◇ *Do Practice Exercises 7 and 8.*

Be careful as you go from a word problem to an expression that represents the word problem because it may be necessary to use parentheses. The next two examples may help you.

5. Evaluate $7 \cdot 5 - (3 + 2)$.

6. Evaluate $(4 + 1)^2 - 8$.

7. Evaluate $7^2 - (36 \div 2)$.

8. Evaluate $2^4 + 3 \cdot (75 \div 25)$.

ANSWERS
5. 30
6. 17
7. 31
8. 25

9. Write an expression for forty minus the quotient of thirty-three divided by eleven, then evaluate it.

ANSWER
9. $40 - 33 \div 11$; 37

EXAMPLE 8

Write an expression for the sum of 15 and the difference between 4 and 1, then evaluate it.

SOLUTION

sum of 15 and the difference between 4 and 1

$15 + (4 - 1)$	Add the difference to 15.
$15 + 3$	contents of parentheses
18	Result. ■

EXAMPLE 9

Write an expression for twelve squared minus the sum of two and sixteen, then evaluate it.

SOLUTION

12 squared minus the sum of 2 and 16

$12^2 - (2 + 16)$	
$12^2 - 18$	contents of parentheses
$144 - 18$	exponents
126	Result. ■

◇ *Do Practice Exercise 9.*

Evaluate.

1. $6 + (9 - 2)$

2. $18 \div 6 - 3$

3. $4 + 9 \div 9$

4. $36 \div 3 - 4$

5. $(2 + 6)^2$

6. $7 - (6 \div 2)$

7. $3^2 + 5^2$

8. $18 \div 3 \cdot 4$

9. $25 \cdot 5 - 10$

10. $9 - 2^2$

11. $6 - 2 \cdot 3$

12. $(3 - 1)^3$

13. $6 + 20^2$

14. $16 \div 4 + 7 \cdot 2$

15. $3 + 7 \cdot 8$

16. $6 \cdot 3 \div 9$

17. $9 + 3 - 8$

18. $10 - 8 + 3$

19. $3 + 5^2 \cdot 2$

20. $9 + (70 - 4^2)$

21. $(21 \div 7)^2$

22. $(7^2 - 2^2) \div 5$

23. $150 \div 5^2$

24. $3 \cdot (48 - 2)$

25. $19 - 6 \div 2$

26. $32 \cdot 5^2 \div 16$

27. $(4 + 18) \div 11$

28. $9 + 7^3$

29. $25^2 \cdot 8$

30. $1{,}800 \div 60 - 4^2$

31. $3^2 + (40 \div 10)^2$

32. $50 \div 5^2 + 19 \cdot 2$

33. $81 + 8 \cdot (12 - 12)$

34. $(4 - 4)^3 + 1$

35. $(2^3 \cdot 3 \div 12)^2$

36. $(7 - 5) \div 2 + (19 - 3^2)$

37. $27^2 \div 1$

38. $180 \div 30 + 9 \div 3$

39. $16 \cdot 5 - 2$

40. $1^3 + 0^2$

Write each as an expression and evaluate.

41. Ten minus the product of seven and one.

42. Seven plus eight times seven.

43. Sixteen added to the product of seven times three.

44. The difference of ten and the quantity two multiplied times five.

45. Seven times eight minus one.

46. The quantity ten plus three, all raised to the second power.

47. Two hundred fifty-six divided by the sum of six plus two.

48. Twenty divided by the difference between five and one.

49. Eight subtracted from the sum of three and five.

50. Fifty divided by the product of five and two.

51. Seven times eight divided by two.

52. Three to the second power plus the product of five and six.

Answer the following questions.

53. Does $3^4 - 6 \cdot 9$ evaluate to the same value as does $3^4 - (6 \cdot 9)$?

54. Does $36 \div 3 - 1$ evaluate to the same value as does $36 \div (3 - 1)$?

55. Does $21 \cdot 7$ evaluate to the same value as does $7 \cdot 21$? What name is used to describe this property of multiplication?

56. Does $35 \div 7 \cdot 9$ evaluate to the same value as does $(35 \div 7) \cdot 9$?

57. Does $(16 \cdot 5 \div 4)$ evaluate to the same value as does $16 \cdot 5 \div 4$?

JOURNAL EXERCISES

1. Explain the concept of and rules of Order of Operations.

2. Create your own Order of Operations problem with several operations and solve it.

3. What are your first thoughts when asked to evaluate an expression such as:

$$\{(9 + 4^2) \div (4 \div 4)^2\}^3$$

Now evaluate this expression.

1 Tests for Divisibility

A number is said to be **divisible** or **evenly divisible** by a second number if the second number divides into it without any remainder. For example, twenty is divisible by five since five divides into twenty with *no remainder*. There are simple, quick tests to find out if a number is divisible by 2, 3, 4, 5, 6, 8, 9, or 10.

Tests for Divisibility

A number is divisible by	
2	if it ends with 0, 2, 4, 6, or 8
3	if the sum of the digits is divisible by 3
4	if the last two digits are divisible by 4
5	if it ends with a 0 or a 5
6	if it is divisible by both 2 and 3
8	if the last three digits are divisible by 8
9	if the sum of the digits is divisible by 9
10	if it ends with a 0

EXAMPLE 1

Find out if 454 is divisible by 2, 5, and/or 10.

SOLUTION

Yes, 454 is divisible by 2 since 454 ends with a 4.

No, 454 is not divisible by 5 since it does not end in either a 0 or a 5.

No, 454 is not divisible by 10 since it does not end in a 0. ■

◇ *Do Practice Exercises 1 and 2.*

EXAMPLE 2

Does 3 divide 213 evenly, that is, without a remainder?

SOLUTION

The digits of 213 are 2, 1, and 3. The sum of these digits is $2 + 1 + 3 = 6$. Using the rule above for divisibility by 3, the number 6 is divisible by 3. Therefore, 213 must be divisible by 3. ■

The tests for divisibility tell you how to find out if a number is divisible by 2, 3, 4, 5, 6, 8, 9, or 10. They do not provide the result of the division. If you wish to find the result of the division, you must divide.

EXAMPLE 3

Is 365 divisible by 3, 4, or 5? If so, divide.

SOLUTION

No, 365 is not divisible by 3 since $3 + 6 + 5 = 14$ which is not divisible by 3.

No, 365 is not divisible by 4 since the last two digits (65) are not divisible by 4.

OBJECTIVE

1 Decide if a number is divisible by 2, 3, 4, 5, 6, 8, 9, or 10.

1. Is 365 divisible by 5? By 10?

2. Does 2 divide evenly into 1,265,324?

ANSWERS
1. yes, no
2. yes

3. Is 715 divisible by 2? By 3?

4. Is 2,321,005 divisible by 4? By 5?

5. Is 22,714 divisible by 2, 3, 4, or 5?

6. Is 322,065 divisible by 6, 8, 9, or 10?

ANSWERS
3. 715 is not divisible by either 2 or 3.
4. 2,321,005 is divisible by 5, but not by 4.
5. 22,714 is divisible by 2, but not by 3, 4, or 5.
6. 322,065 is divisible by 9, but not by 6, 8, or 10.

Yes, 365 is divisible by 5 since it ends with a 5.

$$\begin{array}{r} 73 \\ 5\overline{)365} \\ \underline{35} \\ 15 \\ \underline{15} \\ 0 \end{array}$$ ■

◇ *Do Practice Exercises 3 and 4.*

EXAMPLE 4

Find a six digit number that is divisible by 4.

SOLUTION
Any number that *ends with two digits divisible by 4* is divisible by 4. Thus any number that ends in 00, 04, 08, 12, 16, . . . is divisible by 4. Each of the following six digit numbers is divisible by 4.

603,200   699,204   721,380
799,304   111,124   999,932 ■

EXAMPLE 5

Check to see if 3,609 is evenly divisible by 2, 3, 4, 5, 6, 8, 9, or 10.

SOLUTION
1. 3,609 *is not* divisible by 2 since it does not end in 0, 2, 4, 6, or 8.
2. 3,609 *is* divisible by 3 since the sum of the digits = 3 + 6 + 0 + 9 = 18, which is divisible by 3.
3. 3,609 *is not* divisible by 4 since the last two digits = 09 are not divisible by 4.
4. 3,609 *is not* divisible by 5 since it does not end in 0 or 5.
5. 3,609 *is not* divisible by 6 since it is not divisible by 2.
6. 3,609 *is not* divisible by 8 since 609 is not divisible by 8.
7. 3,609 *is* divisible by 9 since the sum of the digits = 3 + 6 + 0 + 9 = 18, which is divisible by 9.
8. 3,609 *is not* divisible by 10 since it does not end in 0. ■

◇ *Do Practice Exercises 5 and 6.*

EXAMPLE 6

Find a number that is evenly divisible by 3 and by 10.

SOLUTION
In order to be divisible by 3, the sum of the digits must be divisible by 3. In order to be divisible by 10, the number must end with a 0. Therefore, to be divisible by both 3 and 10, the number must end with a 0 and the digits must sum to a number divisible by 3.

For example, 30 and 210 are each divisible by both 3 and 10. ■

name date hour

Use the tests for divisibility to determine if each number is divisible by 2, 3, 4, 5, 6, 8, 9, and/or 10.

1. 28

2. 465

3. 750

4. 236

5. 420

6. 1,000

7. 646

8. 894

9. 560

10. 245

11. 775

12. 880

13. 720

14. 937

15. 1,040

16. 2,231

17. 2,680

18. 3,921

19. 7,243

20. 6,540

21. 4,970

22. 8,957

23. 10,270

24. 22,431

25. 60,241

26. 17,304

27. 31,724

28. 60,994

29. 40,032

30. 672,100

31. 780,021

32. 271,033

33. 365,029

34. 241,312

35. 6,000,000

36. 8,250,325

Evaluate. (Review)

37. $15 + (6 - 1)$

38. $2 \cdot 9 \div 2$

39. $70 \div 35 - 1$

40. $5^2 + 7$

41. $(9 - 1) \cdot 20$

42. $7 \cdot 8 \div 4$

43. $36 \div 3 + 4 \cdot 8$

44. $27 \div 9 \div 3 - 1$

45. 6^2

46. 5^0

Work each problem.

47. Find a ten digit number that is divisible by 3.

48. Find a number greater than 100 that is evenly divisible by both 3 and 5.

49. Find a number ending with 0 that is divisible by both 2 and by 3.

50. Find a number which is divisible by 2, 3, 4, 5, 9, and 10.

51. Find three numbers that can be used in the thousands place in 64 ___ ,302 in order for the number to be divisible by 3.

52. Find a number greater than 100 that ends with a 1 and that is evenly divisible by 9.

53. Is 2,305,204 divisible by 3? By 4?

54. Is 37,261,215 divisible by 8? By 9?

JOURNAL EXERCISES

1. Look up the definition of proof and write it down. Use the rules for Tests for Divisibility and prove that a number divisible by 10 must also be divisible by both 2 and 5. (*Hint:* Look at the last example in the section.)

2. State a rule to find out if a number is divisible by 18. (*Hint:* If a number is divisible by 18, it must be divisible by 2 and by 9.)

OBJECTIVES

1. Know the difference between prime and composite numbers.
2. Find the prime factorization of a number.

1 Prime versus Composite

A **prime number** is a whole number greater than one that has exactly two factors: itself and the number 1. In other words, the only numbers that divide evenly into a prime number are the prime number itself and the number 1. For example, 7 is a prime number since the only whole numbers that divide 7 with no remainder are 7 and 1. On the other hand, the number 6 is a not a prime number since it can be broken apart into the product of 2 times 3.

The first few prime numbers are 2, 3, 5, 7, 11, 13, 17, 19, ... By definition, the number 1 *is not* a prime number.

EXAMPLE 1

Identify the prime numbers:

(a) 31 **(b)** 27 **(c)** 35 **(d)** 97

SOLUTION

(a) 31 is a prime number since its only factors are 1 and 31.
(b) 27 is not a prime number since it is divisible by 1, 3, 9, and 27.
(c) 35 is not a prime number since it is divisible by 1, 5, 7, and 35.
(d) 97 is a prime number since its only factors are 1 and 97. ■

◇ *Do Practice Exercise 1.*

1. Identify the prime numbers:

(a) 19

(b) 40

(c) 41

Every whole number larger than one is either a prime number or a composite number. A **composite number** is a whole number greater than one that has more than two factors. For example, 10 is a composite number with factors 1, 2, 5, and 10.

EXAMPLE 2

Identify the composite numbers:

(a) 37 **(b)** 42 **(c)** 49 **(d)** 63

SOLUTION

(a) 37 is not a composite number since it has no factor other than 1 and 37.
(b) 42 is a composite number with factors 1, 2, 3, 6, 7, 14, 21, and 42.
(c) 49 is a composite number with factors 1, 7, and 49.
(d) 63 is a composite number with factors 1, 3, 7, 9, 21, and 63. ■

◇ *Do Practice Exercise 2.*

2. Identify the prime and composite numbers:

(a) 24

(b) 47

(c) 55

(d) 61

The first few counting numbers are shown below. The prime numbers are circled. On the row below the counting numbers, each composite number is broken into the product of two factors neither of which is the number 1.

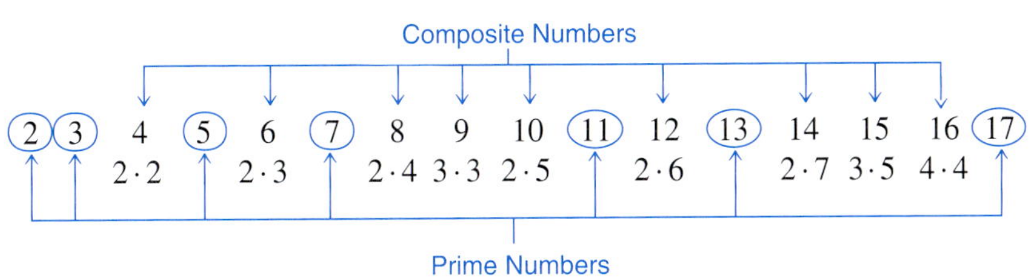

ANSWERS
1. (a) 19 is prime (b) 40 is not prime
(c) 41 is prime
2. (a) 24 is a composite number.
(b) 47 is a prime number.
(c) 55 is a composite number.
(d) 61 is a prime number.

3. List all of the numbers between 20 and 30 and decide which are prime and which are composite.

Prime numbers are currently of great importance to the intelligence community such as the CIA. This is because some secret codes are based on very large prime numbers.

◇ Do Practice Exercise 3.

2 Prime Factorization

Different humans have different fingerprints. In a way, different composite numbers also have different fingerprints. You see, each composite number is composed of or made up of the product of its own unique set of *prime numbers*. For example, 6 equals 2×3 and we know that one 2 times one 3 is always 6. So the "fingerprint" of 6 is one 2 times one 3.

The **prime factorization** of a number is the unique group of primes whose product equals that number. Different numbers have different prime factorizations. The idea that each number has its own unique prime factorization is called the **Fundamental Theorem of Arithmetic.**

Several numbers and their prime factorizations follow.

Number	*Prime Factorization*	
4	$2 \cdot 2 = 2^2$	
7	7	(7 is a prime number.)
8	$2 \cdot 2 \cdot 2 = 2^3$	
12	$2 \cdot 2 \cdot 3 = 2^2 \cdot 3$	
150	$2 \cdot 3 \cdot 5 \cdot 5 = 2 \cdot 3 \cdot 5^2$	
1,010	$2 \cdot 5 \cdot 101$	(101 is a prime number.)

Notice that every number in the column labeled "Prime Factorization" is a prime number. The number 1 is not a prime number and does not occur in this column. Also notice that the prime factorization of a prime number is the prime number itself.

The prime factorization of a number can be found using a **factor tree**. A factor tree is a simple way to keep track of numbers during the process of breaking a composite number into the product of smaller numbers. For example, a factor tree is used to find the prime factorization of 42:

factor tree

42

$6 \cdot 7 = 42$ 42 is written as the product of 6 and 7; 6 is not prime.

$2 \cdot 3 \cdot 7 = 42$ 6 is written as the product of 2 and 3.

The product $2 \cdot 3 \cdot 7$ written on the bottom row of the factor tree is the prime factorization of 42.

Every composite number has a unique prime factorization. The order in which a composite number is broken apart into the product of smaller numbers does not change the prime factorization. For example, the prime factorization of 42 is the same whether 42 is first written as the product of 6 and 7 as above or as the product of 2 and 21 as shown next:

factor tree

42

$2 \cdot 21 = 42$ 42 is written as the product of 2 and 21; 21 is not prime.

$2 \cdot 3 \cdot 7 = 42$ 21 is written as the product of 3 and 7.

ANSWER

3. 21—composite 22—composite
23—prime 24—composite
25—composite 26—composite
27—composite 28—composite
29—prime

As you can see, the final result contains the same prime factors. The prime factorization of 42 is $2 \times 3 \times 7$.

HELPFUL HINT

You have not found the prime factorization of a number until every number in the factorization is a prime number.

In general, repeated factors in a prime factorization will be shown using exponents. Look at the next two examples.

EXAMPLE 3

Find the prime factorization of 36.

SOLUTION

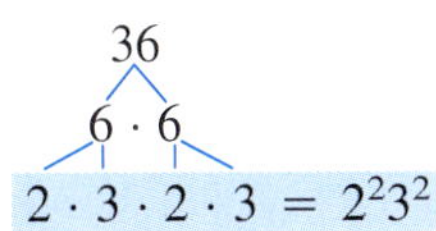

$$36 = 6 \cdot 6 = 2 \cdot 3 \cdot 2 \cdot 3 = 2^2 3^2$$

prime factorization of 36 ■

EXAMPLE 4

Find the prime factorization of 50.

SOLUTION

$$50 = 2 \times 5 \times 5 = 2 \times 5^2$$ ■

◇ *Do Practice Exercise 4.*

Finding the prime factorization of a large number begins with the trial and error process of finding a number that divides evenly into the large number. First, use the tests for divisibility to see if the large number is evenly divisible by 2, 3, 4, 5, 6, 8, 9, or 10. Or, you can try dividing the large number by the prime numbers, beginning with the smaller primes. The large number is a prime number if no prime number divides into it.

EXAMPLE 5

Find the prime factorization of 126.

SOLUTION

We know that 126 is divisible by 3 since $1 + 2 + 6 = 9$ is divisible by 3. So, first divide a 3 out and then continue:

$$126 = 3 \times 42 = 3 \times 6 \times 7 = 3 \times 3 \times 2 \times 7 = 2 \times 3^2 \times 7$$ ■

4. Find the prime factorization of each:

(a) 14

(b) 20

(c) 24

ANSWERS
4. (a) 2×7 (b) $2^2 5$ (c) $2^3 3$

5. Find the prime factorization of two hundred fifty.

EXAMPLE 6

Find the prime factorization of 89.

SOLUTION

Using the tests for divisibility, none of the numbers 2, 3, 4, 5, 6, 8, 9, or 10 divide 89 evenly. Next, try to divide 89 by the primes 7, 11, 13, 17, 19, 23, 29, ... In fact, neither 7 nor 11 divide into 89 evenly so that 89 is a prime number. The prime factorization is the number 89 itself. ■

◇ *Do Practice Exercises 5 and 6.*

6. Find the prime factorization of two hundred sixty-seven.

Prime factors can be used to see if one number divides a second number. A number is divisible by a second if all of the primes in the prime factorization of the divisor *are also in* the prime factorization of the dividend. For example, 30 is divisible by 6 since the prime factors of 6 (2 and 3) are in the prime factorization of 30 ($2 \times 3 \times 5$).

EXAMPLE 7

Use primes to find out if 345 is divisible by 15.

SOLUTION

The prime factorizations of the two numbers are:

prime factorizations

$$345 = 3 \cdot 5 \cdot 23 \quad \text{and} \quad 15 = 3 \cdot 5$$

All of the primes in 15 are also in 345.

7. Use primes to determine if 300 is divisible by 21.

All of the primes in the factorization of 15 (3 and 5) *are also in* the prime factorization of 345. Therefore, 15 divides evenly into 345. ■

EXAMPLE 8

Use prime factors to find out if 150 is divisible by 21.

SOLUTION

$$150 = 2 \times 3 \times 5 \times 5$$
$$21 = 3 \times 7$$

The prime factorization of 21 contains a 7. The prime factorization of 150 does not contain a 7. Therefore, 21 does not divide 150 evenly. ■

◇ *Do Practice Exercise 7.*

ANSWERS

5. 2×5^3

6. 3×89

7. $300 = 2 \times 2 \times 3 \times 5 \times 5$ and $21 = 3 \times 7$
21 does not divide into 300 evenly since there is no 7 in the factorization of 300.

Determine whether each number is prime or composite.

1. 25 **2.** 37 **3.** 42 **4.** 61

5. 40 **6.** 2 **7.** 99 **8.** 175

9. 144 **10.** 101 **11.** 137 **12.** 367

13. 155 **14.** 231 **15.** 707 **16.** 1,053

Find the prime factorization of each.

17. 9 **18.** 15 **19.** 12 **20.** 35

21. 32 **22.** 75 **23.** 66 **24.** 80

25. 75 **26.** 99 **27.** 100 **28.** 64

29. 125 **30.** 81 **31.** 121 **32.** 320

33. 264 **34.** 280 **35.** 205 **36.** 175

37. 210 **38.** 340 **39.** 460 **40.** 130

41. 375 **42.** 1,025 **43.** 2,500 **44.** 10,000

Evaluate. (Review)

45. 3^2 **46.** 5^3 **47.** 1^2 **48.** 12^2

Use prime factors to find out if the division has no remainder.

49. 360 ÷ 30

50. 265 ÷ 15

51. 1,000 ÷ 8

52. 372 ÷ 6

53. 2,760 ÷ 15

54. 5,124 ÷ 32

55. 1,250 ÷ 45

56. 2,850 ÷ 25

Work each problem.

57. List all of the prime numbers up to fifty.

58. List all even numbers (numbers that end in 0, 2, 4, 6, or 8) that are also prime numbers.

59. Is there any number that ends with a zero that is a prime number?

JOURNAL EXERCISES

1. Explain the differences between composite numbers and prime numbers.

2. Explain the steps used to find the prime factorization of a number. How do you know when you have found the prime factorization?

1 Least Common Multiple and Greatest Common Divisor

The **multiples** of a number are the products of that number with the counting numbers: 1, 2, 3, 4, 5, ... Thus the multiples of 6 are $6 \times 1 = 6$, $6 \times 2 = 12$, $6 \times 3 = 18$ and so on.

The first few multiples of 6 and of 10 are

	Multiples
of 6:	6, 12, 18, 24, **30**, 36, 42, 48, 54, **60**, 66, 72, 78, 84, **90**, . . .
of 10:	10, 20, **30**, 40, 50, **60**, 70, 80, **90**, . . .

The numbers in bold print are called **common multiples** of 6 and 10 because they are common to both lists.

Common multiples: 30, 60, 90, . . .

The smallest of all of the common multiples of 6 and 10 is 30, which is called the **least common multiple** or **LCM** of 6 and 10. The LCM of two numbers is the smallest whole number that is evenly divisible by both. Since 30 is the LCM of 6 and 10, it is the smallest number that is evenly divisible by both 6 and 10.

EXAMPLE 1

Find the LCM of 4 and 6.

SOLUTION

Multiples of 4: 4, 8, 12, 16, 20, 24, 28, 32, 36, 40, . . .
Multiples of 6: 6, 12, 18, 24, 30, 36, 42, . . .
Common multiples: 12, 24, 36, . . .
Least Common Multiple: 12 ■

◇ *Do Practice Exercises 1 and 2.*

The **greatest common divisor (GCD)** of two numbers is the largest whole number that divides into both. For example, the GCD of 18 and 27 is 9. This is true since 9 divides into both 18 and 27 and is the largest whole number to do so. The next example shows the process of finding the GCD.

EXAMPLE 2

Find the GCD of 10 and 25.

SOLUTION

Divisors of 10 are: 1, 2, 5, and 10.
Divisors of 25 are: 1, 5, and 25.

Common divisors: 1 and 5

Greatest Common Divisor: 5 ■

◇ *Do Practice Exercise 3.*

OBJECTIVES

1 Use least common multiples and greatest common divisors.
2 Find the least common multiple using prime factors.

1. Find the least common multiple of 5 and 15.

2. Find the LCM of each:

(a) 8 and 10

(b) 12 and 15

3. Find the GCD of each:

(a) 4 and 10

(b) 12 and 15

(c) 30 and 70

ANSWERS
1. 15
2. (a) 40 (b) 60
3. (a) 2 (b) 3 (c) 10

2 Use Primes to Find the Least Common Multiple

The LCM can be found using prime numbers as follows.

Rule: To Find the Least Common Multiple

Step 1 Find the prime factorization of each number. List the primes in columns (of two's, of three's, of five's, and so on).

Step 2 Circle the largest product of identical primes in each column.

Step 3 The Least Common Multiple is the product of all of the circled primes.

EXAMPLE 3

Find the least common multiple of 8 and 10.

SOLUTION

Step 1 The prime factorizations are:

$$8 = 2 \times 2 \times 2 \quad \text{and} \quad 10 = 2 \times 5$$

List the primes in columns of 2s and 5s:

Column contains only 2s.

		2s		5s
8	=	2 × 2 × 2		
10	=	2	×	5

Column contains only 5s.

Step 2 Circle the largest product in each column:

Column contains only 2s.

		2s		5s
8	=	(2 × 2 × 2)		
10	=	2	×	(5)

Column contains only 5s.

Step 3 The LCM is the product of all of primes circled in Step 2.

$$2 \times 2 \times 2 \times 5 = 40$$

The least common multiple of 8 and 10 is 40. This means that the smallest number that both 8 and 10 divide into evenly is 40. There is no number smaller than 40 which *both* 8 and 10 divide into evenly. ■

EXAMPLE 4

Find the least common multiple of 20 and 35.

SOLUTION

Step 1 List the prime factors in columns of identical primes.

		2s	5s	7s
20	=	2 × 2 ×	5	
35	=		5 ×	7

Step 2 Circle the largest product in each column.

20	=	(2 × 2) ×	(5)	
35	=		5 ×	(7)

Step 3 The Least Common Multiple $= 2 \times 2 \times 5 \times 7 = 140$. ■

◇ *Do Practice Exercises 4 and 5.*

The LCM of two numbers contains all of the prime factors of both numbers. For instance, in the last example we saw that 140 is the LCM of 20 and 35. Therefore, the prime factorization of 140 must contain all of the prime factors of 20 *and* all of the prime factors of 35.

$$140 = \underbrace{2 \cdot 2 \cdot 5}_{\text{prime factors of 20}} \cdot 7 \qquad 5 \cdot 7 \text{: prime factors of 35}$$

Since every prime in the prime factorization of 20 is in 140, 20 must divide evenly into 140. Every prime in the prime factorization of 35 is also in 140, thus 35 must also divide evenly into 140. Further, 140 is the smallest number that both 20 and 35 divide into.

The process of finding the least common multiple of three or more numbers follows the same pattern as shown in the next example.

EXAMPLE 5

Find the least common multiple of 4, 10, and 15.

SOLUTION

		2s	3s	5s
4	=	(2 × 2)		
10	=	2 ×		(5)
15	=		(3) ×	5

LCM $= 2 \times 2 \times 3 \times 5 = 60$. ■

◇ *Do Practice Exercise 6.*

EXAMPLE 6

A race car is making a lap every 40 seconds while a second car is making the same lap every 35 seconds. The two cars begin at START at time zero. How long will it be before they are again side by side at the same location on the track?

SOLUTION

A series of diagrams show the cars at the beginning of the race and at various times during the race.

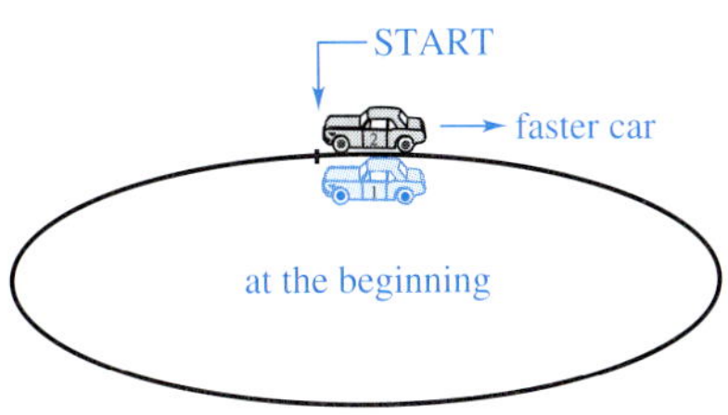

4. Find the LCM of 15 and 9.

5. Find the smallest number which both 20 and 22 divide.

6. Find the LCM of 6, 10, and 18.

ANSWERS
4. 45
5. 220
6. 90

7. One snail makes a lap every 45 minutes and a second snail makes a lap every 50 minutes. Given that they start together, how long will it be until they meet again at the starting place?

ANSWER
7. 450 minutes

Car #2 is back at the starting point every 35 seconds since it makes one complete lap in 35 seconds. Car #1 is slower than Car #2 since it needs 40 seconds to make one complete lap. Car #1 makes less than a complete lap in 35 seconds.

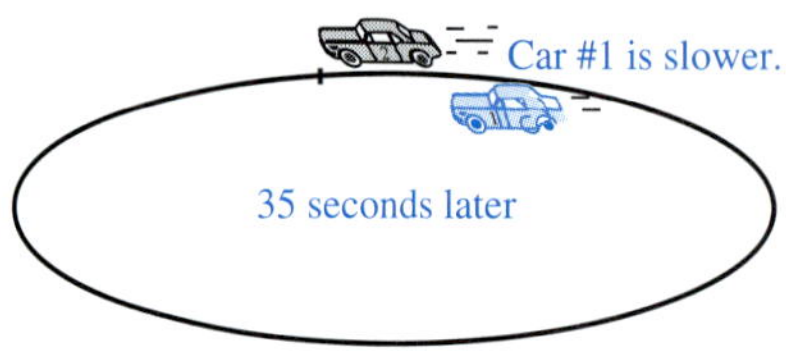

Car #2 has made one complete lap.

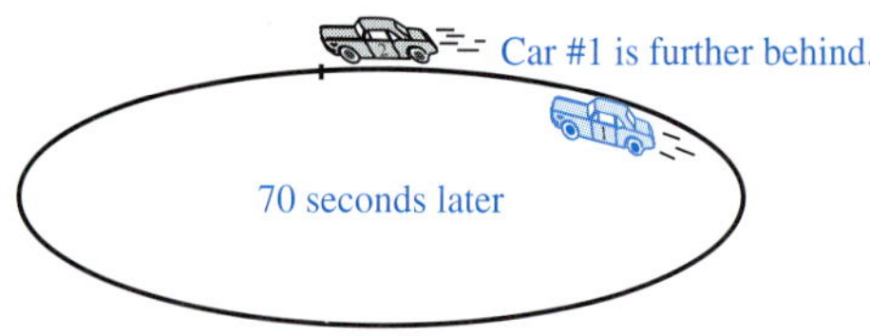

Car #2 has made two complete laps.

Car #1 is slower and falls further and further behind with more 35 second intervals.

Car #2 has traveled farther than car #1.

Finally both cars again meet at the starting point. However, Car #2 will have completed one more lap than Car #1.

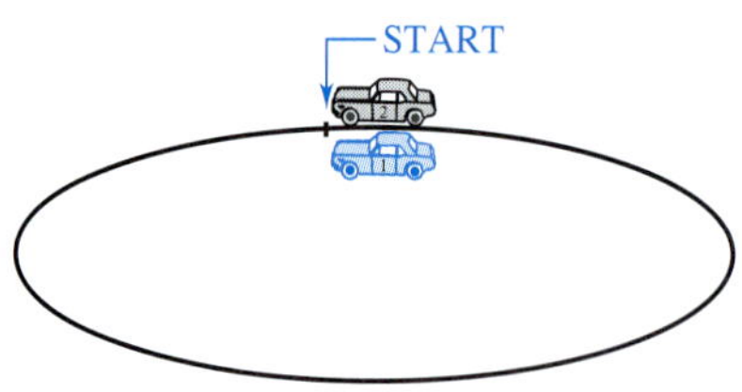

Car #2 has lapped car #1.

Car #2 will be at the starting point ONLY at multiples of 35 seconds, in other words at 35 seconds, 70 seconds, 105 seconds, and so on. Car #1 will be at the starting point ONLY at multiples of 40 seconds, in other words at 40 seconds, 80 seconds, 120 seconds, and so on. BOTH Car #1 and Car #2 will be at the starting point ONLY at multiples of 35 AND 40 seconds. The least amount of time required for both cars to meet again at the starting point is the least common multiple of 35 and 40.

35	=		5 ×	(7)
40	=	(2 × 2 × 2) ×	(5)	

$$\text{LCM} = 2 \times 2 \times 2 \times 5 \times 7 = 280 \text{ seconds}$$

Both cars will again be at the START at 280 seconds. The number of laps that each car makes in 280 seconds is found by dividing the time required for each car to make one lap into 280 seconds.

Car #1	Car #2
$40\overline{)280}$ = 7 laps	$35\overline{)280}$ = 8 laps
280	280
0	0

Car #1 will have made 7 laps and Car #2 will have made 8 laps. ■

◇ *Do Practice Exercise 7.*

name date hour

Find the prime factorization of each number. (Review)

1. 25 **2.** 60 **3.** 85 **4.** 92

5. 120 **6.** 100 **7.** 98 **8.** 150

Determine whether or not the prime factorization of each number is given. (Review)

9. $35 = 7 \cdot 5$ **10.** $45 = 9 \cdot 5$ **11.** $50 = 25 \cdot 2$ **12.** $80 = 2 \cdot 2 \cdot 2 \cdot 2 \cdot 5$

13. $60 = 5 \cdot 3 \cdot 4$ **14.** $100 = 2 \cdot 5 \cdot 10$ **15.** $120 = 2 \cdot 2 \cdot 2 \cdot 3 \cdot 5$ **16.** $250 = 2 \cdot 5 \cdot 5 \cdot 5$

Find the least common multiple.

17. 5, 4 **18.** 16, 20 **19.** 12, 20

20. 10, 25 **21.** 8, 12 **22.** 6, 15

23. 9, 21 **24.** 20, 45 **25.** 50, 60

26. 8, 72 **27.** 12, 27 **28.** 4, 80

29. 8, 14 **30.** 9, 24 **31.** 14, 35

32. 24, 28, 40 **33.** 12, 16, 18 **34.** 3, 35, 10

35. 12, 14, 28

36. 3, 20, 36

37. 10, 20, 30

38. 16, 24, 62

39. 4, 8, 9, 14

40. 3, 6, 7, 12

Find the smallest number evenly divisible by both numbers.

41. 10, 7

42. 6, 18

43. 16, 27

44. 30, 42

45. 20, 21

46. 10, 12

47. 24, 36

48. 40, 65

Find the greatest common divisor.

49. 20, 50

50. 25, 10

51. 18, 36

52. 40, 24

53. 30, 42

54. 35, 49

55. 28, 36

56. 45, 99

57. 40, 65

58. 50, 125

59. 32, 72

60. 40, 140

Work each problem.

61. Two trains leave the station on different tracks, but at the same time and travel in a circle. The trains take 8 hours and 6 hours respectively to make one complete lap. How long after they meet in front of the train station will it be until they meet again in front of the train station? How many laps will the slow train have made during that time?

62. Two satellites orbit the earth. The spy satellite travels along the equator, and the weather satellite travels over the North and South Poles. The spy satellite requires twenty-four hours to make one complete revolution around the earth. The weather satellite requires eighteen hours to make one complete revolution. Given that the satellites meet, how long will it be before they meet again? How many revolutions around the earth will each satellite have made?

63. Find the product of eight times the cube of four.

64. Find the LCM of 6, 12, 15, and 42.

65. Estimate the product of 84 and 48. Then find the exact value.

66. The time required for each of several planets to make one complete revolution around the sun is:

Planet	*Time Required*
Mercury	88 Earth days
Venus	225 Earth days
Earth	365 Earth days

At one moment the three planets are lined up. How long will it be before they are again lined up at the same location?

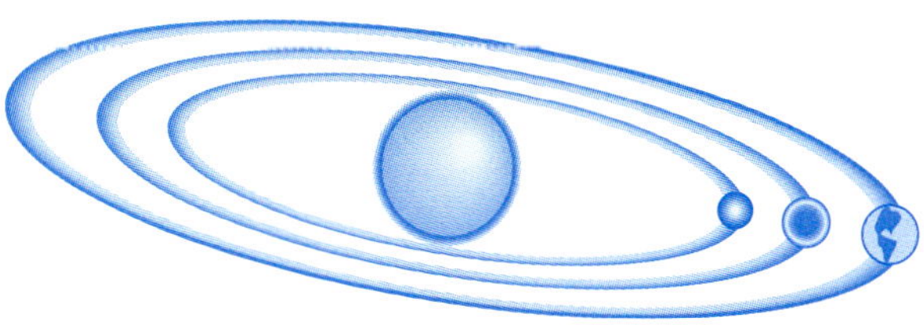

JOURNAL EXERCISES

1. Explain the concept of Least Common Multiple and the process of finding one.
2. Explain the difference between LCM and GCD. Find both the LCM and GCD of 20 and 28.
3. Create your own LCM problem using cars, trains, satellites, or something else and solve it.

1 Tables

Information is frequently given in the form of numbers in a **table**. Tables are usually easy to understand *if* you take the time to read the headings. Look at the following table.

Comparison of the Five Largest U.S. Rivers

River	*Average Discharge**	*Length (miles)*	*Drainage Area (square miles)*
Mississippi	593,000	2,350	1,150,000
St. Lawrence	348,000	1,900	396,000
Ohio	281,000	1,310	203,000
Columbia	265,000	1,240	258,000
Yukon	225,000	1,980	328,000

*Discharges are given in cubic feet per second.
Source: U.S. Geological Survey

The title of the table shows that it contains information about the five largest rivers in the United States. Under the title, the source of the data is given as the United States Geological Survey. Credit must be given when data is taken from another source. The subheadings below the title are:

River	*Average Discharge**	*Length (miles)*	*Drainage Area (square miles)*

The subheadings tell the type of information given for each river. From the table, the length of the Columbia River is found by going down the column titled Length to the row for the Columbia River. The Columbia River is 1,240 miles long. The asterisk (*) after the word Discharge in the subheading of the table is a reference to the footnote, which begins with the * at the bottom of the table. The footnote gives the units used to measure the volume of water being discharged. For example, the Mississippi River discharges 593,000 cubic feet of water every second (on average).

Several factors about the rivers are not given in the table. For example:

1. Where is each river located?
2. Into which ocean does each river empty?

A table cannot provide all of the information about a river. But it does give enough information to compare some of the features of the rivers.

Calculations using the data in a table can be used to answer other questions. For example, the total discharge of all five rivers listed in the table is found by adding the five numbers in the Average Discharge column. The total discharge is 1,712,000 cubic feet of water every second.

◇ *Do Practice Exercise 1.*

OBJECTIVES

1 Work with tables containing data.

2 Work with charts.

1. Use data from the table to answer the questions that follow.

	Semester Costs
Tuition	$650
Books	$145
Supplies	$38

What was the total cost for the semester? Do books or tuition cost more?

ANSWER

1. $833, tuition costs more

2 Charts

Where a table shows numbers, a **chart** is a picture built from numbers. A chart can show a trend that may be hard to see in a table full of numbers. The best way to show data depends on the particular situation. A table is better if accuracy is very important. A chart may be a better way to communicate an idea to an audience who does not need exact numbers. Sometimes, both are used.

Some information about college students is given in a table. The same basic information is then shown in a **pie chart** and a **bar chart**.

Main Source of Financial Support for Students at Comfort Community College*

Type of Support	*Number of Students*	*Percent*
Government grants	600	25%
Parental support	1,200	50%
Scholarships	240	10%
Self-supporting	360	15%
Totals	2,400	100%

*1993–1994 school year

A pie chart, shown next, divides the total into pieces.

Main Source of Financial Support for Students at Comfort Community College*
(2,400 students)

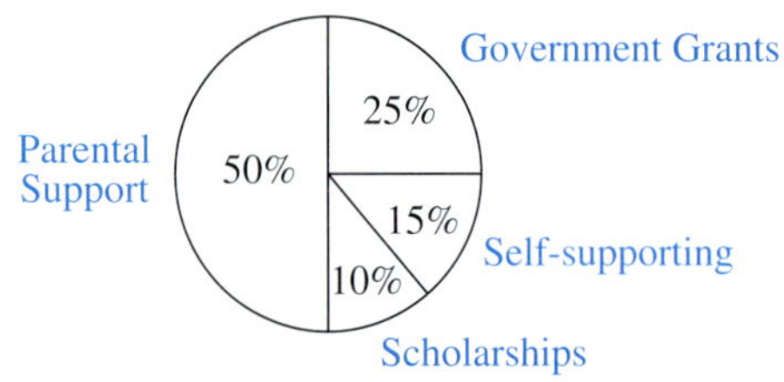

* 1993–1994 school year

Finally, a bar chart is a good way to see a trend.

Main Source of Financial Support for Students at Comfort Community College*

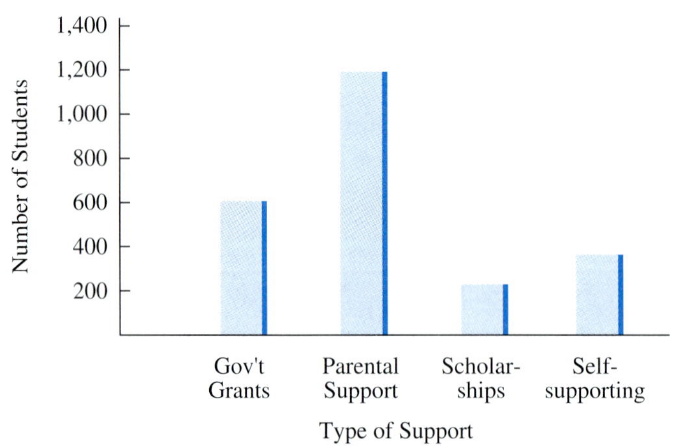

* 1993–1994 school year

The table and both charts contain the same basic information but show it very differently. The table has the most accurate information but the charts show the differences more graphically. From the bar chart, it is difficult to decide if there are 210 students or 250 students on scholarships. Yet, the bar chart quickly shows that more students are supported by their parents than received scholarships.

A chart which provides information using pictures is called a **pictograph**. Data in the following table is also shown in a pictograph.

Cattle on U.S. Farms

Year	*Number of Cattle*
1935	68,846,000
1945	85,573,000
1955	96,592,000
1965	109,000,000
1975	132,028,000
1985	109,749,000

Source: National Agricultural Statistics (U.S.D.A.)

Cattle on U.S. Farms

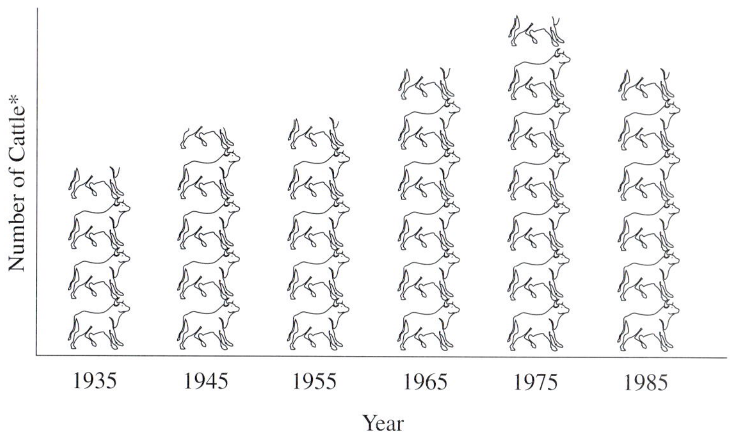

* Each figure represents 20,000,000 head of cattle. One-half of a figure represents 10,000,000 head of cattle.

Source: National Agricultural Statistics (U.S.D.A.)

The pictograph shows that the number of cattle increased from 1935 through 1975 followed by a decrease in 1985. The exact number of cattle on U.S. farms in 1965 cannot be obtained from the pictograph.

◇ *Do Practice Exercise 2.*

2. Show the data in the table of cattle on U.S. farms using a bar chart.

ANSWER

2.

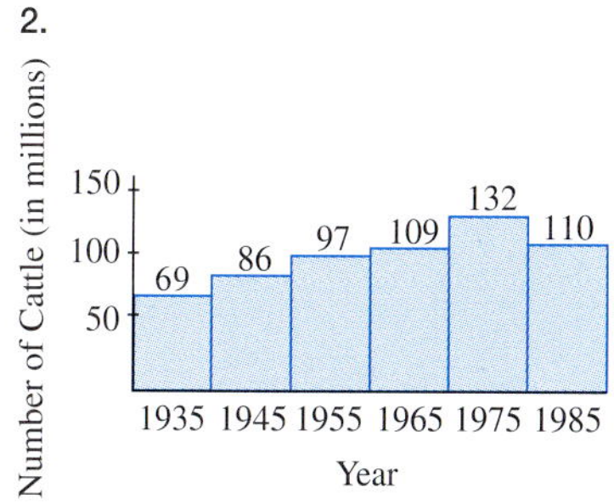

Answer the questions using the information in the tables.

1993–1994 College Algebra Grade Distribution at Comfort Community College

Grade	*Number of Students*	*Percent*
A	40	6%
B	85	13%
C	180	29%
D	215	34%
F	110	18%
Totals	630	100%

1. The table refers to students taking which class?

2. How many students took College Algebra at Comfort Community College during the 1993–1994 school year?

3. How many students passed College Algebra with a grade of A, B, or C?

4. What percent of the students made an F?

5. Does the table tell you where Comfort Community College is located?

6. Does the table tell you the average age of the students who took College Algebra?

1990 Average Starting Salary for Professionals

Profession	*College Degree*	*Average Starting Salary*
Engineer	Bachelor's*	$29,820
Computer programmer	"	$28,331
Accountant	"	$22,838
Teacher	"	$18,850
Retailer	"	$17,035
Engineer	Master's*	$34,776
Accountant	"	$29,700
Medical doctor	Doctorate*	$85,630
Dentist	"	$40,190

*The typical number of years of college required to obtain each degree is: Bachelor's—4 years; Master's—6 years; Doctorate—9 years.
Source: *U.S. News and World Report*

7. What was the average starting salary for accountants with a Master's degree?

8. Was the average starting income higher for medical doctors or dentists?

9. In 1990, what was the additional amount of money that an engineer with a Master's degree could earn compared to an engineer with a Bachelor's degree?

10. What is the annual income of a person working 40 hours per week, 50 weeks per year at $5.00 per hour?

11. What is the difference in income of a starting teacher's salary compared to the individual's salary in problem 10?

12. In 1990, a business hired a computer programmer and a retailing graduate. What additional *monthly* salary expense does the business now have if they pay average salaries?

Answer the questions based on information in the charts.

1993 Profits for a Large Corporation

13. Which source gave the company the largest contribution to its profit?

14. Was profit from manufacturing more than twice the profit from investment income?

15. What was the total profit from all four sources?

16. Do profits from sale of real estate and investment income exceed profits from manufacturing and overseas sales?

Years of Life Expected at Birth (United States)

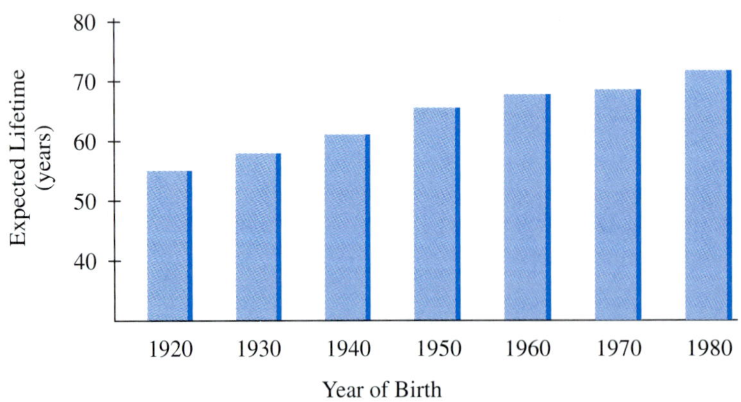

Source: National Center for Health Statistics

17. What is the expected length of the life of a person born in 1920?

18. What has been the trend of expected lifetimes over the past 60 years?

19. A large number of people died at a young age during World War II (1940–1945). What effect would this have on the expected lifetime of people born in 1920?

20. Name some changes in society which might help explain the increase in expected lifetimes.

Deaths Caused by Volcanic Eruptions
(Five Eruptions Causing the Most Fatalities)

Date of Eruption	Volcano	
79 A.D.	Mt. Vesuvius, Italy	
1669	Mt. Etna, Sicily	
1883	Krakatau, Indonesia	
1902	Mt. Pelee, Martinique	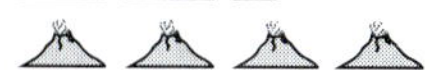
1985	Nevado del Ruiz, Columbia	

Each volcano represents 10,000 deaths.
Source: *The World Almanac and Book of Facts*

21. Which of the volcanic eruptions was thought to have caused the most deaths? How many?

22. How many deaths are thought to have been caused by Mt. Vesuvius?

23. When did Mt. Vesuvius erupt?

24. Which volcanic eruption ejected the most ash and lava, or can this be found using the table?

25. Approximately how many deaths are thought to have been caused by all five volcanos combined? Is this larger or smaller than the 407,316 American casualties during World War II?

Use all of the data to answer the questions that follow.

Quarterly Candy Sales for Carrollton Candy Company

Year	*1st Qtr**	*2nd Qtr*	*3rd Qtr*	*4th Qtr*	*Total*
1991	28,650	26,450	25,900	52,300	133,300
1992	32,100	35,300	31,750	68,000	167,100
1993	43,000	37,650	39,000	?	?

*Qtr = quarter which is three months of the year. Therefore the 1st Qtr is made up of January, February, and March.

Candy Sales for Carrollton Candy Company by Type of Candy*

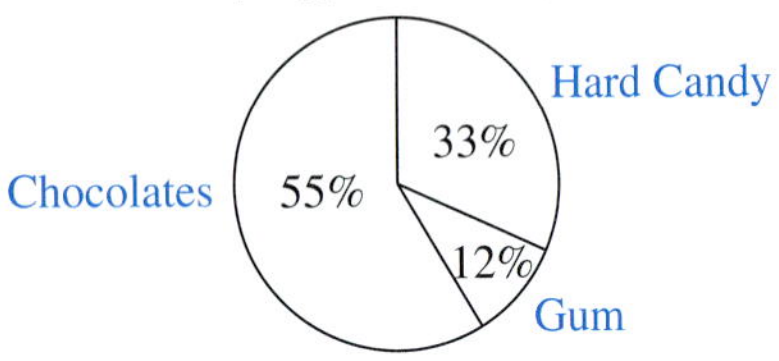

* Data averaged from January 1, 1991, through the first nine months of 1993.

Candy Sales for Carrollton Candy Company with a 4th Quarter 1993 Projection

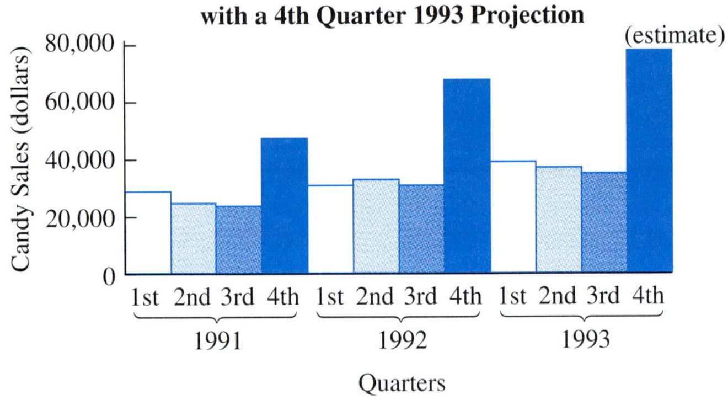

26. What were the sales, to the nearest dollar, for the third quarter of 1992?

27. Which quarter of each year has the highest sales volume?

28. What percent of sales are chocolates?

29. What are the projected sales for the fourth quarter of 1993?

30. What are the projected total sales for 1993?

31. In which quarter and of what year was the data most likely collected and presented in the table and charts?

32. Make a bar chart showing annual sales for 1991, 1992, and 1993 using the answer to problem 30 for 1993 sales.

33. What is the trend of annual sales, up or down?

34. Based on the data available, can you tell whether or not Carrollton Candy Company is making a profit? Explain.

Follow the directions for each problem.

35. Show the following data in a bar chart. Label the chart carefully.

Average Price of Crude Oil

Year	*Average Price per Barrel*	
1977	$ 8	
1979	12	
1981	31	
1983	26	
1985	24	
1987	16	
1989	18	
1991	17	
1993	20	(estimated)

Source: IPAA

36. Show the following data in a table.

Speeds of Various Animals

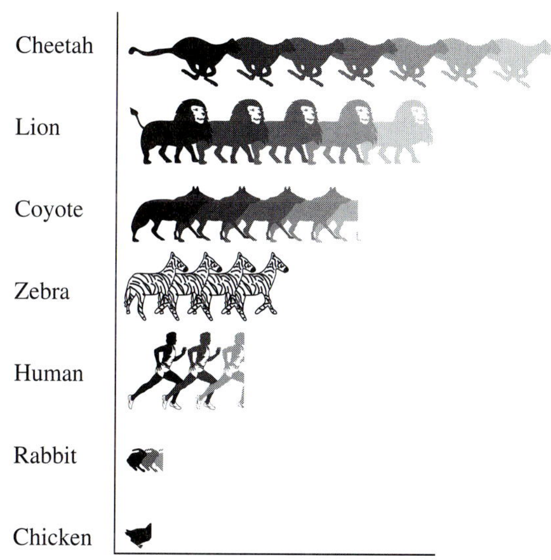

Each figure represents 10 miles per hour
Source: *Natural History Magazine*, March 1974

37. Show the data from the following pie chart in bar chart form.

1993 Sales for XYZ Company, Inc.

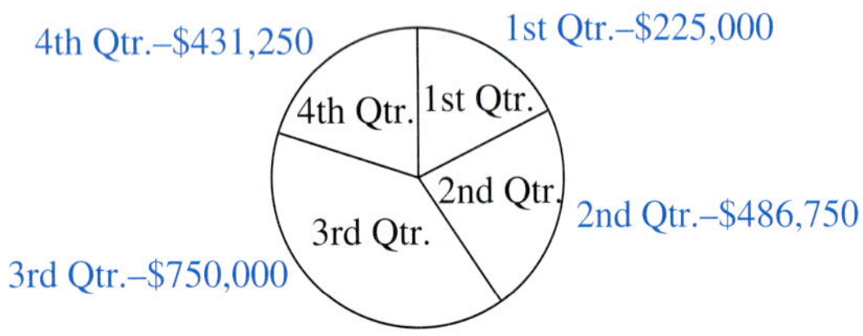

JOURNAL EXERCISES

1. Make up a more or less realistic budget for a family of four including monthly income and expenses and show the data in a table.

2. An individual receives her high school diploma at age 18 and works part time at a fast food restaurant during college. After four years in college she graduates and goes to work full time for a large corporation. After ten years of experience she is promoted to a manager over a small department. Make a bar chart of your best estimates of annual income based on age and position of this individual.

3. Present some data from a newspaper story, from a magazine article, or from a class you are taking in the form of a table, a bar chart, and a pictograph.

1 Number Line

Recall from Chapter 1 that a **number line** is used to represent numbers. A number line has a position for every whole number. Since there is no largest whole number, the number line continues to the right without any end. Points for the numbers 3 and 6 are plotted on the number line.

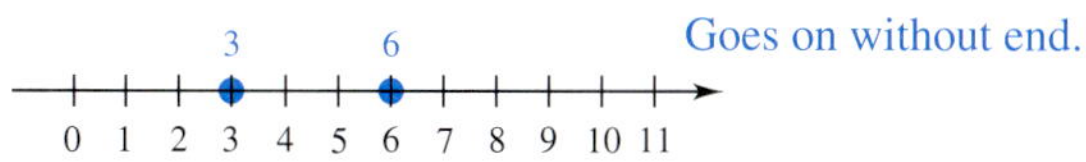

◇ *Do Practice Exercises 1 and 2.*

2 Line Graphs

A number line is used to represent a number in one dimension, or on a line. **Line graphs** are usually drawn on coordinate systems that use two lines which meet as shown:

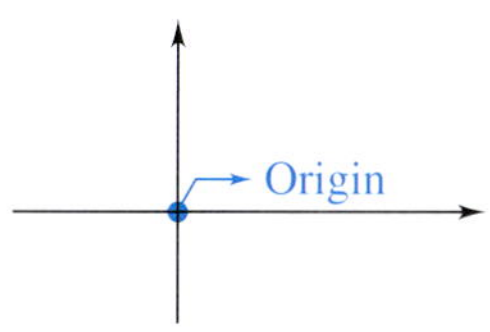

The point at which the two lines meet is called the **origin**. The line going from left to right is called the ***x*-axis**. The line going from top to bottom is called the ***y*-axis**. A scale similar to the one used for the one-dimensional number line is placed on both the x-axis and the y-axis:

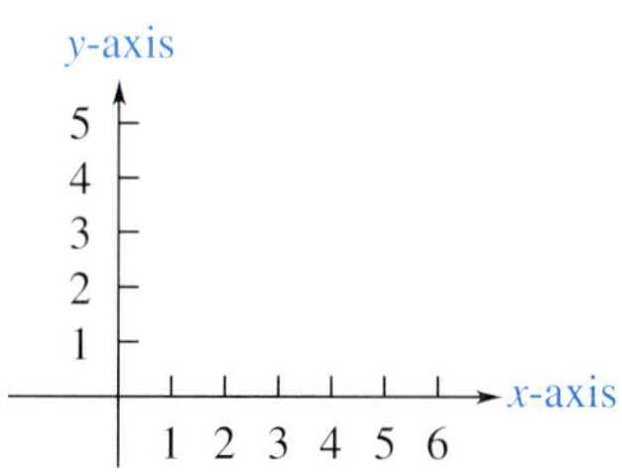

The pair of numbers (1,4) is an example of an ordered pair. The **ordered pair** (1,4) represents *one and only one* location. The first number in (1,4) is a 1 and is the number of units the point is located to the right of the origin, along the x-axis. The second number in (1,4) is a 4 and is the number of units the point is above the origin, along the y-axis. Thus, (1,4) is a point 1 unit to the right of and 4 units above the origin.

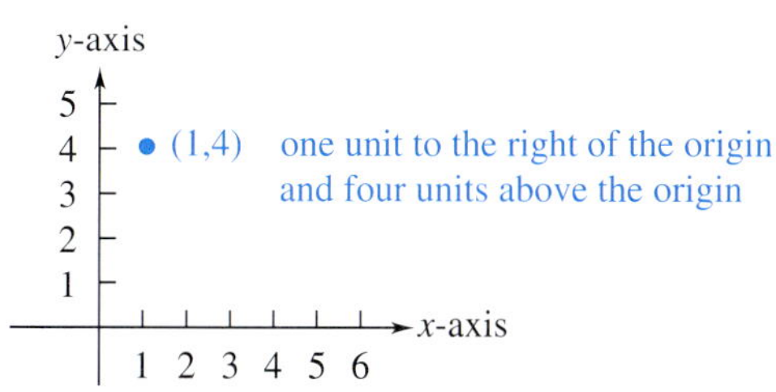

OBJECTIVES

1 Draw and use a number line.

2 Draw and use line graphs.

3 Use a line graph for a real problem.

1. Plot 4 and 10 as points on a number line.

2. Plot 0, 1, and 11 as points on a number line.

ANSWERS

1.

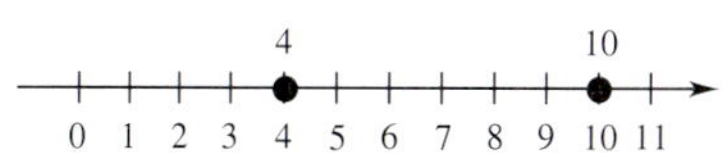

2.

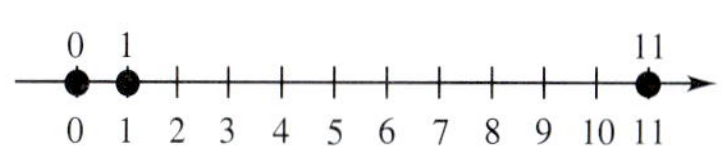

3. Plot (1,5) and (4,1) on a line graph.

4. Plot (0,0), (5,0) and (0,2) on a line graph.

EXAMPLE 1

Locate (0,0).

SOLUTION

The point (0,0) is 0 units to the right of the origin and 0 units above the origin. Thus, (0,0) are the coordinates of the origin.

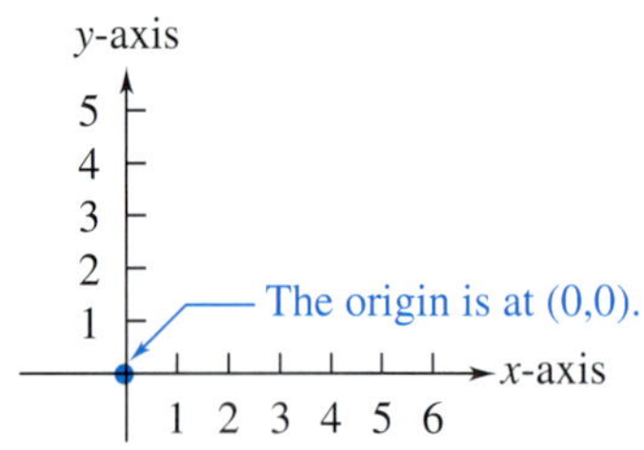

■

EXAMPLE 2

Plot the points (2,3), (4,1), and (6,0) on a line graph.

SOLUTION

(2,3) is two units to the right of and three units above the origin.
(4,1) is four units to the right of and one unit above the origin.
(6,0) is six units to the right of and zero units above the origin, thus (6,0) is on the *x*-axis.

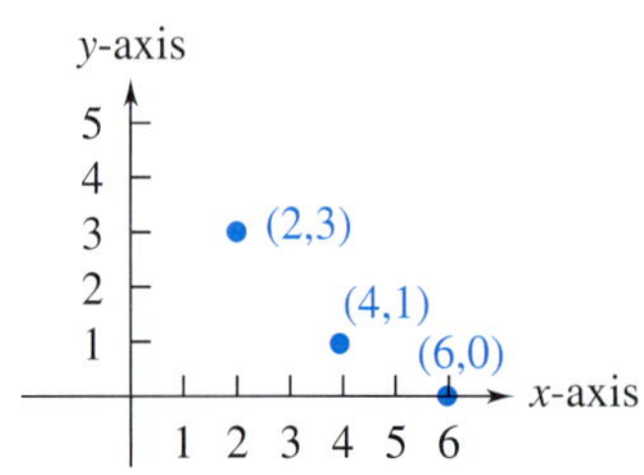

■

◇ ***Do Practice Exercises 3 and 4.***

Every point on a line graph can be written as an ordered pair. The first number of the ordered pair is found by counting the number of units the point is to the right of the origin. The second number of the ordered pair is found by counting the number of units the point is above the origin.

EXAMPLE 3

Find the coordinates of the point.

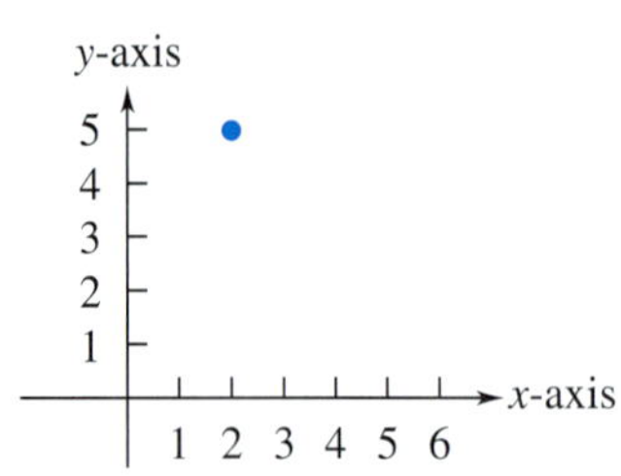

ANSWERS

3.

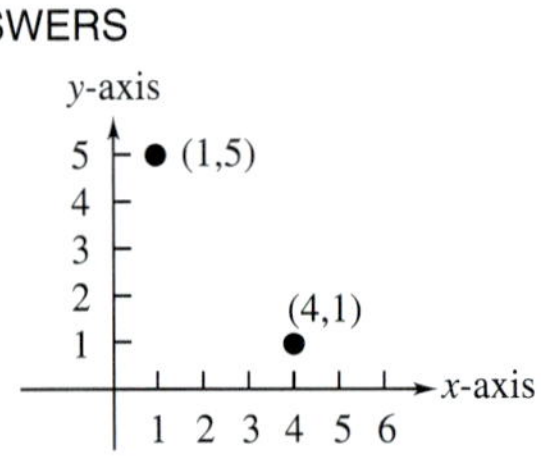

4.

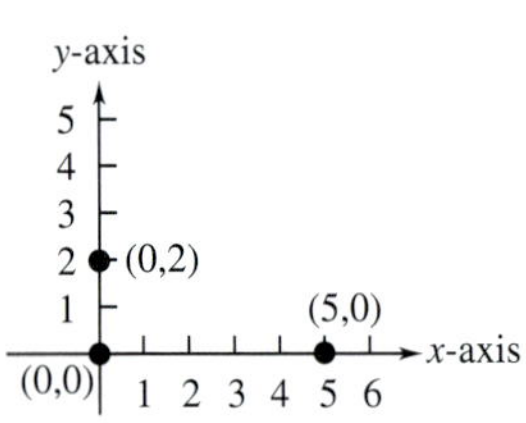

SOLUTION

The point is two units to the right of the origin so the first number of the ordered pair is 2. The point is five units above the origin so the second number of the ordered pair is 5. The ordered pair describing the coordinates of the point is (2,5). ■

3 Applications Using Line Graphs

Line graphs are frequently used to show information. They are used in business, science, and in magazine and newspaper articles. These graphs are effective because they communicate information with a picture. In contrast, it may be hard to see a pattern from a table full of numbers.

A table showing year and corresponding world population is given below. A line graph using the data in the table is then constructed.

World Population

Year	*Population (millions)*
1650	550
1750	725
1850	1,175
1900	1,600
1950	2,565
1990	5,333

Let the x-axis of the coordinate system be the year. The scale on the x-axis must cover all of the years for which there is data, that is, from 1650 through 1990. An appropriate x-axis follows:

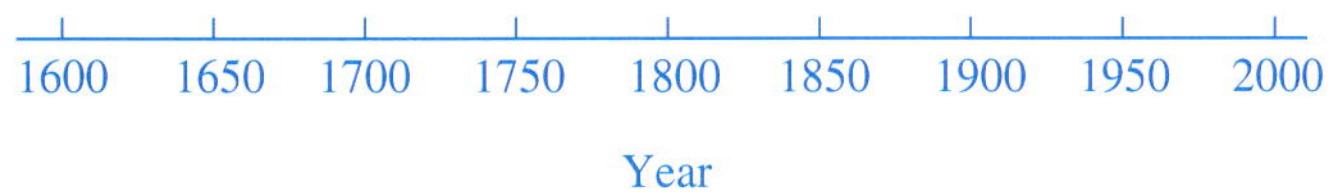

World population will be on the y-axis. The scale used on the y-axis must cover the entire range of population figures, that is, from a low of 550 million to a high of 5,333 million. An appropriate y-axis follows:

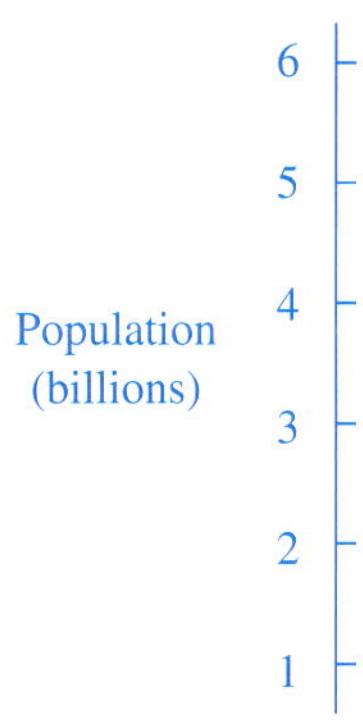

The heading and axes of the graph follow. Notice that in this case, the origin is not the point (0,0). Rather, the origin is (1600,0).

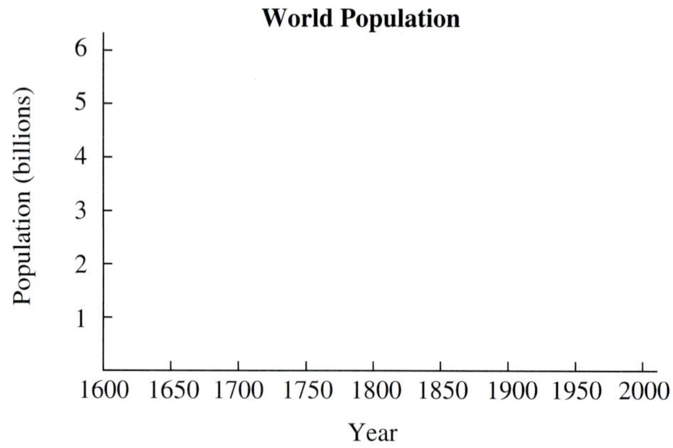

The coordinates of the origin are (1600,0) since 1600 is the first year that appears on the x-axis. The final graph is built by plotting each of the data points from the table and connecting the points by straight lines.

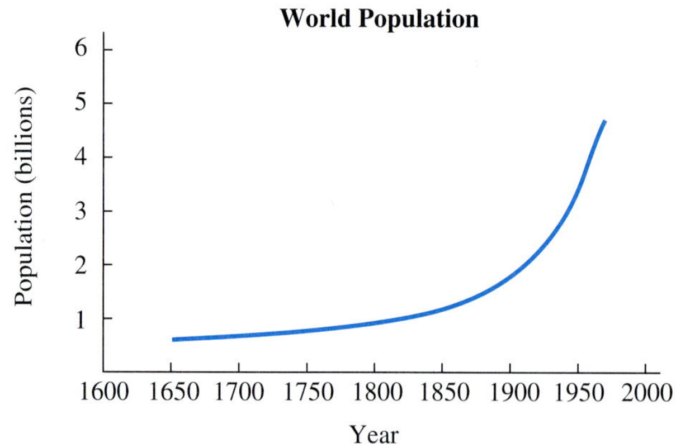

The graph gives a clear picture of a rapidly growing human population. This idea probably comes across more clearly from the graph than it did from the table. The graph shows that world populations levels have been increasing very rapidly since 1900. Most scientists believe that this rapid growth rate of the human population cannot go on forever.

From the graph, it is also possible to estimate world population for the year 1800, a year for which there was no data in the table. First, find the year 1800 on the x-axis. Second, move up to the curve. Finally, move to the left and read the number on the y-axis. This is done for you on the next graph.

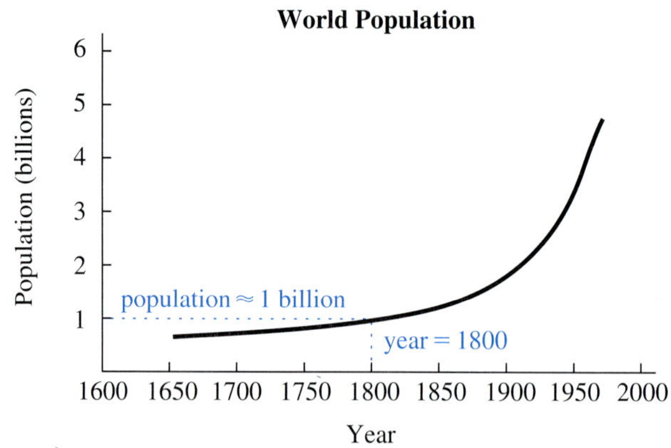

In 1800, the world population was about 1,000,000,000 people.

Draw a one dimensional number line for each and plot the points.

1. 3

2. 4, 1

3. 0, 5, 7

4. 1, 2, 7

5. 2, 4, 15

Find the coordinates of each point.

6.

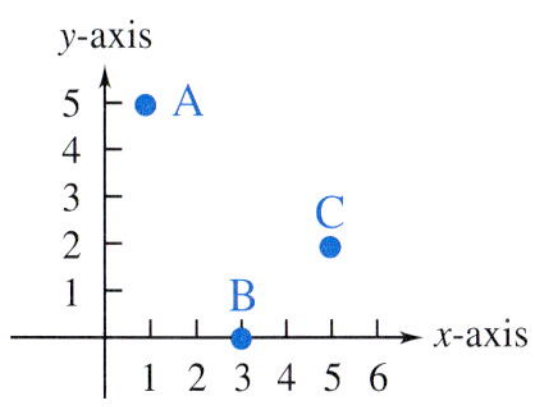

7.

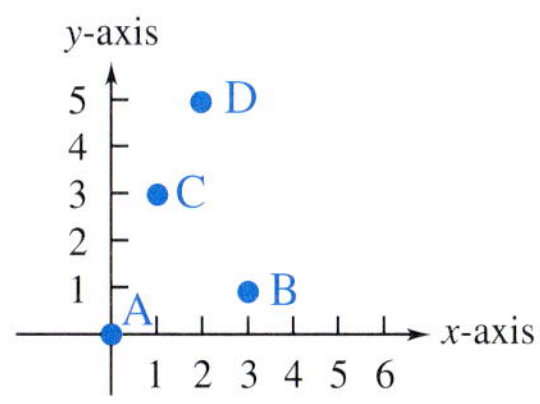

8.

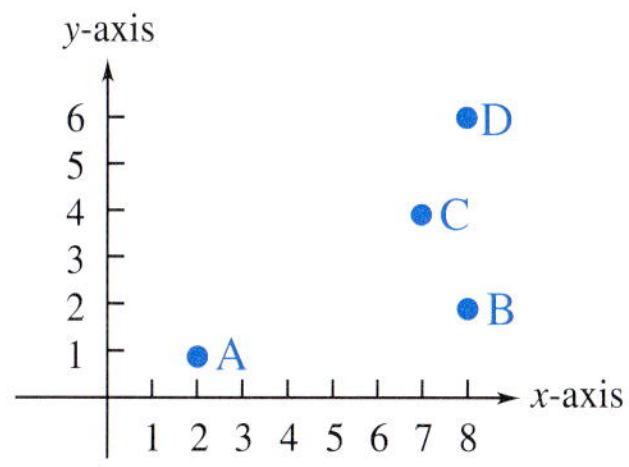

Work each problem.

9. Find the coordinates of the corners of the rectangle in the following figure.

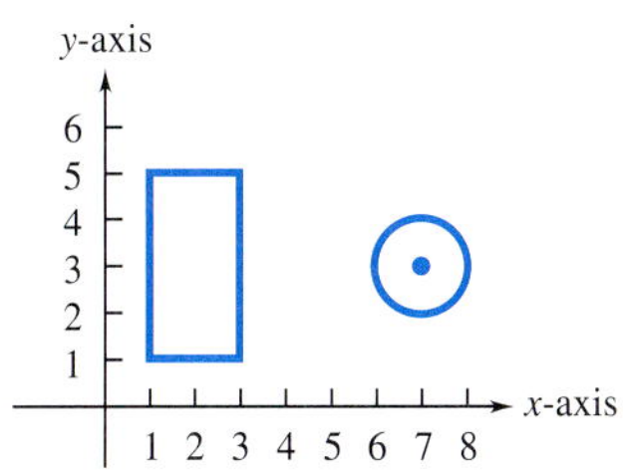

10. Find the coordinates of the center of the circle in problem 9.

11. Draw a line graph and plot (3,7), (1,5), and (2,4).

12. Draw a line graph, plot, and label the following points; A = (3,1), B = (2,7), and C = (3,3).

13. Draw a line graph, plot the corners of a square located at (4,0), (4,3), (7,3), and (7,0), and draw the square.

14. Draw a line graph, then draw a square with sides of length 5 whose lower left corner is at the origin.

15. Draw a line graph, then draw a rectangle with width 5 and height 2 whose lower left corner is at (2,1).

16. Make a bar chart using data from the following table.

Canadian Gold Production

Year	*Production (thousands of troy ounces)*
1983	2,363
1984	2,683
1985	2,815
1986(e)*	3,365
1987(p)*	3,788

*(e) estimated (p) preliminary
Source: *The World Almanac and Book of Facts.*

17. Make a bar chart from the following data. (Review)

Minimum Distance from Sun to Each Planet

Planet	*Minimum Distance (millions of miles)*
Mercury	28
Venus	67
Earth	91
Mars	129
Jupiter	461
Saturn	838
Uranus	1,669
Neptune	2,760
Pluto	2,756

18. Make a graph from the following data.

1990 Federal Estate and Gift Tax

Tax Computed on	*Tax*
$ 20,000	$ 3,800
40,000	8,200
60,000	13,000
80,000	18,200
100,000	23,800

Use data from the graphs to answer the questions that follow.

Annual Oil Production from an Oil Well Drilled in 1985

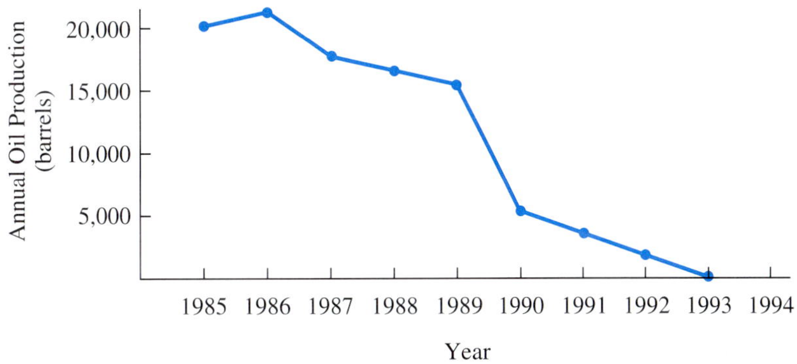

19. What was the production from the oil well during 1989?

20. For how many years did the oil well produce?

21. In what year did production decline the most?

22. Approximately what was the total oil production from the well during its lifetime?

23. If the average selling price of oil from the well was $20 per barrel, what was the approximate total income generated by the well during its lifetime?

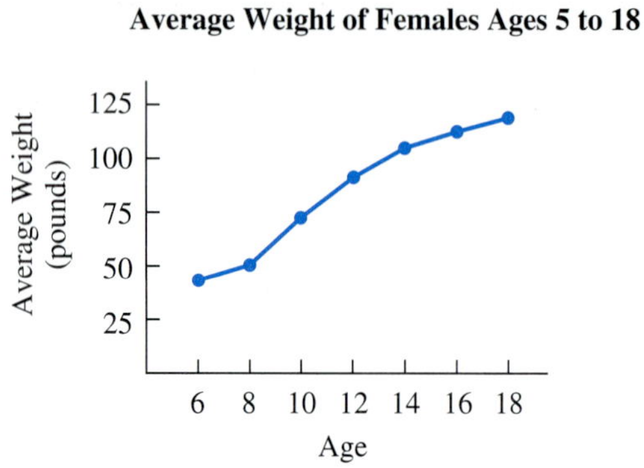

24. What is the approximate average weight of females aged 8?

25. During what ages do females grow the fastest?

26. Do females grow much after age 16?

27. The growth of a girl with a severe medical problem was:

Age (in years):	6	8	10	12	14	16	18
Weight (in pounds):	40	48	60	61	72	105	115

Copy the graph of Average Weight of Females Ages 5 to 18 given above on to your own paper and plot this girl's growth on the same chart. At what age does her weight begin to differ significantly from a normal child's pattern?

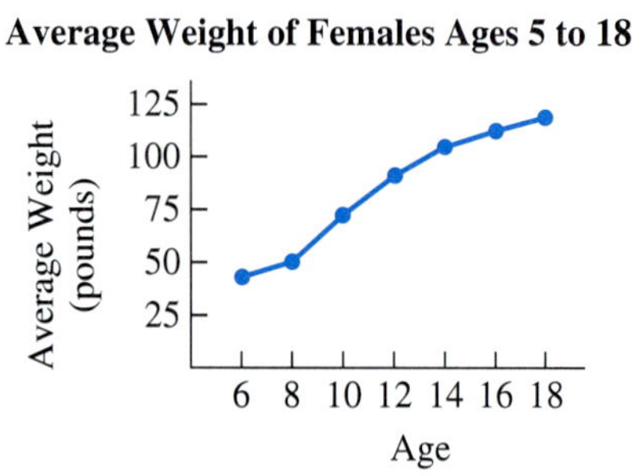

28. The medical problem of the girl was diagnosed and successfully treated. At what age did her weight first recover to near normal?

JOURNAL EXERCISES

1. Find some data of interest to you and graph it. Then identify any trends or patterns.

2. Explain the old saying "A picture is worth a thousand words" as it relates to showing information with a graph versus a table full of numbers. Can you think of any information that cannot be shown in a graph?

CHAPTER 2 REVIEW

SECTION	TOPIC OR TERM	EXAMPLE OR DEFINITION
2.1		
Exponents	Exponent	An exponent is the number of times a value is to be multiplied times itself: $4^2 = 4 \cdot 4$.
	Squared	Squared means that a value is to be raised to the second power: three squared $= 3^2$.
	Cubed	Cubed means that a value is to be raised to the third power: one cubed $= 1^3$.
2.2		
Order of Operations	Order of Operations	The order of operations are the rules for the order in which math operations in an expression are calculated, for example: $3 \times 5 - 1$ $15 - 1$ Multiply first. 14 Result.
2.3		
Tests for Divisibility	Divisible	One number is divisible by a second if the second number divides into the first without remainder.
2.4		
Prime Numbers and Prime Factorization	Prime Number	A prime number is a number whose only factors are itself and 1, such as 17.
	Composite Number	A composite number is a number with factors other than itself and 1, such as 10 whose factors include 2 and 5.
	Prime Factorization	Prime factorization is the unique set of primes that when multiplied together produce the original number. The prime factorization of 55 is 5 times 11.
2.5		
Least Common Multiples	Least Common Multiple	The LCM of two numbers is the smallest number that both divide. The LCM of 8 and 12 is: $8 = \| 2 \cdot 2 \cdot 2 \| \|$ $12 = \| 2 \cdot 2 \cdot \| 3 \|$ LCM $= 2 \cdot 2 \cdot 2 \cdot 3 = 24$
	Greatest Common Divisor	The GCD of two numbers is the largest number that is a factor of both. The GCD of 8 and 20 is: Divisors of 8 are: 1, 2, 4, and 8. Divisors of 20 are: 1, 2, 4, 5, 10, and 20 The GCD is 4.

SECTION	TOPIC OR TERM	EXAMPLE OR DEFINITION
2.6		
Tables and Charts	Chart	Charts are diagrams that show data.
2.7		
Graphs	Line Graph	A line graph uses a line to connect adjacent data points on a coordinate system.

NAPOLEON BONAPARTE'S INVASION OF RUSSIA

Napoleon Bonaparte was born on August 15, 1769. His life was one of many battles. In 1805, he fought the Austrians at Austerlitz. In 1807, he fought the Russians at Friedland. In 1809, he fought the Austrians again, but this time at Wagram. One of Napoleon's greatest tragedies was his 1812 invasion of Russia. He began the invasion in February of 1812 with a "grand army" of 422,000 men. His army faced many battles, forced marches, and starvation.

1. How old was Napoleon when he invaded Russia?

Near Moscow in a city called Borodino, the Russians fought and lost a major battle to the French. One of his soldiers describes the battlefield:

> **Here (center of the battlefield) the corpses lay piled higher ever higher. . . . The ditches were filled to the brim with bodies. . . . Men and horses had been gashed and maimed in every conceivable way. . . .**

Source: *With Napoleon in Russia 1812* by Henrich Vossler. Published by Folio Society of London, 1969.

The Russians let Napoleon and his remaining army of 100,000 enter Moscow without resistance. Napoleon waited in Moscow for their surrender, but they did not. His army began the retreat back to France. The retreat was marked by starvation, more battles, and the horrible cold of the Russian winter. Only 10,000 soldiers of the grand army survived to return to France.

2. Complete the following graph using the above information.

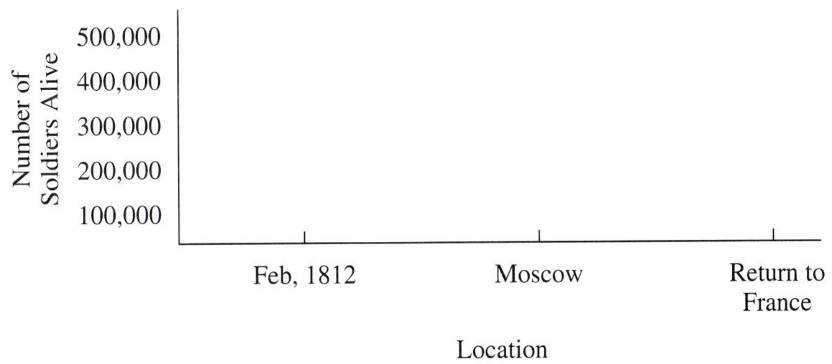

The number of soldiers alive versus location of the invading army is shown in the graph below. The number of soldiers alive is shown by the width of the band. The darker band shows the retreat of the grand army. This graph was taken from one drawn in 1869 by French engineer, Charles Joseph Mitard.

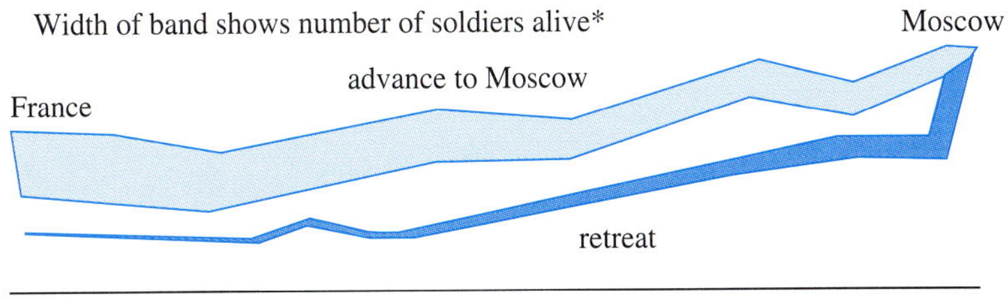

* Army began from France with 422,000 soldiers. 100,000 soldiers remained at Moscow. Only 10,000 soldiers returned alive to France.

3. How many years after the invasion was the graph drawn?

4. Go to a map of Europe and estimate the distance from Paris, France, to Moscow, Russia, and back to Paris, France. How many miles did the army average per day if they walked that distance in 400 days?

Napoleon was exiled to an island where he died in 1821.

5. Complete the ''life event'' chart below for Napoleon showing major battles and events in his life.

Birth

1769

Year

CHAPTER 2 REVIEW PROBLEMS

Evaluate.

1. 7^2
2. 3^4
3. $2^4 3^2$
4. 5^3
5. $2^3 5^2$
6. $100 - 6 \cdot 5$
7. $5 + 9^2$
8. $(10 - 1)^2 \div 3$
9. $4 + 7 \cdot 0$
10. $8 \div 2^2$
11. $64 - 7 \cdot 2 \cdot 3$
12. $3^2 \cdot (6 + 1)$
13. $112 \div 2 - 50 \div 10$
14. $5 \cdot 2^2 - 7$
15. $[(16 \div 4) + 3] \cdot 9 \cdot 6$
16. $32 \div 4 - 3 \cdot 2$

Simplify.

17. $8{,}127^1$
18. 10^0

Write as a base to a power.

19. $5 \times 5 \times 5$
20. $7^2 7^3$

Write each as an expression and evaluate.

21. Find the sum of three squared plus five.
22. Find the sum of two cubed with four squared.
23. Find the product of eight times the sum of two and four.
24. Find the product of six with the quotient of sixty-four divided by sixteen.
25. Find the sum of two squared, four cubed, and eight.
26. Find the product of the first three prime numbers.

Determine if each number is divisible by 2, 3, 4, 5, 6, 8, 9, and/or 10.

27. 320

28. 1,004

29. 689

30. 7,500

Find the prime factorization of each.

31. 18

32. 35

33. 60

34. 75

Find the least common multiple.

35. 6, 8

36. 10, 14

37. 12, 20

38. 15, 25

Find the great common divisors.

39. 20, 28

40. 60, 42

Work each problem.

41. Is the LCM of 6 and 20 greater than 40?

42. What is the smallest number that both 8 and 10 divide into?

43. What is the smallest number that $2 \cdot 3 \cdot 3 \cdot 5$ and $2 \cdot 5 \cdot 7 \cdot 7$ both divide?

44. Find the square of the least common multiple of six and fifteen.

45. Is 141 a prime number or a composite number?

Use the bar chart to answer the three questions that follow.

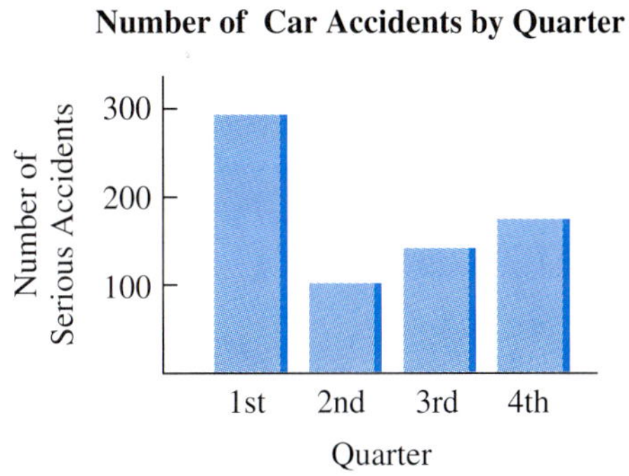

46. How many accidents occurred during the second quarter?

47. The highest number of accidents occurred in which quarter?

48. This bar chart was labeled poorly. Can you think of additional descriptive information that would help us understand the data?

Use data from the table to answer the questions that follow.

Food Preferences for a Group of 200 Students

Preference	*Number*	*Percent*
Beef	76	38%
Chicken	98	49%
Fish	24	12%
Other	2	1%
Totals	200	100%

49. Make a bar chart showing number of students on the y-axis versus food preference on the x-axis.

50. Make a line graph showing number of students on the y-axis versus food preference on the x-axis.

CHAPTER 2 TEST

Evaluate.

1. 7^3

2. $3^2 + 5^3$

3. 8^0

4. $(7 - 1)^2 \div 18$

5. $4 + 30 \div 2$

6. $3 + 7 \cdot 2^3$

7. $3 \cdot (2 - 1)$

8. $(8 \cdot 4 \div 2) - 8$

9. $(14 \div 7)^2$

10. $3 \cdot 5 - 1^2$

Work each problem.

11. Find three squared minus two cubed.

12. Is 742 divisible by 2? By 3?

13. Is 6,450 divisible by 2, 4, and 5?

14. Find the prime factorization of 75.

15. Find the prime factorization of 140.

16. Name the next larger prime number after 37.

17. What is the least common multiple of eight and twelve?

18. What is the LCM of fourteen and sixteen?

19. Is 81 a prime number?

20. Is 141 a composite number?

21. Find the greatest common divisor of 75 and 150.

22. Find the greatest common divisor of 100 and 28.

23. Find seven squared minus four.

24. Sum sixteen with the quotient of seventy divided by ten.

25. Find the smallest number divisible by 2, 3, and 8.

26. List all prime numbers between 20 and 30.

Use data from the table to answer the questions that follow.

Number of Employees at Joseph's Manufacturing Company

Year	*Number of Employees at End of Year*
1989	209
1990	101
1991	70
1992	78
1993	86
1994	92*

*projected

27. In what year were the number of employees at Joseph's Manufacturing Company the highest?

28. In which year did the company lose the most employees?

29. Draw a bar chart showing the same data.

30. Draw a line graph showing the same data.

CUMULATIVE TEST CHAPTERS 1 and 2

Write each number using words.

1. 3,654

2. 41,710

3. 73,215

4. 40,679

5. 2,147,000

6. 1,264,100,001

Evaluate, showing all work.

7. 6×9

8. $27 \div 3$

9. $125 + 27$

10. $74 - 9$

11. $17 \cdot 40$

12. $3{,}309 \div 9$

13. $1{,}724 + 209$

14. $1{,}600 - 327$

15. $\begin{array}{r} 1{,}785 \\ \times \quad 31 \\ \hline \end{array}$

16. $\begin{array}{r} 2{,}754 \\ 382 \\ + \quad 169 \\ \hline \end{array}$

17. $\begin{array}{r} 1{,}962{,}000 \\ - \quad 3{,}200 \\ \hline \end{array}$

18. $17\overline{)2{,}413}$

19. 3^2

20. 4^3

21. $2^5 \cdot 1$

22. $7^2 4^3$

23. $3 + 9^2$

24. $16 - 50 \div 10$

25. $(30 \div 6)^2 \cdot 5$

26. $32 \div 2 + 1$

27. $70 \div 10 \cdot (5 - 2)$

28. $22 + (25 - 2 \times 3)$

29. $(3 - 1) \cdot (8 \div 8 - 1)$

30. $12^2 \cdot 2^3$

Find the prime factorization of each.

31. 60

32. 45

Find the least common multiple.

33. 8, 36

34. 12, 30, 2

Use the chart to answer the three problems that follow.

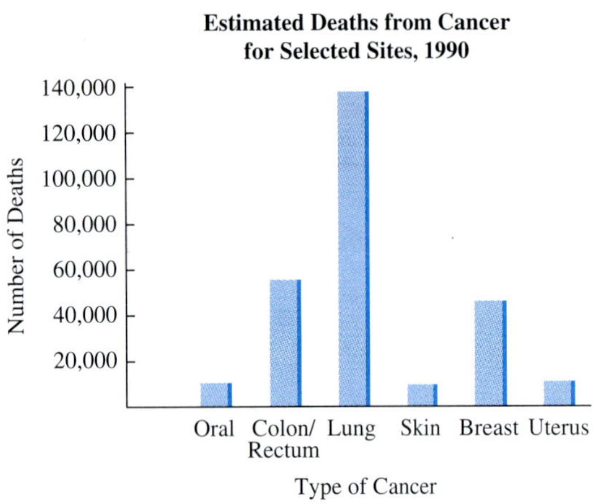

Source: American Cancer Society

35. What was the approximate number of deaths from breast cancer in 1990?

36. What was the approximate total number of deaths from all cancers for the selected sites during 1990?

37. Approximately how many times as many people died from lung cancer as from colon/rectum cancer during 1990?

Work each problem.

38. Find the greatest common divisor of 100 and 36. Find the average of 100 and 36. Are these two numbers equal to one another?

39. A study of a small company showed that an average of four staples are used per hour by each employee. How many staples will one employee use in five eight-hour days?

40. Jill's grandmother died at age 42 from a heart attack and Jill's mother died at age 38 from a heart attack. Jill will also die from a heart attack before reaching age 45. Is this an example of deductive or inductive reasoning?

3 Fraction Notation

■**APPLICATION** *An auto mechanic estimates that one-third of the cars he repairs will be back in his shop within one year for other repairs. If he repairs 321 cars one month, how many of these automobiles will be back in his shop within a year?*

■**SOLUTION** *The answer is found by multiplying the fraction $\frac{1}{3}$ times the 321 cars which were in the shop that month.*

$$\frac{1}{3} \cdot 321 \text{ cars} = 107 \text{ cars}$$

The mechanic estimates that 107 of the 321 cars will return to the shop within a year.

3.1 BASICS

1 Introduction to Fractions

One way to write a part of a whole number is to use a **fraction**. Some examples of fractions are:

A recipe calls for **two-thirds** of a cup of flour.
A doctor prescribed a dosage of **one-half** teaspoon.
One-fourth of the students made an A on the test.

The idea of a fraction will be introduced using geometrical shapes. Both of the squares below are divided into nine equal sized pieces. In each case, seven of the nine pieces or seven-ninths $\left(\frac{7}{9}\right)$ of each geometric figure are shaded.

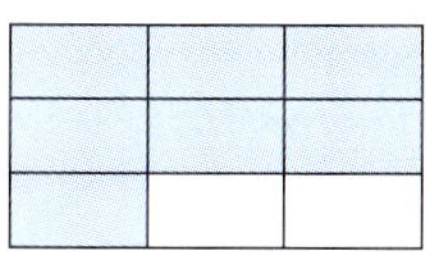

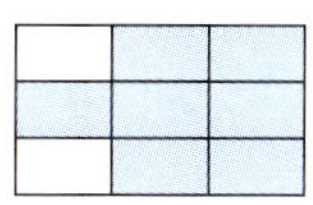

◇ *Do Practice Exercises 1 and 2.*

OBJECTIVES

1 Identify the parts of a fraction.

2 Identify proper and improper fractions.

3 Reduce fractions.

1. Draw a square and divide it into four equal sized pieces. Shade one fourth of the square by shading one of the pieces.

2. What fraction of the following rectangle is shaded?

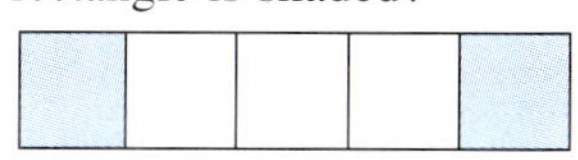

ANSWERS

1. 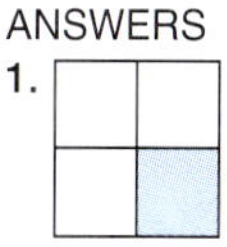

2. $\frac{2}{5}$ths

3. Simplify each:

(a) $\frac{4}{2}$

(b) $\frac{15}{3}$

(c) $\frac{1{,}000}{20}$

A fraction is written as one number on top of another. The number on the top is called the **numerator** and the number on the bottom is called the **denominator**.

$$\frac{5}{8}$$

5 → numerator
8 → denominator

This fraction is read as five-eighths and represents five parts out of eight parts.

A fraction is another way to write a division problem. The fraction $\frac{3}{4}$ has the same meaning as 3 divided by 4, $3 \div 4$, or three parts out of four. Each of the following fractions is shown as a division problem.

$$\frac{1}{3} = 1 \div 3 \qquad \frac{2}{5} = 2 \div 5 \qquad \frac{14}{14} = 14 \div 14 \qquad \frac{6}{3} = 6 \div 3$$

The fraction $\frac{14}{14}$ equals $14 \div 14$ which equals 1. The fraction $\frac{6}{3}$ equals $6 \div 3$ which is 2.

EXAMPLE 1

Simplify the fraction $\frac{20}{5}$.

SOLUTION

$$\frac{20}{5} = 20 \div 5 = 4 \quad \blacksquare$$

◇ *Do Practice Exercise 3.*

A zero can be in the numerator of a fraction since a whole number divides into 0 evenly as we saw in Chapter 1.

$$\frac{0}{12} = 0 \div 12 = 0$$

However, a zero cannot be in the denominator since division by zero is undefined.

$$\frac{15}{0} = 15 \div 0 \qquad \text{which is undefined}$$

Calculator Application

If you did not do so in Chapter 1, use your calculator to divide a whole number such as 15 by 0. What is the result?

Answer: error

ANSWERS
3. (a) 2 (b) 5 (c) 50

The following example shows when zero can and cannot be used.

EXAMPLE 2

Simplify each:

(a) $\frac{0}{15}$ (b) $\frac{5}{0}$

SOLUTION

(a) $\frac{0}{15} = 0$ (b) $\frac{5}{0}$ is undefined ■

◇ *Do Practice Exercise 4.*

2 Proper and Improper Fractions

A **proper fraction** is one in which the numerator is *less than* the denominator. For example, $\frac{2}{3}$ is a proper fraction since 2 is smaller than 3. The fraction $\frac{2}{3}$ equals 2 divided by 3 which is a value less than one.

An **improper fraction** is a fraction in which the numerator is either *equal to or greater than* the denominator. For example, $\frac{12}{5}$ is an improper fraction since 12 is larger than 5. The fraction $\frac{7}{7}$ is also an improper fraction since the numerator is equal to the denominator. The denominator of every improper fraction will divide at least once into the numerator since the denominator is smaller than the numerator.

HELPFUL HINT

Every proper fraction is less than one.
Every improper fraction is one or greater.

Using fraction notation, the number 1 can be written in an infinite number of different ways, such as:

$$\frac{1}{1}, \quad \frac{2}{2}, \quad \frac{17}{17}, \quad \frac{123}{123}, \quad \text{or} \quad \frac{2{,}341}{2{,}341}$$

Each of these is equal to one. Fractions in which the numerators and denominators are equal to each other are called **unit fractions** or **forms of one**. For example, both $\frac{7}{7}$ and $\frac{38}{38}$ are forms of one since both equal one. The fraction $\frac{125}{125}$ is also a form of one.

Some examples of the different types of fractions are:

Proper fractions—The numerator is smaller than the denominator.

$$\frac{1}{4}, \quad \frac{9}{10}, \quad \frac{3}{107}, \quad \frac{0}{5}, \quad \frac{0}{49}$$

4. Simplify each:

(a) $\frac{0}{7}$

(b) $\frac{15}{3}$

(c) $\frac{12}{0}$

ANSWERS
4. (a) 0 (b) 5 (c) undefined

5. Classify each as a proper fraction, an improper fraction, or undefined:

(a) $\frac{5}{4}$

(b) $\frac{9}{10}$

(c) $\frac{0}{8}$

(d) $\frac{15}{0}$

6. Identify the forms of one:
$\frac{3}{4}, \frac{6}{7}, \frac{2}{2}, \frac{9}{5}, \frac{11}{11}$

7. Write $\frac{3}{1}$ and $\frac{1}{1}$ as whole numbers.

8. Write 7 and 2,125 as fractions with a denominator of 1.

ANSWERS
5. (a) improper (b) proper (c) proper (d) undefined
6. $\frac{2}{2}, \frac{11}{11}$
7. 3, 1
8. $\frac{7}{1}, \frac{2,125}{1}$

Improper fractions—The numerator is greater than or equal to the denominator.

$$\frac{6}{5}, \quad \frac{70}{70}, \quad \frac{100}{8}, \quad \frac{7}{1}, \quad \frac{12}{1}$$

Forms of one—The numerator equals the denominator. Actually, these are also improper fractions.

$$\frac{8}{8}, \quad \frac{19}{19}, \quad \frac{265}{265}, \quad \frac{1,000}{1,000}$$

Undefined fractions—Division by zero is not allowed.

$$\frac{7}{0}, \quad \frac{1,000}{0}, \quad \frac{99}{0}$$ ⟵ These have no meaning.

◇ *Do Practice Exercises 5 and 6.*

An improper fraction of special interest occurs when the denominator of the fraction is the number 1. If the denominator is 1, then the fraction is equal to the numerator:

$$\frac{8}{1} = 8 \div 1 = 8 \qquad \frac{12}{1} = 12 \div 1 = 12$$

The reverse is also true. A whole number can be written as a fraction by placing a 1 in the denominator:

$$15 = 15 \div 1 = \frac{15}{1} \qquad 27 = 27 \div 1 = \frac{27}{1}$$

EXAMPLE 3

Simplify $\frac{56}{1}$.

SOLUTION

$$\frac{56}{1} = 56$$ ■

EXAMPLE 4

Write 48 as a fraction with a denominator of 1.

SOLUTION

$$48 = \frac{48}{1}$$ ■

◇ *Do Practice Exercises 7 and 8.*

3 Reducing Fractions

One-half of the circle on the left is shaded and two-fourths of the circle on the right is shaded. In both cases, the amount shaded is the same. Therefore, $\frac{1}{2}$ must equal $\frac{2}{4}$.

$\frac{1}{2}$ shaded

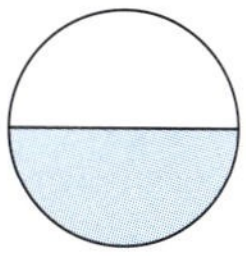

$\frac{2}{4}$ shaded

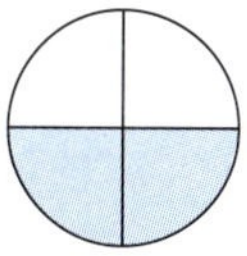

Actually, the fraction $\frac{1}{2}$ can be written in an infinite number of ways:

$$\frac{1}{2} = \frac{2}{4} = \frac{3}{6} = \frac{4}{8} = \frac{5}{10} = \frac{6}{12} = \ldots = \frac{50}{100} = \ldots$$

The fractions $\frac{1}{2}$, $\frac{2}{4}$, $\frac{3}{6}$, and $\frac{50}{100}$ are exactly the same number even though they look different. It is easier to understand:

$$\frac{1}{2} \text{ of the students}$$

than it is to understand:

$$\frac{100}{200} \text{ of the students}$$

even though both mean the same thing. For this reason, fractions are usually written in simplified or reduced form. A fraction is said to be **simplified** or **reduced to lowest terms** when written using the smallest numbers possible in both the numerator and denominator.

Rule: To Reduce a Fraction

Step 1 Find a whole number larger than one that divides evenly into both the numerator and denominator.

Step 2 Divide the numerator by the whole number and replace with the result of the division.

Step 3 Divide the denominator by the whole number and replace with the result of the division.

Step 4 If the numerator and denominator have a common factor greater than 1, go back to Step 1.

9. Reduce each:

(a) $\frac{8}{10}$

(b) $\frac{16}{24}$

(c) $\frac{35}{40}$

ANSWERS

9. (a) $\frac{4}{5}$ (b) $\frac{2}{3}$ (c) $\frac{7}{8}$

EXAMPLE 5

Reduce $\frac{6}{8}$ using the rule.

SOLUTION

Step 1 Both 6 and 8 are evenly divisible by 2.

Step 2 The number 2 divides into the numerator (6) three times. Replace 6 with 3.

Step 3 The number 2 divides into the denominator (8) four times. Replace 8 with 4.

Step 4 There are no common factors in 3 and 4.

Thus $\frac{6}{8}$ reduces to $\frac{3}{4}$ or $\frac{6}{8} = \frac{3}{4}$. ■

EXAMPLE 6

Reduce each:

(a) $\frac{25}{30}$ **(b)** $\frac{105}{165}$

SOLUTION

(a) $\frac{25}{30} = \frac{5}{6}$ **(b)** $\frac{105}{165} = \frac{35}{55} = \frac{7}{11}$ ■

Recall that the greatest common divisor (GCD) of the numerator and denominator is the largest whole number that divides both. The GCD can be used to reduce fractions. For example, the GCD of the numerator and denominator of $\frac{50}{125}$ is 25. Divide both the numerator and denominator by 25:

$$\frac{50}{125} = \frac{2}{5}$$

◇ *Do Practice Exercise 9.*

Forms of one can also be used to reduce fractions. Using this method, the first step is to find the prime factorization of the numerator and of the denominator. Next, identify forms of one and divide them out. Look at the next example.

EXAMPLE 7

Reduce $\frac{70}{154}$ using forms of one.

SOLUTION

The prime factorization of 70 is $2 \cdot 5 \cdot 7$.
The prime factorization of 154 is $2 \cdot 7 \cdot 11$.

$$\frac{70}{154}$$

$$\frac{2 \cdot 5 \cdot 7}{2 \cdot 7 \cdot 11} \quad \text{prime factorizations}$$

$$\frac{2 \cdot 7 \cdot 5}{2 \cdot 7 \cdot 11} \quad \text{Identify forms of one and divide them out.}$$

$$\frac{5}{11} \quad \text{Result.}$$ ■

◇ *Do Practice Exercise 10.*

10. Reduce each using forms of one:

(a) $\frac{27}{30}$

(b) $\frac{42}{49}$

(c) $\frac{50}{90}$

ANSWERS

10. (a) $\frac{9}{10}$ (b) $\frac{6}{7}$ (c) $\frac{5}{9}$

What fraction of each geometric figure is shaded?

1. 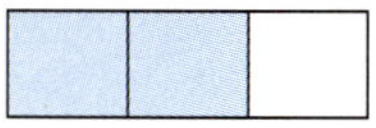

2. 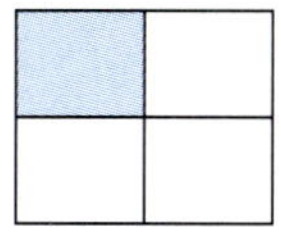

3. 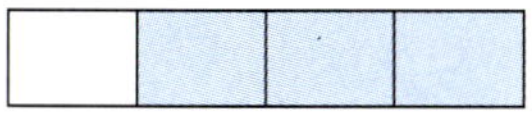

4. 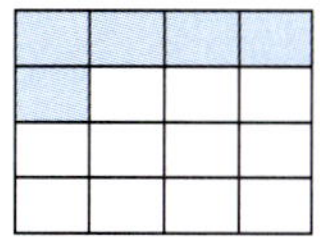

5. 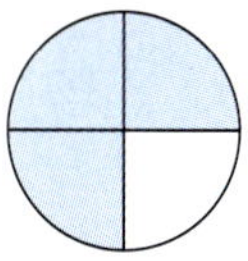

6. 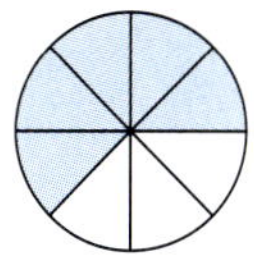

7. 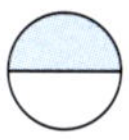

8. 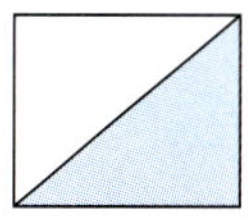

9. 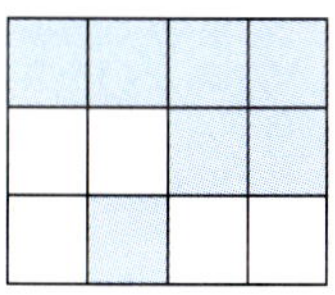

10. 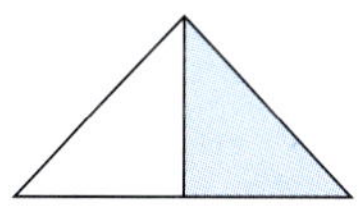

What fraction of each group of geometric figures is shaded?

11. 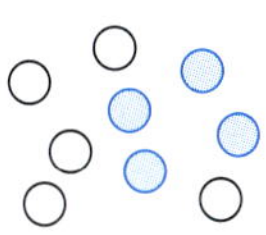

12. 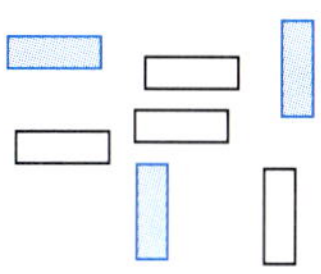

13. 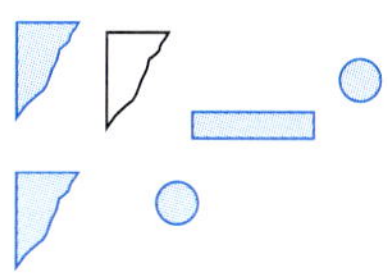

14. 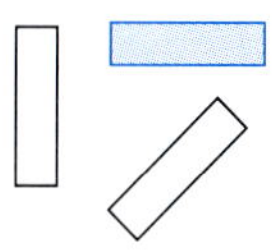

Identify each as a proper or improper fraction and state which fractions are less than one.

15. $\frac{1}{4}, \frac{2}{3}, \frac{3}{5}$

16. $\frac{7}{8}, \frac{0}{5}, \frac{12}{3}, \frac{4}{4}$

17. $\frac{3}{2}, \frac{100}{1}, \frac{17}{20}, \frac{0}{5}$

18. $\frac{9}{2}, \frac{1}{5}, \frac{7}{7}, \frac{0}{10}, \frac{4}{8}$

Which fractions are undefined?

19. $\frac{3}{4}, \frac{0}{9}, \frac{12}{5}, \frac{10}{0}$

20. $\frac{0}{0}, \frac{5}{0}, \frac{0}{7}, \frac{6}{9}$

21. $\frac{2}{3}, \frac{9}{5}, \frac{4}{5}, \frac{0}{8}$

22. $\frac{4}{5}, \frac{7}{0}, \frac{0}{3}, \frac{0}{2}$

Which fractions represent a form of one?

23. $\frac{3}{5}, \frac{8}{8}, \frac{0}{9}$

24. $\frac{4}{6}, \frac{2}{2}, \frac{12}{3}$

25. $\frac{9}{9}, \frac{2}{7}, \frac{1}{1}$

26. $\frac{4}{8}, \frac{7}{7}, \frac{13}{13}$

Evaluate each. (Review)

27. 6^2

28. $15 - 3 \cdot 2$

29. $14 \div 2 \cdot 8$

30. $4 \cdot (5 - 3)$

31. $7 \cdot 8 - 2$

32. $3^2 2^4$

33. 4^0

34. $3^2 + 5 \cdot 4$

Reduce each fraction to lowest terms.

35. $\frac{2}{8}$

36. $\frac{4}{8}$

37. $\frac{3}{15}$

38. $\frac{2}{18}$

39. $\frac{5}{90}$

40. $\frac{5}{20}$

41. $\frac{8}{10}$

42. $\frac{8}{32}$

43. $\frac{300}{400}$

44. $\frac{21}{27}$

45. $\frac{18}{27}$

46. $\frac{95}{100}$

47. $\frac{35}{40}$ **48.** $\frac{34}{90}$ **49.** $\frac{63}{120}$ **50.** $\frac{45}{65}$

51. $\frac{24}{44}$ **52.** $\frac{25}{55}$ **53.** $\frac{18}{48}$ **54.** $\frac{17}{51}$

55. $\frac{37}{111}$ **56.** $\frac{20}{250}$ **57.** $\frac{14}{49}$ **58.** $\frac{64}{800}$

59. $\frac{18}{20}$ **60.** $\frac{50}{250}$ **61.** $\frac{51}{54}$ **62.** $\frac{12}{46}$

63. $\frac{42}{124}$ **64.** $\frac{95}{125}$ **65.** $\frac{112}{240}$ **66.** $\frac{234}{216}$

67. $\frac{55}{121}$ **68.** $\frac{49}{35}$ **69.** $\frac{120}{480}$ **70.** $\frac{121}{165}$

71. $\frac{250}{1{,}000}$ **72.** $\frac{300}{10{,}000}$ **73.** $\frac{2{,}280}{4{,}920}$ **74.** $\frac{1{,}000{,}000}{10{,}000{,}000}$

Find the fraction shaded and reduce to lowest terms.

75. 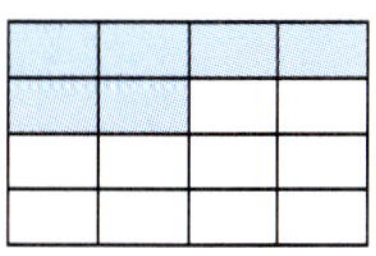

76. 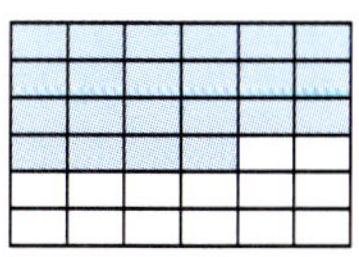

77. 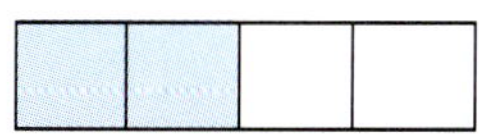

78. 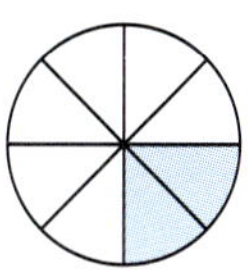

79. 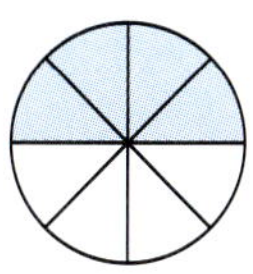

80. 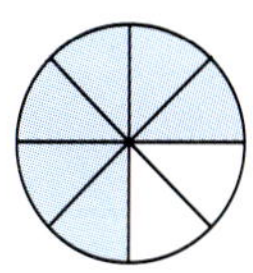

Reduce any fractions to lowest terms if possible.

81. A deck of 52 playing cards contains four aces. What fraction of the deck is aces? Assume two complete decks of 52 cards each are mixed together. What fraction of the two decks is aces?

82. Four of the twenty-four patients that Dr. Jones sees on a particular day have colds. What fraction of the patients have colds?

83. A day is what fraction of a year containing three hundred sixty-five days? Winter (three months) is what fraction of a year?

84. Mrs. O'Leary planted a garden with two rows of peas and six rows of corn. Her cow got into the garden and destroyed all of the corn plants. What fraction of the garden remains?

JOURNAL EXERCISES

1. Explain how greatest common divisors and how forms of one can be used in reducing fractions.

2. Explain the word simplify as it is used in mathematics.

3. In your own words, explain why $\frac{7}{0}$ is undefined. As used here, what is the meaning of the word undefined?

4. In our imagination, there is no largest number. Is there a smallest positive number? (*Hint:* $\frac{1}{1}$, $\frac{1}{2}$, $\frac{1}{3}$, ...) Do you think these types of concepts are important in math? Why, or why not?

5. Define numerator, denominator, proper fraction, improper fraction, and forms of one.

6. Can every whole number be written as a fraction? Can every fraction be written as a whole number? Why, or why not?

1 Equivalent Fractions

Two fractions are **equivalent** if they are equal to one another. For example, $\frac{2}{4}$ and $\frac{1}{2}$ are equivalent since $\frac{2}{4}$ reduces to $\frac{1}{2}$. The following rule can be used to determine if two fractions are equivalent.

Rule: To Determine if Two Fractions Are Equivalent

Step 1 Find the product of the numerator of the first fraction with the denominator of the second fraction.

Step 2 Find the product of the denominator of the first fraction with the numerator of the second fraction.

Step 3 The fractions are equivalent if the two products are equal.

OBJECTIVES

1 Identify equivalent fractions.

2 Raise a fraction to higher terms.

EXAMPLE 1

Is $\frac{6}{9}$ equivalent to $\frac{8}{12}$?

SOLUTION

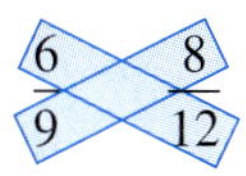

Step 1 Multiply the numerator of $\frac{6}{9}$ times the denominator of $\frac{8}{12}$:

$$6 \times 12 = 72$$

Step 2 Multiply the denominator of $\frac{6}{9}$ times the numerator of $\frac{8}{12}$:

$$9 \times 8 = 72$$

Step 3 The fractions are equivalent, or $\frac{6}{9} = \frac{8}{12}$. ■

EXAMPLE 2

Are $\frac{10}{15}$ and $\frac{15}{20}$ equivalent?

SOLUTION

Step 1 $10 \times 20 = 200$

Step 2 $15 \times 15 = 225$

Step 3 These two products are not equal so that

$$\frac{10}{15} \text{ is not equivalent to } \frac{15}{20}$$ ■

1. Are $\frac{19}{20}$ and $\frac{9}{10}$ equivalent?

2. Does $\frac{3}{4} = \frac{35}{40}$?

3. Is $\frac{13}{25}$ equal to $\frac{65}{125}$?

ANSWERS

1. No, $\frac{19}{20}$ and $\frac{9}{10}$ are not equivalent.
2. No, $\frac{3}{4}$ is not equal to $\frac{35}{40}$.
3. Yes, $\frac{13}{25}$ is equal to $\frac{65}{125}$.

EXAMPLE 3

Are $\frac{0}{10}$ and $\frac{0}{15}$ equivalent?

SOLUTION

$$0 \times 15 = 0$$
$$10 \times 0 = 0$$

They are equivalent since both are equal to 0. ■

◇ *Do Practice Exercises 1 through 3.*

2 Raise a Fraction to Higher Terms

Raising a fraction to higher terms means writing an equivalent fraction using larger numbers in both the numerator and denominator. The idea will first be shown using figures. The fraction $\frac{3}{5}$ is shown as the shaded part of the large square on the left. The question is how many of the small squares in the figure on the right need to be shaded to represent the same fraction.

In other words, what number should be used in place of the "?" so that $\frac{3}{5} = \frac{?}{20}$. This problem is solved in Example 4.

Rule: To Raise a Fraction to Higher Terms

Step 1 Divide the small denominator into the large denominator.

Step 2 The new numerator is found by multiplying the result of the division times the old numerator.

EXAMPLE 4

Find the value for ? which makes $\frac{3}{5} = \frac{?}{20}$ true.

SOLUTION

Step 1 Divide the small denominator (5) into the large denominator (20):

$$20 \div 5 = 4$$

Step 2 Multiply the 4 from Step 1 times the numerator of the fraction $\frac{3}{5}$, that is $4 \times 3 = 12$. The number 12 is the new numerator.

Thus $\frac{3}{5} = \frac{12}{20}$ ■

EXAMPLE 5

$\frac{8}{15} = \frac{?}{60}$

SOLUTION

Step 1 60 divided by 15 is 4.

Step 2 4 times 8 is 32.

Thus $\frac{8}{15} = \frac{32}{60}$ ■

◇ *Do Practice Exercises 4 through 6.*

EXAMPLE 6

Write two fractions equivalent to $\frac{3}{5}$, one with a denominator of 25 and one with a denominator of 65.

SOLUTION

$\frac{3}{5} = \frac{?}{25}$

Step 1 $25 \div 5 = 5$

Step 2 $3 \cdot 5 = 15$

Thus $\frac{3}{5} = \frac{15}{25}$

$\frac{3}{5} = \frac{?}{65}$

Step 1 $65 \div 5 = 13$

Step 2 $13 \cdot 3 = 39$

Thus $\frac{3}{5} = \frac{39}{65}$

or $\frac{3}{5} = \frac{15}{25} = \frac{39}{65}$ ■

◇ *Do Practice Exercise 7.*

EXAMPLE 7

$\frac{4}{5} = \frac{?}{15}$ and $\frac{?}{15} = \frac{4}{5}$ are two different ways to state the exact same problem. Find the value for the unknown.

SOLUTION

Step 1 $15 \div 5 = 3$ Divide the small denominator into the large one.

Step 2 $3 \cdot 4 = 12$ Multiply times the numerator.

Thus, $\frac{12}{15} = \frac{4}{5}$ ■

4. Find the value for ? in $\frac{4}{5} = \frac{?}{40}$.

5. Two-thirds is equal to what number divided by sixty?

6. Write a fraction equivalent to three-fourths that has a denominator of eighty.

7. Find two fractions equivalent to $\frac{3}{4}$, one with a denominator of 20 and one with a denominator of 44.

ANSWERS
4. 32
5. 40
6. $\frac{60}{80}$
7. $\frac{15}{20}$, $\frac{33}{44}$

Do the shaded areas represent equivalent fractions?

1. 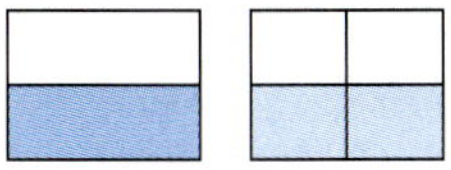

2. 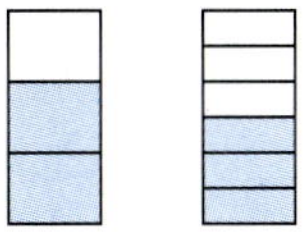

3. 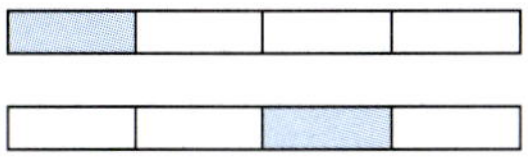

4. 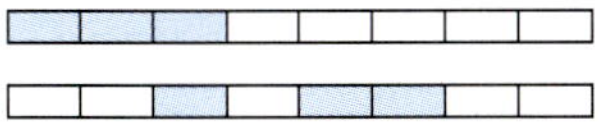

5. 

6. 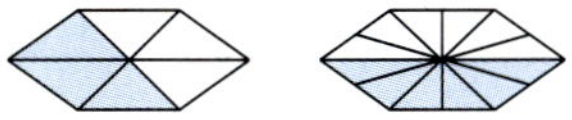

Are the following fractions equivalent?

7. $\frac{2}{3}, \frac{3}{4}$

8. $\frac{1}{2}, \frac{20}{10}$

9. $\frac{1}{4}, \frac{12}{48}$

10. $\frac{25}{75}, \frac{1}{3}$

11. $\frac{7}{9}, \frac{21}{20}$

12. $\frac{75}{5}, \frac{15}{1}$

13. $\frac{50}{30}, \frac{200}{60}$

14. $\frac{9}{10}, \frac{90}{100}$

Reduce, if possible. (Review)

15. $\frac{8}{10}$

16. $\frac{20}{30}$

17. $\frac{5}{55}$

18. $\frac{3}{40}$

19. $\frac{18}{20}$

20. $\frac{16}{90}$

21. $\frac{25}{375}$

22. $\frac{13}{700}$

23. $\frac{81}{425}$

24. $\frac{75}{375}$

25. $\frac{13}{39}$

26. $\frac{650}{900}$

Find the unknown by raising each to higher terms.

27. $\frac{5}{15} = \frac{?}{45}$

28. $\frac{10}{35} = \frac{?}{70}$

29. $\frac{25}{5} = \frac{?}{15}$

30. $\frac{200}{300} = \frac{?}{900}$

31. $\frac{195}{5} = \frac{?}{10}$

32. $\frac{?}{15} = \frac{2}{3}$

33. $\frac{?}{20} = \frac{15}{5}$

34. $\frac{?}{50} = \frac{10}{25}$

35. $\frac{8}{2} = \frac{?}{10}$

36. $\frac{?}{30} = \frac{6}{5}$

37. $\frac{?}{36} = \frac{4}{9}$

38. $\frac{7}{8} = \frac{?}{640}$

39. $\frac{?}{45} = \frac{5}{9}$

40. $\frac{3}{7} = \frac{?}{28}$

41. $\frac{7}{11} = \frac{?}{99}$

42. $\frac{3}{7} = \frac{?}{70}$

43. $\frac{3}{4} = \frac{?}{12}$

44. $\frac{3}{8} = \frac{?}{16}$

45. $\frac{5}{9} = \frac{?}{36}$

46. $\frac{8}{12} = \frac{?}{48}$

47. $\frac{9}{10} = \frac{?}{100}$

48. $\frac{7}{110} = \frac{?}{220}$

49. $\frac{?}{200} = \frac{4}{25}$

50. $\frac{7}{15} = \frac{?}{90}$

51. $\frac{9}{10} = \frac{?}{1{,}000}$

52. $\frac{21}{17} = \frac{?}{85}$

53. $\frac{11}{12} = \frac{?}{144}$

54. $\frac{3}{4} = \frac{?}{440}$

Work each problem.

55. Write a fraction equivalent to $\frac{7}{9}$ that has a denominator of 63.

56. Write a fraction equivalent to $\frac{5}{8}$ that has a denominator of 32.

57. Write two fractions equivalent to $\frac{5}{7}$, one with a denominator of 35 and one with a denominator of 70.

58. Write two fractions equivalent to $\frac{7}{10}$, one with a denominator of 20 and one with a denominator of 50.

59. Two twins are arguing about who had eaten the most pie. Mervin ate one-fourth of a pie and Henry ate two-eighths. Which brother has eaten the most?

60. Write the digits of sixteen million, eight hundred thousand, one and round to the nearest one million.

61. Is $5^0 > 4^1$ a true statement?

62. Is $2^2 < 1^5$?

63. A racer biked one-half of a one mile track in thirty seconds. A second racer completed five-tenths of the same track in the same amount of time. Which of the two is faster?

JOURNAL EXERCISES

1. Is there a largest fraction? Explain.

2. Explain the concepts of reducing a fraction and raising a fraction to higher terms.

1 Multiplication

We will show how to multiply fractions using figures. One-half of the circle on the left is shaded. One-half of the shaded part equals one-half of one-half or one-fourth of the circle on the right. Therefore, one-half times one-half must equal one-fourth.

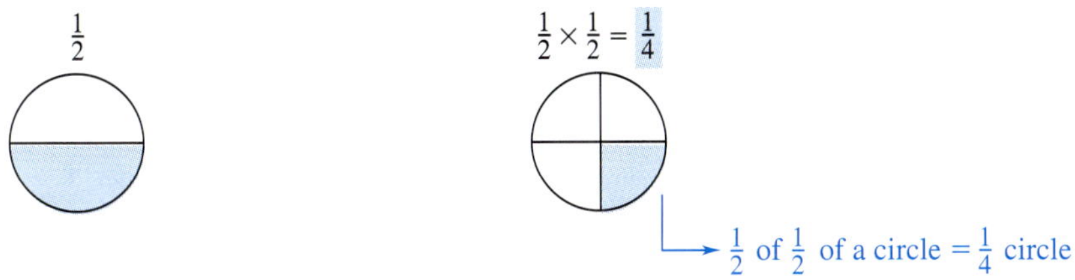

The following rule gives the steps to multiply two fractions.

> **Rule: To Multiply Two or More Fractions**
>
> **Step 1** The new numerator is the product of the numerators.
>
> **Step 2** The new denominator is the product of the denominators.
>
> **Step 3** Reduce the fraction if possible.

EXAMPLE 1

Find the product of $\frac{3}{4}$ and $\frac{5}{7}$.

SOLUTION

Step 1 The new numerator is $3 \times 5 = 15$.

Step 2 The new denominator is $4 \times 7 = 28$.

Step 3 The fraction cannot be reduced.

Multiply numerators.

$$\frac{3}{4} \times \frac{5}{7} = \frac{3 \times 5}{4 \times 7} = \frac{15}{28}$$

Multiply denominators. ■

EXAMPLE 2

Multiply $\frac{3}{4} \cdot \frac{2}{5}$.

SOLUTION

Reduce.

$$\frac{3}{4} \cdot \frac{2}{5} = \frac{3 \cdot 2}{4 \cdot 5} = \frac{6}{20} = \frac{3}{10}$$ ■

OBJECTIVES

1. Multiply fractions.
2. Cross reduce before multiplying.

1. What is the product of $\frac{4}{5}$ and $\frac{2}{3}$?

2. Multiply two-fifths by one-fourth, then reduce.

3. Evaluate $\frac{3}{8} \cdot \frac{9}{10}$.

4. Find $\frac{3}{5} \cdot \frac{1}{6} \cdot \frac{10}{9}$, then reduce.

5. Multiply two-elevenths times four.

6. Find the product of $4 \cdot \frac{3}{5} \cdot \frac{1}{10}$.

ANSWERS

1. $\frac{8}{15}$
2. $\frac{1}{10}$
3. $\frac{27}{80}$
4. $\frac{1}{9}$
5. $\frac{8}{11}$
6. $\frac{6}{25}$

Actually, it may be possible to reduce before multiplying. One method of doing this is to first find the prime factorization of the numerator and the denominator. Then find any forms of one and divide them out. Study the next example.

EXAMPLE 3

Multiply $\frac{2}{3} \cdot \frac{3}{5}$.

SOLUTION

form of one

$$\frac{2}{3} \cdot \frac{3}{5} = \frac{2 \cdot 3}{3 \cdot 5} = \frac{\mathbf{3} \cdot 2}{\mathbf{3} \cdot 5} = \frac{2}{5}$$ ■

◇ *Do Practice Exercises 1 through 3.*

The rule for multiplying fractions can be extended to the multiplication of three or more fractions as shown in the next example.

EXAMPLE 4

Multiply $\frac{2}{3} \cdot \frac{3}{5} \cdot \frac{1}{7}$.

SOLUTION

Identify forms of one and divide them out.

$$\frac{2}{3} \cdot \frac{3}{5} \cdot \frac{1}{7} = \frac{2 \cdot 3 \cdot 1}{3 \cdot 5 \cdot 7} = \frac{\mathbf{3} \cdot 2 \cdot 1}{\mathbf{3} \cdot 5 \cdot 7} = \frac{2}{35}$$ ■

◇ *Do Practice Exercise 4.*

In order to multiply a fraction times a whole number, the whole number must first be *written as a fraction*. Recall that a whole number can be written as a fraction by placing the whole number over the denominator one. For example:

$$7 = \frac{7}{1} \quad \text{or} \quad 45 = \frac{45}{1}$$

EXAMPLE 5

Multiply $\frac{5}{6} \times 12$.

SOLUTION

First write the whole number as a fraction,

$$\frac{5}{6} \cdot 12 = \frac{5}{6} \cdot \frac{12}{1} = \frac{5 \cdot 12}{6 \cdot 1} = \frac{60}{6} = 10$$

then multiply. ■

◇ *Do Practice Exercises 5 and 6.*

2 Cross Reducing

There is another method of reducing fractions BEFORE MULTIPLYING called **cross reducing**. Two fractions that are being multiplied can be cross reduced if either:

1. The numerator of the first fraction and the denominator of the second fraction have a common factor.
2. The denominator of the first fraction and the numerator of the second fraction have a common factor.

In either event, the common factor(s) can be divided out before multiplying numerators and denominators.

The idea of cross reducing will be shown using $\frac{6}{25} \cdot \frac{3}{4}$. The idea will be shown using prime factors.

Find the prime factorizations: $$\frac{6}{25} \cdot \frac{3}{4} = \frac{6 \cdot 3}{25 \cdot 4} = \frac{2 \cdot 3 \cdot 3}{5 \cdot 5 \cdot 2 \cdot 2}$$

Locate forms of one: $$\frac{\mathbf{2} \cdot 3 \cdot 3}{5 \cdot 5 \cdot \mathbf{2} \cdot 2} = \frac{\mathbf{2} \cdot 3 \cdot 3}{\mathbf{2} \cdot 5 \cdot 5 \cdot 2}$$

Divide out forms of one: $$\frac{3 \cdot 3}{5 \cdot 5 \cdot 2} = \frac{9}{50}$$

In this case, a 2 was divided out of the numerator and denominator. The 2 in the numerator came from the 6, the numerator of the first fraction. The 2 in the denominator came from the 4, the denominator in the second fraction.

This 2 came from the 6.

$$\frac{6}{25} \cdot \frac{3}{4} = \frac{2 \cdot 3 \cdot 3}{5 \cdot 5 \cdot 2 \cdot 2}$$

This 2 came from the 4.

Cross reducing is the process of dividing out identical factors that occur in the numerators and denominators of different fractions. It applies ONLY when multiplying fractions. Frequently, cross reduction is shown as follows:

$$\frac{\overset{3}{\cancel{6}}}{25} \cdot \frac{3}{\underset{2}{\cancel{4}}} = \frac{3 \cdot 3}{25 \cdot 2} = \frac{9}{50}$$

EXAMPLE 6

Multiply $\frac{3}{4} \cdot \frac{5}{9}$ by cross reducing first.

SOLUTION

The numerator of the first fraction and the denominator of the second fraction both have the factor 3.

$$\frac{\overset{1}{\cancel{3}}}{4} \cdot \frac{5}{\underset{3}{\cancel{9}}} = \frac{1 \cdot 5}{4 \cdot 3} = \frac{5}{12} \quad \blacksquare$$

◇ *Do Practice Exercises 7 and 8.*

7. Multiply $\frac{5}{8} \cdot \frac{2}{3}$.

8. Find the product of $\frac{1}{2}$ and $\frac{4}{17}$.

ANSWERS

7. $\frac{5}{12}$

8. $\frac{2}{17}$

9. Multiply $\frac{6}{7} \cdot \frac{1}{4} \cdot \frac{3}{8}$.

10. Find the product of $\frac{4}{5}$, 15, and $\frac{1}{3}$.

Sometimes two cross reductions can be done before multiplying. Look at the next example.

EXAMPLE 7

Multiply $\frac{8}{45} \cdot \frac{3}{34}$.

SOLUTION

A 3 can be cross reduced out of 45 and 3:

$$\frac{8}{\underset{15}{\cancel{45}}} \cdot \frac{\overset{1}{\cancel{3}}}{34}$$

and a 2 can be cross reduced out of 8 and 34:

$$\frac{\overset{4}{\cancel{8}}}{\underset{15}{\cancel{45}}} \cdot \frac{\overset{1}{\cancel{3}}}{\underset{17}{\cancel{34}}} = \frac{4 \cdot 1}{15 \cdot 17} = \frac{4}{225} \quad \blacksquare$$

Cross reducing can be extended to the multiplication of three or more fractions as the next example shows.

EXAMPLE 8

Find three-fourths of seven-fifths of eight-fifths.

SOLUTION

The problem is to multiply the three fractions. Both the 4 in the denominator of the first fraction and the 8 in the numerator of the third fraction are evenly divisible by 4. Therefore, they can be cross reduced before multiplying.

$$\frac{3}{\underset{1}{\cancel{4}}} \cdot \frac{7}{5} \cdot \frac{\overset{2}{\cancel{8}}}{5} = \frac{3 \cdot 7 \cdot 2}{1 \cdot 5 \cdot 5} = \frac{42}{25} \quad \blacksquare$$

◇ ***Do Practice Exercises 9 and 10.***

ANSWERS

9. $\frac{9}{112}$

10. 4

name date hour

EXERCISE SET 3.3

Multiply.

1. $\frac{3}{4} \cdot \frac{1}{5}$

2. $\frac{3}{5} \cdot \frac{1}{7}$

3. $\frac{1}{7} \times \frac{3}{8}$

4. $\frac{2}{7} \times \frac{5}{9}$

5. $\frac{9}{8} \times \frac{3}{7}$

6. $\frac{3}{13} \cdot \frac{7}{10}$

7. $\left(\frac{13}{32}\right)\left(\frac{1}{7}\right)$

8. $\left(\frac{3}{7}\right)\left(\frac{4}{7}\right)$

9. $\frac{3}{5} \times \frac{31}{7}$

10. $\frac{4}{5} \cdot \frac{11}{17}$

11. $\frac{4}{5} \cdot \frac{9}{7}$

12. $\frac{3}{4} \cdot \frac{5}{8}$

13. $\frac{5}{6} \cdot \frac{5}{7}$

14. $\frac{3}{7} \cdot \frac{25}{32}$

15. $\frac{6}{7} \cdot \frac{1}{5}$

16. $\frac{5}{9} \cdot \frac{0}{3}$

17. $\frac{1}{5} \cdot \frac{4}{7}$

18. $\frac{3}{11} \cdot \frac{7}{55}$

19. $\frac{3}{8} \cdot \frac{5}{7}$

20. $\frac{9}{20} \cdot \frac{9}{35}$

Reduce if possible. (Review)

21. $\frac{27}{36}$

22. $\frac{75}{225}$

23. $\frac{5}{32}$

24. $\frac{16}{256}$

25. $\frac{65}{70}$

26. $\frac{30}{95}$

Reduce first if possible, then multiply.

27. $\frac{2}{7} \cdot \frac{3}{8}$

28. $\frac{3}{4} \times \frac{2}{1}$

29. $\frac{4}{15} \cdot \frac{3}{7}$

30. $\left(\frac{5}{6}\right)\left(\frac{2}{3}\right)$

31. $\frac{15}{40} \cdot \frac{8}{5}$

32. $\frac{5}{6} \cdot \frac{3}{10}$

33. $\frac{12}{15} \cdot \frac{3}{4}$

34. $\left(\frac{45}{20}\right)\left(\frac{2}{3}\right)$

35. $\frac{5}{7} \times \frac{7}{9}$

36. $\left(\frac{7}{9}\right)\left(\frac{27}{2}\right)$

37. $\left(\frac{3}{5}\right)\left(\frac{5}{4}\right)$

38. $\frac{6}{25} \cdot \frac{7}{36}$

39. $\frac{4}{9} \cdot \frac{3}{6}$

40. $\left(\frac{5}{20}\right)\left(\frac{1}{3}\right)$

41. $\frac{7}{27} \cdot \frac{3}{14}$

42. $\frac{12}{49} \times \frac{7}{8}$

43. $\left(\frac{13}{8}\right)\left(\frac{2}{7}\right)$

44. $\frac{0}{32} \cdot \frac{8}{15}$

45. $\frac{8}{45} \cdot \frac{9}{2}$

46. $\frac{3}{5} \times \frac{5}{3}$

47. $\frac{40}{150} \cdot \frac{15}{20}$

48. $\frac{64}{25} \times \frac{10}{8}$

49. $\left(\frac{46}{75}\right)\left(\frac{10}{23}\right)$

50. $\left(\frac{26}{32}\right)\left(\frac{8}{26}\right)$

51. $\frac{4}{17} \cdot \frac{17}{8}$

52. $\frac{32}{27} \cdot \frac{9}{8}$

53. $\frac{15}{51} \times \frac{17}{45}$

54. $\frac{9}{42} \cdot \frac{14}{45}$

55. $\left(\frac{36}{54}\right)\left(\frac{27}{128}\right)$

56. $\frac{64}{85} \cdot \frac{5}{8}$

Write each integer as a fraction before multiplying.

57. $\frac{3}{4} \times 2$

58. $\frac{3}{10} \cdot 4$

59. $\frac{2}{3} \cdot 9$

60. $\frac{9}{64} \times 8$

61. $\left(\frac{25}{49}\right) \cdot 70$

62. $3 \cdot \frac{1}{2}$

63. $5 \cdot \frac{3}{4}$

64. $\frac{3}{5} \cdot 8$

65. $7 \cdot \frac{2}{3}$

66. $45 \cdot \frac{3}{8}$

Multiply and reduce if possible.

67. $\frac{3}{5} \cdot \frac{5}{7} \cdot \frac{1}{4}$

68. $\left(\frac{7}{8}\right)\left(\frac{6}{5}\right)\left(\frac{3}{7}\right)$

69. $\frac{1}{9} \times 2 \times \frac{5}{4}$

70. $0 \times \frac{2}{3} \cdot \frac{3}{5}$

71. $\frac{1}{3} \times \frac{1}{3} \times \frac{1}{3}$

72. $\frac{2}{5} \times \frac{3}{7} \times 5 \times \frac{7}{9}$

73. $\frac{5}{8} \cdot 0 \cdot \frac{1}{6}$

74. $\frac{3}{4} \cdot 8 \cdot 6$

Work each problem.

75. A freshman class had 102 students. If one-third of the students failed the final exam, how many failed?

76. Jamie's father passed away and left \$100,000 to his heirs. How much will Jamie inherit if she and her four brothers split the inheritance equally?

77. Ownership in producing oil wells is frequently split up by fractions. Mr. Smith owns one thirty-second of a well which produces 640 barrels of oil per month. If oil sells for \$20 per barrel, what monthly revenue will Mr. Smith receive?

78. Cut a piece of paper in half. Take each of the two pieces and cut them in half again. One of these pieces is what fraction of the original sheet of paper?

1–81
odd

79. A farmer plants two-fifths of his two-thousand-acre farm in beans. How many acres does he plant in beans?

80. In the Russian campaign of 1812, thirty-nine out of every forty of Napoleon's four hundred twenty-two thousand troops died. How many troops *survived*?

81. The nine planets make up one one-thousandth of the mass of the solar system and the earth contains about one five-hundredth of the mass of the nine planets. What fraction of the mass of the solar system is contained by the earth?

82. Two-thirds of a manufacturer's nine hundred employees went out on strike. Find the number of employees that went on strike.

JOURNAL EXERCISES

1. Every multiplication problem with whole numbers can be written as a multiplication problem using only fractions. Is this statement true or false? Explain and show examples.

2. The contents of three boxes are:

RED box	300 BLUE marbles
BLUE box	400 RED marbles
GREEN box	empty

Three-fourths of the marbles in the blue box are moved to the green box. Two-thirds of the marbles in the red box are moved to the green box. All of the marbles in the blue box are then switched with those in the red box. Find the number and color of marbles in each of the three boxes. What fraction of the blue marbles end up in the blue box?

3. One-half of your genetic makeup came from your father and one-half from your mother. In turn, one-half of your father's genetic makeup came from his mother and one-half from his father. Tom's great grandfather was a Cherokee Indian. Tom is what fraction Cherokee? Explain.

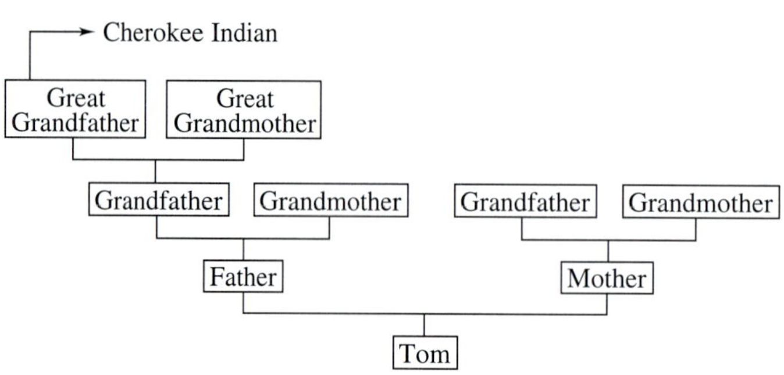

1 Reciprocals

Two fractions are called **reciprocals** or **inverses** if their product is 1. For example, the fractions $\frac{3}{4}$ and $\frac{4}{3}$ are reciprocals since:

$$\frac{3}{4} \cdot \frac{4}{3} = \frac{3 \cdot 4}{4 \cdot 3} = \frac{3 \cdot 4}{3 \cdot 4} = \frac{1}{1} = 1$$

(forms of one)

The reciprocal of a fraction is found by exchanging the numerator and denominator. Several fractions and their inverses are shown below.

Fraction	*Reciprocal*	*Product*
$\frac{2}{3}$	$\frac{3}{2}$	$\frac{2}{3} \cdot \frac{3}{2} = 1$
$\frac{7}{4}$	$\frac{4}{7}$	$\frac{7}{4} \cdot \frac{4}{7} = 1$
$\frac{14}{9}$	$\frac{9}{14}$	$\frac{14}{9} \cdot \frac{9}{14} = 1$

In order to find the reciprocal of a whole number, it is first necessary to change the whole number to a fraction. Recall that a whole number is changed to a fraction by placing the whole number in the numerator and the number 1 in the denominator. The reciprocal is then found by exchanging the numerator and denominator:

Whole Number		*Fraction*	*Reciprocal*	*Product*
5	=	$\frac{5}{1}$	$\frac{1}{5}$	$5 \cdot \frac{1}{5} = 1$
27	=	$\frac{27}{1}$	$\frac{1}{27}$	$27 \cdot \frac{1}{27} = 1$

◇ *Do Practice Exercises 1 and 2.*

2 Division

The division 8 ÷ 4 asks how many 4s are in 8. Similarly, $\frac{4}{5} \div \frac{1}{5}$ is asking how many $\frac{1}{5}$s are in $\frac{4}{5}$?

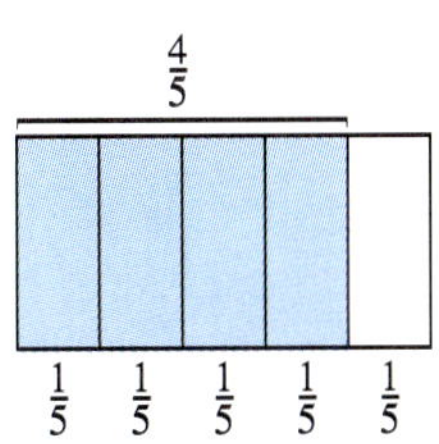

Since there are 4 $\frac{1}{5}$s in $\frac{4}{5}$, $\frac{4}{5} \div \frac{1}{5} = 4$.

OBJECTIVES

1 Find the reciprocal of a fraction.

2 Divide fractions.

1. What is the inverse of $\frac{8}{11}$?

2. Find the reciprocal of 12.

ANSWERS

1. $\frac{11}{8}$

2. $\frac{1}{12}$

3. Divide $\frac{3}{5} \div \frac{4}{9}$.

4. Divide one-eighth by two-thirds.

5. Evaluate $\frac{7}{8} \div \frac{1}{10}$.

ANSWERS

3. $\frac{27}{20}$
4. $\frac{3}{16}$
5. $\frac{35}{4}$

The rule to divide two fractions is given here and is discussed more fully in the section on complex fractions.

Rule: To Divide Two Fractions

Step 1 Replace the second fraction with its reciprocal and change the math operator from division to multiplication.

Step 2 Multiply the remaining fractions and reduce if possible.

EXAMPLE 1

Solve $\frac{2}{3} \div \frac{5}{7}$.

SOLUTION

Step 1 Invert the second fraction and change the division sign to a multiplication sign.

Invert the fraction and

$$\frac{2}{3} \div \frac{5}{7} = \frac{2}{3} \cdot \frac{7}{5}$$

change the division sign to multiplication.

Step 2 Multiply.

$$\frac{2}{3} \cdot \frac{7}{5} = \frac{2 \cdot 7}{3 \cdot 5} = \frac{14}{15}$$ ■

The problem from the last example can be done in one row as follows:

Invert the fraction and

$$\frac{2}{3} \div \frac{5}{7} = \frac{2}{3} \cdot \frac{7}{5} = \frac{2 \cdot 7}{3 \cdot 5} = \frac{14}{15}$$

change the division sign to multiplication.

EXAMPLE 2

Find $\frac{3}{4} \div \frac{5}{2}$.

SOLUTION

Invert

$$\frac{3}{4} \div \frac{5}{2} = \frac{3}{\not{4}_{2}} \cdot \frac{\not{2}^{1}}{5} = \frac{3 \cdot 1}{2 \cdot 5} = \frac{3}{10}$$

Change from division to multiplication. ■

◇ *Do Practice Exercises 3 through 5.*

Be careful when a whole number and a fraction are in the same division problem. Writing the whole number as a fraction *before* dividing will help remove confusion. Look at the next example.

EXAMPLE 3

Divide $\frac{2}{5}$ by 8.

SOLUTION

$\frac{2}{5} \div 8$

$\frac{2}{5} \div \frac{8}{1}$ Write 8 as the fraction $\frac{8}{1}$.

$\frac{2}{5} \cdot \frac{1}{8}$ invert second fraction; change ÷ to ·

$\frac{\overset{1}{\cancel{2}}}{5} \cdot \frac{1}{\underset{4}{\cancel{8}}}$ Cross reduce.

$\frac{1}{20}$ Result. ■

◇ *Do Practice Exercises 6 and 7.*

The numerator of the first fraction in a division problem (the dividend) can be zero. However, the numerator of the second fraction (divisor) cannot contain a zero since division by zero has no meaning. Study the next two examples.

EXAMPLE 4

Find $\frac{0}{10} \div \frac{1}{4}$.

SOLUTION

Invert

$$\frac{0}{10} \div \frac{1}{4} = \frac{0}{10} \cdot \frac{4}{1} = \frac{0 \cdot 4}{10 \cdot 1} = \frac{0}{10} = 0$$

Change from division to multiplication. ■

6. Divide twelve by one-fourth.

7. Find the quotient of three-fourths divided by five.

ANSWERS
6. 48
7. $\frac{3}{20}$

8. Divide $\frac{12}{5}$ by 0.

9. Evaluate $\frac{0}{5} \div \frac{1}{4}$.

ANSWERS
8. undefined
9. 0

EXAMPLE 5

Find $\frac{3}{4} \div \frac{0}{7}$.

SOLUTION

$$\frac{3}{4} \div \frac{0}{7} = \frac{3}{4} \cdot \frac{7}{0} \qquad \text{but } \frac{7}{0} \text{ is undefined}$$

The problem has no answer. ■

◇ *Do Practice Exercises 8 and 9.*

EXAMPLE 6

A large Canadian company owns a Malaysian copper mine which sells three-quarters of a ton of copper to a German firm. The German firm divides the copper into three equal sized shipments. They then send equal amounts to South Africa, Germany, and Japan. How much copper is sent to South Africa?

SOLUTION

The answer is found by dividing three-quarters of a ton by 3:

$$\frac{3}{4} \div 3 = \frac{3}{4} \div \frac{3}{1} = \frac{\overset{1}{\cancel{3}}}{4} \cdot \frac{1}{\underset{1}{\cancel{3}}} = \frac{1 \cdot 1}{4 \cdot 1} = \frac{1}{4} \text{ ton sent to South Africa}$$ ■

For review purposes, a comparison of the multiplication and division of fractions follows.

Multiplication: Cross reduce if possible then multiply the numerators and denominators.

$$\frac{4}{9} \cdot \frac{3}{10} = \frac{\overset{2}{\cancel{4}}}{\underset{3}{\cancel{9}}} \cdot \frac{\overset{1}{\cancel{3}}}{\underset{5}{\cancel{10}}} = \frac{2 \cdot 1}{3 \cdot 5} = \frac{2}{15}$$

Division: Change to a multiplication problem by inverting the fraction following the division sign. Cross reduce if possible, then multiply.

$$\frac{6}{7} \div \frac{3}{5} = \frac{\overset{2}{\cancel{6}}}{7} \cdot \frac{5}{\underset{1}{\cancel{3}}} = \frac{2 \cdot 5}{7 \cdot 1} = \frac{10}{7}$$

Find the inverse of each fraction.

1. $\frac{2}{3}$ **2.** $\frac{3}{4}$ **3.** $\frac{5}{2}$ **4.** $\frac{2}{7}$ **5.** $\frac{7}{3}$

Write each integer as a fraction with a denominator of 1. (Review)

6. 13 **7.** 11 **8.** 2 **9.** 3 **10.** 1

Multiply and reduce to lowest terms. (Review)

11. $\frac{3}{5} \cdot \frac{1}{8}$ **12.** $\frac{4}{7} \cdot \frac{1}{2}$ **13.** $\frac{1}{4} \times \frac{3}{8}$ **14.** $\frac{4}{5} \cdot \frac{5}{6}$

15. $\frac{4}{11} \cdot \frac{3}{20}$ **16.** $\frac{6}{17} \cdot \frac{34}{9}$ **17.** $\frac{3}{4} \cdot \frac{1}{6}$ **18.** $\frac{5}{8} \cdot \frac{2}{5}$

19. $\frac{1}{3} \times \frac{3}{11}$ **20.** $\frac{7}{15} \cdot \frac{2}{3}$ **21.** $\frac{5}{9} \cdot \frac{81}{100}$ **22.** $\frac{16}{21} \times \frac{3}{84}$

Find the product of each fraction with its inverse.

23. $\frac{3}{4} \cdot \frac{4}{3}$ **24.** $\frac{2}{9} \times \frac{9}{2}$ **25.** $\frac{5}{1} \cdot \frac{1}{5}$

26. $\frac{7}{4} \times \frac{4}{7}$ **27.** $\frac{2}{11} \cdot \frac{11}{2}$ **28.** $\frac{1}{40} \cdot 40$

23–67
odd

Divide and reduce to the lowest terms.

29. $\frac{7}{5} \div \frac{1}{8}$

30. $\frac{2}{3} \div \frac{1}{3}$

31. $\left(\frac{1}{7}\right) \div \left(\frac{1}{7}\right)$

32. $\frac{3}{5} \div \frac{2}{5}$

33. $\frac{9}{10} \div \frac{2}{3}$

34. $\frac{7}{15} \div \frac{2}{3}$

35. $\frac{3}{7} \div \frac{5}{8}$

36. $\frac{8}{3} \div \frac{7}{6}$

37. $\frac{4}{7} \div \frac{20}{21}$

38. $\left(\frac{5}{4}\right) \div \left(\frac{1}{7}\right)$

39. $\frac{3}{5} \div \frac{1}{8}$

40. $\frac{6}{11} \div \frac{1}{22}$

41. $\frac{7}{5} \div \frac{3}{8}$

42. $\frac{5}{9} \div \frac{25}{36}$

43. $\frac{3}{8} \div \frac{1}{50}$

44. $\frac{2}{13} \div \frac{26}{25}$

45. $\frac{18}{49} \div \frac{3}{21}$

46. $\left(\frac{25}{12}\right) \div \left(\frac{50}{8}\right)$

47. $\frac{27}{125} \div \frac{9}{75}$

48. $\frac{35}{98} \div \frac{5}{8}$

49. $\frac{12}{45} \div \frac{5}{7}$

50. $\frac{13}{27} \div \frac{26}{9}$

51. $\frac{12}{25} \div \frac{6}{5}$

52. $\frac{64}{81} \div \frac{28}{27}$

53. $\frac{40}{21} \div \frac{30}{42}$

54. $\frac{2}{5} \div 7$

55. $\frac{7}{3} \div 3$

56. $3 \div \frac{4}{5}$

57. $\frac{3}{5} \div \frac{5}{25}$

58. $\left(\frac{16}{3}\right) \div \left(\frac{8}{11}\right)$

59. $0 \div \frac{1}{8}$

60. $\frac{4}{5} \div 5$

61. $7 \div \frac{21}{2}$

62. $18 \div \frac{6}{7}$

Work each problem.

63. Divide $\frac{7}{8}$ by $\frac{2}{3}$.

64. A family of four bought one-half of a watermelon. What fraction of the watermelon does each person get if they share equally?

65. A rancher split her herd of 1,200 head of cattle in half based on age. How many cattle are in the younger group? She then split the younger half into three equal size groups based on quality. How many head were in the young, highest quality group? What fraction of the herd is in this group?

66. A worker inspects one-twentieth of the eight hundred toys coming off an assembly line each day, for quality. Find the number of toys inspected each day.

67. In 1992, California led the United States in number of new business startups. If three twenty-fifths of the nation's two hundred fifty thousand business starts were in California, how many business starts did California have?

68. Fifteen people share equally in lottery winnings of $300,000. How much does each receive?

JOURNAL EXERCISES

1. The process of dividing fractions looks different than the process of dividing whole numbers. Are the two processes different? Change the whole number 60 to a fraction and change the whole number 5 to a fraction. Divide the first fraction by the second. Is the answer the same as that found by dividing the whole number 60 by the whole number 5? Explain.

2. Compare the processes of multiplying and dividing fractions.

1 Common Denominators

Two fractions have a **common denominator** if they have the exact same denominator. For example, the fractions $\frac{1}{12}$ and $\frac{5}{12}$ have the common denominator 12. Fractions that do not have common denominators can be rewritten so that they do have common denominators. The smallest whole number that can be a common denominator for two or more fractions is called the **Least Common Denominator** or **LCD** for short.

Rule: To Rewrite Fractions so that They Have Common Denominators

Step 1 Find the least common multiple of the denominators.

Step 2 Raise each fraction to higher terms using the least common multiple as the new denominator.

EXAMPLE 1

Write an equivalent fraction for $\frac{1}{6}$ and one for $\frac{5}{8}$ using the LCM of the two fractions for denominators.

SOLUTION

Step 1 Find the least common multiple of 6 and 8. (Review Section 2.5.)

$$\begin{array}{rcl|c|c|} 6 & = & 2\times & & 3 \\ 8 & = & 2\times 2\times 2 & & \end{array} \quad \text{LCM} = 2\times 2\times 2\times 3 = 24$$

Step 2 Raise each fraction to higher terms using 24 as the denominator. (Review Section 3.2.)

$$\frac{1}{6} = \frac{?}{24} \qquad \frac{5}{8} = \frac{?}{24}$$

$$\frac{1}{6} = \frac{4}{24} \qquad \frac{5}{8} = \frac{15}{24}$$

Thus, $\frac{1}{6}$ and $\frac{5}{8}$ written as equivalent fractions with common denominators are $\frac{4}{24}$ and $\frac{15}{24}$. ■

◇ *Do Practice Exercises 1 and 2.*

OBJECTIVES

1 Write fractions using least common denominators.

2 Compare two or more fractions.

1. Find the lowest common denominator of $\frac{3}{4}$ and $\frac{5}{18}$.

2. Write an equivalent fraction for $\frac{3}{4}$ and one for $\frac{5}{18}$ using the LCM of the two fractions for denominators.

ANSWERS

1. 36

2. $\frac{3}{4} = \frac{27}{36}, \frac{5}{18} = \frac{10}{36}$

EXAMPLE 2

Write $\frac{2}{3}$, $\frac{3}{5}$, and $\frac{7}{10}$ as equivalent fractions with common denominators.

SOLUTION

Step 1 The least common multiple of 3, 5, and 10 is 30.

Step 2 Raise each fraction to higher terms.

$$\frac{2}{3} = \frac{?}{30} \qquad \frac{3}{5} = \frac{?}{30} \qquad \frac{7}{10} = \frac{?}{30}$$

$$\frac{2}{3} = \frac{20}{30} \qquad \frac{3}{5} = \frac{18}{30} \qquad \frac{7}{10} = \frac{21}{30}$$

Equivalent fractions with a common denominator are $\frac{20}{30}$, $\frac{18}{30}$, and $\frac{21}{30}$ ■

◇ *Do Practice Exercises 3 and 4.*

2 Compare Two or More Fractions

In order to compare the size of two fractions with the same denominator, it is only necessary to compare the numerators. The fraction with the larger numerator is the greater of the two numbers.

EXAMPLE 3

Which fraction is greater $\frac{3}{5}$ or $\frac{4}{5}$?

SOLUTION

Three-fifths and four-fifths are represented as the shaded portions of the two rectangles. The numerator 4 is greater than the numerator 3 so that four-fifths is greater than three-fifths.

$\frac{3}{5} =$

$\frac{4}{5} =$

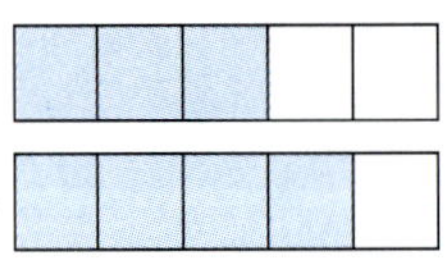

$\frac{4}{5}$ is greater than $\frac{3}{5}$ ■

◇ *Do Practice Exercises 5 and 6.*

EXAMPLE 4

Write $\frac{6}{11}$, $\frac{5}{11}$, and $\frac{9}{11}$ from smallest to largest.

SOLUTION

Since the denominators are identical it is only necessary to compare the numerators. From smallest to largest they are:

$$\frac{5}{11}, \quad \frac{6}{11}, \quad \frac{9}{11}$$

($\frac{5}{11}$ → smallest; $\frac{9}{11}$ → largest) ■

3. Write an equivalent fraction for $\frac{3}{5}$ and one for $\frac{8}{35}$ using the LCM of the two fractions for denominators.

4. Write an equivalent fraction for $\frac{5}{6}$ and one for $\frac{3}{8}$ using the LCM of the two fractions for denominators.

5. Which is greater $\frac{7}{11}$ or $\frac{9}{11}$?

6. Is three-eights greater than five-eighths?

ANSWERS

3. $\frac{3}{5} = \frac{21}{35}$ $\quad \frac{8}{35} = \frac{8}{35}$

4. $\frac{5}{6} = \frac{20}{24}$ $\quad \frac{3}{8} = \frac{9}{24}$

5. $\frac{9}{11}$

6. No. $\frac{5}{8}$ is greater than $\frac{3}{8}$.

The process of comparing fractions that do not have common denominators is different. The two figures below show the difficulty in comparing the fractions $\frac{2}{5}$ and $\frac{1}{3}$ from a geometrical standpoint. It is difficult to tell which fraction is greater by looking at the figures.

$\frac{2}{5}$

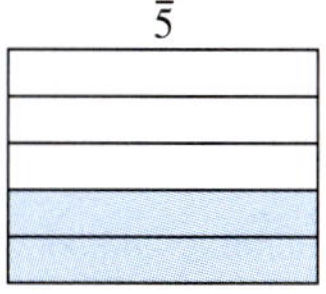

$\frac{1}{3}$

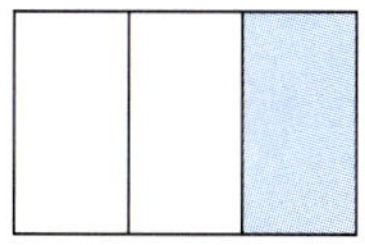

Both fractions can be written in terms of their common denominator 15:

$$\frac{2}{5} = \frac{6}{15} \qquad \frac{1}{3} = \frac{5}{15}$$

The two fractions $\frac{6}{15}$ and $\frac{5}{15}$ can be shown using figures:

$\frac{2}{5} = \frac{6}{15}$

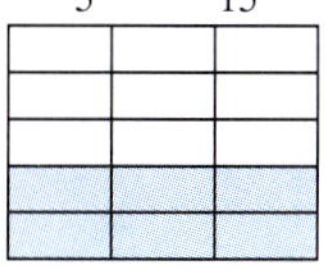

$\frac{1}{3} = \frac{5}{15}$

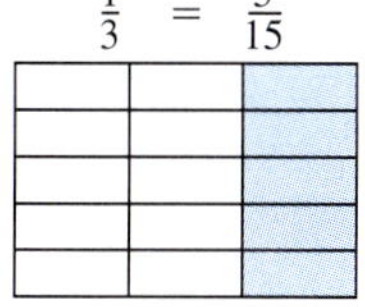

Since $\frac{6}{15}$ is greater than $\frac{5}{15}$, $\frac{2}{5}$ must be greater than $\frac{1}{3}$.

Rule: To Compare the Size of Fractions

Step 1 Rewrite each fraction as an equivalent fraction using the least common denominator.

Step 2 Compare the fractions by comparing the numerators of the equivalent fractions.

EXAMPLE 5

Compare the size of $\frac{2}{9}$ and $\frac{1}{3}$.

SOLUTION

Step 1 Rewrite using the least common denominator of 9.

$$\frac{2}{9} = \frac{?}{9} \qquad \frac{1}{3} = \frac{?}{9}$$

$$\frac{2}{9} = \frac{2}{9} \qquad \frac{1}{3} = \frac{3}{9}$$

Step 2 Compare $\frac{2}{9}$ and $\frac{3}{9}$ by comparing numerators.

$$\frac{3}{9} \text{ is greater than } \frac{2}{9} \qquad \text{thus} \qquad \frac{1}{3} \text{ is greater than } \frac{2}{9} \quad ■$$

Recall from Chapter 1 that the symbol > means greater than and the symbol < means less than.

EXAMPLE 6

Is $\frac{3}{8} > \frac{1}{3}$?

SOLUTION

The common denominator of 8 and 3 is 24. Rewriting each fraction in terms of the common denominator gives us:

$$\frac{3}{8} = \frac{9}{24} \qquad \text{and} \qquad \frac{1}{3} = \frac{8}{24}$$

$$\frac{9}{24} > \frac{8}{24} \qquad \text{so that} \qquad \frac{3}{8} > \frac{1}{3} \quad ■$$

◇ *Do Practice Exercises 7 through 9.*

EXAMPLE 7

Write $\frac{1}{3}$, $\frac{2}{5}$, and $\frac{4}{15}$ from smallest to largest.

SOLUTION

Step 1 Each fraction is written as an equivalent fraction using the least common denominator which is 15.

$$\frac{1}{3} = \frac{?}{15} \qquad \frac{2}{5} = \frac{?}{15} \qquad \frac{4}{15} = \frac{?}{15}$$

$$\frac{1}{3} = \frac{5}{15} \qquad \frac{2}{5} = \frac{6}{15} \qquad \frac{4}{15} = \frac{4}{15}$$

Step 2 From smallest to largest the fractions are $\frac{4}{15}$, $\frac{5}{15}$, and $\frac{6}{15}$. Thus the original fractions from smallest to largest are $\frac{4}{15}$, $\frac{1}{3}$, and $\frac{2}{5}$. ■

◇ *Do Practice Exercise 10.*

7. Is $\frac{3}{4} > \frac{3}{5}$?

8. Is $\frac{1}{2} < \frac{3}{7}$?

9. Is $\frac{2}{3} < \frac{7}{10}$?

10. Write $\frac{2}{3}$, $\frac{5}{8}$, and $\frac{1}{2}$ from smallest to largest.

ANSWERS

7. Yes, $\frac{3}{4} > \frac{3}{5}$.

8. No, $\frac{1}{2} > \frac{3}{7}$.

9. Yes, $\frac{2}{3} < \frac{7}{10}$.

10. $\frac{1}{2}$, $\frac{5}{8}$, $\frac{2}{3}$

name date hour

Find those fractions that have common denominators.

1. $\frac{3}{4}, \frac{1}{4}, \frac{2}{3}$
2. $\frac{5}{18}, \frac{3}{7}, \frac{9}{7}$
3. $\frac{7}{8}, \frac{9}{10}, \frac{3}{8}, \frac{1}{8}$
4. $\frac{107}{13}, \frac{21}{20}, \frac{3}{13}$
5. $\frac{3}{127}, \frac{5}{127}, \frac{2{,}004}{127}$
6. $\frac{2}{3}, \frac{3}{3}, \frac{7}{5}, \frac{8}{3}$
7. $\frac{2}{21}, \frac{3}{21}, \frac{17}{21}$
8. $\frac{5}{64}, \frac{13}{13}, \frac{2}{64}$

Find the least common denominator.

9. $\frac{1}{4}, \frac{3}{8}$
10. $\frac{1}{3}, \frac{1}{6}$
11. $\frac{3}{4}, \frac{5}{6}$
12. $\frac{7}{8}, \frac{2}{3}$
13. $\frac{3}{8}, \frac{13}{16}$
14. $\frac{4}{9}, \frac{7}{30}$
15. $\frac{7}{11}, \frac{3}{17}$
16. $\frac{2}{6}, \frac{3}{8}, \frac{5}{36}$
17. $\frac{7}{20}, \frac{3}{6}, \frac{5}{9}$
18. $\frac{11}{42}, \frac{13}{3}, \frac{1}{7}$

Raise each fraction to higher terms. (Review)

19. $\frac{4}{5} = \frac{?}{10}$
20. $\frac{3}{8} = \frac{?}{24}$
21. $\frac{5}{7} = \frac{?}{49}$
22. $\frac{1}{12} = \frac{?}{36}$
23. $\frac{4}{5} = \frac{?}{40}$
24. $\frac{7}{3} = \frac{?}{30}$
25. $\frac{?}{42} = \frac{5}{6}$
26. $\frac{0}{45} = \frac{?}{90}$
27. $\frac{12}{13} = \frac{?}{39}$
28. $\frac{7}{9} = \frac{?}{189}$

Write the following fractions from smallest to largest.

29. $\frac{1}{3}, \frac{3}{8}$
30. $\frac{3}{8}, \frac{5}{12}$
31. $\frac{4}{10}, \frac{3}{15}$
32. $\frac{15}{8}, \frac{7}{32}$
33. $\frac{6}{7}, \frac{7}{11}$
34. $\frac{3}{4}, \frac{3}{5}$

35. $\frac{4}{9}, \frac{1}{2}$

36. $\frac{5}{8}, \frac{3}{4}$

37. $\frac{3}{7}, \frac{2}{9}$

38. $\frac{5}{8}, \frac{7}{10}$

39. $\frac{25}{27}, \frac{7}{18}$

40. $\frac{1}{20}, \frac{3}{70}$

41. $\frac{3}{32}, \frac{4}{45}$

42. $\frac{5}{47}, \frac{10}{21}$

43. $\frac{7}{8}, \frac{8}{9}$

44. $\frac{7}{8}, \frac{3}{6}, \frac{2}{9}$

45. $\frac{4}{13}, \frac{5}{26}, \frac{7}{13}$

46. $\frac{5}{16}, \frac{12}{15}, \frac{15}{44}$

Write the following fractions from largest to smallest.

47. $\frac{3}{5}, \frac{8}{15}$

48. $\frac{6}{15}, \frac{9}{25}$

49. $\frac{3}{8}, \frac{7}{22}$

50. $\frac{12}{49}, \frac{15}{49}$

51. $\frac{0}{7}, \frac{1}{7}$

52. $\frac{3}{7}, \frac{7}{9}$

53. $\frac{10}{21}, \frac{12}{25}$

54. $\frac{6}{99}, \frac{0}{9}$

55. $\frac{9}{4}, \frac{25}{12}$

56. $\frac{4}{9}, \frac{18}{27}$

57. $\frac{4}{11}, \frac{1}{2}$

58. $\frac{7}{9}, \frac{2}{3}$

59. $\frac{3}{8}, \frac{6}{7}$

60. $\frac{9}{10}, \frac{6}{11}$

61. $\frac{5}{7}, \frac{4}{9}$

62. $\frac{3}{4}, \frac{5}{8}$

63. $\frac{1}{2}, \frac{3}{8}$

64. $\frac{1}{20}, \frac{3}{80}$

65. $\frac{5}{18}, \frac{1}{3}, \frac{2}{5}$

66. $\frac{4}{5}, \frac{6}{7}, \frac{5}{8}$

Work each problem.

67. A sales person used to spend three-fifths of every day with customers but now spends four-sevenths of every day with them. Is she now spending more or less time with customers?

68. It takes Lupe the same amount of time to run five-eighths of a mile as it takes Karen to run seven-tenths of a mile. Which of the two is the faster runner?

69. Jesse makes a recipe calling for three-fifths of a cup of sugar and two-thirds of a cup of flour. Does the recipe call for more sugar than flour?

70. At age 40, one woman out of 1,000 will develop breast cancer. At age 60, one woman out of 500 will develop breast cancer. Is breast cancer more or less likely to develop in 40 year olds compared to 60 year olds?

71. A problem requiring substantial calculation is performed on two different microcomputers. The first computer requires two-thirds of a minute to solve the problem and the second computer requires 42 seconds. Which of the two computers is faster?

72. The following table shows approximate mileages between cities.

From	*To*	*Miles*
New York, New York	Los Angeles, California	2,500
Los Angeles, California	Honolulu, Hawaii	2,600
Paris, France	Cairo, Egypt	2,000
Cairo, Egypt	New Delhi, India	2,800

Passenger A is traveling from New York to Honolulu through Los Angeles. Passenger B is traveling from Paris to New Delhi through Cairo. At Los Angeles and at Cairo, which passenger has completed a higher fraction of their journey?

JOURNAL EXERCISES

1. Of the material on fractions we have covered so far, what do you NOT understand?

2. Was the concept of common denominator needed in order to multiply and divide fractions? Explain the idea of a common denominator in your own words. What were common denominators used for in this section?

3. Define LCD in your own words.

OBJECTIVES

1. Add fractions with common denominators.
2. Subtract fractions with common denominators.

1 Add Fractions with Common Denominators

The addition of two fractions that have a common denominator is shown using circles. The shaded portion of the circle on the left represents the fraction $\frac{1}{5}$ and the shaded portion of the circle in the center represents the fraction $\frac{3}{5}$.

$$\frac{1}{5} \quad + \quad \frac{3}{5} \quad = \quad \frac{4}{5}$$

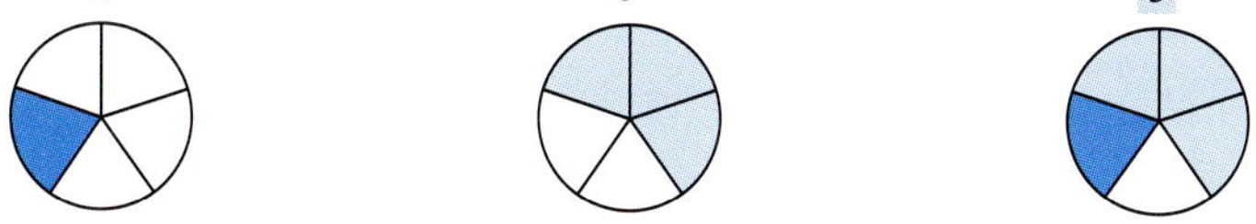

Since both circles have been cut into the same number of equal size pieces, they can be added. The one shaded part in the first circle and the three shaded parts in the second circle have been combined in the circle on the right. Notice that four parts of the last circle are shaded.

Rule: To Add Fractions with Common Denominators

Step 1 The new numerator is the sum of the numerators.

Step 2 The new denominator is the common denominator.

Step 3 Reduce the fraction if possible.

The fractions above will now be added using the rule.

EXAMPLE 1

Add $\frac{3}{5} + \frac{1}{5}$.

SOLUTION

Step 1 Add the numerators: $3 + 1 = 4$.

Step 2 Keep the common denominator 5.

Step 3 Reduce, if possible.

$$\text{Thus } \frac{3}{5} + \frac{1}{5} = \frac{3+1}{5} = \frac{4}{5}$$

Keep the same denominator. ■

EXAMPLE 2

Add $\frac{1}{4} + \frac{1}{4}$.

SOLUTION

Add numerators,

$$\frac{1}{4} + \frac{1}{4} = \frac{1+1}{4} = \frac{2}{4} = \frac{1}{2}$$

but keep the same denominator. ■

1. Add one-eighth to five-eighths.

2. Add $\frac{4}{7} + \frac{2}{7}$.

3. Find the sum of $\frac{1}{12} + \frac{5}{12} + \frac{5}{12}$.

EXAMPLE 3

Find $\frac{3}{20} + \frac{1}{20} + \frac{1}{20}$.

SOLUTION

$$\frac{3}{20} + \frac{1}{20} + \frac{1}{20} = \frac{3+1+1}{20} = \frac{5}{20} = \frac{1}{4}$$

Reduce. ■

EXAMPLE 4

A trailer can hold one-third of a ton. How much weight can the trailer hold if it is reinforced sufficiently to carry an additional one-third ton?

SOLUTION

The answer is found by adding one-third ton to one-third ton.

$$\frac{1}{3} + \frac{1}{3} = \frac{1+1}{3} = \frac{2}{3} \text{ tons}$$ ■

◇ *Do Practice Exercises 1 through 3.*

2 Subtract Fractions with Common Denominators

The subtraction of two fractions that have a common denominator is shown below. The shaded portion of the square on the left represents the fraction $\frac{3}{4}$ and the shaded portion of the square in the center represents the fraction $\frac{1}{4}$.

$$\frac{3}{4} \quad - \quad \frac{1}{4} \quad = \quad \frac{2}{4}\left[=\frac{1}{2}\right]$$

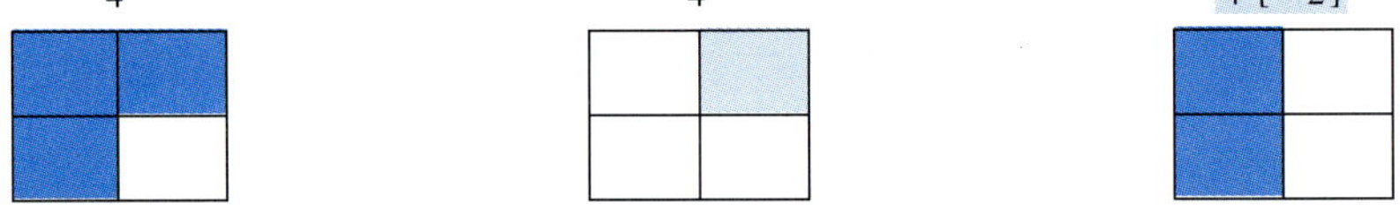

Since both squares have been cut into four equal size pieces, the one shaded part of the square in the center can be subtracted from the three shaded parts in the first square. This has been done in the last square, in which $\frac{1}{2}$ is shaded.

> **Rule: To Subtract Fractions with Common Denominators**
>
> **Step 1** The new numerator is the difference of the numerators.
>
> **Step 2** The new denominator is the common denominator.
>
> **Step 3** Reduce the fraction if possible.

ANSWERS

1. $\frac{6}{8} = \frac{3}{4}$
2. $\frac{6}{7}$
3. $\frac{11}{12}$

EXAMPLE 5

Subtract $\frac{7}{9} - \frac{2}{9}$.

SOLUTION

Step 1 Subtract the numerators: $7 - 2 = 5$.

Step 2 Keep the common denominator 9.

Step 3 The fraction cannot be reduced in this case.

$$\text{Thus } \frac{7}{9} - \frac{2}{9} = \frac{7-2}{9} = \frac{5}{9}$$

Keep the same denominator. ■

EXAMPLE 6

Find $\frac{8}{25} - \frac{3}{25}$.

SOLUTION

Subtract numerators,

$$\frac{8}{25} - \frac{3}{25} = \frac{8-3}{25} = \frac{\overset{1}{\cancel{5}}}{\underset{5}{\cancel{25}}} = \frac{1}{5}$$

but keep the same denominator. ■

EXAMPLE 7

Find the difference between $\frac{7}{10}$ and $\frac{3}{10}$.

SOLUTION

$$\frac{7}{10} - \frac{3}{10} = \frac{7-3}{10} = \frac{\overset{2}{\cancel{4}}}{\underset{5}{\cancel{10}}} = \frac{2}{5} \quad ■$$

◇ *Do Practice Exercises 4 and 5.*

EXAMPLE 8

Three-fourths of a gallon of milk was in the refrigerator. After mowing the yard, Toni drank one-fourth of a gallon. How much was left?

SOLUTION

The answer is found by subtracting $\frac{1}{4}$ of a gallon from $\frac{3}{4}$ of a gallon.

$$\frac{3}{4} - \frac{1}{4} = \frac{3-1}{4} = \frac{2}{4} = \frac{1}{2} \text{ of a gallon remains} \quad ■$$

4. Subtract $\frac{3}{10} - \frac{1}{10}$.

5. Subtract one-fifth from three-fifths.

ANSWERS

4. $\frac{1}{5}$

5. $\frac{2}{5}$

6. Work each of the following:

(a) $\frac{3}{4} \cdot \frac{2}{7}$

(b) $\frac{1}{5} \div \frac{3}{8}$

(c) $\frac{3}{8} + \frac{3}{8}$

(d) $\frac{71}{100} - \frac{63}{100}$

In order to clearly separate the mathematical operations with fractions, an example of each operation studied up to this time is given.

Multiplication $\frac{2}{3} \cdot \frac{4}{5} = \frac{2 \cdot 4}{3 \cdot 5} = \frac{8}{15}$

Division $\frac{1}{5} \div \frac{9}{10} = \frac{1}{\cancel{5}_1} \cdot \frac{\cancel{10}^2}{9} = \frac{1 \cdot 2}{1 \cdot 9} = \frac{2}{9}$

Addition (common denominator) $\frac{3}{4} + \frac{1}{4} = \frac{3 + 1}{4} = \frac{4}{4} = 1$

Subtraction (common denominator) $\frac{9}{7} - \frac{3}{7} = \frac{9 - 3}{7} = \frac{6}{7}$

◇ *Do Practice Exercise 6.*

ANSWERS

6. (a) $\frac{3}{14}$ (b) $\frac{8}{15}$ (c) $\frac{3}{4}$ (d) $\frac{2}{25}$

name date hour

Add fractions by adding the shaded parts.

1. 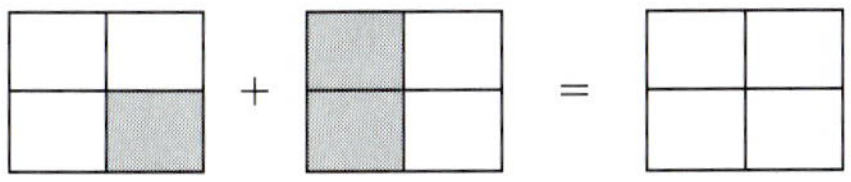

$\frac{1}{4} + \frac{2}{4} = \frac{?}{4}$

2. 

$\frac{1}{8} + \frac{4}{8} = \frac{?}{8}$

Add and reduce if possible.

3. $\frac{3}{5} + \frac{1}{5}$

4. $\frac{2}{7} + \frac{4}{7}$

5. $\frac{3}{9} + \frac{5}{9}$

6. $\frac{3}{13} + \frac{7}{13}$

7. $\frac{4}{15} + \frac{7}{15}$

8. $\frac{7}{15} + \frac{1}{15}$

9. $\frac{1}{7} + \frac{5}{7}$

10. $\frac{4}{15} + \frac{4}{15}$

11. $\frac{7}{20} + \frac{6}{20}$

12. $\frac{3}{8} + \frac{1}{8}$

13. $\frac{2}{5} + \frac{2}{5}$

14. $\frac{1}{4} + \frac{3}{4}$

15. $\frac{6}{9} + \frac{2}{9}$

16. $\frac{1}{12} + \frac{5}{12}$

17. $\frac{4}{13} + \frac{5}{13}$

18. $\frac{2}{12} + \frac{1}{12}$

19. $\frac{3}{20} + \frac{7}{20}$

20. $\frac{5}{9} + \frac{2}{9}$

21. $\frac{2}{40} + \frac{18}{40}$

22. $\frac{3}{10} + \frac{7}{10}$

23. $\frac{17}{56} + \frac{11}{56}$

24. $\frac{5}{49} + \frac{8}{49}$

25. $\frac{11}{100} + \frac{9}{100}$

26. $\frac{7}{96} + \frac{9}{96} + \frac{5}{96}$

Subtract fractions by subtracting the shaded parts.

27. 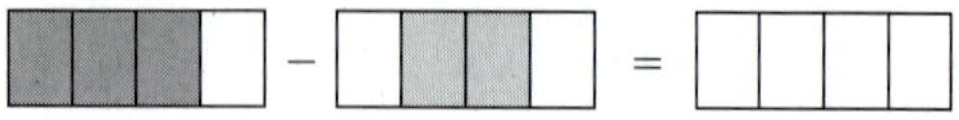

$\frac{3}{4} - \frac{2}{4} = \frac{?}{4}$

28. 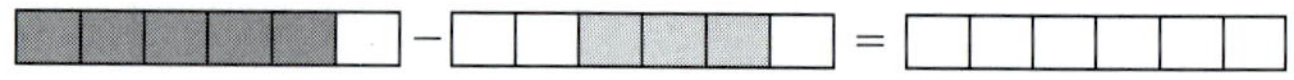

$\frac{5}{6} - \frac{3}{6} = \frac{?}{6}$

Subtract and reduce if possible.

29. $\frac{4}{3} - \frac{2}{3}$

30. $\frac{7}{5} - \frac{3}{5}$

31. $\frac{9}{11} - \frac{3}{11}$

32. $\frac{9}{20} - \frac{7}{20}$

33. $\frac{5}{8} - \frac{5}{8}$

34. $\frac{5}{42} - \frac{1}{42}$

35. $\frac{30}{13} - \frac{4}{13}$

36. $\frac{37}{75} - \frac{12}{75}$

37. $\frac{15}{55} - \frac{4}{55}$

38. $\frac{20}{27} - \frac{8}{27}$

39. $\frac{4}{7} - \frac{1}{7}$

40. $\frac{8}{9} - \frac{1}{9}$

41. $\frac{6}{15} - \frac{3}{15}$

42. $\frac{27}{32} - \frac{1}{32}$

43. $\frac{17}{25} - \frac{12}{25}$

44. $\frac{23}{40} - \frac{3}{40}$

45. $\frac{13}{35} - \frac{4}{35}$

46. $\frac{21}{58} - \frac{7}{58}$

47. $\frac{9}{100} - \frac{3}{100}$

48. $\frac{101}{125} - \frac{76}{125}$

49. $\frac{3}{64} - \frac{1}{64}$

50. $\frac{11}{90} - \frac{1}{90}$

51. $\frac{5}{52} - \frac{5}{52}$

52. $\frac{77}{200} - \frac{27}{200}$

Do the indicated mathematical operation and reduce if possible.

53. $\frac{2}{5} \times \frac{3}{4}$

54. $\frac{5}{7} \div \frac{2}{3}$

55. $\frac{1}{5} + \frac{3}{5}$

56. $\frac{2}{15} - \frac{1}{15}$

57. $\frac{3}{5} \div 5$

58. $\frac{7}{4} - \frac{5}{4}$

59. $\frac{5}{9} + \frac{1}{9}$

60. $\frac{2}{9} \cdot \frac{3}{8}$

61. $5 \div \frac{2}{3}$

62. $\frac{5}{12} - \frac{1}{12}$

63. $\left(\frac{2}{3}\right)\left(\frac{3}{4}\right)$

64. $\frac{1}{2} \div \frac{8}{3}$

65. $\frac{4}{5} - \frac{3}{5}$

66. $\frac{3}{4} \div \frac{1}{20}$

67. $\frac{20}{27} \cdot 9$

68. $\frac{11}{7} - \frac{9}{7}$

69. $33 \div \frac{1}{3}$

70. $\frac{3}{4} \cdot 5$

71. $\frac{8}{9} + \frac{0}{9}$

72. $\left(\frac{0}{12}\right)\left(\frac{0}{12}\right)$

Work each problem.

73. What is the quotient of seven-eighths divided by one-fourth?

74. What is the product of sixty-four and one thirty-second?

75. What is the difference between seven-twelfths and five-twelfths?

76. What is the sum of four-sevenths and two-sevenths?

77. A sale at a furniture store advertises one-third off of everything in the store. A desk normally costing $360 is marked at a price of $280. Has the desk been marked one-third off of the regular price?

78. A water tank containing 10,000 gallons is one-half full. How long will the water last at a consumption rate of 500 gallons per day?

JOURNAL EXERCISES

1. Find an example from a newspaper or magazine article in which fractions are used, cut it out, and attach it to your journal.

2. In your own words, explain the processes used to:

1. Multiply fractions
2. Divide fractions
3. Add fractions with common denominators
4. Subtract fractions with common denominators

OBJECTIVES

1 Add fractions with unlike denominators.
2 Subtract fractions with unlike denominators.

1 Add Fractions with Unlike Denominators

Adding fractions with unlike denominators differs from adding with common denominators. For example, $\frac{1}{3}$ of the square on the left is shaded and $\frac{1}{4}$ of the square in the center is shaded. Both of these shaded areas have been placed in the square on the right. The fraction of the last figure that is shaded cannot be determined by looking at the figures. This is because the square has *not been cut into equal size pieces*. In other words, the fractions do not have common denominators. This problem is worked as Example 1 below.

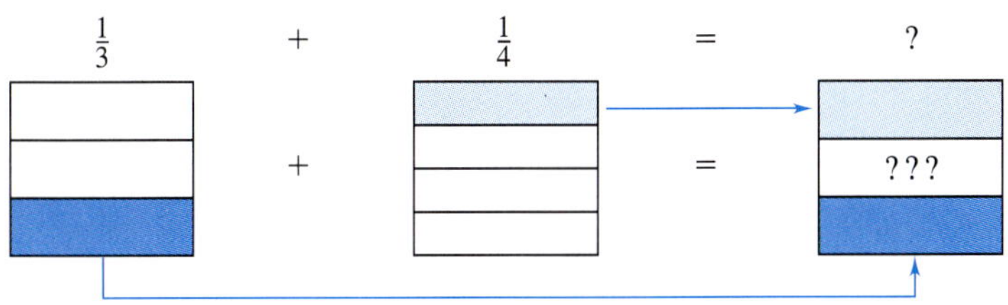

However, the fractions can be added if they are first written using common denominators:

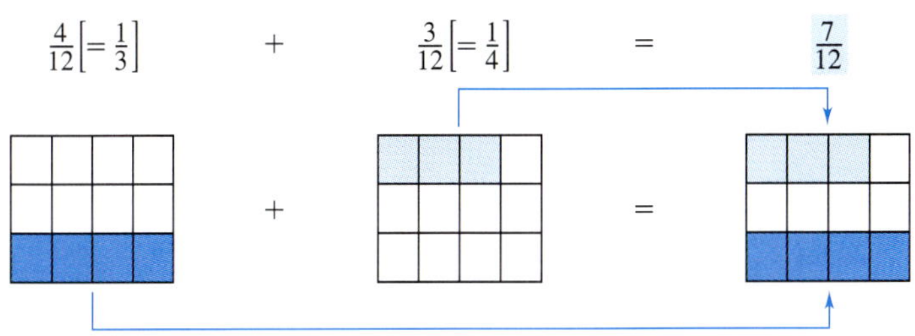

The following rule to add fractions with unlike denominators includes the step of writing each fraction using common denominators.

Rule: To Add Fractions with Unlike Denominators

Step 1 Write each fraction as an equivalent fraction using the least common denominator.

Step 2 Add the fractions by adding the numerators and keeping the common denominator.

Step 3 Reduce, if possible.

EXAMPLE 1

Add $\frac{1}{3} + \frac{1}{4}$.

SOLUTION

Step 1 The least common denominator is 12. Each fraction is written as an equivalent fraction using the common denominator 12:

$$\frac{1}{3} = \frac{4}{12} \quad \text{and} \quad \frac{1}{4} = \frac{3}{12}$$

1. Add $\frac{1}{3} + \frac{1}{5}$.

2. Find the sum of $\frac{2}{6}$ and $\frac{3}{8}$.

Step 2

$$\frac{4}{12} + \frac{3}{12} = \frac{4 + 3}{12}$$

Add numerators.

Keep the same denominator.

Step 3

$$\frac{4}{12} + \frac{3}{12} = \frac{4 + 3}{12} = \frac{7}{12}$$

Try to reduce. ■

HELPFUL HINT

Fractions must have a common denominator before adding.

EXAMPLE 2

Add $\frac{1}{4} + \frac{3}{5}$.

SOLUTION

$\frac{1}{4} + \frac{3}{5}$

$\frac{5}{20} + \frac{12}{20}$ Write using common denominators.

$\frac{5 + 12}{20}$ Add numerators; keep denominator.

$\frac{17}{20}$ Result (cannot be reduced). ■

◇ *Do Practice Exercises 1 and 2.*

EXAMPLE 3

Add $\frac{1}{3} + \frac{1}{4} + \frac{3}{8}$.

SOLUTION

$\frac{1}{3} + \frac{1}{4} + \frac{3}{8}$

$\frac{8}{24} + \frac{6}{24} + \frac{9}{24}$ Write using the common denominator.

$\frac{8 + 6 + 9}{24}$ Add numerators; keep same denominator.

$\frac{23}{24}$ Result. ■

EXAMPLE 4

Add $4 + \frac{2}{3} + \frac{3}{5}$.

ANSWERS

1. $\frac{8}{15}$

2. $\frac{17}{24}$

SOLUTION

$$4 + \frac{2}{3} + \frac{3}{5}$$

$$\frac{4}{1} + \frac{2}{3} + \frac{3}{5} \quad \text{Write 4 as } \frac{4}{1}.$$

$$\frac{60}{15} + \frac{10}{15} + \frac{9}{15} \quad \text{Write using common denominators.}$$

$$\frac{60 + 10 + 9}{15} \quad \text{Add numerators; keep denominator.}$$

$$\frac{79}{15} \quad \text{Result.}$$

◇ *Do Practice Exercises 3 and 4.*

2 Subtract Fractions with Unlike Denominators

Common denominators are also required in order to subtract fractions. The squares below show an attempt to subtract $\frac{1}{5}$ from $\frac{3}{4}$. The solution is not obvious since the squares are cut into different sized pieces:

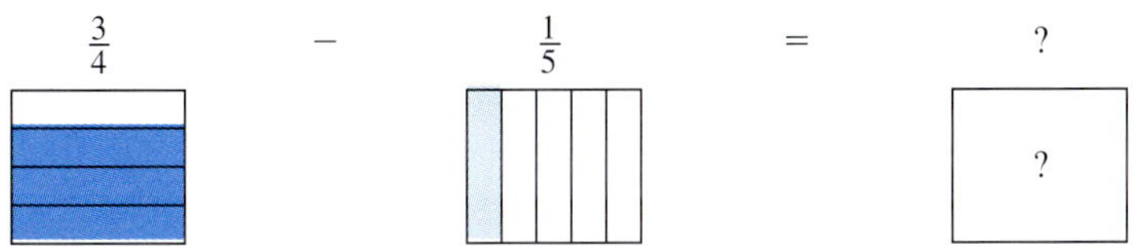

They can be subtracted after writing each fraction using common denominators:

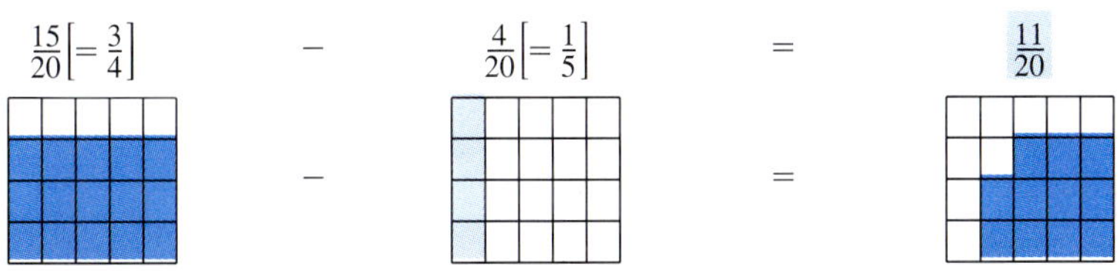

Rule: To Subtract Fractions with Unlike Denominators

Step 1 Write each fraction as an equivalent fraction using the least common denominator.

Step 2 Subtract the fractions by subtracting the numerators and keeping the common denominator.

Step 3 Reduce if possible.

EXAMPLE 5

Subtract $\frac{3}{4} - \frac{1}{5}$.

3. Add $5 + \frac{2}{3}$.

4. Find the sum of 1 with $\frac{1}{8}$.

ANSWERS

3. $\frac{17}{3}$

4. $\frac{9}{8}$

5. Subtract one-eighth from three-fourths.

6. Find the difference between $\frac{3}{5}$ and $\frac{2}{15}$.

ANSWERS

5. $\frac{5}{8}$

6. $\frac{7}{15}$

SOLUTION

Step 1 The least common denominator is 20. Each fraction is rewritten as an equivalent fraction with denominator 20.

$$\frac{3}{4} = \frac{15}{20} \qquad \frac{1}{5} = \frac{4}{20}$$

$$\frac{3}{4} - \frac{1}{5} = \frac{15}{20} - \frac{4}{20}$$

Steps 2 and 3

$$\frac{3}{4} - \frac{1}{5} = \frac{15}{20} - \frac{4}{20} = \frac{15-4}{20} = \frac{11}{20}$$

Keep the same denominator. ■

HELPFUL HINT

Fractions must have a common denominator before subtracting.

EXAMPLE 6

Subtract $\frac{5}{8} - \frac{3}{14}$.

SOLUTION

$\frac{5}{8} - \frac{3}{14}$

$\frac{35}{56} - \frac{12}{56}$ Write using the common denominator.

$\frac{35 - 12}{56}$ Subtract numerators, keep same denominator.

$\frac{23}{56}$ Result. ■

◇ *Do Practice Exercises 5 and 6.*

EXAMPLE 7

Subtract three-fourths from seven.

SOLUTION

$7 - \frac{3}{4}$

$\frac{7}{1} - \frac{3}{4}$ Write 7 as the fraction $\frac{7}{1}$.

$\frac{28}{4} - \frac{3}{4}$ Write using the common denominator.

$\frac{28 - 3}{4}$ Subtract.

$\frac{25}{4}$ Result. ■

◇ *Do Practice Exercises 7 and 8.*

The next example shows an application in which fractions are subtracted from the whole number 1. Study it carefully.

EXAMPLE 8

Aunt Bertha gave one-third of her estate to Max, one-fourth to Karin, and the remainder to Linda. What fraction of the estate goes to Linda?

SOLUTION

The result is found by subtracting $\frac{1}{3}$ and $\frac{1}{4}$ from the entire estate. The hard part may be in recognizing that the entire estate is the number 1, namely 1 estate.

Whole Estate	−	Fraction to Max	−	Fraction to Karin
1	−	$\frac{1}{3}$	−	$\frac{1}{4}$

$\frac{1}{1} - \frac{1}{3} - \frac{1}{4}$ Write 1 as a fraction.

$\frac{12}{12} - \frac{4}{12} - \frac{3}{12}$ Write using common denominator.

$\frac{12 - 4 - 3}{12}$ Subtract numerators.

$\frac{5}{12}$ Result.

Therefore, $\frac{5}{12}$ths of the estate is left for Linda. ■

All of the mathematical operations with fractions are shown next in order to reinforce the different rules used to work the different types of problems.

Multiplication $\frac{3}{4} \cdot \frac{3}{10} = \frac{3 \cdot 3}{4 \cdot 10} = \frac{9}{40}$

Division $\frac{5}{8} \div \frac{7}{2} = \frac{5}{\cancel{8}_4} \cdot \frac{\cancel{2}^1}{7} = \frac{5 \cdot 1}{4 \cdot 7} = \frac{5}{28}$

Addition (common denominator) $\frac{3}{10} + \frac{1}{10} = \frac{3 + 1}{10} = \frac{\cancel{4}^2}{\cancel{10}_5} = \frac{2}{5}$

7. Subtract one-half from 3.

8. Subtract two from eleven-thirds.

ANSWERS

7. $\frac{5}{2}$

8. $\frac{5}{3}$

9. Work each:

(a) $\frac{2}{3} \cdot \frac{11}{16}$

(b) $\frac{3}{4} \div \frac{1}{2}$

(c) $\frac{3}{8} + \frac{5}{8}$

(d) $\frac{3}{14} + \frac{2}{7}$

(e) $\frac{5}{9} - \frac{4}{9}$

(f) $\frac{9}{10} - \frac{3}{4}$

Addition (unlike denominator) $\frac{2}{3} + \frac{3}{4} = \frac{8}{12} + \frac{9}{12} = \frac{8+9}{12} = \frac{17}{12}$

Subtraction (common denominator) $\frac{7}{9} - \frac{1}{9} = \frac{7-1}{9} = \frac{\overset{2}{\cancel{6}}}{\underset{3}{\cancel{9}}} = \frac{2}{3}$

Subtraction (unlike denominator) $\frac{3}{4} - \frac{1}{5} = \frac{15}{20} - \frac{4}{20} = \frac{15-4}{20} = \frac{11}{20}$

◇ *Do Practice Exercise 9.*

ANSWERS

9. (a) $\frac{11}{24}$ (b) $\frac{3}{2}$ (c) 1 (d) $\frac{7}{14} = \frac{1}{2}$ (e) $\frac{1}{9}$ (f) $\frac{3}{20}$

name date hour

Find the least common multiple. (Review)

1. 8, 6

2. 2, 3

3. 4, 6

4. 10, 6

5. 2, 3, 12

6. 6, 8, 10

Raise each fraction to higher terms. (Review)

7. $\frac{3}{4} = \frac{?}{20}$

8. $\frac{2}{3} = \frac{?}{21}$

9. $\frac{14}{9} = \frac{?}{27}$

10. $\frac{?}{40} = \frac{5}{8}$

11. $\frac{?}{100} = \frac{3}{20}$

12. $\frac{4}{5} = \frac{?}{80}$

Find the least common denominator. (Review)

13. $\frac{1}{3}, \frac{1}{4}$

14. $\frac{2}{7}, \frac{1}{8}$

15. $\frac{1}{6}, \frac{3}{4}$

16. $\frac{2}{9}, \frac{5}{6}$

17. $\frac{5}{12}, \frac{1}{6}, \frac{2}{5}$

18. $\frac{1}{3}, \frac{1}{4}, \frac{1}{5}$

Find the least common denominator and write each fraction as an equivalent fraction using the LCD. (Review)

19. $\frac{2}{5}, \frac{1}{3}$

20. $\frac{1}{4}, \frac{2}{3}$

21. $\frac{1}{8}, \frac{5}{6}$

22. $\frac{3}{4}, \frac{5}{12}$

23. $\frac{2}{5}, \frac{1}{25}$

24. $\frac{9}{8}, \frac{5}{18}$

Add or subtract and reduce to lowest terms.

25. $\frac{1}{3} + \frac{3}{4}$

26. $\frac{3}{5} - \frac{1}{2}$

27. $\frac{1}{6} + \frac{2}{3}$

28. $\frac{5}{8} - \frac{1}{4}$

29. $\frac{3}{5} + \frac{3}{35}$

30. $\frac{3}{20} + \frac{1}{4}$

31. $\frac{7}{10} - \frac{2}{3}$

32. $\frac{5}{8} + \frac{1}{4}$

33. $\frac{5}{12} - \frac{1}{3}$

34. $\frac{4}{5} - \frac{1}{100}$

35. $\frac{3}{8} - \frac{1}{7}$

36. $\frac{1}{5} - \frac{1}{8}$

37. $\frac{2}{3} + \frac{1}{9}$

38. $\frac{3}{5} + \frac{4}{25}$

39. $\frac{5}{6} - \frac{1}{3}$

40. $\frac{4}{7} + \frac{3}{14}$

41. $\frac{2}{9} + \frac{5}{24}$

42. $\frac{7}{12} - \frac{3}{8}$

43. $\frac{4}{15} - \frac{1}{18}$

44. $\frac{7}{11} + \frac{3}{11}$

45. $\frac{3}{8} + \frac{17}{24}$

46. $\frac{5}{12} - \frac{1}{8}$

47. $\frac{3}{17} - \frac{5}{34}$

48. $\frac{4}{27} + \frac{5}{27}$

49. $\frac{5}{12} + \frac{2}{9}$

50. $\frac{5}{6} - \frac{1}{75}$

51. $\frac{3}{13} - \frac{5}{54}$

52. $\frac{3}{20} + \frac{5}{36}$

53. $\frac{7}{25} + \frac{5}{75} + \frac{3}{50}$

54. $\frac{1}{10} + \frac{1}{21} + \frac{1}{15}$

55. $\frac{1}{5} + \frac{1}{6} + \frac{7}{30}$

56. $\frac{3}{8} + \frac{1}{5} + \frac{1}{10}$

Work each problem and reduce to lowest terms.

57. $\frac{3}{4} \cdot \frac{1}{2}$

58. $\frac{3}{4} \div \frac{2}{3}$

59. $\frac{4}{7} + \frac{3}{7}$

60. $\frac{3}{4} \div 3$

61. $\frac{2}{3} - \frac{1}{3}$

62. $\frac{3}{4} + \frac{1}{5}$

63. $\frac{1}{5} \times \frac{7}{4}$

64. $\frac{5}{6} - \frac{1}{8}$

65. $7 - \frac{2}{3}$

66. $\frac{4}{5} \cdot 0$

67. $\frac{2}{3} - \frac{1}{8}$

68. $\frac{3}{5} \cdot \frac{1}{9}$

69. $\frac{6}{25} + \frac{1}{5}$

70. $\frac{3}{7} - \frac{2}{7}$

71. $\frac{3}{4} \div \frac{3}{8}$

72. $2 + \frac{1}{5}$

73. $\frac{7}{5} - \frac{1}{4}$

74. $14 \div \frac{1}{7}$

75. $\left(\frac{5}{9}\right)\left(\frac{3}{20}\right)$

76. $\left(\frac{9}{10}\right) - \left(\frac{1}{5}\right)$

Work each problem.

77. Find the sum of three-fifths and four-twentieths.

78. What number equals five-ninths minus four twenty-sevenths?

79. Find the product of four-elevenths with the whole number twelve.

80. One thousand insects are born on an experimental agricultural plot. Ninety-nine out of every 100 do *not* make it to the adult stage. How many of the 1,000 reach adulthood?

81. In one poor country, 35 out of every 100 newborn babies do not survive to age 5. How many out of every 2,000 newborns do survive to age 5, on average?

82. Divide three-eighths by two-thirds, then add the result to three-halves. Is the result a proper or an improper fraction? Is it greater than or less than one?

83. Three major car companies in North America built seven million passenger cars in 1992. Of these, approximately one-seventh were built by Chrysler and two-sevenths were built by Ford. What fraction was built by the remaining car company, General Motors?

84. One-fourth of the world's population is from China. Three-twentieths of the world's population is from India. What fraction of the world's population is neither from China nor from India?

JOURNAL EXERCISES

1. A child claims to spend her time as follows:

Activity	*Time*	*As a Fraction of a Year*
Sleeping, Bathing	10 hours/24 hour day	$\frac{5}{12}$
Eating	2 hours/24 hour day	$\frac{1}{12}$
Weekends	2 days/7 day week	$\frac{2}{7}$
Summer Vacation	3 months/12 month year	$\frac{1}{4}$

She uses this data to claim that there is no time left over for school. Add the fractions in the last column together. Is the sum larger than one year? What is the flaw in the child's reasoning?

2. Explain the differences between adding with common denominators versus adding with unlike denominators.

3. Which mathematical operations require that a common denominator be calculated? Which do not? Why?

1 Raise a Fraction to a Power

Exponents apply to fractions in the same way they apply to whole numbers. An exponent tells how many times a number is used as a factor in multiplication. The following examples review exponents as applied to whole numbers then progress to exponents applied to fractions.

Exponents Applied to

Whole Numbers	*Fractions*
$6^2 = 6 \cdot 6 = 36$	$\left(\frac{1}{2}\right)^2 = \frac{1}{2} \cdot \frac{1}{2} = \frac{1 \cdot 1}{2 \cdot 2} = \frac{1}{4}$
$4^3 = 4 \cdot 4 \cdot 4 = 64$	$\left(\frac{2}{3}\right)^3 = \frac{2}{3} \cdot \frac{2}{3} \cdot \frac{2}{3} = \frac{2 \cdot 2 \cdot 2}{3 \cdot 3 \cdot 3} = \frac{8}{27}$

An exponent at the top right hand corner of a fraction with no parentheses might apply *only* to the numerator of the fraction and NOT to the entire fraction. If the exponent applies to the fraction, avoid confusion by enclosing the fraction in parentheses and placing the exponent outside of the parentheses. Look at the next example.

EXAMPLE 1

Evaluate $\left(\frac{2}{5}\right)^3$ and $\frac{5^2}{8}$.

SOLUTION

Use $\frac{2}{5}$ as a factor three times.

$$\left(\frac{2}{5}\right)^3 = \frac{2}{5} \cdot \frac{2}{5} \cdot \frac{2}{5} = \frac{2 \cdot 2 \cdot 2}{5 \cdot 5 \cdot 5} = \frac{8}{125}$$

Applies only to the numerator.

$$\frac{5^2}{8} = \frac{5 \cdot 5}{8} = \frac{25}{8}$$

■

◇ *Do Practice Exercises 1 through 3.*

2 Complex Fractions

A fraction in which either the numerator and/or the denominator contain a fraction is called a **complex fraction**. Each of the following is a complex fraction:

$$\frac{\frac{2}{3}}{7}, \quad \frac{3}{\frac{4}{\frac{1}{2}}}, \quad \frac{5 - \frac{1}{4}}{\frac{4}{5}}, \quad \frac{\frac{4}{5}}{\frac{1}{7}}$$

The numerator is a fraction or the denominator is a fraction.

The complex fraction $\frac{\frac{3}{4}}{\frac{1}{2}}$ has a numerator which is the fraction $\frac{3}{4}$ and a

OBJECTIVES

1 Raise a fraction to a power.

2 Simplify complex fractions.

1. Find $\left[\frac{3}{7}\right]^2$.

2. Evaluate $\left[\frac{1}{4}\right]^3$.

3. Raise three-eights to the second power.

ANSWERS

1. $\frac{9}{49}$

2. $\frac{1}{64}$

3. $\frac{9}{64}$

denominator which is the fraction $\frac{1}{2}$. A fraction implies division, therefore the complex fraction $\frac{\frac{3}{4}}{\frac{1}{2}}$ is the numerator $\frac{3}{4}$ divided by the denominator $\frac{1}{2}$ or $\frac{3}{4} \div \frac{1}{2}$:

$$\frac{\frac{3}{4}}{\frac{1}{2}}$$

numerator $\frac{3}{4}$; denominator $\frac{1}{2}$

$$\frac{3}{4} \div \frac{1}{2}$$ Divide numerator by denominator.

$$\frac{3}{\cancel{4}_2} \cdot \frac{\cancel{2}^1}{1}$$ invert fraction, ÷ to ·
Cross reduce.

$$\frac{3 \cdot 1}{2 \cdot 1}$$

$$\frac{3}{2}$$ Result.

EXAMPLE 2

Simplify $\frac{\frac{2}{3}}{\frac{4}{7}}$.

SOLUTION

$$\frac{\frac{2}{3}}{\frac{4}{7}} = \frac{2}{3} \div \frac{4}{7} = \frac{\cancel{2}^1}{3} \cdot \frac{7}{\cancel{4}_2} = \frac{1 \cdot 7}{3 \cdot 2} = \frac{7}{6}$$ ■

EXAMPLE 3

Simplify $\frac{\frac{3}{4}}{\frac{5}{8}}$.

SOLUTION

$$\frac{\frac{3}{4}}{\frac{5}{8}}$$

$$\frac{3}{4} \div \frac{5}{8}$$ Divide numerator by denominator.

$\frac{3}{\cancel{4}_1} \cdot \frac{\cancel{8}^2}{5}$ Change to a multiplication problem and reduce.

$\frac{6}{5}$ Result. ■

◇ *Do Practice Exercise 4.*

Recall from Section 3.4 that two fractions can be divided by inverting the second fraction and changing the division problem to a multiplication problem. We will now justify this rule using $\frac{3}{5}$ divided by $\frac{1}{2}$ as an example. First, the division can be written as a complex fraction:

$$\frac{3}{5} \div \frac{1}{2} = \frac{\frac{3}{5}}{\frac{1}{2}}$$

The first objective is to remove the fraction $\frac{1}{2}$ from the denominator of the original complex fraction above. This can be done by using the inverse of $\frac{1}{2}$ which is $\frac{2}{1}$. Notice that the complex fraction

$$\text{new complex fraction} \quad \frac{\frac{2}{1}}{\frac{2}{1}} = \frac{2}{1} \div \frac{2}{1} = 2 \div 2 = 1$$

is equal to 1. It does not change a number to multiply it times 1. Therefore, the complex fraction that equals one can be multiplied times the original complex fraction without changing the result:

$$\frac{\frac{3}{5}}{\frac{1}{2}} \cdot 1 = \frac{\frac{3}{5}}{\frac{1}{2}} \cdot \frac{\frac{2}{1}}{\frac{2}{1}}$$

Now multiply numerators and denominators:

$$\frac{\frac{3}{5} \cdot \frac{2}{1}}{\frac{\cancel{1}}{\cancel{2}} \cdot \frac{\cancel{2}}{\cancel{1}}} = \frac{\frac{3}{5} \cdot \frac{2}{1}}{1} = \frac{3}{5} \cdot \frac{2}{1}$$

Multiply numerators.

Multiply denominators.

We have shown that:

$$\frac{3}{5} \div \frac{1}{2} = \frac{3}{5} \cdot \frac{2}{1}$$

In other words, we have shown that the division of two fractions can be changed to a multiplication problem by inverting the second fraction.

4. Simplify:

(a) $\frac{\frac{5}{8}}{\frac{3}{4}}$

(b) $\frac{\frac{3}{5}}{\frac{1}{6}}$

ANSWERS

4. (a) $\frac{5}{6}$ (b) $\frac{18}{5}$

5. Simplify:

(a) $\dfrac{8}{\frac{2}{3}}$

(b) $\dfrac{\frac{4}{5}}{20}$

The next example is slightly more complex since the numerator is a fraction and the denominator is a whole number. It is important to first change any whole number to a fraction, then work the problem as above.

EXAMPLE 4

Simplify $\dfrac{\frac{6}{11}}{4}$.

SOLUTION

$$\dfrac{\frac{6}{11}}{4}$$

$\dfrac{6}{11} \div 4$ Divide the numerator by the denominator.

$\dfrac{6}{11} \div \dfrac{4}{1}$ Write 4 as a fraction.

$\dfrac{\overset{3}{\cancel{6}}}{11} \cdot \dfrac{1}{\underset{2}{\cancel{4}}}$ Change to a multiplication problem.

$\dfrac{3}{22}$ Result. ■

◇ *Do Practice Exercise 5.*

The final example in this section is one in which the numerator is an expression. In this case, first simplify the numerator. Then, simplify the complex fraction.

EXAMPLE 5

Simplify $\dfrac{3 - \frac{1}{4}}{\frac{2}{3}}$.

SOLUTION

First, simplify the numerator:

numerator → $3 - \dfrac{1}{4}$

$\dfrac{3}{1} - \dfrac{1}{4}$ Write 3 as a fraction.

$\dfrac{12}{4} - \dfrac{1}{4}$ Write using common denominator.

$\dfrac{12 - 1}{4}$ Subtract.

→ $\dfrac{11}{4}$

ANSWERS

5. (a) 12 (b) $\dfrac{1}{25}$

Now, replace the expression in the numerator with the value it is equal to and simplify.

$\dfrac{\frac{11}{4}}{\frac{2}{3}}$ ⟵The expression is replaced with the fraction it is equal to.

$\dfrac{11}{4} \div \dfrac{2}{3}$ Divide the numerator by the denominator.

$\dfrac{11}{4} \cdot \dfrac{3}{2}$ Change to a multiplication problem.

$\dfrac{33}{8}$ Result. ■

◇ *Do Practice Exercise 6.*

6. Simplify:

(a) $\dfrac{1 - \frac{1}{3}}{\frac{4}{5}}$

(b) $\dfrac{3 + \frac{1}{2}}{\frac{9}{10}}$

ANSWERS

6. (a) $\frac{5}{6}$ (b) $\frac{35}{9}$

name date hour

Work each problem and reduce if possible.

1. $\frac{2}{3} \cdot \frac{5}{6}$

2. $\frac{3}{4} + \frac{5}{4}$

3. $\frac{1}{2} \div \frac{4}{5}$

4. $\frac{6}{5} - \frac{3}{5}$

5. $\frac{2}{3} + \frac{4}{5}$

6. $\left(\frac{2}{3}\right)^2$

7. $\left(\frac{3}{4}\right)^3$

8. $\left(\frac{1}{2}\right)^2$

9. $\left(\frac{4}{5}\right)^3$

10. $\left(\frac{1}{4}\right)^2$

11. $\left(\frac{8}{13}\right)^2$

12. $\left(\frac{2}{5}\right)^2$

13. $\frac{3}{4} + \frac{7}{8}$

14. $3 - \frac{4}{5}$

15. $\left(\frac{7}{8}\right) \cdot \left(\frac{0}{6}\right)$

16. $\left(\frac{7}{10}\right)^2$

17. $\left(\frac{5}{6}\right)^3$

18. $\left(\frac{1}{5}\right)^3$

19. $\left(\frac{5}{9}\right)^2$

20. $\frac{1}{2} - \frac{1}{3}$

21. $\left(\frac{2}{3}\right)^3$

22. $\frac{4}{3} \div \frac{1}{2}$

23. $\frac{5}{6} - \frac{1}{15}$

24. $\frac{7}{8} + \frac{9}{8}$

25. $\frac{5}{8} + \frac{3}{8}$

26. $\frac{3}{4} \div 12$

27. $\left(\frac{3}{20}\right)^2$

28. $\left(\frac{1}{4}\right)^3$

29. $\frac{4}{7} \cdot \frac{2}{3}$

30. $\frac{6}{7} + \frac{3}{35}$

31. $\frac{2}{3} - \frac{2}{3}$

32. $\left(\frac{3}{5}\right)^1$

33. $\frac{4}{7} \div \frac{1}{3}$

34. $\frac{4}{15} + \frac{1}{8}$

35. $\frac{1}{9} \div 7$

36. $\left(\frac{4}{9}\right)^2$

37. $\frac{7}{8} + \frac{1}{8}$

38. $\frac{2}{3} \cdot \frac{3}{4}$

39. $\frac{7}{8} - \frac{1}{16}$

40. $\left(\frac{1}{2}\right)^3$

Simplify each complex fraction.

41. $\dfrac{\frac{3}{4}}{\frac{2}{3}}$

42. $\dfrac{\frac{1}{4}}{\frac{6}{7}}$

43. $\dfrac{\frac{4}{7}}{8}$

44. $\dfrac{\frac{5}{6}}{\frac{7}{3}}$

45. $\dfrac{\frac{3}{5}}{\frac{1}{9}}$

46. $\dfrac{\frac{0}{7}}{\frac{1}{20}}$

47. $\dfrac{\frac{2}{5}}{6}$

48. $\dfrac{9}{\frac{1}{10}}$

49. $\dfrac{\frac{4}{3}}{\frac{1}{5}}$

50. $\dfrac{\frac{6}{7}}{\frac{3}{5}}$

51. $\dfrac{\frac{1}{100}}{\frac{1}{1000}}$

52. $\dfrac{\frac{7}{3}}{\frac{49}{4}}$

53. $\dfrac{\frac{1}{4}}{\frac{2}{3}}$

54. $\dfrac{\frac{1}{2}}{\frac{3}{4}}$

55. $\dfrac{1}{\frac{3}{5}}$

56. $\dfrac{\frac{1}{7}}{\frac{4}{9}}$

57. $\dfrac{\frac{3}{10}}{\frac{5}{8}}$

58. $\dfrac{\frac{5}{8}}{5}$

59. $\dfrac{\frac{1}{2}}{\frac{7}{4}}$

60. $\dfrac{\frac{2}{3}}{4}$

61. $\dfrac{2 - \frac{2}{3}}{\frac{1}{2}}$

62. $\dfrac{1 + \frac{1}{3}}{\frac{4}{5}}$

Work each problem.

63. Square four-fifths and multiply the result times nine.

64. Cube four-thirds and add the result to one-eighth.

JOURNAL EXERCISES

1. Math notation can sometimes look complicated. Jot down your thoughts as you simplify:

$$\frac{\frac{4}{5} - \frac{\frac{3}{5} - \frac{1}{2}}{\frac{2}{3}}}{5 - \frac{1}{2}}$$

Hint: Break the problem into parts. First, simplify the numerator:

$$\text{numerator} \quad \frac{4}{5} - \frac{\frac{3}{5} - \frac{1}{2}}{\frac{2}{3}}$$

Then simplify the denominator:

$$\text{denominator} \quad 5 - \frac{1}{2}$$

Finally, simplify the complex fraction that remains.

2. Explain square and cube as they apply to fractions. Then square the following complex fraction:

$$\left[\frac{\frac{3}{4}}{\frac{2}{3}}\right]$$

1 Order of Operations

The order of operations rules for fractions are identical to those for whole numbers. The purpose of this section is to:

1. Reemphasize the order of operations and

2. Provide practice in working with fractions.

Rule: To Evaluate an Expression

Step 1 Evaluate the contents of the parentheses and above and below the fraction bar.

Step 2 Evaluate exponents.

Step 3 Evaluate multiplication and division in the order they occur when moving from left to right across the expression.

Step 4 Evaluate addition and subtraction in the order they occur when moving from left to right across the expression.

OBJECTIVE

1 Apply the Order of Operations to expressions containing fractions.

EXAMPLE 1

Evaluate $\frac{1}{2} + \frac{2}{3} \cdot \frac{6}{11}$.

SOLUTION

$$\frac{1}{2} + \frac{2}{3} \cdot \frac{6}{11}$$

$$\frac{1}{2} + \frac{2}{\cancel{3}_1} \cdot \frac{\cancel{6}^2}{11} \qquad \text{Cross reduce.}$$

$$\frac{1}{2} + \frac{4}{11} \qquad \text{Multiply.}$$

$$\frac{11}{22} + \frac{8}{22} \qquad \text{Write using common denominators.}$$

$$\frac{11 + 8}{22} \qquad \text{Add.}$$

$$\boxed{\frac{19}{22}} \qquad \text{Result.}$$

■

Notice that the first row of the expression in Example 1 is the most complicated. *Only one* mathematical operation is calculated in going from one row to the next. The remainder of each row is identical to the row above and the result is on the last row.

EXAMPLE 2

Evaluate $\frac{3}{4} - \left(\frac{1}{2} \div \frac{4}{5}\right) \cdot \frac{1}{2}$

1. Evaluate $\frac{3}{4} - \frac{1}{2} \cdot \frac{2}{3}$.

2. Evaluate $\left[\frac{4}{5}\right]^2 - \frac{1}{5}$.

3. Evaluate $\left[\frac{3}{5}\right]^2 \div \frac{1}{2}$.

ANSWERS

1. $\frac{5}{12}$

2. $\frac{11}{25}$

3. $\frac{18}{25}$

SOLUTION

$\frac{3}{4} - \left(\frac{1}{2} \div \frac{4}{5}\right) \cdot \frac{1}{2}$	
$\frac{3}{4} - \left(\frac{1}{2} \cdot \frac{5}{4}\right) \cdot \frac{1}{2}$	Change ÷ to · inside the parentheses.
$\frac{3}{4} - \frac{5}{8} \cdot \frac{1}{2}$	contents of parentheses
$\frac{3}{4} - \frac{5}{16}$	Multiply.
$\frac{12}{16} - \frac{5}{16}$	Write using common denominators.
$\frac{12 - 5}{16}$	Subtract.
$\frac{7}{16}$	Result. ■

◇ *Do Practice Exercises 1 and 2.*

EXAMPLE 3

Evaluate $\left(1 - \frac{3}{5}\right)^2 \cdot \frac{2}{3}$.

SOLUTION

$\left(1 - \frac{3}{5}\right)^2 \cdot \frac{2}{3}$	
$\left(\frac{1}{1} - \frac{3}{5}\right)^2 \cdot \frac{2}{3}$	Contents of parentheses, write 1 as a fraction.
$\left(\frac{5}{5} - \frac{3}{5}\right)^2 \cdot \frac{2}{3}$	Write using a common denominator.
$\left(\frac{2}{5}\right)^2 \cdot \frac{2}{3}$	Subtract fractions in parentheses.
$\frac{4}{25} \cdot \frac{2}{3}$	exponents
$\frac{8}{75}$	Result. ■

◇ *Do Practice Exercise 3.*

Work the following and explain each step with a comment.

1. $\frac{3}{4} + \frac{7}{8}$

2. $\frac{4}{9} \cdot \frac{12}{5}$

3. $\frac{7}{3} \div \frac{4}{7}$

4. $\frac{6}{7} - \frac{1}{7}$

5. $\frac{2}{3} \cdot \frac{1}{2} + \frac{1}{5}$

6. $\frac{3}{4} \div \frac{6}{5} \div \frac{2}{5}$

7. $\left(\frac{1}{2} + \frac{3}{4}\right) \div 7$

8. $\frac{4}{5} + 6 \div \frac{1}{7}$

9. $\frac{3}{4} + \frac{1}{2} - \frac{1}{6}$

10. $\left(\frac{4}{5}\right)\left(\frac{2}{3}\right)\left(\frac{3}{20}\right)$

11. $\left(\frac{1}{2}\right)^2 - \left(\frac{1}{3}\right)^2$

12. $\left(\frac{1}{2} - \frac{1}{3}\right)^2$

13. $\frac{1}{9} \cdot \frac{1}{16} \div \frac{1}{4}$

14. $\left(\frac{9}{4}\right)^2 \cdot 1$

15. $\left(\frac{3}{2} + \frac{6}{2}\right) \div \frac{1}{16}$

16. $\frac{2}{3} \cdot 8^2$

17. $\frac{3}{5} \div \frac{1}{2} - \frac{1}{2}$

18. $\frac{1}{2} - \frac{1}{3} \cdot \frac{1}{6}$

19. $\frac{5}{8} \cdot 7 \cdot \frac{2}{3}$

20. $\frac{4}{5} \cdot \frac{1}{7} - \frac{1}{30}$

21. $\frac{7}{4} \div \frac{2}{3} \div \frac{1}{5}$

22. $\frac{7}{4} \div \left(\frac{2}{3} \div \frac{1}{5}\right)$

23. $\frac{1}{4} \div \left[3 - \frac{1}{2}\right]$

24. $\frac{2}{3} \div \left[\frac{1}{2}\right]^2$

25. $\frac{5}{9} - \frac{1}{2} + \frac{1}{3}$

26. $\frac{3}{4} \div \frac{8}{5}$

27. $\frac{1}{8} + \left[\frac{1}{6}\right]^2$

28. $\frac{19}{20} - \frac{4}{5} + \frac{1}{3} \cdot \frac{6}{5}$

29. $\frac{4}{7} + \frac{5}{7} \div \frac{2}{3}$

30. $\left(2 \div \frac{1}{4}\right)^3 \div \frac{2}{3}$

Write each as an expression, solve, and reduce if possible.

31. Find the product of one-fifth and one-fourth then divide by five-thirds.

32. Find the sum of two-thirds and three-fifths.

33. What is the product of sixth-sevenths and two-fifths?

34. Find the difference between three-fourths and one-fifth.

35. Divide three-eighths by four-fifths.

36. Add three-fifths to the product of two-thirds and one-tenth.

37. Subtract one-seventh from the quotient of one-half divided by four-thirds.

38. Multiply two-fifths times the product of one-third and two.

39. One-half is how much more than one-third?

40. By how much is one-fourth less than three-fourths?

41. Divide seven by the difference of five-thirds and one-third.

42. Find the product of one-fourth, two-fifths, and one-sixth.

43. Divide seven-eighths by the sum of two-thirds and one-ninth.

44. Is two-fifths greater than the sum of one-eighth and one-tenth?

45. Four-fifths is how much less than nine-tenths?

46. Multiply the sum of two-fifths and one-fifth times the quotient of one-fourth divided by four.

47. Find the product of one-sixth with the square of two-fifths.

JOURNAL EXERCISES

1. Write down your thoughts as you solve:

$$\frac{1}{4} + \left[\frac{7}{8} - \left(\frac{1}{3} + \frac{1}{2}\right)\right]^2$$

2. Why do you think that the authors of this book suggest you work math problems in a step-by-step manner, with an explanation for each step? Do you feel comfortable doing one tiny step at a time? Why, or why not?

3. Create your own complicated Order of Operations problem using fractions and solve it.

CHAPTER 3 REVIEW

SECTION	TOPIC OR TERM	EXAMPLE OR DEFINITION
3.1		
Basics	Numerator	The numerator is the number on the top in a fraction.
	Denominator	The denominator is the number on the bottom in a fraction.
	Proper Fraction	A proper fraction is one in which the numerator is less than the denominator.
	Improper Fraction	An improper fraction is one in which the numerator is greater than or equal to the denominator.
	Form of One	A form of one is a fraction in which the numerator is equal to the denominator.
	Reduce	Reducing is the process of writing an equivalent fraction with smaller numbers in the numerator and denominator.
3.2		
Raising Fractions to Higher Terms	Equivalent	Equivalent fractions are equal to one another. For example, are $\frac{3}{7}$ and $\frac{15}{35}$ equivalent? $\frac{3}{7} \stackrel{?}{=} \frac{15}{35}$ $\quad 3 \times 35 = 105$, $7 \times 15 = 105 \rightarrow$ equivalent
	Raising Fractions to Higher Terms	$\frac{7}{12} = \frac{?}{60}$ $60 \div 12 = 5, \quad 5 \times 7 = 35$ $? = 35 \quad$ so that $\frac{7}{12} = \frac{35}{60}$
3.3		
Multiplication	Multiply Fractions	$\frac{3}{7} \times \frac{4}{11} = \frac{3 \times 4}{7 \times 11} = \frac{12}{77}$
	Cross Reduce	Cross reducing is the process of reducing across two or more fractions that are being multiplied. For example, $\frac{3}{8} \cdot \frac{2}{5}$ can be cross reduced before multiplying: $\frac{3}{\cancel{8}_{4}} \cdot \frac{\cancel{2}^{1}}{5} = \frac{3 \cdot 1}{4 \cdot 5} \cdot \frac{3}{20}$

SECTION	TOPIC OR TERM	EXAMPLE OR DEFINITION
3.4		
Division	Reciprocals	Reciprocals are two fractions whose product equals 1.
	Divide Fractions	$\frac{4}{5} \div \frac{7}{8} = \frac{4}{5} \cdot \frac{8}{7} = \frac{4 \cdot 8}{5 \cdot 7} = \frac{32}{35}$
3.5		
Comparing Fractions	Common Denominator	A common denominator is a denominator common to 2 or more fractions.
	Compare Fractions	Is $\frac{3}{5} > \frac{2}{3}$? Write using common denominator of 15. $\frac{3}{5} = \frac{9}{15}$ and $\frac{2}{3} = \frac{10}{15}$ $\frac{9}{15}$ is not greater than $\frac{10}{15}$. Therefore, $\frac{3}{5}$ is not greater than $\frac{2}{3}$.
3.6		
Adding and Subtracting with Common Denominators	Adding Fractions with Common Denominators	$\frac{1}{5} + \frac{2}{5} = \frac{(1+2)}{5} = \frac{3}{5}$
	Subtracting Fractions with Common Denominators	$\frac{7}{8} - \frac{3}{8} = \frac{(7-3)}{8} = \frac{4}{8} = \frac{1}{2}$
3.7		
Adding and Subtracting with Unlike Denominators	Adding Fractions with Unlike Denominators	$\frac{1}{4} + \frac{2}{3} = \frac{3}{12} + \frac{8}{12} = \frac{(3+8)}{12} = \frac{11}{12}$
	Subtracting Fractions with Unlike Denominators	$\frac{1}{2} - \frac{1}{5} = \frac{5}{10} - \frac{2}{10} = \frac{(5-2)}{10} = \frac{3}{10}$

SECTION	TOPIC OR TERM	EXAMPLE OR DEFINITION
3.8 Exponents and Complex Fractions	Complex Fractions	A complex fraction is a fraction in which the numerator or the denominator or both are themselves fractions. Simplify $\frac{\frac{3}{4}}{\frac{1}{2}}$. $\frac{3}{4} \div \frac{1}{2}$ Numerator ÷ denominator. $\frac{3}{\cancel{4}_2} \cdot \frac{\cancel{2}^1}{1}$ Multiply; cross reduce. $\frac{3}{2}$ Result.
3.9 Order of Operations	Order of Operations	$\frac{1}{2} - \frac{1}{2} \times \frac{1}{4}$ $\frac{1}{2} - \frac{1}{8}$ Multiply first. $\frac{4}{8} - \frac{1}{8}$ common denominators $\frac{(4 - 1)}{8}$ Subtract. $\frac{3}{8}$ Result.

APPLICATION WORD PROBLEM

DRILLING FOR OIL

The cost of drilling a well to search for crude oil or natural gas can be very large. The chances of drilling a dry hole and losing all of the money invested is usually high. In order to reduce the risk to each investor, the ownership of a well is frequently split into several pieces. Then, each investor will lose only a portion of the cost should the well turn out to be dry.

The estimated dry hole cost of a well named the #1 Goodluck is $60,000. Four individuals agree to take the risk by investing in the #1 Goodluck. The four investors and the portion of the well owned by each are:

Individual	*Ownership*
Paul Jones	$\frac{1}{2}$
Sharon Feinstein	$\frac{1}{4}$
Jerry Walker	$\frac{1}{8}$
Tommy Flores	$\frac{1}{8}$

1. Do the four ownership interests add up to complete ownership of the well, or is another investor needed?

2. Which investor will lose the most money if the well is a dry hole?

3. Which investor will make the most money if the well is an excellent producer?

4. How much money will each of the four investors lose if the well is a dry hole costing exactly $56,000?

Assume the #1 Goodluck is drilled and turns out to be a producing oil well. In that event, the dry hole costs no longer apply since the well is not a dry hole, but a producer.

5. How much money must each investor spend if the total cost of a completed and producing well is $100,000?

CHAPTER 3 REVIEW PROBLEMS

Work each problem.

1. $\frac{1}{2} \cdot \frac{1}{3}$ **2.** $\frac{2}{7} + \frac{4}{7}$ **3.** $\frac{5}{8} - \frac{4}{8}$ **4.** $\frac{3}{4} \div \frac{1}{3}$

5. $\left(\frac{7}{8}\right)\left(\frac{1}{5}\right)$ **6.** $\frac{2}{3} \div \frac{1}{4}$ **7.** $\frac{5}{8} \cdot \frac{1}{6}$ **8.** $\left(\frac{1}{3}\right) + \left(\frac{3}{4}\right)$

9. $\frac{2}{5} - \frac{1}{8}$ **10.** $\frac{7}{8} \div \frac{4}{5}$ **11.** $\frac{5}{6} - \frac{1}{9}$ **12.** $\frac{1}{5} + \frac{4}{15}$

13. $\frac{9}{7} - \frac{1}{7}$ **14.** $\frac{3}{20} - \frac{1}{35}$ **15.** $\frac{80}{100} - \frac{8}{10}$ **16.** $\frac{1}{9} \times \frac{2}{49}$

17. $\left(\frac{2}{15}\right) \div \left(\frac{3}{8}\right)$ **18.** $\frac{3}{5} \div \frac{2}{3}$ **19.** $\frac{7}{4} \cdot \frac{1}{5}$ **20.** $\frac{3}{1000} - \frac{3}{10000}$

Reduce the following.

21. $\frac{6}{18}$ **22.** $\frac{15}{20}$ **23.** $\frac{49}{35}$ **24.** $\frac{60}{85}$ **25.** $\frac{20}{25}$

26. $\frac{125}{1000}$ **27.** $\frac{27}{36}$ **28.** $\frac{70}{95}$ **29.** $\frac{150}{630}$ **30.** $\frac{325}{410}$

Work each and reduce if possible.

31. $\frac{3}{4} \cdot \frac{8}{27}$ **32.** $\frac{5}{8} \div \frac{25}{4}$ **33.** $\frac{5}{7} + \frac{9}{7}$ **34.** $\frac{6}{5} - \frac{3}{50}$

35. $\frac{7}{8} \times 4$ **36.** $\frac{5}{6} + \frac{3}{8}$ **37.** $\frac{14}{9} \div \frac{1}{3}$

38. $\frac{25}{30} \div 5$ **39.** $\frac{29}{49} - \frac{19}{42}$ **40.** $\frac{1}{8} + \frac{5}{12} + \frac{1}{9}$

Arrange the following fractions from smallest to largest.

41. $\frac{2}{3}, \frac{5}{6}$ **42.** $\frac{3}{5}, \frac{7}{8}, \frac{3}{10}$ **43.** $\frac{1}{2}, \frac{1}{3}, \frac{1}{4}$ **44.** $\frac{4}{6}, \frac{5}{6}, \frac{3}{6}$

45. $\frac{2}{3}, \frac{3}{4}, \frac{5}{8}$

46. $\frac{3}{8}, \frac{1}{3}, \frac{2}{5}$

47. $\frac{7}{9}, \frac{6}{7}, \frac{0}{6}$

48. $\frac{6}{5}, \frac{7}{4}, \frac{11}{40}$

Work each using rules for Order of Operation.

49. $\frac{1}{2} + \frac{3}{4} \div \frac{2}{3}$

50. $\frac{7}{4} \div \frac{3}{4} \cdot \frac{1}{8}$

51. $\frac{3}{8} \div \left(\frac{1}{6} + \frac{2}{3}\right)$

52. $\frac{7}{9} \cdot \left(\frac{3}{4} - \frac{1}{8}\right)$

53. $\frac{6}{7} \div \left(\frac{3}{5} - \frac{3}{8}\right)$

54. $\left(\frac{4}{7} + \frac{1}{49}\right)\left(\frac{2}{3} \cdot \frac{3}{8}\right)$

55. $\left(\frac{2}{3}\right)^2 - \frac{1}{9}$

56. $\left(\frac{1}{8} \div \frac{8}{9}\right)$

57. $\frac{7}{8} - \left(1 - \frac{1}{3}\right)^2$

58. $\left(\frac{2}{3}\right)^2 - \left(\frac{1}{4}\right)^2$

Work each problem.

59. Square three-eights.

60. Find the quotient of two-thirds divided by one-third.

61. Add one-fifth and three-eighths.

62. Find the product of sixth-sevenths and four-fifths.

63. Simplify $\frac{\frac{4}{9}}{\frac{3}{5}}$.

64. Laurie Zimms donates one-sixth of her annual salary of $45,000 to charity. How much does she donate to charity?

65. Lee ran three-fifths of a mile in the same amount of time that Joan ran five-eighths of a mile. Who ran faster?

CHAPTER 3 TEST

Reduce each fraction.

1. $\frac{10}{24}$ **2.** $\frac{24}{30}$ **3.** $\frac{225}{250}$ **4.** $\frac{40}{380}$

Work each and reduce if possible.

5. $\frac{3}{8} \cdot \frac{4}{27}$ **6.** $\frac{7}{18} \div \frac{49}{3}$ **7.** $\frac{9}{7} - \frac{1}{14}$

8. $\frac{4}{5} - \frac{3}{5}$ **9.** $\frac{7}{9} + \frac{4}{18}$ **10.** $\frac{9}{5} \div \frac{3}{8}$

11. $\left(\frac{5}{8}\right)\left(\frac{64}{25}\right)$ **12.** $\frac{4}{5} + \frac{7}{20}$ **13.** $\frac{7}{4} - \frac{14}{8}$

14. $\frac{7}{4} \div \frac{2}{3} \cdot \frac{4}{5}$ **15.** $\frac{3}{5} + \frac{1}{3} \div \frac{9}{2}$ **16.** $\frac{4}{5} - \frac{1}{3} \cdot \frac{2}{5}$

17. $\frac{4}{7} \cdot \left(\frac{2}{3} - \frac{1}{8}\right) \cdot \frac{2}{7}$ **18.** $\left(\frac{4}{5}\right)^2 - \frac{1}{8}$ **19.** $\frac{2}{3} + \frac{1}{2} \cdot 7$

20. $\left(\frac{1}{3}\right)^2 \cdot \left(\frac{1}{4} - \frac{1}{4}\right)$ **21.** $25^1 \div \frac{1}{8}$ **22.** $\frac{3}{4} \div \left(\frac{1}{2}\right)^2$

Work each problem.

23. Find the sum of three-fifths and three-eighths.

24. Subtract one-eighth from three-fourths.

25. Divide two-thirds by fourteen.

26. Multiply four-ninths times one-sixth.

27. Simplify $\frac{\frac{3}{8}}{\frac{1}{2}}$.

28. John and Paul each ate one-eighth of a pizza. How much is left?

29. An item originally marked $140 is on sale for one-fourth off. What is the sales price of the item?

30. Five-eighths of a group of eight hundred people have brown eyes. How many *do not* have brown eyes?

Mixed Numbers, Ratios, and Proportions

4

APPLICATION *A bank wishes to fertilize their 15,000 square foot lawn. The lawn requires $3\frac{1}{2}$ pounds of fertilizer for every 1,000 square feet. How much fertilizer do they need?*

SOLUTION *There are 15 one thousands in 15,000 square feet. Therefore, the answer is found by multiplying the mixed number $3\frac{1}{2}$ times 15.*

$$3\frac{1}{2} \times 15$$

$$\frac{7}{2} \times \frac{15}{1} \qquad \text{Change both to fractions.}$$

$$\frac{105}{2} \qquad \text{Multiply fractions.}$$

$$52\frac{1}{2} \qquad \text{Pounds of fertilizer are needed.}$$

4.1 MIXED NUMBERS

OBJECTIVES

1. Identify mixed numbers.
2. Write a mixed number as a fraction.
3. Write an improper fraction as a mixed number.

In the last chapter, we worked with fractions. In this chapter we will be looking at some of the ways that fractions are used.

1 Mixed Numbers

A **mixed number** is the sum of a whole number and a proper fraction. It is called mixed because two different types of numbers are being added or "mixed" together. For example:

whole number ← 3, fraction → $\frac{1}{2}$

$$3 + \frac{1}{2} \qquad \text{three plus one-half.}$$

$$22 + \frac{3}{4} \qquad \text{twenty-two plus three-fourths}$$

In practice the addition sign is dropped when writing mixed numbers. The word "and" is read between the whole number part and the fraction part of each mixed number.

Drop the addition sign.

$$3 + \frac{1}{4} = 3\frac{1}{4} \qquad \text{Read as three AND one-fourth.}$$

→ mixed number

1. Write the sum of five plus two-thirds as a mixed number.

2. Write $85\frac{3}{10}$ using words.

3. Write $21\frac{1}{2}$ as the sum of a whole number and a proper fraction.

4. Plot seven and one-half on a number line.

5. Plot $3\frac{2}{3}$ on a number line.

The figure below shows the idea of the mixed number $3\frac{1}{4}$. Three complete squares are shaded in addition to one-fourth of the fourth square.

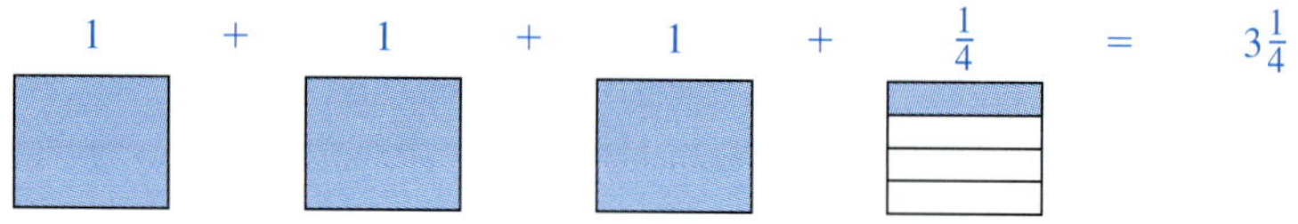

EXAMPLE 1

Read $4\frac{9}{10}$ and write it as the sum of an integer and a fraction.

SOLUTION

Four and nine-tenths $= 4 + \frac{9}{10}$ ■

◇ *Do Practice Exercises 1 through 3.*

The mixed number $2\frac{1}{3}$ equals $2 + \frac{1}{3}$ which is larger than 2 but smaller than 3. The fraction $\frac{1}{3}$ is one part out of three. Thus, $2\frac{1}{3}$ is two plus one part out of three. On a number line, it is represented as follows:

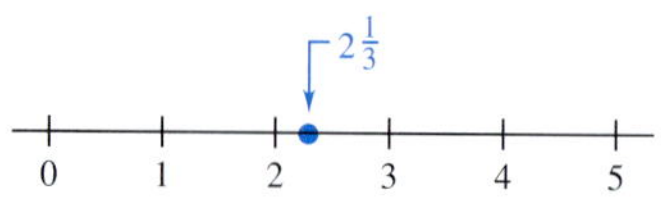

EXAMPLE 2

Plot five and three-fourths on a number line.

SOLUTION

The mixed number $5\frac{3}{4} = 5 + \frac{3}{4}$ is larger than 5 but smaller than 6. It is five plus three-fourths of the way to six.

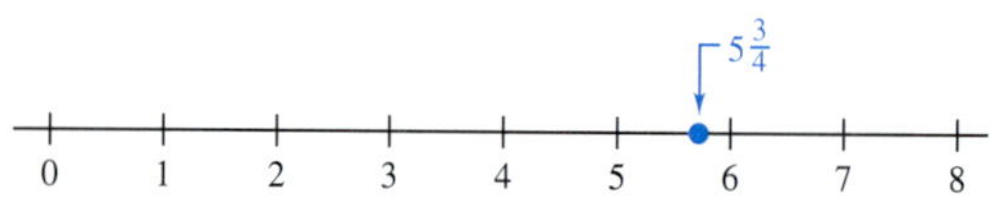

■

◇ *Do Practice Exercises 4 and 5.*

Recall from Chapter 3 that every proper fraction is less than one and every improper fraction is one or greater. Every mixed number is the sum of a whole number and a fraction, so that every mixed number must be greater than one. Here are some examples of proper fractions, improper fractions, and mixed numbers.

ANSWERS

1. $5\frac{2}{3}$
2. eighty-five and three tenths
3. $21 + \frac{1}{2}$
4.

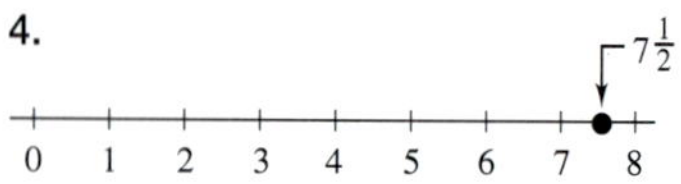

5.

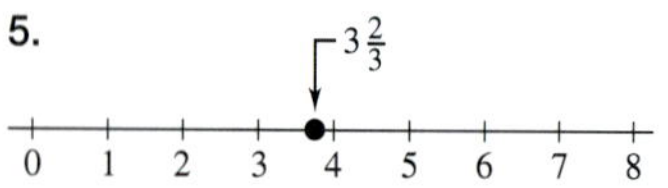

less than one

Proper fractions: $\frac{1}{2}$, $\frac{3}{5}$, $\frac{2}{7}$, $\frac{19}{42}$

Improper fractions: $\frac{3}{3}$, $\frac{17}{2}$, $\frac{91}{5}$, $\frac{101}{101}$ greater than or equal to one

Mixed numbers: $4\frac{1}{2}$, $3\frac{2}{3}$, $1\frac{1}{2}$, $50\frac{1}{8}$

Actually, each mixed number can be written as an improper fraction and each improper fraction can be written as a mixed number. In the next two parts of this section, we will learn how to change a number from one form to the other.

2 From a Mixed Number to an Improper Fraction

The next example uses the properties of a mixed number and the rules of adding fractions to change a mixed number into a fraction. This example should help you understand the rule that follows.

EXAMPLE 3

Change $4\frac{2}{3}$ into an improper fraction.

SOLUTION

$4\frac{2}{3}$

$4 + \frac{2}{3}$ Write as a sum.

$\frac{4}{1} + \frac{2}{3}$ Write 4 as a fraction.

$\frac{12}{3} + \frac{2}{3}$ Write using common denominators.

$\frac{12 + 2}{3}$ Add.

$\frac{14}{3}$ Result. ■

◇ *Do Practice Exercises 6 and 7.*

Notice in the last example that the result was found by:

1. Multiplying the whole number times the denominator of the fraction (to write using common denominators) and,
2. Adding this value to the numerator of the fraction (to add fractions with common denominators).

6. Change $5\frac{3}{4}$ to an improper fraction.

7. Write $4\frac{1}{2}$ as an improper fraction.

ANSWERS

6. $\frac{23}{4}$

7. $\frac{9}{2}$

8. Change two and four-sevenths to an improper fraction.

9. Change $3\frac{2}{3}$ to an improper fraction.

These same steps are included in the rule that follows.

> **Rule: To Change a Mixed Number to an Improper Fraction**
>
> **Step 1** Multiply the whole number times the denominator.
>
> **Step 2** Add the product to the numerator of the fraction. This value is the new numerator.
>
> **Step 3** Keep the same denominator.

EXAMPLE 4

Change seven and three-fifths to an improper fraction.

SOLUTION

$$7\frac{3}{5}$$

3 → numerator
5 → denominator
7 → whole number

Step 1	$7 \times 5 = 35$	Multiply the whole number times the denominator.
Step 2	$35 + 3 = 38$	Add the product to the numerator.
Step 3	5	Keep the same denominator.

Thus $7\frac{3}{5} = \frac{38}{5}$. ■

EXAMPLE 5

Change ten and three-fourths to an improper fraction.

SOLUTION

$$10\frac{3}{4}$$

$$\frac{10 \cdot 4 + 3}{4}$$

Multiply the whole number times the denominator.
Add the product to the numerator.
Keep the same denominator.

$$\frac{40 + 3}{4}$$

$$\frac{43}{4}$$ Result. ■

EXAMPLE 6

Change six and four-fifths to an improper fraction.

SOLUTION

$$6\frac{4}{5} = \frac{6 \cdot 5 + 4}{5} = \frac{30 + 4}{5} = \frac{34}{5}$$ ■

◇ *Do Practice Exercises 8 and 9.*

ANSWERS
8. $\frac{18}{7}$
9. $\frac{11}{3}$

3 From an Improper Fraction to a Mixed Number

The next example uses the rules of fractions to change an improper fraction to a mixed number. This example may help you understand the steps of the rule that follows.

EXAMPLE 7

Change seven-thirds to a mixed number.

SOLUTION

$$\frac{7}{3} = \frac{6+1}{3} = \frac{6}{3} + \frac{1}{3} = 2 + \frac{1}{3} = 2\frac{1}{3} \quad ■$$

◇ *Do Practice Exercises 10 and 11.*

The following rule shows the steps required to change an improper fraction to a mixed number.

Rule: To Change an Improper Fraction to a Mixed Number

Step 1 Divide the denominator into the numerator.

Step 2 The quotient is the whole number.
The remainder is the numerator.
The denominator remains the same.

EXAMPLE 8

Change $\frac{23}{4}$ into a mixed number.

SOLUTION

Step 1 Divide the denominator into the numerator.

$$\begin{array}{r} 5 \\ 4\overline{)23} \\ \underline{20} \\ 3 \end{array}$$

Step 2 Identify the various parts of the mixed number.

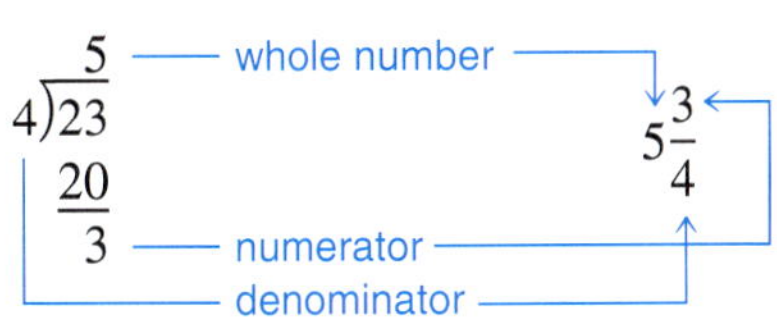

Thus $\frac{23}{4} = 5\frac{3}{4}$. ■

10. Change five-halves into a mixed number.

11. Change seventeen-thirds into a mixed number.

ANSWERS

10. $2\frac{1}{2}$

11. $5\frac{2}{3}$

12. Change nineteen-fourths to a mixed number.

13. Change $\frac{31}{5}$ to a mixed number.

ANSWERS
12. $4\frac{3}{4}$
13. $6\frac{1}{5}$

EXAMPLE 9

Change fifteen-sevenths to a mixed number.

SOLUTION

$$\begin{array}{r} 2 \\ 7\overline{)15} \\ \underline{14} \\ 1 \end{array}$$

2 → whole number
denominator → 7
1 → numerator

$$\frac{15}{7} = 2\frac{1}{7}$$

◇ *Do Practice Exercises 12 and 13.*

State whether each fraction is proper or improper. (Review)

1. $\frac{2}{3}$ **2.** $\frac{7}{4}$ **3.** $\frac{5}{8}$

4. $\frac{9}{9}$ **5.** $\frac{250}{12}$ **6.** $\frac{23}{25}$

Write each as the sum of a whole number and a proper fraction.

7. $5\frac{7}{8}$ **8.** $4\frac{2}{3}$ **9.** $41\frac{2}{3}$ **10.** $1{,}000\frac{12}{13}$

Write each as a mixed number.

11. $5 + \frac{1}{2}$ **12.** $2 + \frac{3}{4}$ **13.** $1 + \frac{7}{8}$ **14.** $4 + \frac{4}{5}$

Find the whole numbers on either side of the mixed number given.

Example: $4\frac{5}{8}$ Answer: $4\frac{5}{8}$ is between 4 and 5.

15. $3\frac{3}{4}$ **16.** $2\frac{5}{8}$ **17.** $18\frac{1}{3}$ **18.** $90\frac{1}{5}$

Draw a number line and plot each number as a point.

19. $3\frac{1}{4}$ **20.** $7\frac{2}{3}$

21. $4\frac{1}{2}$ **22.** $2\frac{3}{4}$

Find the mixed number that represents the number of squares shaded.

23. 

24. 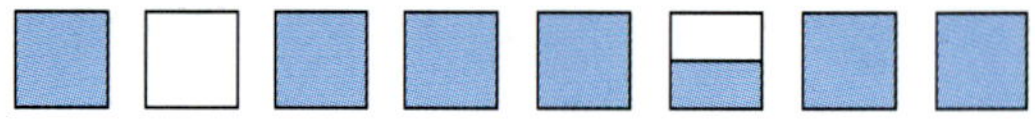

Change from a mixed number to an improper fraction.

25. $3\frac{2}{5}$

26. $1\frac{1}{8}$

27. $4\frac{3}{7}$

28. $5\frac{1}{2}$

29. $9\frac{2}{7}$

30. $10\frac{1}{4}$

31. $8\frac{7}{8}$

32. $21\frac{1}{5}$

33. $12\frac{7}{10}$

34. $9\frac{4}{5}$

35. $4\frac{1}{3}$

36. $5\frac{2}{7}$

37. $6\frac{5}{8}$

38. $23\frac{3}{4}$

39. $9\frac{2}{5}$

40. $11\frac{1}{12}$

41. $12\frac{5}{12}$

42. $15\frac{5}{8}$

43. $25\frac{3}{4}$

44. $1{,}020\frac{17}{20}$

Change from an improper fraction to a mixed number.

45. $\frac{3}{2}$

46. $\frac{8}{5}$

47. $\frac{10}{9}$

48. $\frac{12}{5}$

49. $\frac{15}{11}$

50. $\frac{37}{7}$

51. $\frac{41}{8}$

52. $\frac{17}{13}$

53. $\frac{43}{20}$

54. $\frac{46}{9}$

55. $\frac{53}{2}$

56. $\frac{15}{2}$

57. $\frac{8}{1}$

58. $\frac{73}{3}$

59. $\frac{1{,}012}{9}$

60. $\frac{121}{5}$

61. $\frac{32}{7}$

62. $\frac{49}{9}$

63. $\frac{82}{19}$

64. $\frac{45}{44}$

Complete the table below. The numbers in each column are equal to one another. The first column has been done for you.

Whole number + Fraction	$3 + \frac{4}{5}$	**65.**	$10 + \frac{1}{2}$	**69.**	**71.**
Mixed number	$3\frac{4}{5}$	$2\frac{7}{8}$	**67.**	**70.**	$7\frac{1}{8}$
Improper fraction	$\frac{19}{5}$	**66.**	**68.**	$\frac{11}{4}$	**72.**

Work each problem.

73. Write $1{,}085{,}025\frac{2}{3}$ using words.

74. Write $3{,}000{,}001\frac{1}{5}$ using words.

75. Write eight-thirds as a mixed number.

76. Draw and shade three and two-fifths squares.

77. Draw and shade three and one-half circles.

JOURNAL EXERCISES

1. Explain the differences between proper fractions, improper fractions, and mixed numbers.

2. In your own words, explain the process of changing a mixed number to an improper fraction and the process of changing an improper fraction to a mixed number.

3. A mixed number is made up of a whole number along with the numerator and denominator of a proper fraction. Which of these three numbers can never be zero? Explain.

1 Simplify Mixed Numbers

Sometimes the fraction portion of a mixed number can be reduced. In this case, it is best to go ahead and reduce the fraction.

EXAMPLE 1

Simplify $7\frac{4}{6}$.

SOLUTION

$$7\frac{4}{6} = 7 + \frac{4}{6} = 7 + \frac{2}{3} = 7\frac{2}{3}$$

Reduce. ■

◇ *Do Practice Exercises 1 and 2.*

After doing a mathematical operation, the fraction portion of a mixed number may be an improper fraction. If this occurs, change the fraction so that it is a proper fraction and adjust the whole number accordingly. For example:

$$3\frac{4}{3} = 3 + \frac{4}{3} = 3 + 1\frac{1}{3} = 3 + 1 + \frac{1}{3} = 4\frac{1}{3}$$

EXAMPLE 2

Simplify $4\frac{5}{3}$.

SOLUTION

The number $4\frac{5}{3}$ has not been written in correct form since the fraction $\frac{5}{3}$ is greater than one.

$4\frac{5}{3}$	
$4 + \frac{5}{3}$	Write the mixed number as a sum.
$4 + 1\frac{2}{3}$	improper fraction to a mixed number
$4 + 1 + \frac{2}{3}$	Write the mixed number as a sum.
$5 + \frac{2}{3}$	Add whole numbers.
$5\frac{2}{3}$	Result. ■

◇ *Do Practice Exercises 3 and 4.*

OBJECTIVES

1 Simplify mixed numbers.

2 Compare the size of mixed numbers.

1. Simplify $14\frac{6}{8}$.

2. Simplify $3{,}241\frac{21}{30}$.

3. Simplify $7\frac{6}{5}$.

4. Simplify $14\frac{11}{3}$.

ANSWERS

1. $14\frac{3}{4}$
2. $3{,}241\frac{7}{10}$
3. $8\frac{1}{5}$
4. $17\frac{2}{3}$

5. Is $8\frac{4}{5}$ greater than $9\frac{2}{3}$?

6. Which is greater: $5\frac{99}{100}$ or $4\frac{3}{8}$?

ANSWERS

5. no. $8\frac{4}{5} < 9\frac{2}{3}$

6. $5\frac{99}{100} > 4\frac{3}{8}$

2 Compare the Size of Mixed Numbers

If two mixed numbers have different whole number parts, then the mixed number with the larger whole number part is greater.

EXAMPLE 3

Which is greater: $3\frac{4}{5}$ or $5\frac{1}{4}$?

SOLUTION

Both numbers are placed on a number line to help visualize the process. The mixed number $3\frac{4}{5}$ is between 3 and 4 whereas the mixed number $5\frac{1}{4}$ is between 5 and 6.

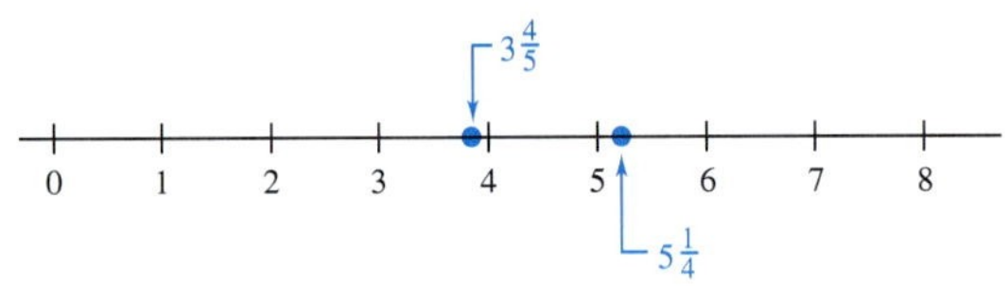

$5\frac{1}{4}$ is greater than $3\frac{4}{5}$ or $5\frac{1}{4} > 3\frac{4}{5}$ ■

◇ *Do Practice Exercises 5 and 6.*

If two mixed numbers have the same whole number part, the greater of the two is the one with the larger fraction.

EXAMPLE 4

Which is greater: $5\frac{2}{3}$ or $5\frac{3}{4}$?

SOLUTION

Both $5\frac{2}{3}$ and $5\frac{3}{4}$ are between 5 and 6. The greater of the two is found by comparing the fractions $\frac{2}{3}$ and $\frac{3}{4}$. These two fractions are compared by writing each as an equivalent fraction with lowest common denominators. (Review Section 3.5.)

$$\frac{2}{3} = \frac{8}{12} \qquad \frac{3}{4} = \frac{9}{12}$$

↑ common denominators

The numerator 9 is greater than the numerator 8, therefore:

$$\frac{9}{12} \text{ is greater than } \frac{8}{12} \quad \text{or} \quad \frac{3}{4} \text{ is greater than } \frac{2}{3}$$

so that $5\frac{3}{4} > 5\frac{2}{3}$. ■

◇ *Do Practice Exercises 7 and 8.*

The size of a mixed number can be compared to the size of an improper fraction by changing both numbers to fractions and comparing the fractions. Alternatively, the size of a mixed number and an improper fraction can be compared by changing both numbers to mixed numbers and comparing the mixed numbers.

EXAMPLE 5

Which is greater: $\frac{23}{6}$ or $3\frac{7}{8}$?

SOLUTION

This problem can be solved in either of two ways.

Compare as Fractions	*Compare as Mixed Numbers*
$\frac{23}{6} \quad 3\frac{7}{8}$	$\frac{23}{6} \quad 3\frac{7}{8}$
$\frac{23}{6} \quad \frac{31}{8}$	$3\frac{5}{6} \quad 3\frac{7}{8}$
$\frac{92}{24} \quad \frac{93}{24}$	$3\frac{20}{24} \quad 3\frac{21}{24}$
$\frac{93}{24}$ is greater than $\frac{92}{24}$	$3\frac{21}{24}$ is greater than $3\frac{20}{24}$
$\frac{31}{8}$ is greater than $\frac{23}{6}$	$3\frac{7}{8}$ is greater than $\frac{23}{6}$
$3\frac{7}{8}$ is greater than $\frac{23}{6}$	

■

7. Which is greater: $8\frac{1}{2}$ or $8\frac{3}{5}$?

8. Which is smallest:

$12\frac{2}{3}$, $12\frac{3}{4}$, or $12\frac{4}{7}$?

ANSWERS

7. $8\frac{3}{5}$

8. $12\frac{4}{7}$

9. Order 18, $15\frac{1}{2}$, and $15\frac{2}{3}$ from smallest to largest.

10. Arrange $8\frac{7}{8}$, $8\frac{4}{5}$, and $8\frac{9}{10}$ from smallest to largest.

ANSWERS

9. $15\frac{1}{2}$, $15\frac{2}{3}$, 18

10. $8\frac{4}{5}$, $8\frac{7}{8}$, $8\frac{9}{10}$

EXAMPLE 6

Write $\frac{12}{5}$, $2\frac{1}{3}$, and $\frac{9}{10}$ from smallest to largest.

SOLUTION

$$\frac{12}{5},\ 2\frac{1}{3},\ \frac{9}{10}$$

$$\frac{12}{5},\ \frac{7}{3},\ \frac{9}{10}$$ Write each as a fraction.

$$\frac{72}{30},\ \frac{70}{30},\ \frac{27}{30}$$ Write using a common denominator.

From smallest to largest: $\frac{27}{30}$, $\frac{70}{30}$, $\frac{72}{30}$

From smallest to largest: $\frac{9}{10}$, $\frac{7}{3}$, $\frac{12}{5}$

From smallest to largest: $\frac{9}{10}$, $2\frac{1}{3}$, $\frac{12}{5}$ ■

◇ *Do Practice Exercises 9 and 10.*

name date hour

Draw a number line and plot each number. (Review)

1. $5, 3\frac{1}{2}$

2. $4\frac{2}{3}, 10\frac{1}{2}$

3. $1\frac{1}{2}, 4\frac{1}{3}$

4. $1\frac{1}{4}, 1\frac{3}{4}$

5. $6\frac{1}{3}, 6\frac{2}{3}$

6. $5, 10\frac{1}{2}$

7. $3, 2\frac{2}{5}$

8. $5\frac{1}{3}, 5\frac{3}{5}$

Simplify each number.

9. $4\frac{9}{12}$

10. $5\frac{12}{7}$

11. $3\frac{6}{5}$

12. $23\frac{40}{12}$

13. $2\frac{20}{12}$

14. $5\frac{3}{3}$

15. $6\frac{40}{3}$

16. $7\frac{5}{2}$

17. $9\frac{13}{5}$

18. $2\frac{7}{7}$

List from smallest to largest.

19. $4\frac{1}{2}$, $5\frac{2}{3}$

20. 12, $7\frac{4}{5}$

21. 8, $4\frac{1}{2}$

22. $\frac{4}{5}$, $\frac{2}{3}$

23. $6\frac{1}{2}$, $6\frac{3}{5}$

24. $7\frac{1}{2}$, 8

25. $2\frac{4}{5}$, $2\frac{1}{2}$

26 $9\frac{3}{4}$, $\frac{37}{3}$

27. $3\frac{4}{5}$, $2\frac{1}{5}$

28. $3\frac{3}{8}$, $3\frac{6}{25}$

29. $5\frac{1}{2}$, $2\frac{2}{3}$

30. $\frac{13}{20}$, $\frac{20}{13}$

31. $1\frac{1}{2}$, $\frac{10}{7}$, $1\frac{3}{8}$

32. $2\frac{4}{5}$, 2, $\frac{17}{7}$

List from largest to smallest.

33. $\frac{7}{3}$, $2\frac{1}{2}$

34. $3\frac{1}{2}$, $7\frac{3}{5}$

35. $6\frac{2}{3}$, $6\frac{3}{5}$

36. $4\frac{5}{8}$, $3\frac{3}{4}$

37. $\frac{25}{6}$, 5

38. $7\frac{1}{2}$, $7\frac{2}{3}$

39. $2\frac{3}{8}$, 2

40. $\frac{34}{5}$, $7\frac{1}{2}$

41. $6\frac{2}{9}$, $7\frac{2}{5}$

42. $4\frac{2}{3}$, $5\frac{1}{4}$, $5\frac{2}{5}$

43. 6, $7\frac{1}{2}$, $6\frac{3}{4}$

44. $5\frac{1}{2}$, $5\frac{9}{10}$

45. $2\frac{1}{4}$, $\frac{18}{5}$, $7\frac{1}{2}$

46. $8\frac{1}{2}$, 8, $8\frac{3}{4}$

47. $4\frac{7}{8}$, $4\frac{9}{10}$, $4\frac{1}{3}$

Work each problem.

48. Which is larger, four and one-half or four and two-fifths?

49. Which is smaller, seventeen and seven-eighths or seven?

50. A blue car makes a lap every one and one-fourth minutes. A red car makes the same lap every one and two-fifths minutes. Which car is faster?

51. A recipe for four people requires:

$2\frac{1}{4}$ cups of sugar $\quad$ $2\frac{1}{2}$ cups of flour

2 eggs $\quad$ 1 teaspoon of salt

Is more sugar or flour needed?

52. Two groups of students make the following grades:

	Group 1	*Group 2*
Student 1	85	83
Student 2	75	70
Student 3	81	92

Write the average grade for each group as a mixed number. Which group has the higher average?

53. Four different stocks were selling for $42\frac{1}{2}$, $38\frac{3}{4}$, $25\frac{1}{2}$, and $40\frac{2}{3}$. List the stocks from least expensive to most expensive.

JOURNAL EXERCISES

1. Explain the process of comparing mixed numbers to fractions and to whole numbers.

2. Is there any mixed number that cannot be written as an improper fraction? Is there any improper fraction that cannot be written as a mixed number? Why do both mixed numbers and improper fractions exist?

1 Add, Subtract, Multiply, and Divide Mixed Numbers

One way to add, subtract, multiply, and divide mixed numbers is to:

1. First, change each mixed number to an improper fraction.
2. Then, work the problem using the rules of fractions.
3. Finally, it is best to change the answer back to a mixed number if possible.

It is not necessary to learn any new rules to solve mixed numbers in this fashion since you already know how to work with fractions. An alternative way to add and subtract mixed numbers will be shown at the end of this section.

EXAMPLE 1

Multiply $3\frac{1}{2} \cdot 4\frac{3}{5}$.

SOLUTION

$3\frac{1}{2} \cdot 4\frac{3}{5}$

$\frac{7}{2} \cdot \frac{23}{5}$ Change each to a fraction.

$\frac{161}{10}$ Multiply.

$16\frac{1}{10}$ Result. ■

EXAMPLE 2

Add $4\frac{1}{3} + 5\frac{3}{4}$.

SOLUTION

$4\frac{1}{3} + 5\frac{3}{4}$

$\frac{13}{3} + \frac{23}{4}$ Change each to a fraction.

$\frac{52}{12} + \frac{69}{12}$ Write using a common denominator.

$\frac{121}{12}$ Add numerators.

$10\frac{1}{12}$ Result. ■

◇ *Do Practice Exercise 1.*

OBJECTIVES

1 Add, subtract, multiply, and divide mixed numbers.

2 Apply the order of operations to whole numbers, fractions, and mixed numbers.

3 Use a second method for adding and subtracting mixed numbers.

1. Work each:

(a) $4\frac{1}{2} + 2\frac{3}{5}$

(b) $5\frac{1}{4} \cdot 2\frac{1}{3}$

(c) $4\frac{1}{5} \div 6\frac{2}{3}$

(d) $12\frac{4}{5} - 3\frac{1}{4}$

ANSWERS

1. (a) $7\frac{1}{10}$ (b) $12\frac{1}{4}$ (c) $\frac{63}{100}$ (d) $9\frac{11}{20}$

2 Order of Operations

The same basic process can be used to simplify an expression with more than two mixed numbers: first change each mixed number to a fraction, then solve using the rules of fractions. The Order of Operations is repeated for your convenience.

Rule: To Evaluate an Expression

Step 1 Evaluate the contents of the parentheses and above and below the fraction bar.

Step 2 Evaluate exponents.

Step 3 Evaluate multiplication and division in the order they occur when moving from left to right across the expression.

Step 4 Evaluate addition and subtraction in the order they occur when moving from left to right across the expression.

EXAMPLE 3

Simplify $4\frac{1}{2} \cdot 2\frac{1}{5} + \frac{3}{5}$.

SOLUTION

$$4\frac{1}{2} \cdot 2\frac{1}{5} + \frac{3}{5}$$

$$\frac{9}{2} \cdot \frac{11}{5} + \frac{3}{5}$$ Write each as a fraction.

$$\frac{99}{10} + \frac{3}{5}$$ Multiplication before addition.

$$\frac{99}{10} + \frac{6}{10}$$ Write using a common denominator.

$$\frac{99 + 6}{10}$$ Add.

$$\frac{105}{10}$$

$$10\frac{5}{10}$$ Write as a mixed number.

$$10\frac{1}{2}$$ Result $\left(\frac{5}{10} \text{ reduces to } \frac{1}{2}\right)$. ■

EXAMPLE 4

Evaluate $8\frac{1}{2} \div 5 - \frac{5}{8}$.

SOLUTION

$8\frac{1}{2} \div 5 - \frac{5}{8}$

$\frac{17}{2} \div \frac{5}{1} - \frac{5}{8}$ Write each as a fraction.

$\frac{17}{2} \cdot \frac{1}{5} - \frac{5}{8}$ Invert second fraction, ÷ to ·.

$\frac{17}{10} - \frac{5}{8}$ Multiply.

$\frac{68}{40} - \frac{25}{40}$ Write using a common denominator.

$\frac{43}{40}$ Subtract.

$1\frac{3}{40}$ Result. ■

◇ *Do Practice Exercises 2 and 3.*

Exponents are applied to mixed numbers in the same way they are applied to whole numbers and fractions. An example is given of the exponent 2 applied to a whole number, to a fraction, and to a mixed number.

Whole numbers $4^2 = 4 \cdot 4 = 16$

Fractions $\left(\frac{1}{3}\right)^2 = \frac{1}{3} \cdot \frac{1}{3} = \frac{1 \cdot 1}{3 \cdot 3} = \frac{1}{9}$

Mixed numbers $\left[3\frac{1}{5}\right]^2 = \left[3\frac{1}{5}\right] \cdot \left[3\frac{1}{5}\right] = \frac{16}{5} \cdot \frac{16}{5} = \frac{256}{25} = 10\frac{6}{25}$

EXAMPLE 5

Evaluate $\left(3\frac{3}{4} + 1\frac{1}{2}\right)^2 \div 2$.

2. Evaluate $4\frac{1}{5} + \frac{3}{5} - \frac{1}{8}$.

3. Evaluate $\left(2\frac{1}{2} + 4\right) \div \frac{2}{3}$.

ANSWERS

2. $4\frac{27}{40}$

3. $9\frac{3}{4}$

4. Evaluate $\left(4\frac{1}{2}\right)^2 + 2^2$.

5. Evaluate $8 \div \left[3\frac{1}{5} - \left(\frac{1}{2}\right)^2\right]$.

ANSWERS

4. $24\frac{1}{4}$

5. $2\frac{42}{59}$

SOLUTION

$\left(3\frac{3}{4} + 1\frac{1}{2}\right)^2 \div 2$

$\left(\frac{15}{4} + \frac{3}{2}\right)^2 \div \frac{2}{1}$ Write each as a fraction.

$\left(\frac{15}{4} + \frac{6}{4}\right)^2 \div \frac{2}{1}$ Write using a common denominator.

$\left(\frac{21}{4}\right)^2 \div \frac{2}{1}$ Add inside parentheses.

$\frac{441}{16} \div \frac{2}{1}$ Exponent: $\left[\frac{21}{4}\right]^2 = \frac{21}{4} \cdot \frac{21}{4} = \frac{441}{16}$.

$\frac{441}{16} \cdot \frac{1}{2}$ Invert second fraction, ÷ to ·.

$\frac{441}{32}$

$13\frac{25}{32}$ Result. ■

◇ *Do Practice Exercises 4 and 5.*

3 A Different Way to Add and Subtract Mixed Numbers

Each mixed number can be written as a sum of a whole number plus a fraction. Therefore, the sum of two mixed numbers is actually the sum of two whole numbers plus the sum of two fractions. This idea can be used in adding or subtracting two mixed numbers. Study the next example.

EXAMPLE 6

Add $4\frac{1}{2} + 2\frac{2}{5}$.

SOLUTION

$4\frac{1}{2} + 2\frac{2}{5}$

$4 + \frac{1}{2} + 2 + \frac{2}{5}$ Write each mixed number as a sum.

$4 + 2 + \frac{1}{2} + \frac{2}{5}$ Commutative property of addition

$6 + \frac{5}{10} + \frac{4}{10}$ Write fractions using common denominators.

$6 + \frac{5 + 4}{10}$ Add fractions.

$6\frac{9}{10}$ Result. ■

Sometimes, the addition of two mixed numbers is written with one mixed number above the other. The steps used to solve the problem are the same. The fractions are written using a common denominator, the fractions are added, and finally the whole numbers are added.

$$\begin{array}{rcr} 4\frac{1}{2} & = & 4\frac{5}{10} \\ +2\frac{2}{5} & = & +2\frac{4}{10} \\ \hline & & 6\frac{9}{10} \end{array}$$

common denominators

Subtraction can be done in a similar fashion as shown in the next example.

EXAMPLE 7

Subtract $14\frac{3}{4} - 5\frac{2}{3}$.

SOLUTION

First, rewrite the fractions using the common denominator 12, then subtract the fractions. Finally, subtract the whole numbers.

$$\begin{array}{rcr} 14\frac{3}{4} & = & 14\frac{9}{12} \\ -\ 5\frac{2}{3} & = & -\ 5\frac{8}{12} \\ \hline & & 9\frac{1}{12} \end{array}$$ ■

◇ *Do Practice Exercises 6 and 7.*

When subtracting the whole numbers 100 − 15, it is necessary to borrow:

$$\begin{array}{r} {}^{9\,1} \\ \not{1}\not{0}0 \\ 15 \\ \hline 85 \end{array}$$

Similarly, it may be necessary to borrow before subtracting mixed numbers. Look carefully at the next example.

EXAMPLE 8

Subtract $7\frac{1}{4} - 2\frac{3}{4}$.

SOLUTION

$$\begin{array}{r} 7\frac{1}{4} \\ -2\frac{3}{4} \\ \hline \end{array}$$

6. Add $3\frac{3}{5} + 2\frac{1}{8}$.

7. Subtract $4\frac{2}{3} - 3\frac{1}{4}$.

ANSWERS

6. $5\frac{29}{40}$

7. $1\frac{5}{12}$

8. Subtract $4\frac{1}{3} - 3\frac{3}{4}$.

9. Subtract $5\frac{1}{4} - 2\frac{2}{3}$.

The fraction $\frac{3}{4}$ cannot be subtracted from $\frac{1}{4}$ since $\frac{1}{4}$ is TOO SMALL. Therefore, it is necessary to borrow one from the 7 making it a 6. Use the one borrowed to increase the size of the fraction:

$$7\frac{1}{4} = 7 + \frac{1}{4} = 6 + 1 + \frac{1}{4} = 6 + 1\frac{1}{4} = 6\frac{5}{4}$$

Therefore:

$$\begin{array}{rcr} 7\frac{1}{4} & = & 6\frac{5}{4} \\ \underline{-2\frac{3}{4}} & = & \underline{-2\frac{3}{4}} \\ & & 4\frac{2}{4} = 4\frac{1}{2} \end{array}$$ ■

◇ *Do Practice Exercises 8 and 9.*

ANSWERS

8. $\frac{7}{12}$

9. $2\frac{7}{12}$

name date hour

Evaluate each.

1. $2\frac{1}{2} \cdot 3\frac{3}{4}$

2. $4\frac{1}{4} + 1\frac{2}{3}$

3. $3\frac{3}{4} - 1\frac{1}{2}$

4. $2\frac{2}{3} \div 1\frac{4}{5}$

5. $5\frac{3}{8} - 3\frac{1}{2}$

6. $3\frac{1}{2} \div 3\frac{5}{8}$

7. $\left[5\frac{3}{4}\right]\left[1\frac{2}{5}\right]$

8. $5\frac{2}{3} + 2\frac{1}{2}$

9. $3\frac{1}{4} \div 1\frac{7}{8}$

10. $5\frac{1}{4} + 3\frac{1}{2}$

11. $6\frac{2}{5} - 1\frac{3}{4}$

12. $7\frac{1}{2} \cdot 4\frac{1}{3}$

13. $5\frac{1}{3} \cdot 6$

14. $7\frac{1}{2} \div 4\frac{2}{3}$

15. $5\frac{1}{3} + 3\frac{3}{4}$

16. $2\frac{1}{2} - 1\frac{5}{8}$

17. $8\frac{2}{3} \div 3\frac{3}{4}$

18. $4\frac{1}{2} + 5\frac{1}{4}$

19. $7\frac{2}{3} - 6\frac{3}{5}$

20. $7\frac{1}{2} \cdot 7\frac{1}{2}$

21. $4\frac{2}{3} \cdot 1\frac{4}{5}$

22. $5\frac{1}{4} - 3\frac{1}{2}$

23. $4\frac{7}{8} \div 1\frac{3}{5}$

24. $9\frac{1}{2} + 9\frac{1}{2}$

25. $6\frac{2}{3} + 3\frac{1}{4}$

26. $5\frac{1}{2} \cdot 2\frac{1}{2}$

27. $8\frac{1}{2} \div 4\frac{1}{4}$

28. $3\frac{3}{8} - 2\frac{2}{3}$

Evaluate each using the Order of Operations.

29. $1\frac{1}{2} \cdot 1\frac{1}{3} + 2\frac{5}{6}$

30. $5\frac{1}{2} - \left[2\frac{1}{2}\right]\left[1\frac{3}{5}\right]$

31. $4\frac{3}{9} + 1\frac{1}{2} \div 1\frac{3}{4}$

32. $4\frac{4}{5} - 2\frac{5}{6} \cdot 1\frac{1}{4}$

33. $5\frac{4}{5} + \left(5\frac{1}{4} \div 1\frac{3}{4}\right)$

34. $\left(7\frac{1}{4}\right)^2 \div 3\frac{3}{4} + 7$

35. $\frac{7}{3} \div 2\frac{1}{6} \cdot 1\frac{3}{5}$

36. $2\frac{1}{2} \cdot 3 \div \frac{2}{5}$

37. $6\frac{5}{8} + \left(0 \cdot 2\frac{3}{5}\right)$

38. $3 + \left(7\frac{1}{2}\right)^2$

39. $\left(4\frac{5}{6} + 5\right) - 1\frac{2}{5}$

40. $12\frac{3}{4} - \left(3\frac{1}{4} - 2\frac{3}{20}\right)$

Work each problem.

41. Find the average of three and one-fourth, four and two-thirds, and two and four-fifths.

42. Find the average of seven and one-third and three and one-fourth.

43. Jim bought 20 shares of stock at $\$44\frac{1}{2}$ for each share. What was the total cost?

44. Janie purchased 100 shares of stock at $\$12\frac{1}{4}$ for each share. What was the total cost?

45. Divide two thousand and four-fifths by twenty-eight and one-third.

46. Three fishermen wish to share a $15\frac{1}{2}$ pound fish that one of them caught. How much will each receive if they share equally?

47. Two hunters wished to share $153\frac{1}{2}$ pounds of bear meat from a bear they killed during hunting season. How much meat will each receive?

48. A company will deliver $6\frac{1}{2}$ cubic yards of concrete for $377. What is the price per cubic yard?

49. A woman bought $12\frac{1}{2}$ yards of one material an $8\frac{1}{4}$ yards of a second. How many yards of material did she buy?

50. Tom found $9\frac{1}{3}$ yards of rope and bought 15 additional yards. How many yards of rope does he now have?

51. An individual bought and sold the stock listed below. How much money did the investor lose?

Company	*Number of Shares*	*Purchase Price*	*Sale Price*
Diapers, Inc.	100	$30	$\$28\frac{5}{8}$

52. A rancher wishes to fence the three **square** pastures shown in the diagram. Write an expression for the amount of fencing material needed to fence the outer edges of all three pastures. Then evaluate the expression.

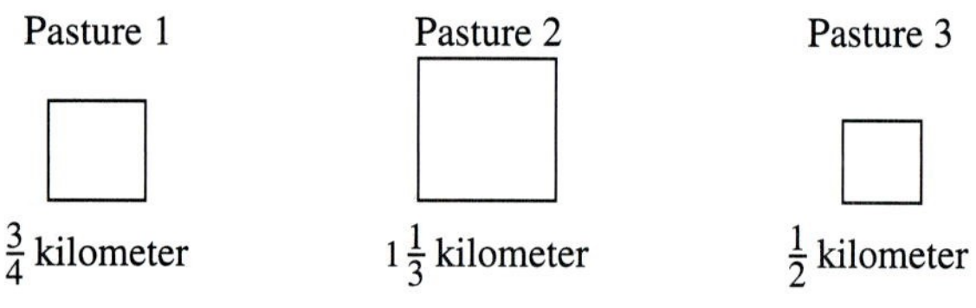

53. A patient received $3\frac{1}{2}$ liters of intravenous fluid in 24 hours. Find the flow rate in liters per hour.

JOURNAL EXERCISES

1. Compare the two different methods for adding mixed numbers given in this section. Work one problem both ways.

2. Can every mixed number be written as a fraction? If so, why do both mixed numbers and fractions exist?

1 Ratios

OBJECTIVE

1 Identify and work with ratios.

A **ratio** is a comparison between two quantities. For example, 401 male students and 504 female students were taking freshman history classes at one university during one semester. The comparison of 401 male students to 504 female students is a ratio. It can be written in three different ways:

1. 401 male students to 504 female students
2. 401 male students : 504 female students
3. $\dfrac{\text{401 male students}}{\text{504 female students}}$

When using the third method above, the first number of the ratio becomes the numerator and the second number becomes the denominator.

Ratios are used to compare quantities with the same units. **Rates** are used to compare quantities with different units, or different labels. Examples of both ratios and rates follow.

Ratios	*Rates*
$\dfrac{\text{401 people}}{\text{509 people}} = \dfrac{401}{509}$	$\dfrac{\text{60 minutes}}{\text{1 hour}}$
$\dfrac{\text{3 weeks}}{\text{5 weeks}} = \dfrac{3}{5}$	$\dfrac{\text{31 feet}}{\text{2 hours}}$
$\dfrac{\text{10 dollars}}{\text{49 dollars}} = \dfrac{10}{49}$	$\dfrac{\text{6 ounces}}{\text{1 gallon}}$

Notice that the units can be left off in ratios since they are identical. The units cannot be removed for rates, since they differ.

A **unit rate** is a rate with a denominator of 1, for example:

$$\frac{\text{45 miles}}{\text{1 hour}}, \quad \frac{\text{3,000 gallons}}{\text{1 acre}}, \quad \frac{\text{8,000 pounds}}{\text{1 truck load}}$$

EXAMPLE 1

Write a ratio (or rate) for each:

(a) 7 deer to 200 acres
(b) 3 boys to 2 girls
(c) 3 apples out of every 20 apples

SOLUTION

(a) $\dfrac{\text{7 deer}}{\text{200 acres}}$

(b) $\dfrac{\text{3 boys}}{\text{2 girls}}$

(c) $\dfrac{\text{3 apples}}{\text{20 apples}} = \dfrac{3}{20}$ ■

1. Write 3 children per 1 family as a ratio.

2. Write 5 dogs per 1 city block as a ratio.

3. Write twelve people to three pizzas as a ratio.

4. Write sixty hospital days to fifteen patients as a ratio.

EXAMPLE 2

What is the ratio of rectangles to triangles?

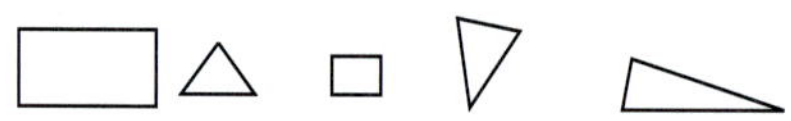

SOLUTION

$$\frac{2 \text{ rectangles}}{3 \text{ triangles}} \quad \blacksquare$$

◇ *Do Practice Exercises 1 and 2.*

The fraction portion of each ratio should be reduced to lowest terms for ease of reading, as shown in the next example.

EXAMPLE 3

An audience is made up of two hundred children and one thousand adults. What is the ratio of children to adults?

SOLUTION

Children to adults implies number of children on top and number of adults on bottom.

Reduces to

$$\frac{200 \text{ children}}{1{,}000 \text{ adults}} = \frac{1 \text{ child}}{5 \text{ adults}} \quad \blacksquare$$

◇ *Do Practice Exercises 3 and 4.*

If possible, identical units should be used in the numerator and denominator of a ratio. This makes it easier to understand.

EXAMPLE 4

Tim has three nickels and two quarters. What is the ratio of the amount of money in nickels versus quarters?

SOLUTION

The first step is to change three nickels to fifteen cents and two quarters to fifty cents. This way both the numerator and denominator of the ratio have the same unit.

$$\frac{3 \text{ nickels}}{2 \text{ quarters}} = \frac{15 \text{ cents in nickels}}{50 \text{ cents in quarters}} = \frac{3 \text{ cents}}{10 \text{ cents}} = \frac{3}{10} \quad \blacksquare$$

ANSWERS

1. $\frac{3 \text{ children}}{1 \text{ family}}$

2. $\frac{5 \text{ dogs}}{1 \text{ city block}}$

3. $\frac{4 \text{ people}}{1 \text{ pizza}}$

4. $\frac{4 \text{ hospital days}}{1 \text{ patient}}$

EXAMPLE 5

One military jet travels up to 1,200 miles in an hour. At maximum speed, NASA's space shuttle would travel 18,000 miles in one hour. Compare the speeds using a ratio of miles traveled in one hour.

SOLUTION

$$\begin{array}{r} \text{shuttle} \\ \text{jet} \end{array} \quad \frac{18{,}000 \text{ miles}}{1{,}200 \text{ miles}} = \frac{15}{1}$$

At maximum speed, the space shuttle flies 15 times faster than the military jet. ■

◇ *Do Practice Exercises 5 and 6.*

5. An automobile traveled seventy miles in two hours. Write its speed as a ratio.

6. The supersonic jet, the Concord, flew two thousand four hundred miles in three hours. Write its average speed as a ratio.

ANSWERS

5. $\dfrac{35 \text{ miles}}{1 \text{ hour}}$

6. $\dfrac{800 \text{ miles}}{1 \text{ hour}}$

name date

EXERCISE SET 4.4

Write each as a ratio (or rate). If possible, use common units and reduce.

1. 6 men to 5 boys

2. 4 horses to 16 cows

3. Three dollars out of every one hundred dollars

4. 2 cars to 1 family

5. 50 students to 2 classrooms

6. 25 students to 35 desks

7. 20 cents to 50 cents

8. 30 cents to 1 dollar

9. 400 miles to 25 gallons

10. 3 pennies to 3 nickels

11. 4 quarters to 1 ten dollar bill

12. 12 inches to 7 feet

13. 1 mile to 3 laps

14. 5 quarters to 50 cents

15. 100 students to 10 teachers

16. 3 children to 1 family

17. 1,500 cattle to 3 ranchers

18. 200 metal fence posts to 50 wooden fence posts

19. 45 minutes to 2 hours

20. 3 hours to 20 minutes

21. 6 books for 2 classes

22. 120 miles in 2 hours

23. $48 for 4 concert tickets

24. 180 pounds to 60 inches

25. 3 cups of milk to 5 cups of flour

26. 35 lawyers per 1,000 people

27. 3 wrecks per 100 drivers

28. 2 cars per 1 family

29. $1,850 tuition for 2 semesters

30. 1,200 calories per 8 slices of pizza

Work each problem.

31. A college has two thousand female students and one thousand eight hundred male students. What is the ratio of female to male students? What is the ratio of male to female students?

32. What is the ratio of female students to the total number of students in problem 31 above?

33. The mass of the earth is about 81 times the mass of the moon. Use a dictionary and write out the definition of ''mass.'' Then write the mass of the moon to the mass of the Earth as a ratio.

34. In humans, twelve pair of nerves begin at the bottom of the brain, whereas thirty-one pair of nerves begin on the spinal chord along the backbone. What is the ratio of the number of nerves beginning on the spinal chord to those beginning at the bottom of the brain?

35. The ratio of sand to cement in concrete is 3 to 1. What does this mean? How many parts of sand are in 4 parts of concrete?

36. Three out of every five people in the world live in Asia. Write this information as a ratio.

37. Africa is approximately 3 times the size of Europe. Write this information as a ratio without using any units. Which continent is smaller?

38. A cat has had two litters of kittens:

Litter Number	*Color of Kittens*		
	Black	*Gray*	*Black and White*
1	2	1	1
2	1	2	0

Using the information for both litters combined, find the ratio of black kittens to total number of kittens.

39. The diameter of the sun is 864,000 miles and the diameter of the earth is 7,926 miles. Round the diameter of the sun to the nearest one hundred thousand miles and that of the earth to the nearest one thousand miles, then compare using a ratio.

40. Jerome read a 250-page novel in 4 hours. Write a rate for the number of pages to hours.

41. A stock is selling at thirty dollars per share. Last year the company earned five dollars per share. Find a ratio for the stock price to the earnings per share. This ratio is the *price to earnings* ratio commonly used by stock market analysts.

JOURNAL EXERCISES

1. The ratio of the mass of the sun to the mass of all of the planets combined is approximately 70:1. Explain the meaning of this data.

2. At sea level, the ratio of the speed of light to the speed of sound is almost one million to one. Which travels faster, light or sound? Using two people located two city blocks apart, how could you demonstrate which travels faster?

3. In 1990, the birth rate in Tunisia was 27.5 births per one thousand people. What does this mean?

4. In 1990, crude oil reserves for North America were ninety billion barrels with most of those in Mexico. The oil reserves for the Middle East were six hundred eighty-five billion barrels. Round both figures to the nearest one hundred billion and write a ratio for proven oil reserves in the Middle East to those in North America. In the decades to come, how might the information in this ratio affect your life? Explain.

EXAMPLE 3

Use both Method I and Method II to find out if:

$$\frac{5}{8} \stackrel{?}{=} \frac{25}{40}$$

SOLUTION

	Method I	*Method II*	
	$\frac{5}{8} \stackrel{?}{=} \frac{25}{40}$	Product of extremes: Product of means:	$5 \cdot 40 = 200$ $8 \cdot 25 = 200$
Reduce fractions.	$\frac{5}{8} \stackrel{?}{=} \frac{\cancel{25}^{5}}{\cancel{40}_{8}}$	The two products are equal, so the proportion must be true.	
They are equal.	$\frac{5}{8} = \frac{5}{8}$		■

◇ *Do Practice Exercises 6 and 7.*

A justification for Method II will be given in the next example. The example shows the steps to remove the denominator from each side of the equation. It illustrates an important concept so study it carefully.

EXAMPLE 4

Does $\frac{6}{7} \stackrel{?}{=} \frac{3}{4}$?

SOLUTION

The 4 in the denominator is removed from the right side of the equation by multiplying both sides times 4.

$$\frac{6}{7} \stackrel{?}{=} \frac{3}{4}$$

$$\frac{6}{7} \cdot 4 \stackrel{?}{=} \frac{3}{\cancel{4}} \cdot \cancel{4}$$ Multiply *both sides* times 4. Divide out forms of one.

$$\frac{6}{7} \cdot 4 \stackrel{?}{=} 3$$

6. Does $\frac{4}{5} = \frac{70}{100}$?

7. Does $\frac{3}{8} = \frac{240}{640}$?

ANSWERS
6. no
7. yes

Now, remove the 7 in the denominator of the left hand side by multiplying both sides of the equation by 7.

$$\frac{6}{7} \cdot 4 \cdot 7 \stackrel{?}{=} 3 \cdot 7$$ Multiply both sides times 7.

$$\frac{6}{\not{7}} \cdot \not{7} \cdot 4 \stackrel{?}{=} 3 \cdot 7$$ Commutative property
Divide out forms of one.

$$6 \cdot 4 \stackrel{?}{=} 3 \cdot 7$$ $6 \cdot 4$: product of extremes; $3 \cdot 7$: product of means

$$24 \stackrel{?}{=} 21$$

The proportion is false since 24 does not equal 21. ■

The last example showed that checking to see if a proportion is true is the same as asking if the product of the extremes equals the product of the means. However, this is the basic idea behind Method II.

Cross multiplication is the process of finding the product of the extremes and the product of the means. In the sections that follow, cross multiplication will be shortened to look as follows:

$$\frac{6}{7} \stackrel{?}{=} \frac{3}{4}$$

$$6 \cdot 4 \stackrel{?}{=} 7 \cdot 3$$ Cross multiply.

Each time you see this, you should recall that this is the result of multiplying both sides of the equation by the two denominators and dividing out forms of one.

Identify the extremes and the means.

1. $\frac{3}{4} = \frac{75}{100}$

2. $\frac{2}{3} = \frac{40}{60}$

3. $\frac{5}{7} = \frac{25}{35}$

4. $\frac{5}{8} = \frac{40}{64}$

5. $\frac{4}{5} = \frac{800}{1{,}000}$

6. $\frac{3}{7} = \frac{30}{70}$

Determine whether the proportion is true *or* false.

7. $\frac{5}{8} = \frac{10}{10}$

8. $\frac{1}{3} = \frac{20}{35}$

9. $\frac{4}{5} = \frac{40}{50}$

10. $\frac{2}{3} = \frac{60}{90}$

11. $\frac{7}{5} = \frac{140}{110}$

12. $\frac{4}{5} = \frac{17}{21}$

13. $\frac{3}{8} = \frac{27}{72}$

14. $\frac{5}{12} = \frac{20}{36}$

15. $\frac{9}{10} = \frac{9{,}000}{10{,}000}$

16. $\frac{19}{16} = \frac{20}{25}$

17. $\frac{7}{8} = \frac{49}{56}$

18. $\frac{7}{2} = \frac{9}{4}$

19. $\frac{4}{1} = \frac{45}{10}$

20. 6:10 = 48:80

21. 3:5 = 7:15

22. $\frac{2}{7} = \frac{6}{21}$

23. 2:9 = 100:500

24. 16:9 = 64:36

25. $\frac{1}{3} = \frac{5}{20}$

26. $\frac{3}{2} = \frac{4}{3}$

27. $\frac{30}{100} = \frac{7}{20}$

28. $\frac{3}{8} = \frac{90}{260}$

29. $\frac{4}{9} = \frac{60}{135}$

30. $\frac{3}{7} = \frac{24}{56}$

Work each problem.

31. The number and gender of employees at two different companies are:

	Gender	
	M	*F*
Company A	100	120
Company B	50	60

Write the ratio of female workers to male workers for each company. Are the ratios equal?

32. Two satellites orbit the earth. Satellite A makes 1 revolution every 20 hours. Write this as a ratio. Satellite B makes 3 revolutions every 72 hours. Write this as a ratio. Are the two ratios equal?

33. Using the data that follows, find a ratio for profit to sales for each year. Are the ratios equal? Find the year with the highest profit to sales ratio by comparing the size of the two fractions.

	1992	*1993*
Sales	1,000,000	1,200,000
Profit	100,000	90,000

JOURNAL EXERCISES

1. Explain cross multiplication in your own words. Work a problem showing every step (see Example 4).

2. What is the difference between a ratio and a proportion? Give examples of both.

OBJECTIVE

1 Solve proportions.

1 Solving Proportions

A proportion may contain one unknown value and that value can be found. For example, the numerator of the fraction on the left hand side of the following proportion is unknown.

$$\frac{?}{4} = \frac{6}{8}$$

The unknown can be represented using a question mark as above or by using a variable, say x:

$$\frac{x}{4} = \frac{6}{8}$$

The problem is then to solve for the value of x that makes the proportion true.

Rule: To Find the Unknown in a Proportion

Step 1 Write the proportion using a variable for the unknown.

Step 2 Cross multiply.

Step 3 Divide both sides of the equation by the number being multiplied times the variable.

EXAMPLE 1

Solve $\frac{x}{5} = \frac{80}{100}$ for the unknown value x.

SOLUTION

Step 1 $\frac{x}{5} = \frac{80}{100}$

Step 2 $100 \cdot x = 5 \cdot 80$ Cross multiply. ($100 \cdot x$: product of extremes; $5 \cdot 80$: product of means)

$100x = 400$

Step 3 The unknown is being multiplied by 100.
Divide both sides of the equation by 100.

$$\frac{100x}{100} = \frac{400}{100}$$ Divide both sides by 100.

$$\frac{\cancel{100}^{1}x}{\cancel{100}_{1}} = \frac{\cancel{400}^{4}}{\cancel{100}_{1}}$$ Reduce.

$$x = \frac{4}{1}$$

$x = 4$ Result.

1. Solve $\frac{y}{40} = \frac{85}{200}$ for y.

2. Find the unknown in $\frac{x}{7} = \frac{1{,}000}{1{,}400}$.

ANSWERS
1. $y = 17$
2. $x = 5$

The x in the original equation can be replaced with 4 and the proportion is true:

$$\frac{4}{5} = \frac{80}{100} \quad ■$$

EXAMPLE 2

Solve $\frac{y}{10} = \frac{56}{70}$.

SOLUTION

$$\frac{y}{10} = \frac{56}{70}$$

$$70 \cdot y = 10 \cdot 56 \qquad \text{Cross multiply.}$$

$$70y = 560$$

$$\frac{70y}{70} = \frac{560}{70} \qquad \text{Divide both sides by 70.}$$

$$y = 8 \quad ■$$

◇ *Do Practice Exercises 1 and 2.*

It does not make any difference where in the proportion the variable is located or which letter is used for the variable, the solution process is the same. Look at the next two examples.

EXAMPLE 3

Solve $\frac{100}{a} = \frac{20}{3}$ for a.

SOLUTION

$$\frac{100}{a} = \frac{20}{3}$$

$$3 \cdot 100 = 20 \cdot a \qquad \text{Cross multiply.}$$

$$300 = 20a$$

$$\frac{300}{20} = \frac{20a}{20} \qquad \text{Divide both sides by 20.}$$

$$15 = a \quad \text{or} \quad a = 15 \quad ■$$

EXAMPLE 4

Solve $\frac{7}{30} = \frac{b}{90}$ for b.

SOLUTION

$$\frac{7}{30} = \frac{b}{90}$$

$$7 \cdot 90 = 30 \cdot b$$ Cross multiply.

$$630 = 30b$$

$$\frac{\overset{21}{\cancel{630}}}{\underset{1}{\cancel{30}}} = \frac{\cancel{30}b}{\cancel{30}}$$ Divide both sides by 30.

$$21 = b$$ ■

◇ *Do Practice Exercises 3 and 4.*

3. Solve $\frac{90}{b} = \frac{18}{11}$ for the unknown.

4. Solve $\frac{3}{7} = \frac{165}{Y}$ for Y.

ANSWERS
3. $b = 55$
4. $Y = 385$

Which proportions are true? (Review)

1. $\frac{4}{5} = \frac{16}{20}$

2. $\frac{5}{8} = \frac{20}{30}$

3. $\frac{2}{3} = \frac{9}{6}$

4. $\frac{1}{4} = \frac{25}{100}$

5. $\frac{6}{10} = \frac{120}{200}$

6. $\frac{5}{30} = \frac{1}{5}$

7. $\frac{3}{7} = \frac{21}{49}$

8. $\frac{5}{8} = \frac{25}{100}$

Solve for the unknown.

9. $\frac{x}{5} = \frac{40}{50}$

10. $\frac{y}{3} = \frac{10}{15}$

11. $\frac{a}{4} = \frac{75}{100}$

12. $\frac{2}{3} = \frac{x}{24}$

13. $\frac{3}{5} = \frac{60}{z}$

14. $\frac{5}{b} = \frac{25}{30}$

15. $\frac{8}{y} = \frac{25}{30}$

16. $\frac{3}{5} = \frac{33}{x}$

17. $\frac{b}{4} = \frac{50}{100}$

18. $\frac{7}{8} = \frac{c}{56}$

19. $\frac{2}{3} = \frac{20}{y}$

20. $\frac{5}{x} = \frac{15}{27}$

21. $\frac{d}{9} = \frac{15}{45}$

22. $\frac{20}{x} = \frac{80}{100}$

23. $\frac{3}{14} = \frac{21}{y}$

24. $\frac{1}{40} = \frac{x}{8}$

25. $\frac{16}{28} = \frac{4}{y}$

26. $\frac{35}{60} = \frac{x}{12}$

27. $\frac{3}{a} = \frac{24}{40}$

28. $\frac{8}{b} = \frac{60}{30}$

29. $\frac{x}{15} = \frac{66}{90}$

30. $\frac{60}{42} = \frac{x}{14}$

31. $\frac{a}{40} = \frac{200}{320}$

32. $\frac{40}{z} = \frac{320}{600}$

JOURNAL EXERCISES

1. When you are solving proportions, you are solving equations. Explain.

2. In solving proportions, why are both sides of the equation divided by the number being multiplied times the variable? In other words, why divide by 8 in the following problem?

$$\frac{3}{8} = \frac{y}{42}$$

$$3 \cdot 42 = 8 \cdot y \quad \text{Cross multiply.}$$

$$\frac{126}{8} = \frac{8 \cdot y}{8} \quad \text{Why divide by 8?}$$

1 Word Problems

Proportions are a very useful tool for solving certain word problems. The first step in any word problem is to read the problem carefully. Then determine whether or not proportions are the appropriate tool for solving the problem.

OBJECTIVE

1 Solve word problems using proportions.

EXAMPLE 1

A carton containing eggs is square with six eggs along each side. How many eggs does it hold?

SOLUTION

This problem is solved using multiplication, *not* by using a proportion.

$$6 \times 6 = 36 \text{ eggs}$$ ■

Recall that a proportion is two ratios that have been set equal to each other. Thus, two ratios must occur in a problem *before* a proportion can even be set up. Not only that, the two ratios must be equal to each other.

EXAMPLE 2

In 8 minutes Ting jogged 4 laps. If she continues to jog at the same speed, how long will it take her to jog 10 laps?

SOLUTION

Let the variable x = the time required for her to jog 10 laps. Then the two ratios are:

$$\frac{8 \text{ minutes}}{4 \text{ laps}} \quad \text{and} \quad \frac{x \text{ minutes}}{10 \text{ laps}}$$

These two ratios can be set equal to each other since she continues to jog at the same speed.

$$\frac{8 \text{ minutes}}{4 \text{ laps}} = \frac{x \text{ minutes}}{10 \text{ laps}}$$

$$\frac{8}{4} = \frac{x}{10} \qquad \text{Remove units.}$$

$$8 \cdot 10 = 4x \qquad \text{Cross multiply.}$$

$$4x = 80$$

$$\frac{\overset{1}{\cancel{4}}x}{\underset{1}{\cancel{4}}} = \frac{\overset{20}{\cancel{80}}}{\underset{1}{\cancel{4}}} \qquad \text{Divide both sides by 4.}$$

$$x = 20 \text{ minutes to jog 10 laps}$$ ■

1. A patient is to receive 2 milligrams of a drug for every 40 pounds of body weight. How many milligrams are needed for a person weighing 160 pounds?

EXAMPLE 3

Four pounds of raw fish can be bought for \$8. How many pounds of raw fish can be bought for \$240?

SOLUTION

Let x = the number of pounds of raw fish that can be bought for \$240. Then:

$$\frac{4 \text{ pounds}}{\$8} = \frac{x \text{ pounds}}{\$240} \quad \text{Ratios are equal.}$$

$$\frac{4}{8} = \frac{x}{240} \quad \text{Remove the units.}$$

$$4 \cdot 240 = 8x \quad \text{Cross multiply.}$$

$$960 = 8x$$

$$\frac{\overset{120}{\cancel{960}}}{\cancel{8}} = \frac{\cancel{8}x}{\cancel{8}} \quad \text{Divide both sides by 8.}$$

$$120 = x$$

Therefore, 120 pounds of raw fish can be bought for \$240. ■

◇ *Do Practice Exercise 1.*

It is *very important* to arrange the ratios of a proportion in one of two ways before setting the ratios equal to one another. Both methods are shown next; either method may be used. First, note that **units** refers to the unit of measurement such as feet or gallons. **Labels** refer to the object being measured such as gallons of alcohol or gallons of ammonia. The difference between the two is shown in Example 5.

METHOD I: Write the proportion so that the same units and labels are in each of the numerators and the same units and labels are in each of the denominators.

EXAMPLE 4

Two inches on a map represents thirty miles. How many miles are represented by six inches?

SOLUTION

As given in the problem, the two ratios are:

$$\frac{2 \text{ inches}}{30 \text{ miles}} \qquad \frac{\text{How many miles?}}{6 \text{ inches}}$$

Using Method I, the proportions *must be* written so that the same unit is in the numerator of each ratio and the same unit is in the denominator of each ratio. Thus, the second ratio above must be inverted before setting the ratios equal to one another.

$$\frac{2 \text{ inches}}{30 \text{ miles}} \qquad \frac{6 \text{ inches}}{\text{How many miles?}} \longleftarrow \text{inverted}$$

Now inches are in both numerators and miles are in both denominators. Let x equal the unknown number of miles and set the two ratios equal to each other.

ANSWER
1. 8 milligrams

$$\frac{2}{30} = \frac{6}{x}$$ Drop units.

$$2 \cdot x = 30 \cdot 6$$ Cross multiply.

$$2x = 180$$

$$\frac{\not{2}x}{\not{2}} = \frac{\overset{90}{\not{1}\not{8}\not{0}}}{\underset{1}{\not{2}}}$$ Divide both sides by 2.

$$x = 90$$

Therefore, 6 inches on the map equals 90 miles. ■

METHOD II: Write the proportion so that the units and labels in the numerator and denominator of each ratio are the same. Write the other ratio with the same units and labels in the numerator and denominator, and in the same order.

EXAMPLE 5

A chemical plant must mix eight hundred gallons of a particular chemical. Two hundred gallons of the chemical requires fifty gallons of alcohol. How much alcohol is needed to mix the eight hundred gallons?

SOLUTION

Ratios in which numerators and denominators have the same units and labels are:

(unit → gallons; label → chemical)

$$\frac{\text{800 gallons of chemical}}{\text{200 gallons of chemical}}, \quad \frac{\text{50 gallons of alcohol}}{\text{How much alcohol?}}$$

These ratios can be set equal to each other only if the two numerators "go together." The numerator of the first ratio is the total volume of chemical. Therefore, the numerator of the second ratio must also be for the total volume, or in this case for the amount of alcohol in the total volume. The second ratio must be inverted before setting the two ratios equal to one another.

$$\frac{\text{800 gallons of chemical}}{\text{200 gallons of chemical}} = \frac{\text{How much alcohol?}}{\text{50 gallons of alcohol}}$$

$$\frac{800}{200} = \frac{x}{50}$$ Drop units and labels.

$$\frac{4}{1} = \frac{x}{50}$$ Reduce.

$$4 \cdot 50 = 1 \cdot x$$ Cross multiply.

$$200 = x$$

200 gallons of alcohol are needed to mix 800 gallons of the chemical. ■

◇ *Do Practice Exercise 2.*

EXAMPLE 6

A six foot individual casts a nine foot shadow. How tall is a tree that casts a sixty foot shadow?

2. A molecule of carbon dioxide is made up of one part carbon and two parts oxygen. How many parts of oxygen are needed to combine with two hundred parts of carbon to make two hundred molecules of carbon dioxide?

ANSWER
2. 400 parts of oxygen

3. How would you change $\frac{\text{600 cents}}{\text{3 children}}$ so that it can be set equal to the ratio $\frac{\text{how many dollars}}{\text{1 child}}$?

ANSWER
3. First change 600 cents to 6 dollars, then proceed.

SOLUTION
Draw a sketch, if it helps.

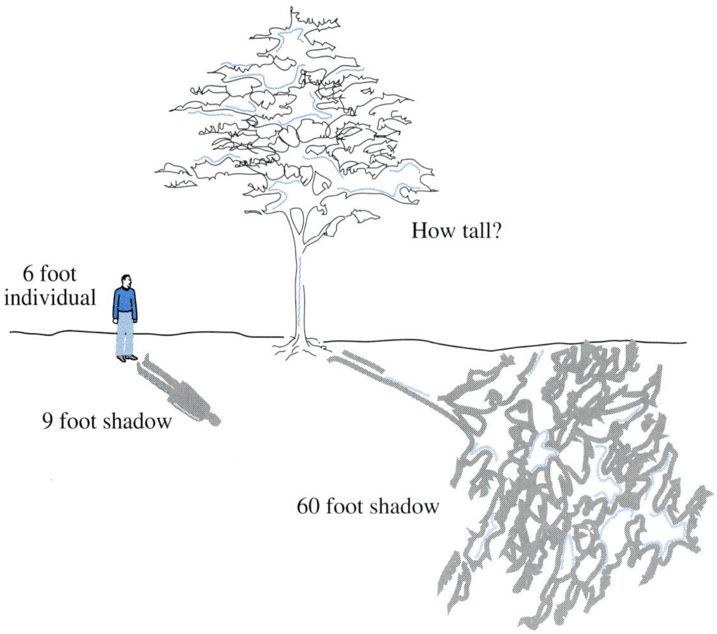

The ratios are written using Method I with the ''tall'' label on top of each ratio and ''length of shadow'' label in the denominator.

$$\frac{\text{6 foot person}}{\text{9 foot shadow}} = \frac{\text{How tall is the tree?}}{\text{60 foot shadow}}$$

$$\frac{6}{9} = \frac{x}{60} \quad \text{Drop units and labels.}$$

$$\frac{2}{3} = \frac{x}{60} \quad \text{Reduce.}$$

$$2 \cdot 60 = 3 \cdot x \quad \text{Cross multiply.}$$

$$120 = 3x$$

$$\frac{120}{3} = \frac{3x}{3} \quad \text{Divide by 3.}$$

$$40 = x \quad \text{Result.}$$

The tree is 40 feet tall. ■

Sometimes two ratios cannot be set equal to each other in a proportion because the units are different. For example, the following two ratios cannot be set equal to one another, since one numerator is in dollars and the other is in cents.

$$\frac{\text{40 dollars}}{\text{3 children}} \qquad \frac{\text{How many cents?}}{\text{1 child}}$$

In this case, it is necessary to first change \$40 to 4,000 cents. The ratios can then be set equal to one another.

◇ ***Do Practice Exercise 3.***

Solve each, showing all work.

1. If the total cost of three shirts is thirty dollars, how much would seven of these shirts cost?

2. If two pair of shoes sell for eighteen dollars, how much would five pairs of shoes cost?

3. If it takes a contractor five hours to do two small jobs, how many of these small jobs can be done in a forty-hour work week?

4. If a typist can type twenty pages in two hours, how many hours will it take to type seventy-five pages. Round your answer to the nearest hour.

5. A bricklayer completed one-third of a fence in five hours. How long will it take to finish the fence?

6. How many books are in seven boxes if each box has eighteen books?

7. An author wrote 120 pages in twenty days. At that rate, how many days does the author need to write a 570-page book?

8. A quality control inspector found twelve defective items out of a sample of sixty. Assuming the same ratio applies, how many defective items are in a group of eighteen hundred items?

9. Two hundred trees out of the seven hundred trees in a five-acre tract of land are ready to be logged. Estimate the number of trees that are ready to be logged out of a hundred-acre tract.

10. A paint store sold four hundred gallons of paint in eight months. At that rate, how much paint can the manager expect to sell in three months?

11. A tray of donuts uses six cups of flour and two and one-half cups of sugar. How much flour is needed to make forty trays of donuts?

12. An automobile uses 2 gallons of gasoline to travel 50 miles. How much gas is needed to travel 420 miles? Round the answer to the nearest gallon.

13. Five parts of an alloy are made up of 3 parts niobium and 2 parts strontium. How many kilograms of strontium does it take to make eighty kilograms of alloy?

14. A ten story building casts an eighty foot shadow at the same time that a second builing casts a two hundred and forty foot shadow. How tall is the second building in feet, if the average height of each story is twelve feet?

15. A mixture is made from three parts alcohol and five parts of other solutions. How much alcohol is needed to make forty gallons of the mixture?

16. The temperature of the earth increases by approximately 1°F for every 150 feet in depth from the surface. What would you expect the temperature to be at a depth of 18,000 feet if the temperature at the surface is 90°F?

17. Twenty-five pounds of a particular fertilizer is needed for every two thousand square feet of lawn. How many square feet can be fertilized with one thousand pounds of fertilizer? About how many acres is this if an acre is forty-three thousand five hundred and sixty square feet?

18. A car is traveling at thirty miles per hour. The car is increasing its speed by five miles per hour every three seconds. How fast will it be traveling in fifteen seconds?

19. Electricity and gas for the Ramirez's home averages sixty dollars per month. What is the total utility bill for the year? What fraction of the Ramirez's combined $21,600 income does this equal?

20. An average human has approximately six million cubic millimeters of blood. The average number of three different types of blood cells found in each cubic millimeter of human blood is:

Cell Type	*Average Number of Cells*	*Cell Function*
Red blood cells	5,000,000	Carry oxygen and carbon dioxide
Platelets	350,000	Help stop bleeding
White blood cells	7,000	Fight bacteria

What is the ratio of white blood cells to red blood cells? Approximately how many white blood cells are in an average human? Assume the above figures are for a 150 pound person and estimate the number of red blood cells in a 200 pound person. What assumption did you make?

JOURNAL EXERCISES

Make up and solve your own proportion word problem regarding the following.

1. The speed with which a car travels and the distance it travels.

2. The height of a person, a tree, and the length of their shadows.

3. The changing of a recipe from one that feeds four people to one that feeds ten people.

4. The estimating of annual sales for a company given sales for eight months.

5. A subject of your choice.

CHAPTER 4 REVIEW

SECTION	TOPIC OR TERM	EXAMPLE OR DEFINITION
4.1		
Mixed Numbers	Mixed Number	A mixed number is a number that is the sum of a whole number and a fraction: $4\frac{1}{2}$ or $3\frac{4}{7}$
	From a Mixed Number to a Fraction	$6\frac{2}{3} = \frac{6 \cdot 3 + 2}{3} = \frac{18 + 2}{3} = \frac{20}{3}$
	From a Fraction to a Mixed Number	$\frac{7}{5}$ $\quad \begin{array}{r} 1 \\ 5\overline{)7} \\ \underline{5} \\ 2 \end{array}$ $\quad \frac{7}{5} = 1\frac{2}{5}$
4.2		
Simplifying and Comparing Mixed Numbers	Simplify	Simplification is the process of writing a number in a simpler way, for example: $5\frac{7}{3} = 5 + \frac{7}{3} = 5 + 2\frac{1}{3} = 7\frac{1}{3}$
	Compare Mixed Numbers	$12\frac{1}{3} > 3\frac{1}{4}$ Since $12 > 3$ $4\frac{1}{8} < 4\frac{3}{5}$ since $\frac{1}{8} < \frac{3}{5}$
4.3		
Order of Operations	Solve Problems	$2\frac{1}{3} \div 1\frac{3}{4}$ $\frac{7}{3} \div \frac{7}{4}$ Change to fractions. $\frac{\overset{1}{\cancel{7}}}{3} \cdot \frac{4}{\underset{1}{\cancel{7}}}$ from ÷ to · $\frac{4}{3} = 1\frac{1}{3}$ Result.

SECTION	TOPIC OR TERM	EXAMPLE OR DEFINITION
4.4		
Ratios	Ratio	A ratio is a comparison of two values that have the same units and labels. For example, 3:5 also written as $\frac{3}{5}$
	Rate	A rate is a comparison of two values that have different units or labels. For example, $\frac{\$300}{7 \text{ weeks}}$.
	Unit Rate	A unit rate is a rate that has a denominator of 1. For example, $\frac{45 \text{ miles}}{1 \text{ hour}}$.
4.5		
Proportions	Proportion	Proportions are two ratios or rates that equal one another.
	Extremes and Means	extreme → $\frac{5}{6} = \frac{25}{30}$ ← mean; mean → 6, 30 ← extreme
4.6		
Solving Proportions	Solving Proportions	$\frac{3}{7} = \frac{60}{x}$ $3x = 7 \cdot 60$ Cross multiply. $3x = 420$ $\frac{3x}{3} = \frac{420}{3}$ Divide by 3. $x = 140$

FLOWER SHOP

Beatrice owns a small flower shop. The total income from the shop averages $6,000 per month. The average monthly expenses are shown below:

Item	*Monthly Expense*
Supplies	$2,100
Rent (includes utilities)	1,400
Payroll	700

1. What are her total monthly expenses?

2. How much profit does she make per month?

3. What does her hourly wage amount to if she works 200 hours per month?

Beatrice plans to move her business to a larger building. She believes that she can increase monthly sales to $8,000 per month.

4. Use a proportion to find her new monthly cost of supplies. Assume the cost of supplies is proportional to sales.

The building Beatrice will be moving *from* has 800 square feet and the building she will be moving *to* has 1,000 square feet.

5. Use a proportion to find the rent on the new building. Assume rent is proportional to the size of the space.

6. Find the sum of all expenses at the new location if payroll costs remain the same.

7. Based on these figures, will she have more profit at the new location? How much more?

CHAPTER 4 REVIEW PROBLEMS

Change from an improper fraction to a mixed number.

1. $\frac{10}{3}$

2. $\frac{12}{5}$

3. $\frac{15}{2}$

4. $\frac{16}{7}$

5. $\frac{24}{13}$

6. $\frac{35}{6}$

Change from a mixed number to a fraction.

7. $4\frac{1}{2}$

8. $3\frac{2}{3}$

9. $5\frac{3}{4}$

10. $2\frac{7}{8}$

11. $8\frac{3}{5}$

12. $4\frac{2}{9}$

Draw a number line and plot each as a point.

13. $6, 4\frac{1}{2}$

14. $5\frac{1}{2}, 5\frac{3}{4}$

15. $7, 10\frac{1}{4}$

16. $2\frac{1}{2}, 1\frac{3}{5}$

Evaluate each.

17. $4\frac{1}{2} + 2\frac{2}{3}$

18. $3\frac{4}{5} \cdot 2\frac{5}{8}$

19. $5\frac{3}{8} \div 4\frac{2}{5}$

20. $5\frac{3}{4} - 2\frac{3}{8}$

21. $2\frac{3}{10} - 5\frac{3}{5} \div 4\frac{2}{3}$

22. $4\frac{2}{7} \div 3\frac{4}{5} + 4\frac{2}{3}$

23. $5\frac{1}{2} + 2\frac{1}{4} \cdot 5$

24. $\left\{3\frac{3}{5} \div 2\frac{1}{4}\right\}^2 + 3\frac{2}{5}$

Write each as a ratio (or rate) and reduce, if possible.

25. Four cents to three dollars

26. Six cars to three families

27. Twelve hundred students to forty faculty

28. Eighteen million red blood cells to three cubic milliliters

Determine whether each proportion is true *or* false.

29. $\frac{3}{7} = \frac{27}{63}$

30. $\frac{14}{35} = \frac{2}{5}$

31. $\frac{5}{4} = \frac{30}{20}$

32. $\frac{3}{4} = \frac{25}{35}$

33. $\frac{1}{3} = \frac{30}{90}$

34. $\frac{6}{8} = \frac{21}{28}$

35. $\frac{2}{3} = \frac{8}{10}$

36. $\frac{12}{7} = \frac{10}{2}$

Solve for the unknown.

37. $\frac{3}{5} = \frac{x}{35}$

38. $\frac{5}{y} = \frac{30}{48}$

39. $\frac{b}{7} = \frac{20}{70}$

40. $\frac{5}{7} = \frac{60}{y}$

41. $\frac{7}{r} = \frac{35}{50}$

42. $\frac{12}{25} = \frac{x}{200}$

43. $\frac{12}{20} = \frac{3}{y}$

44. $\frac{32}{x} = \frac{4}{5}$

45. $\frac{48}{60} = \frac{y}{15}$

46. $\frac{75}{125} = \frac{3}{y}$

47. $\frac{50}{250} = \frac{a}{5}$

48. $\frac{2}{x} = \frac{20}{60}$

Work each problem.

49. What is the average of $4\frac{1}{2}$, $2\frac{1}{3}$, and $5\frac{3}{4}$?

50. Three hours (180 minutes) of television programming contains forty-five minutes of commercials. What is the ratio of commercials to total TV programming?

51. A company has two hundred fifty microcomputers and five hundred employees. What is the ratio of microcomputers to employees?

52. A doctor ordered three milliliters of a drug for every twenty pounds of body weight. What dosage is needed for a patient weighing one hundred eighty pounds?

53. A recipe for chocolate chip cookies requires seventy-two chocolate chips per dozen cookies. How many chocolate chips are required to make one hundred thirty-two cookies?

54. Four whales were spotted in a one hundred square mile area. How many whales could be expected in twelve hundred square miles? What assumption must you make?

55. A company payroll is $21,000 per month for twelve employees. What is their total payroll after hiring three new employees at the same average salary?

56. A chemical is made up of three parts of solution A and one part of solution B. How many gallons of solution A are needed to make one hundred gallons of the chemical?

57. An investor earned $200 interest on an investment of $2,600. At the same rate of return, how much must be invested to earn $700 interest?

58. Joe and Carol went on a long journey of approximately one thousand two hundred miles. They drove one-third of the distance in two days. At that rate, how many days will it take to complete the journey?

59. Teresa, June, and Vicky are sisters. Their aunt died and left one-fifth of her estate to Teresa. The remainder of the estate is to be split equally between June and Vicky. What fraction of the estate will June inherit?

60. At a particular mine, an average of fourteen tons of rock must be extracted and crushed to yield 4 pounds of gold. How much rock must be extracted in order to find 200 pounds of gold?

CHAPTER 4 TEST

Change to an improper fraction.

1. $4\frac{4}{5}$

2. $7\frac{3}{4}$

3. $100\frac{3}{20}$

4. $15\frac{3}{7}$

Change to a mixed number.

5. $\frac{25}{6}$

6. $\frac{34}{4}$

7. $\frac{700}{12}$

8. $\frac{490}{17}$

Write each as a ratio.

9. 3 cars with bad brakes for every 100 cars on the road

10. \$250 taxes on \$1,000 income

11. 800 head of goats on 2,500 acres

12. 1 good driver out of every 20 drivers

Evaluate each.

13. $5\frac{1}{3} + 4\frac{2}{5}$

14. $6\frac{3}{8} \cdot 4\frac{3}{4}$

15. $7\frac{3}{5} - 2\frac{7}{8}$

16. $4\frac{3}{4} \div 5\frac{1}{2}$

17. $3\frac{2}{3} + \frac{4}{5}$

18. $6 \cdot 9\frac{1}{2}$

19. $4\frac{3}{4} \div \left(1\frac{2}{5} + \frac{3}{4}\right)$

20. $3\frac{2}{5} + 4\frac{4}{5} \cdot 1\frac{1}{6}$

Determine whether the proportion is true *or* false.

21. $\frac{5}{7} = \frac{15}{20}$

22. $\frac{2}{3} = \frac{8}{12}$

Solve for the unknown.

23. $\frac{5}{6} = \frac{x}{66}$

24. $\frac{y}{3} = \frac{24}{36}$

25. $\frac{3}{40} = \frac{36}{a}$

26. $\frac{7}{b} = \frac{490}{779}$

Work each problem.

27. A small rodent was observed to eat for seventy-two minutes out of every two hours (120 minutes). Write a ratio for the amount of time eating to total amount of time.

28. A hard disk for a personal computer rotates at the rate of three thousand six hundred revolutions per minute. How many revolutions will it make in an hour?

29. Aziz ran the following distances: Monday $3\frac{1}{2}$ miles, Tuesday $2\frac{3}{4}$ miles, and Wednesday $5\frac{1}{5}$ miles. Find the average number of miles run per day.

30. Seven defectives were found in a sample of two hundred. How many defectives would you expect in a batch of fifteen thousand?

CUMULATIVE TEST CHAPTERS 3 AND 4

Identify each proper fraction.

1. $\frac{3}{4}, \frac{8}{7}, \frac{9}{2}$

2. $\frac{1}{3}, \frac{17}{2}, \frac{12}{23}$

Find each undefined fraction.

3. $\frac{5}{0}, \frac{12}{1}, \frac{0}{3}$

4. $\frac{2}{3}, \frac{9}{0}, \frac{0}{5}$

Change each to a mixed number.

5. $\frac{12}{5}$

6. $\frac{75}{12}$

7. $\frac{35}{6}$

8. $\frac{41}{15}$

Change each to a fraction.

9. $7\frac{5}{7}$

10. $12\frac{7}{9}$

11. $30\frac{1}{12}$

12. $75\frac{1}{8}$

Evaluate.

13. $\frac{1}{2} + \frac{3}{4}$

14. $6 - \frac{5}{49}$

15. $\frac{2}{3} \cdot \frac{7}{9}$

16. $5\frac{1}{2} \div 7\frac{1}{4}$

17. $62 \cdot 43$

18. $9^2 + 4^3$

19. $\frac{2}{3} + \frac{4}{5}$

20. $\frac{2}{3} \div 4\frac{5}{9}$

Solve for the unknown.

21. $\frac{x}{3} = \frac{30}{45}$

22. $\frac{70}{y} = \frac{63}{54}$

23. $\frac{3}{18} = \frac{1}{t}$

24. $\frac{64}{80} = \frac{8}{y}$

25. $\frac{30}{75} = \frac{s}{5}$

26. $\frac{a}{80} = \frac{3}{10}$

Work each problem.

27. Write a ratio for the number of faculty (100) to the number of students (2,400) at a small college.

28. Magnolia Woods Elementary School has twelve teachers and three hundred seventy-two children. Write a ratio for the number of children per teacher.

29. A particular washing machine requires 7 gallons of water for each load. How much total water is needed to wash 3 loads per week for a year?

30. A human brain is estimated to contain one hundred billion cells. Write the digits of this number.

31. Add two and two-thirds to the product of seven and one-fifth with four and three-eighths.

32. Find the average of $3\frac{3}{5}$, $7\frac{1}{2}$, and $2\frac{2}{3}$.

33. Find the average of the following three numbers:

fraction:	three-fourths
whole number:	twelve
mixed number:	five and one-half

34. One thousand gallons of a solvent costs a manufacturer three thousand eight hundred dollars. What would be the cost of eighteen hundred gallons?

35. Sound travels three thousand three hundred feet in three seconds. How long will it take to travel one hundred miles given that 1 mile = 5,280 feet? Round your answer to the nearest whole number.

36. Round 6,280,320 to the nearest ten thousand then to the nearest one million.

37. Find two-thirds of one-half of seven thousand two hundred.

38. A fence is needed to fence in and *separate* two pastures shaped as drawn. All distances are in feet.

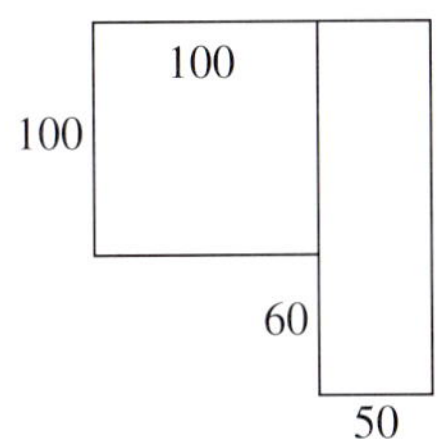

Find the amount of fencing material needed.

39. A boxer weighs $168\frac{3}{4}$ pounds. How many pounds must he lose each day to make the 158 pound fight limit for a fight scheduled to occur in six days?

40. Every time a baseball is thrown it falls to the ground. The next time a baseball is thrown it will also fall to the ground. Is this an example of inductive or deductive logic?

5

Decimals

APPLICATION *Sales at a small corner grocery store average $523.69 per day. What are the total sales for a month in which the store is open 30 days?*

SOLUTION *The answer is found by multiplying the average daily sales times the number of days.*

$$\$523.69 \times 30 = \$15,710.70$$

Total sales for the month are estimated to be $15,710.70.

5.1 BASICS

OBJECTIVES

1. Identify place values in decimal numbers.
2. Write decimal numbers using expanded notation.

In Chapter 3 we saw that a proper fraction is one way to write a number that is smaller than one. In this chapter, we will see a different way to write the same set of numbers, those smaller than one. This notation is referred to as **decimal notation** because it uses a **decimal point**. Numbers larger than one can also be written using decimal notation. You are already familiar with decimal notation because you use it every time you buy or sell something.

1 Place Values

In Chapter 1, we learned to read large numbers by separating each group of three digits with commas and naming place values. The next example reviews that process for you.

EXAMPLE 1

Read 6,025,331.

SOLUTION

6, 0 2 5, 3 3 1

- 1 → ones place
- 3 → tens place
- 3 → hundreds place
- 5 → thousands place
- 2 → ten thousands place
- 0 → hundred thousands place
- 6 → millions place

The number is six million, twenty-five thousand, three hundred thirty-one. ■

◇ *Do Practice Exercise 1.*

Every whole number ends with a decimal point, although the decimal point is frequently not written. The number 6,025,331 from the last example can also be written with a decimal point:

$$6{,}025{,}331 = 6{,}025{,}331.$$

(→ decimal point)

The location of the decimal point is important because the naming of *all* of the place values is based upon the location of the decimal point. Digits to the left of the decimal point indicate the number of ones, or of tens, or of hundreds, and so on. Therefore, any number with a digit of 1 or larger to the left of the decimal point is **greater than one**. For example, each of the following is greater than one.

(> : greater than)

$$17. > 1$$
$$183. > 1$$
$$1{,}965{,}402. > 1$$

◇ *Do Practice Exercise 2.*

Numbers can also be written to the right of the decimal point where they indicate a value **less than one**. The first position to the right of the decimal place is the tenths place. The number one tenth can be written using either fraction notation or decimal notation:

$$\frac{1}{10} = 0.1$$

one tenth ($\frac{1}{10}$ → fraction notation) = 0.1 one tenth (→ decimal notation)

The numbers .1 and 0.1 are two different ways to write the exact same number. It is common to place a 0 in front of the decimal point for any number that has no digits to the left of the decimal point. For example:

0.3 or 0.02 or 0.905

The decimal numbers that may occur in the tenth place are shown next along with a geometric representation of each.

1. Read 4,025,961,000.

2. Find the decimal point for each:

(a) 0

(b) 1

(c) 56

(d) 1,624

(e) 247,321

ANSWERS
1. four billion, twenty-five million, nine hundred sixty-one thousand
2. (a) 0. (b) 1. (c) 56. (d) 1,624. (e) 247,321.

Decimal Notation		*Fraction Notation*	*Geometric Representation*
0.1	=	$\frac{1}{10}$	
0.2	=	$\frac{2}{10}$	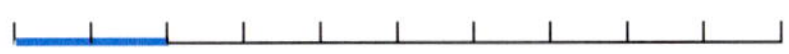
0.3	=	$\frac{3}{10}$	
0.4	=	$\frac{4}{10}$	
0.5	=	$\frac{5}{10}$	
⋮		⋮	
0.9	=	$\frac{9}{10}$	

◇ *Do Practice Exercises 3 and 4.*

The place values on the right hand side of the decimal point are as follows:

Name	*Decimal Notation*		*Fraction Notation*
tenths	0.1	=	$\frac{1}{10}$
hundredths	0.01	=	$\frac{1}{100}$
thousandths	0.001	=	$\frac{1}{1{,}000}$
ten-thousandths	0.0001	=	$\frac{1}{10{,}000}$
hundred-thousandths	0.00001	=	$\frac{1}{100{,}000}$
millionths	0.000001	=	$\frac{1}{1{,}000{,}000}$
⋮	⋮		⋮

Every place value to the left of the decimal point except for the ones place has a corresponding place value on the right side of the decimal point. The only difference is that the name of every place value on the right side of the equation ends in the letters "th." The place values of the number 101,234.98765 are as follows:

tens place ← 101,234.98765 → tenths place

hundreds place ← 101,234.98765 → hundredths place

thousands place ← 101,234.98765 → thousandths place

ten thousands place ← 101,234.98765 → ten-thousandths place

hundred thousands place ← 101,234.98765 → hundred-thousandths place

3. Write seven tenths using both decimal notation and fraction notation.

4. Write $\frac{8}{10}$ using words.

ANSWERS

3. 0.7, $\frac{7}{10}$

4. eight tenths

5. Name the place values in 71,025.038.

6. Write the decimal form for each:

(a) $\frac{7}{10}$

(b) $\frac{9}{100}$

(c) $\frac{3}{1,000}$

ANSWERS

5. 71,025.038

ten thousands ← 7; thousands ← 1; hundreds ← 0; tens ← 2; ones ← 5; 0 → tenths; 3 → hundredths; 8 → thousandths

6. (a) 0.7 (b) 0.09 (c) 0.003

Calculator Application

What is the smallest place value that your calculator can show?

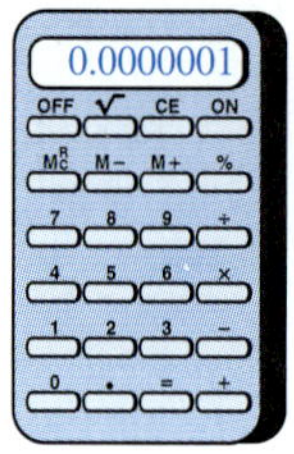

Notice that *every place value to the right of the decimal* ends in the letters "th." Also notice that no commas are written to the right of the decimal point, and the first position to the right of the decimal point is the tenths place.

EXAMPLE 2

Name the place values in 4,521.0132.

SOLUTION

4 , 5 2 1 . 0 1 3 2

- 4 → thousands
- 5 → hundreds
- 2 → tens
- 1 → ones
- 0 → tenths
- 1 → hundredths
- 3 → thousandths
- 2 → ten-thousandths ■

EXAMPLE 3

Write the decimal form for each fraction:

(a) $\frac{3}{10}$ **(b)** $\frac{6}{100}$ **(c)** $\frac{7}{1,000}$

SOLUTION

(a) $\frac{3}{10} = 0.3$

(b) $\frac{6}{100} = 0.06$

(c) $\frac{7}{1,000} = 0.007$ ■

◇ *Do Practice Exercises 5 and 6.*

2 Expanded Notation

Expanded notation is the writing of a number as the sum of its simpler, component parts. For example, twenty-one cents can be written using decimal notation, fraction notation, or using expanded notation with either decimals or fractions:

Decimal Notation		*Expanded Decimal Notation*		*Fraction Notation*		*Expanded Fraction Notation*
0.21	=	0.2 + 0.01	=	$\frac{21}{100}$	=	$\frac{2}{10} + \frac{1}{100}$

EXAMPLE 4

Write 0.6821 in expanded notation using fractions.

SOLUTION

$$0.6821 = \frac{6}{10} + \frac{8}{100} + \frac{2}{1{,}000} + \frac{1}{10{,}000}$$

■

EXAMPLE 5

Write $\frac{9}{10} + \frac{3}{100}$ using decimal notation.

SOLUTION

$$\frac{9}{10} + \frac{3}{100} = 0.93$$

■

◇ *Do Practice Exercises 7 and 8.*

Two decimal numbers can be compared using expanded notation. Look carefully at the next example.

EXAMPLE 6

Is $0.25 > 0.3$?

SOLUTION

$$0.25 = \frac{2}{10} + \frac{5}{100} = \frac{25}{100}$$

$$0.3 = \frac{3}{10} = \frac{30}{100}$$

$\frac{25}{100}$ is not greater than $\frac{30}{100}$

Therefore, $0.25 < 0.3$. ■

◇ *Do Practice Exercises 9 and 10.*

7. Write 0.379 using expanded fraction notation.

8. Write $\frac{1}{10} + \frac{5}{100} + \frac{7}{1{,}000} + \frac{1}{10{,}000}$ using decimal notation.

9. Is 0.6 greater than 0.15?

10. Is $0.22 < 0.5$?

ANSWERS

7. $\frac{3}{10} + \frac{7}{100} + \frac{9}{1{,}000}$
8. 0.1571
9. yes, $0.6 > 0.15$
10. yes, $0.22 < 0.5$

Write each number using words. (Review)

1. 6,225

2. 485,290

3. 12,297,000

4. 19,000,001

5. 7,004,125

6. 3,000,000,010

Name each place value. (Review)

7. 1,091

8. 45,120

9. 960,104

10. 2,045,000

11. 69,275,321

12. 81,271,369,123

Write the digits of each number and place the decimal point correctly.

13. Two thousand, seventy-five

14. Nine hundred one thousand, six hundred twelve

15. Eight million, seventy-four

16. Ninety trillion, four hundred million

Find the digit in the tenths place.

17. 35.2 **18.** 4.681 **19.** 3,621.71 **20.** 4,000.001

Find the digit in the hundredths place.

21. 36.04 **22.** 0.921 **23.** 50.091 **24.** 6,000.01

Find the digit in the thousandths place.

25. 0.083 **26.** 9.214 **27.** 16.0125 **28.** 2,000.8341

Find the digit in the indicated place value.

	Place Value
29. 6,021.3	tenths
30. 0.026	hundredths
31. 10.537	tens
32. 6.057	thousandths
33. 0.0001	hundredths
34. 57,221	thousandths
35. 401.104	hundredths
36. 5,125.07	tenths
37. 327.091	thousandths
38. 64,207.811	hundredths
39. 97.00451	ones
40. 2,000.03	hundredths

	Place Value
41. 103,655.075	hundreds
42. 81,942.085	tenths
43. 95,000.005	thousandths
44. 64,200.81	thousands
45. 78,908.656	thousandths
46. 265.9617	hundredths
47. 3,012.0	thousands
48. 2,025,037.1	one
49. 6,049.037	thousandths
50. 0.00005	hundred thousandths
51. 77,000,000.0001	ten thousandths
52. 2,104,350.2618	one

Write in expanded fraction notation and expanded decimal notation.

53. 0.45

54. 0.512

55. 0.4

56. 0.125

57. 0.1691

58. 0.2500

59. 0.6104

60. 0.0001

Find the greater number.

61. 20, 40

62. 0.79, 0.172

63. 0.6, 0.7

64. 0.004, 0.1

65. 8.0, 10

66. 0.999, 0.99

67. 9.12, 9.967

68. 7,124.1, 6,000.999

Work each problem.

69. Write both seven tenths and three hundredths as fractions, then subtract.

70. Jamie ran three tenths of a mile. Did she run less than or more than 0.5 miles?

71. Does 0 = 0.0?

72. Write 0.25 in expanded notation using decimal notation and again using fraction notation in which the fractions are reduced.

73. Does 9.7 = 9.70?

74. Identify all of the place values in 54,321.1234.

JOURNAL EXERCISES

1. Is the number three tenths a fraction, a decimal, or can it be either one? Explain.

2. Is 0.0001 smaller than 0.001? Is 0.00000001 a number? In our imagination, is there a smallest decimal number? Explain.

1 Read Decimal Numbers

Decimal numbers are widely used, from describing an amount of money on a check to the diameter of a cylinder in an engine. For this reason, it is important to be able to read and write decimal numbers.

OBJECTIVES

1 Read decimal numbers.
2 Write decimal numbers.
3 Identify and work with trailing zeros.

Rule: To Read a Decimal Number

Step 1 Read the whole number part of the decimal number.

Step 2 Replace the decimal point with the word "and."

Step 3 Read the number to the right of the decimal point as a whole number.

Step 4 Name the place value of the LAST DIGIT to the right of the decimal point.

EXAMPLE 1

Read 1.17.

SOLUTION

Step 1 1 → one

Step 2 . → AND

Step 3 17 → seventeen

Step 4 → hundredths

The 7 is located in the hundredths place position. Therefore the number is:

1.17 = one and seventeen hundredths ■

Using the decimal number from the last example, the quantity $1.17 is written as one dollar and seventeen cents. Cents is used in place of hundredths when referring to currency (money).

EXAMPLE 2

Read 4,500.331.

SOLUTION

Step 1 4,500 → four thousand five hundred

Step 2 . → AND

Step 3 331 → three hundred thirty-one

Step 4 → thousandths

Thus the number is:

Four thousand five hundred and three hundred thirty-one thousandths

The number ends with thousandths since the final digit, the 1, falls in the thousandths place value. ■

1. Read 51.87.

2. Read 180.213.

3. Read 4,001.025.

4. Read 15,000.0001.

ANSWERS
1. fifty-one and eighty-seven hundredths
2. one hundred eighty and two hundred thirteen thousandths
3. four thousand one and twenty-five thousandths
4. fifteen thousand and one ten-thousandth

EXAMPLE 3

Read 0.3506.

SOLUTION
The number is:

three thousand five hundred six ten-thousandths ■

HELPFUL HINT

Recall that the word "and" is not used when reading whole numbers. Instead, the word "and" represents the decimal point in decimal notation.

◇ *Do Practice Exercises 1 and 2.*

A whole number such as 1,000,000 requires zeros to hold place values. Similarly, some decimal numbers need zeros on the right hand side of the decimal point to hold place values. These zeros are not mentioned when reading the number. The only place value mentioned on the right hand side of the decimal point is the one in which the last digit falls. Look carefully at the next example.

EXAMPLE 4

Read 3,000,005.000652.

SOLUTION

Step 1	three million five	left side of decimal point
Step 2	and	decimal point
Step 3	six hundred fifty-two	right side of decimal point
Step 4	millionths	place value of last digit

or,

three million five	and	six hundred fifty-two	millionths
↓	↓	↓	↓
whole number	decimal point	decimal number	place value of last digit ■

◇ *Do Practice Exercises 3 and 4.*

2 Write Decimal Numbers

The digits of a decimal number can be found from the written number by using the same basic process, in reverse. Recall that the word "and" indicates the position of the decimal point. Everything to the left of "and" is the whole number portion. Everything to the right of "and" is the decimal portion. Study the next two examples.

EXAMPLE 5

Write the digits of eight and five tenths.

SOLUTION

eight and five tenths → 8.5 ■

EXAMPLE 6

Write the digits of fifty-four and thirty-two hundredths.

SOLUTION

fifty-four and thirty-two hundredths → 54.32 ■

Similarly, fifty-four dollars and thirty-two cents is written as $54.32.

EXAMPLE 7

The speed of a lion is two and eighteen hundredths times the speed of a human sprinter. Write the decimal number. What are the chances a person can outrun a lion?

SOLUTION

Two and eighteen hundredths is 2.18. The chances that a human can outrun a lion are very low, since the lion is more than twice as fast as a human. ■

◇ *Do Practice Exercise 5.*

Zeros may be needed on the right hand side of the decimal point to hold place values. Just remember, the last digit of a decimal number *must* fall in the place value mentioned. Look at the next example.

EXAMPLE 8

Write seven thousand two and four thousandths as a decimal number.

SOLUTION

seven thousand two 7,002
and
four ?4
thousandths 004
→ insert 2 zeros

7,002.004

Two zeros were placed to the right of the decimal point, before the 4. The 4 must end up in the thousandths place. ■

Actually, a number may require several zeros to hold place values as shown in the next example.

EXAMPLE 9

Write eight billion forty thousand and one hundred twelve ten-thousandths as a decimal number.

5. Write the digits of each:

(a) twelve and seven tenths

(b) eighteen and fourteen hundredths

(c) seventy-five and three hundred twenty-two thousandths

ANSWERS
5. (a) 12.7 (b) 18.14 (c) 75.322

6. Write the digits of four thousand and one hundredth.

7. Write twenty and seven thousandths as a decimal number.

8. Does 1,000 = 1,000.000?

9. Does 1,000 = 1,000.001?

10. Does 0.24 = .24?

ANSWERS
6. 4,000.01
7. 20.007
8. yes
9. no
10. yes

SOLUTION

eight billion forty thousand — 8,000,040,000
and
one hundred twelve — ?112
ten-thousandths — 0112

One zero was inserted, so the 2 falls in the ten-thousandths place value.

8,000,040,000.0112 ■

◇ *Do Practice Exercises 6 and 7.*

3 Trailing Zeros

Trailing zeros are zeros that are located to the right of the decimal point and that are at the end of a number. For example, the last three zeros of the following number are trailing zeros.

1,600.0507000

(not trailing zeros: the zeros in 0507; trailing zeros: the final 000)

The addition or removal of trailing zeros does not change the value of a number. Thus:

1,600.507 = 1,600.5070 = 1,600.507000
(trailing zeros)

The three numbers above are equal to one another. The trailing zeros simply state that there are no ten-thousandths, no hundred-thousandths, and no millionths.

Although seldom seen, leading zeros also do not change the value of a number. **Leading zeros** are zeros that occur in front of the whole number. Look at the next example.

EXAMPLE 10

Does 65.09 = 0000065.090000000?

SOLUTION
The two numbers are equal if they have the same numbers in the same place values.

```
          6 5 . 0 9
0 0 0 0 0 6 5 . 0 9 0 0 0 0 0 0 0
```

- no ten-thousandths
- no thousandths
- nine hundredths
- no tenths
- five ones
- six tens
- no hundreds

Thus 65.09 = 0000065.090000000 ■

◇ *Do Practice Exercises 8 through 10.*

State which numbers are greater than one. (Review)

1. 65.0 **2.** 0.054 **3.** 7.001 **4.** 0.00025 **5.** 12.25

6. .99999 **7.** 6.254 **8.** $\frac{3}{4}$ **9.** $5\frac{2}{7}$ **10.** 0.008

Name each place value. (Review)

11. 10.3 **12.** 5.07 **13.** 25

14. 6,003.02 **15.** 0.0123 **16.** 0.0005

Write each number using words.

17. 5.29 **18.** 7.1

19. 0.45 **20.** 6.03

21. 17.52 **22.** 1.03

23. 7.25 **24.** 7.0025

25. 75 **26.** 8.03

27. 12.4 **28.** 16.95

29. 21.005 **30.** 17.92

31. 100.01 **32.** 302.5

33. 362.85 **34.** 971.002

35. 1,040.002 **36.** 3,220.06

37. 650.003 **38.** 0.0007

39. 8,221,324.35 **40.** 5,000,601.003

Write the digits of each number.

41. Eighteen hundredths

42. Seven and two tenths

43. Twelve and four hundredths

44. Sixty-eight

45. Four and seventy-one hundredths

46. Sixty-eight thousand

47. Sixty-eight ten-thousandths

48. Ninety-four and ninety-six hundredths

49. Four hundred twelve and three thousandths

50. Two and one hundredth

51. Six hundred and four thousandths

52. Ninety-nine and nine tenths

53. Seven hundred fifty and twelve thousandths

54. Eight thousand and three hundredths

55. Ninety-two thousand, four hundred

56. Forty-seven million

57. Four thousandths

58. Six thousand and twelve hundredths

59. Four hundred thousand and nine hundredths

60. Nine million forty-five and three tenths

61. One millionth

62. Two hundred million and forty-five thousandths

63. Sixty billion four hundred and six hundredths

64. Sixteen billion sixty-one and two hundred seven thousandths

Work each problem.

65. Write \$27.05 using words.

66. Write \$561.28 using words.

67. Write \$1,340.00 using words.

68. Write the digits of eight hundred dollars and forty cents.

69. Write the digits of one thousand ten dollars and three cents.

70. Write the digits of eight thousand dollars and ninety cents.

71. A gun commonly used by criminals is the 38 special. The number 38 refers to the diameter of the barrel that is 0.38 inches. Write the number using words.

72. An insurance company had to pay a claim of eight million three hundred seventy dollars and sixty cents as their part of the cost of a bridge that collapsed. Write the digits of this number.

73. The eggs of one species of wasp average 0.025 inches in diameter. Write this number using words.

JOURNAL EXERCISES

1. A micrometer (a tool) measures distances to the nearest hundredth of an inch. Write the digits of one hundredth. One part has a diameter of 0.001 inch and another part has a diameter of 0.002 of an inch. Can the micrometer be used to tell them apart? Why, or why not?

2. Human hair grows at the rate of approximately 0.00000001 miles per hour. Write this number using words. We usually talk about the speed of an automobile or airplane in miles per hour. Can miles per hour be used to describe the rate at which hair grows? Can you think of a better scale than miles per hour for hair growth?

OBJECTIVES

1 Add decimal numbers.
2 Subtract decimal numbers.
3 Round decimal numbers.

1 Addition

As you have already learned, the first step in adding whole numbers is to line up place values. Then add columns of identical place values. For example:

Thousands place value
Hundreds place value
Tens place value
Ones place value

$$\begin{array}{r} 3{,}164 \\ +\ \ \ \ 102 \\ \hline 3{,}266 \end{array}$$

The process of adding decimal numbers is basically the same. First, line up identical place values, then add columns of identical place values.

Rule: To Add Decimal Numbers

Step 1 Write the decimal numbers above each other with the decimal points lined up.

Step 2 Add trailing zeros, if necessary, so that both numbers have the same number of digits after the decimal point.

Step 3 Bring the decimal point down.

Step 4 Add as if whole numbers, beginning at the right most column.

EXAMPLE 1

Add 6.2 to 5.34.

SOLUTION

Step 1 $\begin{array}{r} 6.2 \\ +\ 5.34 \\ \hline \end{array}$ Line up the decimal points.

Step 2 $\begin{array}{r} 6.20 \\ +\ 5.34 \\ \hline \end{array}$ Add a trailing zero.

Step 3 $\begin{array}{r} 6.20 \\ +\ 5.34 \\ \hline \end{array}$ Bring the decimal point down.

Step 4 $\begin{array}{r} 6.20 \\ +\ 5.34 \\ \hline 11.54 \end{array}$ Add.

Thus $6.2 + 5.34 = 11.54$ ■

Carrying is used if the numbers in a place value add to 10 or more. This idea is shown in the next example.

1. Add 4.27 + 0.9.

2. Add twenty-one and seven hundredths to thirteen and three tenths.

EXAMPLE 2

Add 17.351 + 6.9.

SOLUTION

```
  17.351
+  6.9          Line up place values.
--------
  17.351
+  6.900        Add trailing zeros.
--------
       .        Bring decimal point down.

  1 1
  17.351
+  6.900        Add.
--------
  24.251
```

■

The process of adding three or more decimal numbers follows the same pattern, as shown in the next example.

EXAMPLE 3

Add \$3.71 + \$2.60 + \$5.04.

SOLUTION

```
  $3.71
   2.60         Line up place values.
+  5.04         Bring decimal point down.
-------
  $  .

   3.71
   2.60
+  5.04         Add.
-------
 $11.35
```

■

◇ Do Practice Exercises 1 and 2.

Always find the decimal point for each number *before* adding. The next two examples show the importance of this idea by adding a whole number to a decimal number.

EXAMPLE 4

Add 12 to 14.6.

SOLUTION

Note that 12 = 12. = 12.0.

```
  12.0          Line up place values.
+ 14.6          Add trailing zero.
------
     .          Bring decimal point down.

  12.0
+ 14.6
------
  26.6          Result.
```

■

ANSWERS
1. 5.17
2. 34.37

EXAMPLE 5

One sign of a kidney malfunction is that a patient may retain water. For this reason, nurses frequently measure the liquids taken in by a patient and also measure the urine excreted by the patient. Mr. Jones took in 2 liters of fluid one day and 1.3 liters the following day. What is the total amount of fluid taken in by Mr. Jones?

SOLUTION

The solution is found by adding 2 liters to 1.3 liters. First, find the decimal point associated with 2:

$$2 = 2.$$

then align the decimal points and add:

→ A trailing zero was added.

```
  2.0      first day
+ 1.3      second day
  3.3 liters total
```

Calculator Application

A new car without an air conditioner and with the small engine is priced at $12,500. The air conditioner is an extra $995.50 and the larger engine is an extra $1,100. Find the total cost including air conditioner, large engine, and a sales tax of $1,167.64.

Answer: $15,763.14

◇ *Do Practice Exercises 3 and 4.*

2 Subtraction

The steps to subtract two decimal numbers is very similar to those of adding decimal numbers. The only difference is that place values are subtracted rather than added.

Rule: To Subtract Two Decimal Numbers

Step 1 Write the decimal numbers above each other with the decimal points lined up.

Step 2 Add trailing zeros, if necessary, so that both numbers have the same number of digits after the decimal point.

Step 3 Bring the decimal point down.

Step 4 Subtract as if whole numbers, beginning at the right most column.

3. Add 16 + 0.05.

4. Find the sum of forty-two and four hundredths with four and one tenth.

ANSWERS
3. 16.05
4. 46.14

5. Subtract 96.53 − 14.2.

6. Subtract seven and fifty-five hundredths from one hundred.

EXAMPLE 6

Subtract 6.9 − 2.4.

SOLUTION

Step 1
$$\begin{array}{r} 6.9 \\ -2.4 \\ \hline \end{array}$$
Line up decimal points.

Step 2
$$\begin{array}{r} 6.9 \\ -2.4 \\ \hline \end{array}$$
Trailing zeros are not required.

Step 3
$$\begin{array}{r} 6.9 \\ -2.4 \\ \hline \end{array}$$
Bring the decimal point down.

Step 4
$$\begin{array}{r} 6.9 \\ -2.4 \\ \hline 4.5 \end{array}$$
Subtract. ■

EXAMPLE 7

Subtract 2.007 from 61.05.

SOLUTION

Step 1
$$\begin{array}{r} 61.05 \\ -\ \ 2.007 \\ \hline \end{array}$$
Line up decimal points.

Step 2
$$\begin{array}{r} 61.050 \\ -\ \ 2.007 \\ \hline \end{array}$$
Add trailing zeros as needed.

Step 3
$$\begin{array}{r} 61.050 \\ -\ \ 2.007 \\ \hline \end{array}$$
Bring the decimal point down.

Step 4
$$\begin{array}{r} {\scriptsize 4\,1}\ \ \\ 61.0\not{5}0 \\ -\ \ 2.007 \\ \hline 59.043 \end{array}$$
Subtract as if whole numbers.
Borrow as needed.
Result. ■

◇ *Do Practice Exercises 5 and 6.*

HELPFUL HINT

The next time you purchase a few items, use your calculator to find the total before going to the cashier.

Borrowing with decimal numbers is similar to that of borrowing with whole numbers, as shown in the next example.

ANSWERS
5. 82.33
6. 92.45

EXAMPLE 8

An item costs $12 plus a sales tax of $0.96. What will the change be if purchased with a twenty dollar bill?

SOLUTION

First add to find the total cost:

```
  $12.00
+    .96
  $12.96
```

Then subtract to find the change:

```
   1 9 91
  $20.00
-  12.96
   $7.04
```

The change should be $7.04. ■

◇ *Do Practice Exercise 7.*

7. An item costs $3.49 plus a tax of $0.23. What would the change be from a five dollar bill?

3 Rounding

Recall that rounding is used to find an approximate value for a number, one with fewer place values of accuracy. The process of rounding a decimal number is the same as the rounding of whole numbers which was covered in Chapter 1. The rule is repeated here for convenience.

Rule: To Round a Number to a Desired Place Value

Step 1 Find the desired place value.

Step 2 Look at the digit immediately to the right.

Step 3 If this digit is five or larger, increase the digit in the desired place value. Otherwise do not change the digit in the desired place value.

Step 4 Replace all digits to the right of the desired place value with zeros.

In the case of decimal numbers, it is also customary to drop any trailing zeros located to the right of the decimal point.

EXAMPLE 9

Round 1,230.657 to the nearest tenth.

SOLUTION

Step 1 1,230.657 → tenths place value

→ number to the right of tenths

Step 2 1,230.657

Step 3 5 is large, thus round 6 tenths up to 7 tenths

Step 4 1,230.700 or 1,230.7 after dropping trailing zeros ■

ANSWER

7. $1.28 in change

8. Round 1,684.278 to the nearest tenth.

9. Round seventy and thirty-nine thousandths to the nearest hundredth.

10. Round 724.61 to the nearest one hundred (not hundredth).

ANSWERS
8. 1,684.3
9. 70.04
10. 700

EXAMPLE 10

Round 9.81 to the nearest ones place value.

SOLUTION

Step 1 9.81 (9 → ones place value)

Step 2 9.81 (8 → number to the right of ones)

Step 3 8 is large, thus round 9 ones up to 10 ones

Step 4 10.00 or 10 after dropping trailing zeros ■

◇ *Do Practice Exercises 8 through 10.*

Calculator Application

Some calculators round off decimal numbers by allowing you to specify the number of digits to the right of the decimal point. Does your calculator have this feature?

Write each number using words. (Review)

1. 6.02

2. 50.3

3. 7,420.02

4. 40.005

5. 6,221.99

6. 2,021,000.75

Write the digits of each number. (Review)

7. Seven and four tenths

8. Fourteen and four thousandths

9. Sixty-five

10. Seventy-four thousandths

11. One thousand fifty-one and thirty-five ten-thousandths

12. Two thousand and six tenths

Round each to the indicated place value.

	to the nearest
13. 1,340.52	tenth
14. 25.625	hundredth
15. 57.178	one
16. 750.27	ten
17. 3,721.5	one thousand
18. 0.006	hundredth
19. 54,332	ten thousand
20. 312.06	tenth

	to the nearest
21. 415.81	one
22. 916.07	hundredth
23. 4,250.075	tenth
24. 1,422.68	one thousand
25. 3,000.9	thousand
26. 4,500.68	tenth
27. 9,681.008	hundredth
28. 10,080.99	hundred
29. 25,000.069	tenth
30. 6,245,221.1365	thousandth

Evaluate each.

31. 6.4 − 2.5

32. 1.9 + 12

33. 7.3 + 0.01

34. 7.3 + 2.5

35. 15.7 − 6.1

36. 7.4 + 1.03

37. 1.05 + .02

38. 6.21 − 4

39. 3.07 + 900

40. 12 + 17

41. 8.4 − 0.12

42. 1,040 − 2.1

43. 17.01 − 4.5

44. 3.6 + 2.13

45. 1.9 + 2.04 + 5

46. 3.21 − 1.7

47. 6.54 − 2.08

48. 3.21 + 6.9

49. 9.9 + 12

50. 13 − 11.9

51. 51 − 2.07

52. 12 − 6.02

53. 7.4 + 15.003

54. 99.99 − 0.003

55. $170 - 4.2$

56. $12 + 3.1 + 0.09$

57. $1{,}007 - 975.9$

58.
$$\begin{array}{r} 16.1 \\ 3.24 \\ +\ 4.005 \\ \hline \end{array}$$

59.
$$\begin{array}{r} 17{,}034.213 \\ -\ \ \ \ 66.07 \\ \hline \end{array}$$

60.
$$\begin{array}{r} 1{,}740.21 \\ 21.035 \\ +\ 145.\ \ \ \ \\ \hline \end{array}$$

First estimate by rounding to the nearest whole number and solving. Then find the exact answer.

61. $6.5 + 3.2$

62. $7.8 - 1.2$

63. $5.1 + 4.65$

64. $65.3 + 2{,}041$

65. $7{,}053.6 - 21.4$

66. $50 - 2.07$

67. $23.7 - 16.09$

68. $19 + 12.8$

69. $16.4 + 4.85$

70. $12.2 - 3.7$

Work each problem.

71. Rosa is paid \$4.50 per hour. Use addition to find her wages for three hours.

72. Items costing \$4.65, \$3.75, and \$19.99 were purchased with a fifty dollar bill. The change given was \$20.61. Is this the correct change? If not, what is the correct change?

73. A tax of forty-two cents is added to an item costing six dollars and nine cents. What is the change from a twenty dollar bill?

74. A garden is shaped like a square that measures eight and four tenths meters on each side. How many meters of fencing material is needed to keep a dog out of the garden?

75. Sum 5.41 and 19, then subtract this quantity from 31.5.

76. Bernice's hair boutique had the following income and expenses last month:

Week	*Income*	*Expenses*
1	$1,420.35	$1,865.25
2	$1,840.26	$ 345.00
3	$1,369.50	$ 827.70
4	$2,050.85	$1,540.75

Find the total income and expenses. How much profit did the shop make?

JOURNAL EXERCISES

1. A salesperson must keep records of auto mileage for a company car. She maintains a log in which she writes down the odometer reading before and after work. The car is not driven again until the next day. Find her daily mileage for all three days shown.

Date	*Before Work*	*After Work*
9/10	26,341.2	26,379.9
9/11	26,379.9	26,351.2
9/12	26,451.2	26,550.7

There is a mistake with at least one of the numbers, find it. How do you know it is a mistake?

2. The amount of natural gas needed to heat a home was:

December	71.5 cubic feet
January	80.3 cubic feet
February	102.3 cubic feet

Which month would you guess to be the coldest? What assumptions are you making? Would your answer change if you found out that the furnace had been turned off for two weeks during January? Explain.

3. In your own words, compare the process of adding decimal numbers to that of adding whole numbers.

4. Ferris Automotive had a checking account balance of $3,680.21 on August 1. During the first week of August, they made the following deposits and wrote the following checks:

Date	*Deposit*	*Checks*
August 2	$ 384.27	$ 18.99
"		172.43
August 5		1,243.71
"		1,902.12
"		812.14
August 6	1,743.21	625.00
"		90.00

What is the account balance at the end of the week? Check their daily balances and determine whether or not the account balance ever fell below zero. Contact a bank and find out their policy regarding checks that bring an account balance below zero and describe your findings.

1 Multiplication

Two decimal numbers can be multiplied by first changing them to fractions, then multiplying the fractions as shown in the first example.

EXAMPLE 1

Multiply 0.7 times 0.3 by first changing each to a fraction, then multiplying.

SOLUTION

$$0.7 \cdot 0.3$$

$$\frac{7}{10} \cdot \frac{3}{10} \quad \text{Change each to a fraction.}$$

$$\frac{7 \cdot 3}{10 \cdot 10} \quad \text{Multiply numerators and denominators.}$$

$$\frac{21}{100} \quad \text{or} \quad 0.21$$ ■

◇ *Do Practice Exercises 1 and 2.*

Notice in Example 1 that 7 and 3 were multiplied (in the numerator) to get 21. This value was then divided by 100, resulting in the movement of the decimal point two places to the left. Effectively, the decimal numbers were multiplied without regard to the decimal point. Then the decimal point was moved to the left. This idea is used in the following rule.

Rule: To Multiply Two Decimal Numbers

Step 1 Write one number above the other.

Step 2 Ignore the decimal points and multiply as if they are whole numbers.

Step 3 Count the number of digits to the right of the decimal place in both factors.

Step 4 Move the decimal point to the left the same number of spaces.

OBJECTIVES

1 Multiply decimal numbers.

2 Multiply decimal numbers by powers of ten.

1. Multiply 0.5×0.7.

2. Find the product of three hundredths with seven tenths.

ANSWERS
1. 0.35
2. 0.021

3. Multiply 1.2 times 12.1.

4. Multiply three and four hundredths times eight and two tenths.

EXAMPLE 2

Multiply 6.41 times 3.1.

SOLUTION

Step 1
$$\begin{array}{r} 6.41 \\ \times\ \ 3.1 \\ \hline \end{array}$$
Write one above the other.

Step 2
$$\begin{array}{r} 6.41 \\ \times\ \ 3.1 \\ \hline 641 \\ 1923\ \\ \hline 19871 \end{array}$$
Ignore decimal points and multiply.

Step 3
$$\begin{array}{r} 6.41 \\ \times\ 3.1 \\ \hline \end{array}$$
6.41 → 2 digits to the right of decimal place plus
3.1 → 1 digit to the right of decimal place makes 3 digits to right of decimal place.

Step 4 1 9.8 7 1 Move decimal to left 3 places.

Thus $6.41 \cdot 3.1 = 19.871$ ■

EXAMPLE 3

Find the product of 2.3 and 0.3.

SOLUTION

Step 1
$$\begin{array}{r} 2.3 \\ \times 0.3 \\ \hline \end{array}$$
Write one above the other.

Step 2
$$\begin{array}{r} 2.3 \\ \times 0.3 \\ \hline 69 \end{array}$$
Multiply.

Step 3 2 digits to the right of decimal places

Step 4 .6 9 Move decimal point to the left 2 places.

0.69 ■

EXAMPLE 4

Evaluate $16 \cdot 3.0002$.

SOLUTION

$$\begin{array}{r} 3.0002 \\ \times\ \ \ \ 16 \\ \hline 180012 \\ 30002\ \\ \hline 480032 \end{array}$$
3.0002 → There are 4 digits to the right of the decimal point.

The decimal point is moved four places to the left: 48.0032 ■

◇ *Do Practice Exercises 3 and 4.*

ANSWERS
3. 14.52
4. 24.928

Sales tax is a very common use of decimal numbers. Look at the next example.

EXAMPLE 5

A farmer bought a tractor for \$27,400. The sales tax in his area was found by multiplying 0.0825 times the cost of the tractor. Find the sales tax.

SOLUTION

$$\begin{array}{r} \$27{,}400 \\ \times\ 0.0825 \\ \hline 137000 \\ 54800 \\ 219200 \\ \hline 2{,}260.5000 \end{array}$$

The sales tax on the tractor is \$2,260.50. ■

Calculator Application

In one city, a sales tax is calculated by multiplying 0.075 times the cost of an item. How much sales tax applies to a new automobile costing \$15,645? What is the total cost of the automobile including sales tax?

Answers: \$1,173.38; \$16,818.38

Exponents have the same meaning for decimal numbers as they have for whole numbers and fractions. Recall that a quantity raised to the second power is squared and a quantity raised to the third power is cubed. Look at the next example.

EXAMPLE 6

Find the square and cube of 0.2.

SOLUTION

Square: $(0.2)^2 = 0.2 \cdot 0.2 = 0.04$

Cube: $(0.2)^3 = 0.2 \cdot 0.2 \cdot 0.2 = 0.008$ ■

◇ *Do Practice Exercises 5 through 7.*

5. Find $(0.9)^2$.

6. Find the cube of three tenths.

7. Square sixteen hundredths.

ANSWERS
5. 0.81
6. 0.027
7. 0.0256

8. Multiply $624 \times 1{,}000$.

9. Multiply $32.4 \cdot 100$.

10. Multiply two thousand and one hundredth times ten.

2 Multiplication by a Power of Ten

Recall from Chapter 1 that a **power of ten** is ten raised to some power, such as:

			power of ten
$10^0 =$	1	$=$	1
$10^1 =$	10	$=$	10
$10^2 =$	$10 \cdot 10$	$=$	100
$10^3 =$	$10 \cdot 10 \cdot 10$	$=$	$1{,}000$
$10^4 =$	$10 \cdot 10 \cdot 10 \cdot 10$	$=$	$10{,}000$

We saw that a whole number can be multiplied times a power of ten by counting the number of zeros in the power of ten and attaching the same number of zeros to the end of the whole number. For example:

$$764 \cdot 100 = 76{,}400$$

764 → whole number; 100 → power of ten; 76,400 → Add 2 zeros.

The process of multiplying a decimal number times a power of ten is identical. The pattern is shown below:

$$\begin{array}{r} 3.14 \\ \times \quad 1 \\ \hline 3.14 \end{array} \qquad \begin{array}{r} 3.14 \\ \times \quad 10 \\ \hline 31.4 \end{array} \qquad \begin{array}{r} 3.14 \\ \times \; 100 \\ \hline 314 \end{array} \qquad \begin{array}{r} 3.14 \\ \times 1{,}000 \\ \hline 3{,}140 \end{array}$$

powers of ten

The decimal place in the product is moved to the right one additional place for each additional zero in the power of ten. A decimal number can be multiplied by a power of ten by:

1. First count the number of zeros in the power of ten.
2. Then move the decimal point to the *right* the same number of places and insert zeros as needed.

The next example shows this process.

EXAMPLE 7

Multiply $65.8 \times 100{,}000$.

SOLUTION

The number 100,000 has 5 zeros. Move the decimal point in 65.8 to the right 5 places filling in zeros as needed.

$$65.80000.$$

Move decimal point five places.

Thus $65.8 \times 100{,}000 = 6{,}580{,}000.$ ■

◇ ***Do Practice Exercises 8 through 10.***

ANSWERS
8. 624,000
9. 3,240
10. 20,000.1

Numbers such as 0.1, 0.01, 0.001, and 0.0001, are also powers of ten. This will be more fully explained when working with negative exponents in algebra. However, the pattern below suggests that the power of ten occurs in the denominators.

$$0.1 = \text{one tenth} = \frac{1}{10} = \frac{1}{10^1}$$

$$0.01 = \text{one hundredth} = \frac{1}{100} = \frac{1}{10^2}$$

$$0.001 = \text{one thousandth} = \frac{1}{1{,}000} = \frac{1}{10^3}$$

$$0.0001 = \text{one ten-thousandth} = \frac{1}{10{,}000} = \frac{1}{10^4}$$

$$\vdots \qquad \vdots \qquad \vdots \qquad \vdots$$

Multiplication by one tenth is the same as dividing by ten as shown below:

$$27.6 \times 0.1 = 27.6 \times \frac{1}{10} = \frac{27.6}{1} \times \frac{1}{10} = \frac{27.6}{10} = 2.76$$

The end result is to move the decimal point to the left one place. Multiplication by one hundredth is the same as dividing by one hundred or moving the decimal point to the left two places, and so forth.

Multiplying a decimal number times a power of ten that is less than one is done by:

1. First count the number of zeros in the power of ten.
2. Then move the decimal point to the *left* the same number of places and insert zeros as needed.

EXAMPLE 8

Multiply 475.6 × 0.0001.

SOLUTION

The 1 in 0.0001 is four places to the right of the decimal point. Multiply by moving the decimal place in 475.6 to the left four places, adding zeros as needed.

$$0.0475.6$$

475.6 × 0.0001 = 0.04756 ■

HELPFUL HINT

Multiplication by a power of ten greater than one results in a larger number thus the decimal point moves to the *right*. Multiplication by a number less than one results in a smaller number thus the decimal point moves to the *left*.

◇ *Do Practice Exercises 11 through 13.*

11. Multiply 625 · 0.01.

12. Multiply 3,050 by 0.001.

13. Multiply twenty-seven times one-millionth.

ANSWERS
11. 6.25
12. 3.05
13. 0.000027

Work each showing all steps.

1. $3.8 \cdot 2.1$ **2.** $5.7 \cdot 8$ **3.** 9.04×2.5 **4.** $3.2 \cdot 4.06$

5. $7.1 \cdot 3.4$ **6.** $(9)(9.4)$ **7.** $5.8\,(9.7)$ **8.** $6.3 \cdot 9$

9. $4.4 \cdot 5.6$ **10.** $(3.4)^2$ **11.** 5.21×0 **12.** $6.43 \cdot 1$

13. $12.2 \cdot 4$ **14.** $15 \cdot 3.9$ **15.** $6.8 \cdot 8.7$ **16.** $4.2 \cdot 0.03$

17. $(10.9)(1.2)$ **18.** $8.8 \cdot 0.45$ **19.** 12.4×13.15 **20.** $6.9 \cdot 4.01$

21. $6.5 \cdot 0.01$ **22.** $7.03 \cdot 1.4$ **23.** $12 \cdot 1.8$ **24.** 16×0.35

25. $7.9 \cdot 80$ **26.** $3.4 \cdot 6.09$ **27.** 12.75×1 **28.** $9.4 \cdot 1.001$

29. $0.08 \cdot 80.09$ **30.** $111.01 \cdot 4.8$ **31.** $12.7 \cdot 4.03$ **32.** $10.81 \cdot 14.5$

Work each showing all steps. (Some are review problems.)

33. 6.4×3.1 **34.** $5.7 + 2.801$ **35.** $7.9 - 2$ **36.** $(5.7)(3.05)$

37. $6.5 \cdot 7.3$

38. $9.1 \cdot 0.005$

39. $3{,}024 + 0.02$

40. 75×4.1

41. $17.21 - 6.99$

42. 7.04×8.1

43. $6 \cdot 0.003$

44. $(2.79)(81.5)$

45. 8.7×13.12

46. 4×5.9

47. $3.1 \cdot (0.03)$

48. $(2.45)(17.5)$

49. $(7.24) + 3.51$

50. $(2.1)^2$

51. $(3.01)^2$

52. $21 - 6.04$

53. $3.2 - 0.09$

54. $4.2 \cdot 1.8$

55. $4.5 + 2.7$

56. 3.6×0.05

57. $\begin{array}{r} 7.813 \\ \times \quad 1.5 \\ \hline \end{array}$

58. $\begin{array}{r} 6.214 \\ +2.1 \\ \hline \end{array}$

59. $\begin{array}{r} 5.91 \\ -4.07 \\ \hline \end{array}$

60. $\begin{array}{r} 4.31 \\ \times \ 0.1 \\ \hline \end{array}$

Multiply by the power of ten.

61. $\begin{array}{r} 7{,}243.5 \\ \times \qquad .1 \\ \hline \end{array}$

62. $\begin{array}{r} 642.75 \\ \times \quad 100 \\ \hline \end{array}$

63. $\begin{array}{r} \$451.25 \\ \times \qquad 10 \\ \hline \end{array}$

64. $\begin{array}{r} 6{,}005.21 \\ \times \qquad .01 \\ \hline \end{array}$

65. $\begin{array}{r} 2{,}431.05 \\ \times \quad 1{,}000 \\ \hline \end{array}$

66. $\begin{array}{r} 751.09 \\ \times \ .0001 \\ \hline \end{array}$

67. $\begin{array}{r} 16.005 \\ \times \ 1{,}000 \\ \hline \end{array}$

68. $\begin{array}{r} 1{,}021.0005 \\ \times \qquad 0.001 \\ \hline \end{array}$

Write each number using words. (Review)

69. 645.12

70. 7,240.008

71. 1,912.0375

72. 35,000.01

Write the digits of each number. (Review)

73. Seven hundredths

74. Four and sixteen thousandths

75. Twelve and forty-seven thousandths

76. Six million and six tenths

Work each problem.

77. Multiply three hundredths times five and one tenth.

78. Add three to fourteen and seventy-two thousandths.

79. Square twelve hundredths.

80. Multiply six and four tenths times three and four hundredths and add the product to seventeen.

81. Subtract two and three tenths from seven and four tenths then multiply the difference times twenty and four hundredths.

82. An automobile is purchased with one thousand dollars down and a payment of one hundred twenty dollars and thirty cents per month for forty months. Find the total cost of the automobile.

83. A sales tax is calculated by multiplying 0.08 times the sales price. What is the total cost, including tax, of an item priced at $12.00?

84. An investment returns an *annual* interest calculated by multiplying 0.095 times the investment. What would one year's interest be on an investment of one thousand five hundred dollars?

85. An individual earns \$6.30 per hour for regular time and 1.5 times \$6.30 for overtime. What is the total income for a sixty hour workweek if everything above forty hours is paid at the overtime wage rate?

86. The cost of shipping a particular item is \$3.15 for the first one, and \$.03 for each item thereafter. Find the total cost to ship fifty of these items.

87. It costs \$55.00 per day plus ten cents per mile to rent an automobile. What is the cost of renting the car for a day during which it was driven sixty-five miles?

88. A snail moves about twenty-five feet per hour or five thousandths miles per hour. A military supersonic (faster than sound) jet travels at approximately four hundred thousand times the speed of the snail. How fast does the jet travel in feet per hour and in miles per hour?

JOURNAL EXERCISES

1. An individual plans to walk a mile. After walking one-half of a mile he writes the time in a book. After walking one-half of the remaining one-half mile distance he writes that time in the book. In fact, each time he walks one-half of the remaining distance he writes the time in the book. Draw a sketch describing the situation. How many entries will have been made in the book by the time the individual finishes the mile?

2. Explain the differences and similarities between multiplying whole numbers and multiplying decimal numbers.

3. A house costs \$65,000. The house is purchased with a down payment of \$3,600 and 360 payments of \$555.12 each. Find the total of the down payment plus the monthly payments and round to the nearest ten thousand dollars. Compare the rounded figure for the total payments to the original price using a ratio. Why are they so different?

1 Division

Division with whole numbers is accomplished using four basic steps:

1. Divide.
2. Multiply.
3. Subtract.
4. Bring down the next digit.

The division continues until there are no more digits to bring down. The answer to the division problem is the quotient plus any remainder. Division with decimals is very similar except that two extra steps are needed before dividing decimal numbers. The idea behind these two extra steps will now be explained.

A division problem, even one with decimals, can be written as a fraction. For example:

$$15.54 \div 4.2 = \frac{15.54}{4.2}$$

It is easier to keep track of decimal points if the denominator of the fraction (divisor) is a whole number. The denominator 4.2 can be made into a whole number if it is multiplied times 10. If the denominator is multiplied by 10, the numerator must also be multiplied by 10 in order for the fraction to remain the same. Thus we have:

$$15.54 \div 4.2 = \frac{15.54}{4.2} = \frac{15.54}{4.2} \times \underbrace{\frac{10}{10}}_{\text{form of one}} = \frac{155.4}{42}$$

(form of one: $\frac{10}{10}$; now a whole number: 42)

Notice that the decimal points of both 15.54 and 4.2 were moved to the right one place. Also notice that the denominator (divisor) is now a whole number. The result of the process is the same as the first two steps of the rule that follows the first example.

EXAMPLE 1

What would you multiply $\frac{5.9}{7.21}$ by to get an equivalent fraction with a whole number in the denominator?

SOLUTION
The divisor, 7.21, must be multiplied by 100 to become a whole number. Therefore, it is necessary to multiply the original fraction times $\frac{100}{100}$:

$$\frac{5.9}{7.21} \times \frac{100}{100} = \frac{590}{721}$$ ■

◇ *Do Practice Exercises 1 and 2.*

OBJECTIVES

1 Divide using decimal numbers.

2 Divide by powers of ten.

1. What would you multiply $\frac{26.241}{3.25}$ by to get an equivalent fraction with a whole number in the denominator?

2. What would you multiply $\frac{7.054}{5.001}$ by to get an equivalent fraction with a whole number in the denominator?

ANSWERS

1. multiply by $\frac{100}{100}$

2. multiply by $\frac{1,000}{1,000}$

EXAMPLE 3

Divide 54 by 2.25.

SOLUTION

```
2.25)54

2.25.)54.00.      Move decimal point two places.

        24. ——→ Result.
  225)5400.       Divide.
      450
       900
       900
         0
```

◇ *Do Practice Exercises 3 and 4.*

The first two steps of the rule *are not needed* if the divisor is already a whole number as shown in the next example.

EXAMPLE 4

Three pair of socks are on sale for $4.59. What is the average cost per pair?

SOLUTION

The answer is found by dividing 3 into 4.59. No decimal points need to be moved since the divisor 3 is already a whole number.

```
   $1.53
  3)4.59
    3          Divide.
    15
    15
      9
      9
      0
```

Thus the average cost per pair is $1.53 ■

◇ *Do Practice Exercises 5 and 6.*

Calculator Application

A college bought 350 student desks for $23,992.50. What was the average cost per desk?

Answer: $68.55

As we have seen before, a division may go on without end. If this occurs, the division can be stopped after learning several digits. The number of digits needed depends on the place value of accuracy that is desired.

3. Divide 15.75 ÷ 2.5.

4. Divide 17.86 by 3.8.

5. Divide 98.4 ÷ 8.

6. Find the quotient of twenty-one and forty-eight hundredths divided by twelve.

ANSWERS
3. 6.3
4. 4.7
5. 12.3
6. 1.79

Rule: To Divide Decimal Numbers

Step 1 Move the decimal point in the divisor to the right until the divisor is a whole number.

Step 2 Move the decimal point in the dividend to the right the same number of places.

Step 3 Place the decimal point in the quotient directly above the decimal point in the dividend.

Step 4 Divide.

Step 5 Add trailing zeros to the dividend as needed.

EXAMPLE 2

Divide 15.54 by 4.2.

SOLUTION

$$4.2\overline{)15.54}$$

Step 1 4.2 — Move decimal point in divisor one place.

Step 2 15.5.4 — Move decimal point in dividend one place.

Step 3 $42\overline{)155.4}$ (with decimal point above) — Place decimal point in quotient.

Step 4 Divide as if dividing whole numbers.

```
         3.7
   42)155.4
      126
       294
       294
         0
```

15.54 ÷ 4.2 = 3.7 ■

The division from the previous example can be checked for accuracy by multiplying the quotient times the divisor.

```
   3.7
 × 4.2
    74
  148
 15.54
```

15.54 is the dividend, therefore the division was done correctly.

7. Divide 26.9 ÷ 3.

8. Divide three by eleven.

EXAMPLE 5

Evaluate 10.1 ÷ 3.

SOLUTION

$$
\begin{array}{r}
3.366\ \ldots \\
3\overline{)10.100} \\
\underline{9} \\
1\ 1 \\
\underline{9} \\
20 \\
\underline{18} \\
20 \\
\underline{18} \\
2
\end{array}
$$

■

In the last example, the division continues without any end, so that the answer can be rounded off as follows.

If Rounded to the Nearest	*Digits Needed*	*Answer*
tenth	3.36	3.4
hundredth	3.366	3.37
thousandth	3.3666	3.367
ten-thousandth	3.36666	3.3667

Depending on the number of digits desired, any of several answers would be correct for the last example. Alternatively, the result could have been written as:

$$3.3666\ \ldots \quad \text{or} \quad 3.3\overline{6}$$

where the bar over the $\overline{6}$ means that the 6 repeats without end. A number such as $3.3\overline{6}$ is called a **repeating decimal**.

◇ *Do Practice Exercises 7 and 8.*

ANSWERS

7. $8.96666\ldots = 8.9\overline{6}$

8. $0.272727\ldots = 0.\overline{27}$

2 Division by Powers of Ten

Division of a number by a power of ten greater than 1 results in a *smaller* number thus the decimal number must be moved to the left. For example:

$$16.4 \div 100 = \frac{16.4}{100} = 0.164$$

$$2{,}123.05 \div 1{,}000 = \frac{2{,}123.05}{1{,}000} = 2.12305$$

Division of a number by a power of ten less than 1 results in a *larger* number thus the decimal point must be moved to the right. For example:

Use rule to divide fractions.

$$27.3 \div 0.01 = 27.3 \div \frac{1}{100} = 27.3 \cdot \frac{100}{1} = 2{,}730$$

$$562.47 \div 0.001 = 562.47 \div \frac{1}{1{,}000} = 562.47 \cdot \frac{1{,}000}{1} = 562{,}470$$

EXAMPLE 6

Divide each:

(a) $27 \div 0.001$
(b) $35.2 \div 10$
(c) $69.05 \div 0.01$

SOLUTION

(a) 27,000 Decimal moved to the right 3 places.
(b) 3.52 Decimal moved to the left 1 place.
(c) 6,905 Decimal moved to the right 2 places. ■

◇ *Do Practice Exercises 9 and 10.*

9. Divide $17 \div 1{,}000$.

10. Evaluate $1.85 \div 0.01$.

ANSWERS
9. 0.017
10. 185

name date hour

Work each showing all steps. If you would like, check your answer with a calculator.

1. 6.9 ÷ 3

2. 12.3 ÷ 0.3

3. 4.5 ÷ 2.25

4. 22.448 ÷ 1.84

5. 12.15 ÷ 4.05

6. 25.50 ÷ 2.5

7. 23.45 ÷ 0.5

8. 45 ÷ 0.8

9. 96.6 ÷ 4.2

10. 2.54 ÷ 0.8

11. 16.8 ÷ 0.02

12. 28.2 ÷ 12

13. 0.78 ÷ 5.2

14. 12.09 ÷ 0.03

15. 36 ÷ 0.5

16. 0.47 ÷ 0.005

17. 22.0 ÷ 5.5

18. 11.165 ÷ 3.85

19. 9.87 ÷ 4.7

20. 0.152 ÷ 1.9

21. 0.276 ÷ 0.04

22. 4.875 ÷ 1.5

23. 6,834 ÷ 3.4

24. 1.725 ÷ 57.5

Find the quotient to the nearest tenth.

25. 17.3 ÷ 4.5

26. 6.1 ÷ 0.03

27. 9.9 ÷ 12

28. 805.7 ÷ 3

29. 65 ÷ 0.25

30. 471 ÷ 5.1

31. 16.5 ÷ 2.4

32. 3.25 ÷ 9.1

33. 14.7 ÷ 2.05

34. 75.36 ÷ 0.34

Find the quotient to the nearest hundredth.

35. $16.2\overline{)0.413}$

36. $5.9\overline{)5.991}$

37. $0.3\overline{)7.21}$

38. $7.4\overline{)0.005}$

39. $6\overline{)12.231}$

40. $17\overline{)12.005}$

41. 165 ÷ 7.4

42. 0.321 ÷ 0.04

43. 7.85 ÷ 4.21

44. 0.1 ÷ 0.3

Divide by the power of ten.

45. $16.5 \div 10$ **46.** $2.71 \div 100$ **47.** $4.73 \div 0.01$ **48.** $724 \div 0.001$

49. $2.95 \div 1{,}000$ **50.** $6{,}241.35 \div 0.1$ **51.** $16.08 \div 0.1$ **52.** $175 \div 100$

Evaluate, showing all work. (Some are review problems.)

53. $5.2 \cdot 3$ **54.** $4.1 + 12$ **55.** $6.4 - 3.01$

56. $12 \div 0.25$ **57.** $5.3 - 2.4$ **58.** $6.71 \cdot 0.2$

59. $5.9 \cdot 7$ **60.** $25.92 \div 8.1$ **61.** $4 + 6.1$

62. $19.5 - 0.07$ **63.** $(16)(8.1)$ **64.** $14.13 - 4.71$

65. 9.24 + 159

66. 0.512 ÷ 16

67. 16 + 17.5

68. 7.5 ÷ 30

69. 0.265 ÷ 1.06

70. 14 · 3.72

Work each problem.

71. Find the average of seven, four and two hundredths, and twelve and four thousandths. Round the answer to the nearest tenth.

72. Fourteen items cost forty dollars and eighty-one cents. That includes a tax of two dollars and thirty-one cents. What is the average cost of each item not including taxes?

73. A home is to be paid for with three hundred sixty monthly payments. The sum of the payments is equal to one hundred eighty-nine thousand, one hundred forty-four dollars. Find the monthly payment.

74. The sales tax in one city is found by multiplying 0.075 times the cost of the items. What is the total tax on a new minivan listed at eighteen thousand five hundred dollars?

75. For a total of one thousand nine hundred five dollars, Sal installed carpet costing fifteen dollars per square yard. How many square yards of carpet did Sal buy?

JOURNAL EXERCISES

1. In your own words, explain why decimal places may need to be moved before dividing decimal numbers. Try to find a way to divide decimal numbers without first moving the decimal point.

2. What are repeating decimals? Does every division result in repeating decimals? Identify several fractions that contain repeating decimals when written as decimal numbers.

Every decimal number can be written as a fraction and every fraction can be written as a decimal number although the decimal number may have an infinite number of digits. A mixed number is simply a different way to write an improper fraction and can also be written as a decimal number. In this section we are going to learn how to change a number from one notation form to another.

1 From Fraction Notation to Decimal Notation

A fraction can be thought of as a division problem. A fraction can be changed into a decimal number by dividing the denominator into the numerator.

EXAMPLE 1

Write the fraction $\frac{3}{4}$ as a decimal number.

SOLUTION

$4\overline{)3}$ — $\frac{3}{4} = 3 \div 4$

$4\overline{)3.000}$ — Place decimal point. Add trailing zeros.

$$\begin{array}{r} .75 \\ 4\overline{)3.000} \\ \underline{2\,8} \\ 20 \\ \underline{20} \\ 0 \end{array}$$ Divide.

Thus $\frac{3}{4} = 0.75$ ■

EXAMPLE 2

Change $\frac{1}{8}$ to decimal notation.

SOLUTION

$$\frac{1}{8} = 1 \div 8 = 0.125$$ ■

EXAMPLE 3

Change $\frac{7}{10}$ to decimal notation.

SOLUTION

$$\frac{7}{10} = 7 \div 10 = 0.7$$ ■

◇ *Do Practice Exercises 1 through 3.*

OBJECTIVES

1 Change from fraction notation to decimal notation.

2 Change from decimal notation to fraction notation.

3 Change from mixed number notation to decimal notation.

4 Change from decimal notation to mixed number notation.

1. Write $\frac{11}{20}$ using decimal notation.

2. Change fifteen sixteenths to decimal notation.

3. Write $\frac{5}{8}$ as a decimal number.

ANSWERS
1. 0.55
2. 0.9375
3. 0.625

Some fractions and their decimal equivalents occur frequently enough to be memorized. Several common ones are shown below.

$\frac{1}{2} = 0.5$	$\frac{1}{4} = 0.25$	$\frac{1}{5} = 0.2$	$\frac{1}{8} = 0.125$
	$\frac{3}{4} = 0.75$	$\frac{2}{5} = 0.4$	$\frac{3}{8} = 0.375$
		$\frac{3}{5} = 0.6$	$\frac{5}{8} = 0.625$
		$\frac{4}{5} = 0.8$	$\frac{7}{8} = 0.875$

It is common to round the decimal equivalent to a desired place value of accuracy for any fraction that does not have an exact decimal representation. For example, find the decimal equivalent of the fraction $\frac{1}{3}$ to the nearest one hundredth:

$$
\begin{array}{r}
.333\ldots \\
3\overline{)1.000} \\
\underline{9} \\
10 \\
\underline{9} \\
10 \\
\underline{9} \\
1
\end{array}
$$

The digits go on without end.

The fraction 3 rounded to the nearest one hundredth is 0.33.

2 From Decimal Notation to Fraction Notation

Finding the fraction that is equivalent to a decimal number is as easy as reading the decimal number. Recall that the last two steps of the rule to read a decimal number are to:

Step 3 Read the number to the right of the decimal point as a whole number.

Step 4 Name the place value of the LAST DIGIT to the right of the decimal point.

For example, the decimal number 0.57 is read as fifty-seven hundredths. Therefore, the fractional equivalent of 0.57 is 57 out of 100, or:

$$0.57 = \frac{57}{100}$$

The steps are given in the following rule.

Rule: To Change From Decimal Notation to Fraction Notation

Step 1 Write everything to the right of the decimal point in the numerator.

Step 2 Place the correct power of ten in the denominator. The correct power of ten is the place value of the last digit.

Step 3 Reduce the fraction if possible.

EXAMPLE 4

Write 0.375 as a fraction.

SOLUTION

Step 1 $\dfrac{375}{?}$ Place 375 in the numerator.

Step 2 $\dfrac{375}{1{,}000}$ The 5 in 0.375 is in thousandths place so put 1,000 in the denominator.

Step 3 $\dfrac{375}{1{,}000} = \dfrac{3}{8}$ Reduce.

Thus $0.375 = \frac{3}{8}$. ■

EXAMPLE 5

Change 0.06 to fraction notation.

SOLUTION

$$0.06 = \frac{6}{100} = \frac{3}{50}$$ ■

EXAMPLE 6

Convert 0.12 to fraction notation.

SOLUTION

$$0.12 = \frac{12}{100} = \frac{3}{25}$$ ■

◇ *Do Practice Exercises 4 through 6.*

4. Write 0.50 using fraction notation and reduce.

5. Write sixty-five hundredths as a fraction and reduce.

6. Write forty-five thousandths as a fraction and reduce.

ANSWERS

4. $\frac{1}{2}$

5. $\frac{13}{20}$

6. $\frac{9}{20}$

7. Write $4\frac{5}{8}$ using decimal notation.

8. Write $1{,}000{,}004\frac{4}{5}$ using decimal notation.

9. Write $7\frac{1}{100}$ using decimal notation.

ANSWERS
7. 4.625
8. 1,000,004.8
9. 7.01

3 From Mixed Number Notation to Decimal Notation

A mixed number has a whole number portion and a fraction portion. The whole number part of the mixed number becomes the whole number part of the decimal number. The fraction part of the mixed number becomes the decimal part of the decimal number.

EXAMPLE 7

Write $4\frac{3}{5}$ as a decimal number.

SOLUTION

The whole number part of $4\frac{3}{5}$ is 4 and becomes the whole number part of the decimal number. The fraction part of $4\frac{3}{5}$ is changed to a decimal by dividing the denominator of the fraction into the numerator:

$$\begin{array}{r} .6 \\ 5\overline{)3.0} \\ \underline{3\,0} \\ 0 \end{array}$$

Since $\frac{3}{5}$ is equal to 0.6:

$$4\frac{3}{5} = 4.6$$

■

EXAMPLE 8

Write $3\frac{1}{2}$ using decimal notation.

SOLUTION

$$3\frac{1}{2} = 3 + \frac{1}{2} = 3 + 0.5 = 3.5$$ ■

EXAMPLE 9

Write $12\frac{3}{8}$ using decimal notation.

SOLUTION

$$12\frac{3}{8} = 12 + \frac{3}{8} = 12 + 0.375 = 12.375$$ ■

◇ *Do Practice Exercises 7 through 9.*

4 From Decimal Notation to Mixed Number Notation

Every mixed number has a whole number. Thus, only decimal numbers that have a whole number can be changed into a mixed number. The whole number part of the decimal number becomes the whole number part of the mixed number. The decimal point is dropped and everything to the right of the decimal point is used to find the fraction part of the mixed number.

EXAMPLE 10

Write 7.625 as a mixed number.

SOLUTION

$$7.625 = 7\frac{625}{1000} = 7\frac{5}{8}$$ ■

EXAMPLE 11

Write 1,500.75 as a mixed number.

SOLUTION

$$1{,}500.75 = 1{,}500\frac{75}{100} = 1{,}500\frac{3}{4}$$ ■

EXAMPLE 12

Change 4.6 to a mixed number, then to an improper fraction.

SOLUTION

$$4.6 = 4\frac{6}{10} = 4\frac{3}{5} = \frac{4 \cdot 5 + 3}{5} = \frac{23}{5}$$ ■

◇ *Do Practice Exercises 10 through 12.*

10. Write 12.5 as a mixed number, then as an improper fraction.

11. Write one hundred seventy and seven hundredths as a mixed number.

12. Write 2,124.95 as a mixed number.

ANSWERS

10. $12\frac{1}{2}$, $\frac{25}{2}$

11. $170\frac{7}{100}$

12. $2{,}124\frac{19}{20}$

Change from fraction notation to mixed number notation. (Review)

1. $\frac{22}{3}$ **2.** $\frac{7}{3}$ **3.** $\frac{12}{5}$ **4.** $\frac{61}{8}$

Change from mixed number notation to fraction notation. (Review)

5. $5\frac{1}{4}$ **6.** $7\frac{3}{5}$ **7.** $8\frac{1}{12}$ **8.** $9\frac{5}{7}$

Change from fraction notation to decimal notation.

9. $\frac{1}{4}$ **10.** $\frac{1}{8}$ **11.** $\frac{1}{2}$ **12.** $\frac{3}{8}$

13. $\frac{5}{8}$ **14.** $\frac{3}{4}$ **15.** $\frac{1}{3}$ **16.** $\frac{9}{40}$

17. $\frac{3}{80}$ **18.** $\frac{7}{20}$ **19.** $\frac{37}{100}$ **20.** $\frac{27}{50}$

21. $\frac{29}{20}$ **22.** $\frac{77}{25}$ **23.** $\frac{8}{5}$ **24.** $\frac{14}{10}$

Change from decimal notation to fraction notation, and reduce if possible.

25. 0.50 **26.** 0.75 **27.** 0.2 **28.** 0.125

29. 0.85 **30.** 0.60 **31.** 0.25 **32.** 0.7

33. 0.875 **34.** 0.35 **35.** 0.14 **36.** 0.244

37. 0.45 **38.** 0.375 **39.** 0.405 **40.** 0.55

Change from mixed number notation to decimal notation.

41. $4\frac{1}{2}$ **42.** $5\frac{1}{4}$ **43.** $3\frac{3}{4}$ **44.** $7\frac{3}{5}$

45. $12\frac{13}{100}$ **46.** $15\frac{7}{25}$ **47.** $3\frac{7}{20}$ **48.** $18\frac{11}{20}$

49. $21\frac{4}{5}$ **50.** $38\frac{17}{25}$ **51.** $47\frac{1}{80}$ **52.** $35\frac{1}{4}$

53. $17\frac{3}{5}$ **54.** $21\frac{9}{20}$ **55.** $99\frac{1}{100}$ **56.** $1{,}023\frac{3}{10}$

Change each to mixed number notation and reduce if possible.

57. 6.25 **58.** 7.4 **59.** 8.55 **60.** 9.75

61. 3.35 **62.** 12.2 **63.** 20.08 **64.** 6.15

65. 14.375 **66.** 16.875 **67.** 10.125 **68.** 9.05

69. 13.1 **70.** 25.08 **71.** 37.42 **72.** 59.006

Complete the following table. The first row has been done for you.

Decimal Notation	*Mixed Number*	*Fraction Notation*
4.375	$4\frac{3}{8}$	$\frac{35}{8}$
73.	$7\frac{3}{5}$	**74.**
5.12	**75.**	**76.**
77.	**78.**	$\frac{61}{4}$
35.2	**79.**	**80.**

JOURNAL EXERCISES

1. Explain the major differences between fraction, decimal, and mixed number notations.
2. List any questions you have regarding fraction, decimal, and/or mixed number notation.
3. Explain the process of converting from mixed number notation to decimal notation and from decimal notation to mixed number notation.
4. Explain the process of converting from decimal notation to fraction notation and from fraction notation to decimal notation.

1 Order of Operations

The same Order of Operations rules apply to decimals as apply to whole numbers, fractions, and mixed numbers.

Rule: To Evaluate an Expression

Step 1 Evaluate the contents of the parentheses and above and below the fraction bar.

Step 2 Evaluate exponents.

Step 3 Evaluate multiplication and division in the order in which they occur when moving from the left to the right across the expression.

Step 4 Evaluate addition and subtraction in the order in which they occur when moving from the left to the right across the expression.

OBJECTIVES

1 Apply the Order of Operations to problems with decimals, fractions, and mixed numbers.

2 Solve more word problems using decimal notation.

EXAMPLE 1

Evaluate $0.6 \cdot 2.5 \div (1.3 + 0.2)$.

SOLUTION

$0.6 \cdot 2.5 \div (1.3 + 0.2)$

$0.6 \cdot 2.5 \div 1.5$ Contents of parentheses

$1.5 \div 1.5$ Multiply first.

1 Result. ■

EXAMPLE 2

Evaluate $(6.9 \cdot 4 - 1) \div 0.5$.

SOLUTION

$(6.9 \cdot 4 - 1) \div 0.5$

$(27.6 - 1) \div 0.5$ Multply in parentheses.

$26.6 \div 0.5$ Subtract, drop parentheses.

53.2 Result. ■

EXAMPLE 3

Evaluate $17.4 - 3.2 \cdot 2^2$.

SOLUTION

$17.4 - 3.2 \cdot 2^2$

$17.4 - 3.2 \cdot 4$ exponents

$17.4 - 12.8$ Multiply before subtracting.

4.6 Result. ■

1. Evaluate $4.1 \cdot 2 + 7.53$.

2. Evaluate $2 + 5 \cdot 1.1 \cdot 2$.

HELPFUL HINT

Calculators are a good tool for solving expressions containing only decimal numbers. However, you must still use the rules for Order of Operations, even though you are using a calculator.

◇ *Do Practice Exercises 1 and 2.*

One way to solve an expression containing both fractions and decimals is to first change each number to a fraction, then solve using the rules of fractions. A second way to solve an expression containing both fractions and decimals is to first change each number to a decimal, then solve using the rules of decimals. Either method is acceptable. Look at the next two examples.

EXAMPLE 4

Evaluate $0.4 \div 0.01 + \frac{1}{4}$.

SOLUTION

$$0.4 \div 0.01 + \frac{1}{4}$$

$$0.4 \div 0.01 + 0.25 \qquad \text{Change } \frac{1}{4} \text{ to decimal notation.}$$

$$40 + 0.25 \qquad \text{Divide.}$$

$$40.25 \qquad \text{Result.}$$

■

EXAMPLE 5

$\left(\frac{1}{2}\right)^2 + 0.5 \cdot 1.8 + 2\frac{3}{4}$

SOLUTION

$$\left(\frac{1}{2}\right)^2 + 0.5 \cdot 1.8 + 2\frac{3}{4}$$

$$\frac{1}{4} + 0.5 \cdot 1.8 + 2\frac{3}{4} \qquad \text{exponent}$$

$$\frac{1}{4} + \frac{1}{2} \cdot 1\frac{8}{10} + 2\frac{3}{4} \qquad \text{Write using fraction notation.}$$

$$\frac{1}{4} + \frac{1}{2} \cdot 1\frac{4}{5} + 2\frac{3}{4} \qquad \text{Simplify } 1\frac{8}{10} \text{ to } 1\frac{4}{5}.$$

$$\frac{1}{4} + \frac{1}{2} \cdot \frac{9}{5} + \frac{11}{4} \qquad \text{Mixed number to fraction}$$

$$\frac{1}{4} + \frac{9}{10} + \frac{11}{4} \qquad \text{Multiply.}$$

ANSWERS
1. 15.73
2. 13

$$\frac{5}{20} + \frac{18}{20} + \frac{55}{20}$$ Write using common denominator.

$$\frac{5 + 18 + 55}{20}$$ Add.

$$\frac{78}{20}$$

$$3\frac{18}{20}$$ Change to a mixed number.

$$3\frac{9}{10}$$ Result. ■

◇ *Do Practice Exercises 3 and 4.*

3. Evaluate $(4.2)^2 - 3\frac{1}{2} \cdot 2$

4. Evaluate $13.8 \div 6 + \frac{5}{4}$

Calculator Application

Evaluate $7 \div 0.02 \cdot 1.4 \div 3.5$ using a calculator.

Answer: 140

2 Word Problems

The next example may help you relate to the width of North America.

EXAMPLE 6

An automobile used 16 gallons of gasoline to travel 320 miles. Find the miles per gallon. Using this gas mileage calculate how many gallons it will take to drive the approximate 3,000 miles across North America. What is the total cost of the gasoline required, assuming the price is $1.29 per gallon?

SOLUTION

Step 1 Find the average number of miles per gallon.

Step 2 Find the number of gallons needed to travel the distance.

Step 3 Find the cost of gasoline for the trip.

```
    20  miles per gallon
16)320
   32
    00
     0
     0
```

The number of gallons needed to drive across North America is found by dividing the number of miles by the average miles per gallon, or $3{,}000 \div 20$.

```
    150  gallons of gasoline
20)3000
   20
   100
   100
     0
     0
```

ANSWERS

3. 10.64 OR $10\frac{16}{25}$

4. 3.55 OR $3\frac{11}{20}$

The cost of 150 gallons of gasoline is found by multiplying 150 times 1.29 cents per gallon for gasoline.

$$\begin{array}{r} 150 \\ \times 1.29 \\ \hline 1350 \\ 300 \\ 150 \\ \hline \$193.50 \end{array} = \text{cost of gasoline} \quad \blacksquare$$

The last example can also be written as one expression and solved using the order of operations. The pieces of overall problem are:

Gas mileage = 320 ÷ 16

Number of gallons of gas needed = 3,000 ÷ gas mileage

Total cost = number of gallons of gas needed · $1.29

Now, put the pieces together:

$$\underbrace{3{,}000 \div \overbrace{(320 \div 16)}^{\text{gas mileage}} \cdot \overbrace{1.29}^{\text{price of gas}}}_{\text{number of gallons of gas needed}}$$

$3{,}000 \div (320 \div 16) \cdot 1.29$

$3{,}000 \div 20 \cdot 1.29$ — contents of parentheses

$150 \cdot 1.29$ — from left to right

$193.50

Decimal numbers are also found in proportions as shown in the next example.

EXAMPLE 7

A mixture requires 1.5 gallons of Chemical X for every 8 gallons of Chemical Y. How many gallons of Chemical X are needed to make 600 gallons of Chemical Y?

SOLUTION

The two ratios as given are:

$$\frac{1.5 \text{ gallons of X}}{8 \text{ gallons of Y}} \qquad \frac{? \text{ gallons of X}}{600 \text{ gallons of Y}}$$

These two ratios can be set equal to one another since they have the same labels in the numerators and denominators. Let A = amount of Chemical Y needed, then

$$\frac{1.5}{8} = \frac{A}{600}$$

$1.5 \cdot 600 = 8 \cdot A$ — Cross multiply.

$900 = 8A$

$\frac{900}{8} = \frac{8A}{8}$ — Divide both sides by 8.

$112.5 = A$ — Result.

112.5 gallons of Chemical X are needed. ■

Evaluate. (Review)

1. $6.5 + 12.203$

2. $17.9 \cdot 1.41$

3. $16.8 \div 2$

4. $19 - 3.45$

5. $175 + 0.05$

6. $(16.2)(0.87)$

7. $21 \div 0.05$

8. $21.3 - 17.08$

9. $12.1 + 13.75$

10. $21(3.27)$

11. $21.4 \div 2.5$

12. $19.7 - 3.01$

Evaluate the expressions.

13. $6.04 + 5.1 \cdot 3$

14. $7.5 \cdot 1.4 \div 0.25$

15. $(7 + 0.02) \div (1.8 \cdot 2)$

16. $7 \cdot 9 \div 0.5$

17. $(4.4 \times 2)^2$

18. $(7.8 - 3.4) \cdot 1.55$

19. $(16 \div 0.02) - 40.5$

20. $(300 \div 15) + (2.7 + 5)$

21. $7.4 \cdot 3.6 \div 0.18$

22. $(16 - 4.02)^2 - 3.7$

23. $\frac{1}{2} + 0.25 \cdot 5.7$

24. $4\frac{1}{3} + 2\frac{1}{2} \cdot 3\frac{1}{5}$

25. $15\frac{1}{4} - 8.7 \cdot 1.2$

26. $\frac{5}{4} \cdot 0.35 - 0.05$

27. $6.2 - \frac{3}{4} \cdot 7\frac{1}{2}$

28. $5\frac{1}{6} \cdot 1.2 \cdot \frac{3}{5}$

29. $3.4 \cdot 2.1 + (7.66 - 4)$

30. $10.5 \div 25 \cdot 3.5$

Write each as an expression, then solve.

31. Six items cost an average of $6.49 apiece. What is the total cost?

32. Find the average of 2.5 and 3.4.

33. A motorcycle travels two hundred seventy miles on four and one-half gallons of gas. If gasoline costs one dollar and twenty cents per gallon, how much would it cost to drive the motorcycle two thousand four hundred miles?

34. An item sells for two dollars and fifty cents plus a tax calculated as eight hundredths times the sales price. Find the cost, including tax, for five of these items.

35. A mixture requires a ratio of 4 parts of zinc to 7 parts of sulphur. How many pounds of zinc are required for a mixture containing 100 pounds of sulphur?

36. One blend of crude oil at a refinery called for 3.2 parts of a heavy crude to 7 parts of a light crude. Find the amount of heavy crude required for a mixture containing 3,500 gallons of light crude.

37. A stereo system costing $800 is bought with no down payment and 36 monthly payments of $27.75 each. What is the sum of the monthly payments? What is the total interest charge to finance the stereo system?

38. Multiply one thousand times one thousandth.

39. One alligator at a zoo eats two and one-half pounds of raw meat each day. How much meat is needed to feed the alligator for thirty days?

40. Erlinda is paid $4.50 per hour plus a bonus of 35 cents for each item she sews in that hour. What is her total pay for an eight hour shift in which she completes 43 items?

JOURNAL EXERCISES

1. An average person breathes 35 pounds of air each day. Assume one thousandth of the 35 pounds is made up of pollutants such as auto emissions. How many pounds of pollutant does a person breathe in one year (365 days)? Could this possibly be a health factor? Explain.

2. At a particular interest rate, sixty thousand dollars in mortgage debt ends up costing $189,554.46 if paid in 360 monthly installments. What is the monthly payment? How much interest is paid over the 30 year period? Are these numbers realistic?

CHAPTER 5 REVIEW

SECTION	TOPIC OR TERM	EXAMPLE OR DEFINITION
5.1		
Basics	Place Value	Place value is the idea that the value of a digit depends on its location in a number. For example, the 4 in 25.04 is in the hundredths place.
	Expanded Notation	Expanded notation is the writing of a number as the sum of its place value pieces. For example, $0.37 = 0.3 + 0.07 = \frac{3}{10} + \frac{7}{100}$
5.2		
Reading and Writing Decimal Numbers	Trailing Zeros	Trailing zeros are zeros located to the right of the decimal point and at the end of the number. For example, 85.2000 = 85.2. (000 → trailing zeros)
	Leading Zeros	Leading zeros are zeros that occur in front of a number. For example, 002.5 = 2.5.
5.3		
Addition, Subtraction, and Rounding	Addition of Decimal Numbers	$\begin{array}{r} 5.14 \\ +3.70 \\ \hline 8.84 \end{array}$ Line up place values. Add trailing zeros. Add.
	Subtraction of Decimal Numbers	$\begin{array}{r} \overset{8\ 1}{\not{9}.05} \\ -1.32 \\ \hline 7.73 \end{array}$ Line up place values. Add trailing zeros. Subtract.
	Rounding	65.28 rounded to the nearest tenth is 65.3.
5.4		
Multiplication	Multiplication of Decimal Numbers	$\begin{array}{r} \overset{2}{6.4} \\ \times\ 0.5 \\ \hline 3.20 \end{array} = 3.2$ 6.4: one digit to the right of the decimal place; 0.5: one digit to the right of the decimal place; 3.20: two digits to the right of the decimal place
	Power of Ten	A power of ten is a 1 followed by 1 or more zeros, such as 10, 100, 1000, and so on.

SECTION	TOPIC OR TERM	EXAMPLE OR DEFINITION
5.5		
Division	Division of Decimal Numbers	$3.57 \div 5.1$ $5.1\overline{)3.57}$ = 0.7; $357 - 357 = 0$
5.6		
Decimals, Fractions, and Mixed Numbers	Fraction to Decimal	$\frac{3}{4} = 3 \div 4 = 0.75$
	Decimal to Fraction	$0.35 = \frac{35}{100} = \frac{7}{20}$
	Mixed Number to Decimal	$4\frac{3}{5} = 4 + \frac{3}{5} = 4 + 0.6 = 4.6$
	Decimal to Mixed Number	$18.7 = 18 + 0.7 = 18 + \frac{7}{10} = 18\frac{7}{10}$
5.7		
Order of Operations and Word Problems	Order of Operations	$3.4 - (0.5)^2$ $3.4 - 0.25$ exponent 3.15 result.

INVENTORY

Henry is in charge of inventory at a hardware store. Inventory is a word used to describe all of the items in the store at one time. A partial list of the inventory in the store on July 1 and the number of each item on hand is:

Item	*Number in Store*
Lawn mower	12
Step ladder	10
Hand saw	15
Electric cord	7

Henry uses the *reorder point* method to determine when to reorder a particular item. The idea behind the reorder point method is to reorder an item whenever the number in stock falls either to or below a certain number called the *reorder point*. For example, the reorder point shown in the table below for lawn mowers is 10. Whenever the number of lawn mowers in the store drops to 10 or lower, Henry automatically orders additional lawn mowers.

Furthermore, Henry orders 5 lawn mowers each time he orders mowers. Why 5 lawn mowers and not 6 or 8? Both this number and the reorder point were set by management. His job is to watch the inventory of each item and reorder the correct amount when the number in stock drops to the reorder point. The reorder points and the order quantities are shown in the table below.

Item	*Reorder Point*	*Number to Order*	*Cost Each*
Lawn mower	10	5	$69.25
Step ladder	8	5	$42.00
Hand saw	12	10	$ 6.40
Electric cord	5	5	$ 8.45

As another example, Henry automatically reorders 10 hand saws when the number in the store drops to 12 or lower. The cost to the store for each hand saw is $6.40.

The following items are sold during July.

Item	*Number Sold*
Lawn mower	5
Step ladder	3
Hand saw	7
Electric cord	1

1. What is the inventory at the end of July? In other words, how many of each item remain in the store at the end of July?

2. Which items fell below their reorder points during July and must be reordered?

3. Make a table of the items that need to be reordered showing the following:

Item	*Number Ordered*	*Price Each*	*Total Cost*

Assume the items listed in question 3 are ordered, paid for, and delivered to the store all on the last day of July.

4. What is the number of each item in the store on August 1?

5. What is the amount of money invested in all of the items in the store on August 1? In other words, what is the value of the inventory on August 1?

6. Make a list of five different businesses in your neighborhood that carry inventory and list at least three inventory items for each business.

CHAPTER 5 REVIEW PROBLEMS

Write each number using words.

1. 16.75 **2.** 210.375 **3.** 1,405.08 **4.** 6,241.57

Write the digits of each number using decimal notation.

5. Seven and four thousandths

6. Fifty and twenty-one hundredths

7. Two thousand seven and seven tenths

8. Seventy-five ten-thousandths

Round to the indicated place value.

	to the nearest
9. 7.51	tenth
10. 12.407	hundredth
11. 17.51	ones
12. 751.5197	thousandth

Evaluate each showing all work.

13. $6.9 + 3.15$ **14.** $16 \cdot 2.4$ **15.** 7.5×1.04

16. $3.5(6.9)$ **17.** $0.86 \div 2.5$ **18.** $12.7 - 4.03$

19. $12 + 3.47$

20. $3.5 + 2.17$

21. $0.304 \div 0.02$

22. $17.6 - 12.93$

23. $5 + 4.93$

24. $(6)(4.29)$

25. $29.16 \div 1.2$

26. $6.93 + 2.1$

27. $7.4 \cdot 3.57$

28. 17.5×0.007

29. $19.7 - 0.003$

30. 4.5×0.05

Change each to fraction notation or to mixed number notation as applicable.

31. 0.7

32. 21.5

33. 0.035

34. 0.01

35. 14.12

36. 1,050.65

Change each to a decimal number.

37. $3\frac{4}{5}$

38. $\frac{3}{5}$

39. $12\frac{7}{8}$

40. $75\frac{3}{4}$

41. $27\frac{3}{8}$

42. $\frac{75}{4}$

Evaluate each showing all work.

43. $6.5 \cdot 3.45 + 0.07$

44. $(2.5)^2 - 1.07$

45. $75 - 6.08 \div 3.2$

46. $(16 - 9) \cdot 1.4$

47. $34.8 \div 3 - 7$

48. $7 \cdot 3.5 + (1.4)(2.1)$

49. $1.4 \cdot 3.05 \div 0.03$

50. $2^2 + 5.1$

Work each problem.

51. Find the sum of seven thousandths with four and five tenths.

52. Find the product of one hundredth with one hundredth.

53. Subtract six and two hundredths from twelve.

54. Divide seventy-three by two tenths.

55. Find the product of two thousandths and three and one-fourth.

56. Subtract two squared from four and one tenth.

57. An automobile was purchased with a down payment of $1,200 and forty-eight payments of $195.27 each. What is the total cost?

58. An automobile gets 15.5 miles per gallon of gasoline. The automobile can travel the one hundred miles from city A to city B exactly three and one-half times on one tank of gas. How many gallons of gas does the tank hold? Round your answer to the nearest tenth of a gallon.

59. Wolfgang ran five-eighths of a mile and Joan ran sixty-five hundredths of a mile. Who ran farther?

60. Write eleven-fourths using fraction notation, mixed number notation, and decimal notation.

CHAPTER 5 TEST

Write each number using words.

1. 605.27

2. 512.8

3. 7,243.05

4. 1,000,000.08

Write each using decimal notation.

5. Six and fourteen thousandths

6. Seventy-five hundredths

7. Two thousand one and four tenths

8. Eighty thousand one and five hundredths

Evaluate.

9. $6.3 + 12$

10. $2.4 - 0.05$

11. $(7.3)(2.05)$

12. $0.174 \div 0.025$

13. $3.7 \div 1000$

14. $6.4 \cdot 0.91$

15. $9.9 + 2.4 \cdot 3.5$

16. $16 - 4.23 \div 3$

17. $\frac{1}{2} \cdot 0.52 + 8.7$

18. $(7.3 + 2.52) \div 2.5$

19. $3 \cdot (1.1)^2$

20. $4.5 \cdot \frac{11}{5} - 1$

Change each to decimal notation.

21. $\frac{3}{4}$

22. $\frac{5}{8}$

Change each to fraction notation and reduce.

23. 0.4

24. 0.14

Work each problem.

25. Round 7.984 to the nearest hundredth.

26. The total cost of six items is $45.54. Find the average cost of each of the items.

27. Find the average of 2.1, $\frac{8}{5}$, and $1\frac{1}{4}$.

28. Write the fraction seventeen-fourths as a mixed number and as a decimal number.

29. Write 4.7, $4\frac{1}{5}$, 0.499, and $\frac{20}{3}$ from smallest to largest.

30. Business cards can be printed for nine tenths of a cent each. How many business cards can be printed for twenty-seven dollars?

6

Percent

APPLICATION *A bank lends an individual \$800 for a year at 12% simple interest. What interest must be paid at the end of one year?*

SOLUTION *The interest is found by multiplying percent times loan amount times one year.*

Interest = Percent · Loan amount · Time

Interest = 12% · \$800 · 1

Interest = 0.12 · 800 · 1   Change 12% to 0.12.

Interest = \$96   Multiply.

The interest charge for the one year loan is \$96.

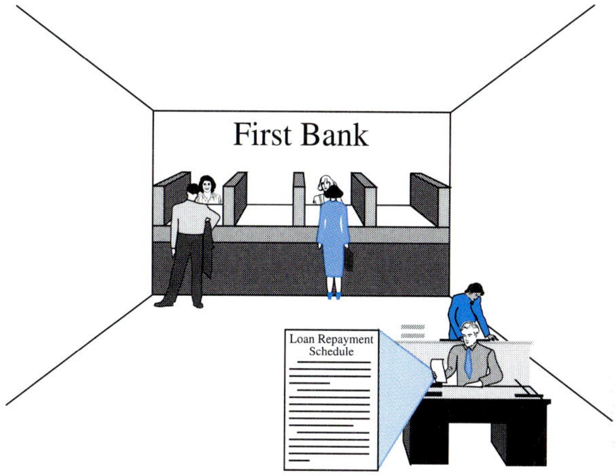

6.1 BASICS

OBJECTIVES

1. Know the meaning of percent.
2. Solve percent problems using the definition of percent.
3. Solve percent problems using proportions.

1 Basics of Percent

The word **percent** means "out of one hundred." The symbol % may be used in place of the word percent. Thus 25% is read as twenty-five percent and means 25 out of every 100.

$$25\% = \text{twenty-five } \textit{percent} = 25 \text{ out of } 100 = \frac{25}{100} = 0.25$$

Similarly, $62\% = \text{sixty-two percent} = 62 \text{ out of } 100 = \frac{62}{100} = 0.62$

EXAMPLE 1

Write each as a fraction with a denominator of 100, then write using decimal notation:

(a) 63%   **(b)** 25%   **(c)** 10%   **(d)** 100%

SOLUTION

(a) $63\% = \frac{63}{100} = 0.63$

(b) $25\% = \frac{25}{100} = 0.25$

(c) $10\% = \frac{10}{100} = 0.10 = 0.1$

(d) $100\% = \frac{100}{100} = 1.00 = 1$ ■

Percents are commonly used. Some examples in which you may have seen them include:

Example	*Meaning*
8% sales tax	\$8 tax on each \$100 in sales
7% unemployment	7 unemployed out of every 100 workers
6% commission	\$6 commission for each \$100 in sales
12% interest rate	\$12 interest for each \$100 owed
83% test score	83 correct out of each 100 problems

The idea of percent can be shown using a glass of water. The glass on the left is empty or 0% full. Moving to the right, the following glasses are 25% full, 50% full, and finally 100% full.

$0=\frac{0}{100}=0\%$ $\quad \frac{1}{4}=\frac{25}{100}=25\%$ $\quad \frac{1}{2}=\frac{50}{100}=50\%$ $\quad 1=\frac{100}{100}=100\%$

A glass that is 125% full is one in which 125 parts out of every 100 parts is full. The glass is more than full

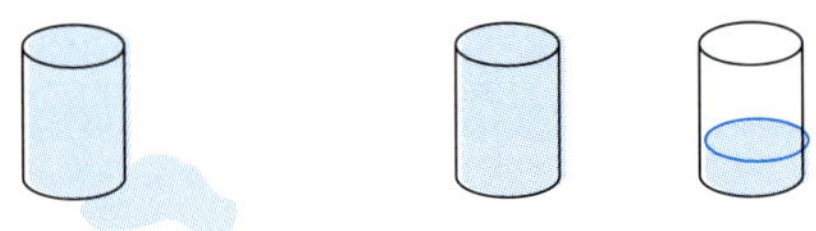

$\frac{125}{100} = 125\%$ or $100\% + 25\% = 125\%$

One hundred percent (100%) is the ''total'' or ''complete'' amount. Any percentage smaller than 100% is less than the whole and any percentage greater than 100% is more than the whole. Zero percent (0%) represents the number 0.

In this chapter we will learn how to work with percents. First, we will use the idea that percent is the number out of 100 to solve problems. Then we will use proportions to solve percent problems.

2 Solve Percent Problems Using the Definition of Percent

Percents can be found using the idea that percent means the number out of each 100. Look at the next two examples.

EXAMPLE 2

What percent of the rectangle is shaded?

18 shaded out of 200

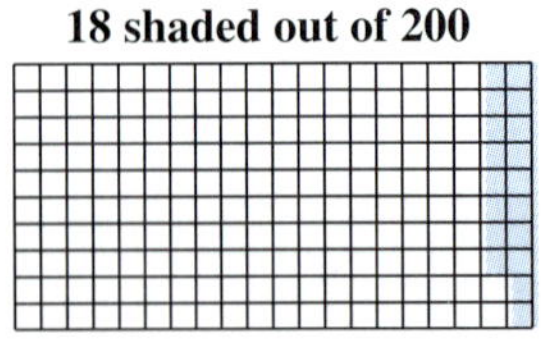

SOLUTION

The number of shaded squares is 18 and the total number of squares is 200. The ratio of the number of squares shaded to the total number of squares is

$$\frac{18 \text{ shaded}}{200 \text{ squares}} \quad \text{which reduces to} \quad \frac{9 \text{ shaded}}{100 \text{ squares}}$$

9 out of every 100 squares are shaded or 9% is shaded. ■

EXAMPLE 3

What percent of the rectangle is shaded?

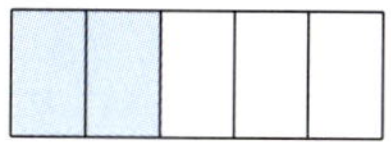

SOLUTION

Two parts out of five parts are shaded. First, rewrite this fraction with a denominator of 100:

$$\frac{2}{5} = \frac{?}{100}$$

$$\frac{2}{5} = \frac{40}{100} \quad \text{Divide 5 into 100 and multiply by 2.}$$

40% of the rectangle is shaded. ■

◇ *Do Practice Exercise 1.*

EXAMPLE 4

A real estate commission on a \$100,000 sale was \$7,000. What percent is this?

SOLUTION

The idea here is to write \$7,000 out of \$100,000 as some number out of 100.

$$\frac{\$7{,}000}{\$100{,}000} = \frac{7}{100} = 7\%$$

Seven out of one hundred equals 7 percent so that the real estate commission is 7%. ■

◇ *Do Practice Exercises 2 and 3.*

3 Solve Percent Problems Using Proportions

Percent problems can also be solved using proportions which were covered in Chapter 4. The steps are shown in the next few examples.

1. What percent of each geometric figure is shaded?

(a)

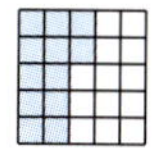

(b)

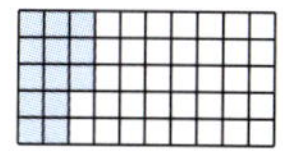

(c)

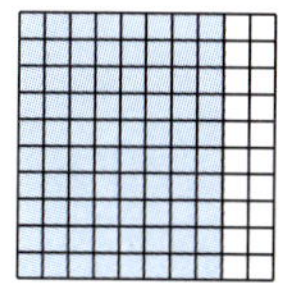

(d)

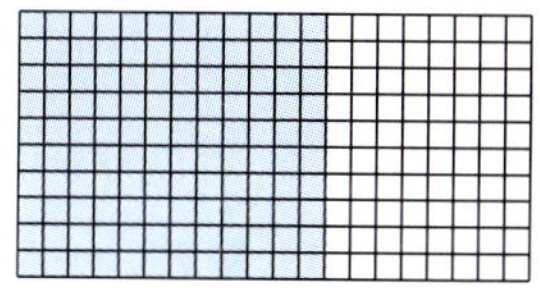

2. The sales tax is \$12 for every \$200 in sales. What percent is the sales tax?

3. Lisa had 47 correct answers out of 100 questions on her test. What percent correct did she have?

ANSWERS
1. (a) 48% (b) 26% (c) 80% (d) 60%
2. 6% sales tax
3. 47% correct

4. Find each:

(a) 75% of 40

(b) 30% of 150

(c) 10% of 90

(d) 0% of 62

(e) 1% of 600

(f) 125% of 72

ANSWERS
4. (a) 30 (b) 45 (c) 9 (d) 0 (e) 6 (f) 90

EXAMPLE 5

Find 20% of 85.

SOLUTION

There are two ratios in this problem. The first ratio is given by 20% which is 20 out of 100. The second ratio is some portion of 85. Let x equal the unknown amount.

First ratio: $\frac{20}{100}$ (= 20%)

Second ratio: $\frac{x}{85}$

The two ratios can be set equal to one another, forming a proportion:

This ratio represents 20%. (pointing to $\frac{20}{100}$)

$$\frac{20}{100} = \frac{x}{85}$$

The proportion can be solved using techniques from Chapter 4.

$$\frac{20}{100} = \frac{x}{85}$$

$$20 \cdot 85 = 100 \cdot x \qquad \text{Cross multiply.}$$

$$1700 = 100x$$

$$\frac{1700}{100} = \frac{100x}{100} \qquad \text{Divide both sides by 100.}$$

$$17 = x \qquad \text{solution to proportion}$$

Therefore, 20% of 85 is 17. ■

EXAMPLE 6

Find 40% of 120.

SOLUTION

$$\text{First ratio} = 40\% = \frac{40}{100}$$

$$\text{Second ratio} = \frac{x}{120} \qquad \text{if } x \text{ is the unknown}$$

$$\frac{40}{100} = \frac{x}{120} \qquad \text{set the ratios equal}$$

$$40 \cdot 120 = 100x \qquad \text{Cross multiply.}$$

$$4{,}800 = 100x$$

$$\frac{4{,}800}{100} = \frac{100x}{100} \qquad \text{Divide both sides by 100.}$$

$$48 = x$$

40% of 120 is 48. ■

◇ ***Do Practice Exercise 4.***

EXAMPLE 7

A mechanic joked that 30% of the cars in his neighborhood break down each month. If this is true and if his neighborhood has 400 cars, how many break down each month?

SOLUTION

Let N = the number of cars in the neighborhood that break down each month, then:

$$\text{First ratio:} \quad 30\% = \frac{30}{100}$$

$$\text{Second ratio:} \quad \frac{N}{400}$$

$$\frac{30}{100} = \frac{N}{400}$$

$$30 \cdot 400 = 100N \qquad \text{Cross multiply.}$$

$$12{,}000 = 100N$$

$$\frac{12{,}000}{100} = \frac{100N}{100} \qquad \text{Divide both sides by 100.}$$

$$120 = N$$

If the mechanic is correct, 120 of the 400 cars in the neighborhood break down each month. ■

One hundred percent is the total amount. Sometimes a percent of the total is given and the question asks about the remaining portion, as shown in the next example.

EXAMPLE 8

Forty percent of the employees of a local company have a college degree. What percent of the employees *do not* have a college degree?

5. Twenty percent of an employee's pay is withheld for taxes. What percent does the employee actually receive?

6. Sixty-five percent of the students in a difficult class passed. What percent failed? Write this percent as a ratio.

ANSWERS
5. The employee receives 80%.
6. 35% failed;
$$\frac{35 \text{ fail}}{100 \text{ students}} = \frac{7 \text{ fail}}{20 \text{ students}}$$

SOLUTION

An employee either has, or does not have, a college degree. The percent that have a college degree plus the percent that do not have a college degree equals all, or 100%, of the employees:

Have a college degree	Do not have a college degree
100% of employees	

Since 40% of the total employees have a college degree, then 100% − 40% do not have a college degree.

$$\begin{array}{r} 100\% \\ -\ 40\% \\ \hline 60\% \end{array}$$ do not have a college degree.

Check the result by adding:

$$\begin{array}{r} 40\% \\ +\ 60\% \\ \hline 100\% \end{array}$$

have a college degree
do not have a college degree
= all of the employees ■

◇ *Do Practice Exercises 5 and 6.*

name date hour

EXERCISE SET 6.1

Find the percent of each figure that is shaded.

1. 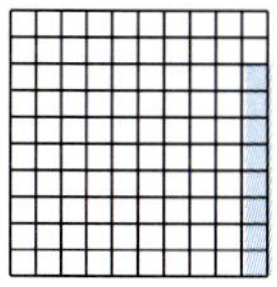

2. 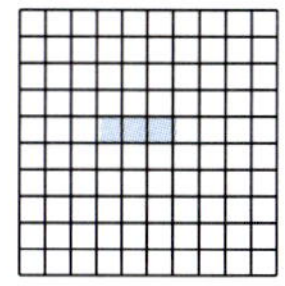

3. 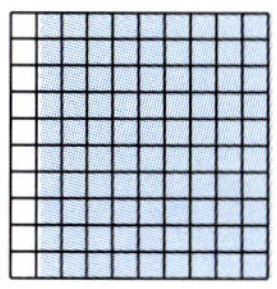

4. 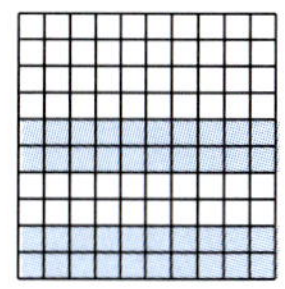

5. 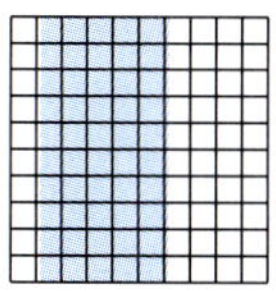

6. 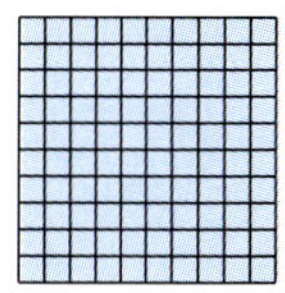

7. 

8. 

9. 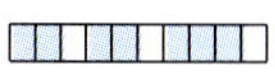

10. 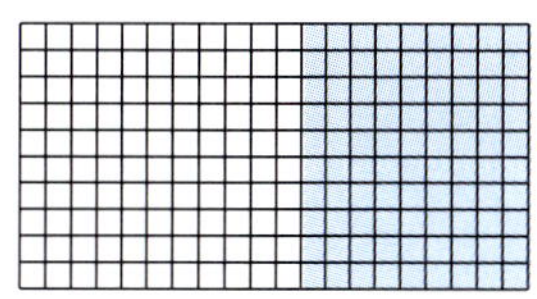

11. 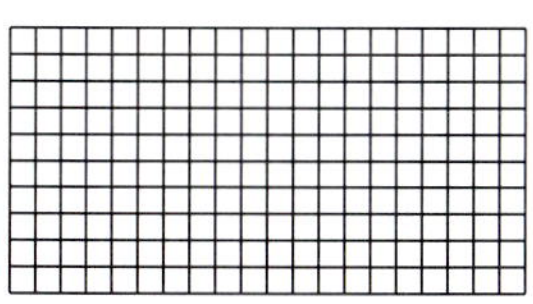

12. 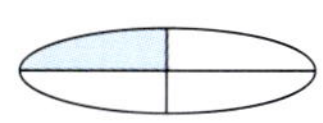

13. 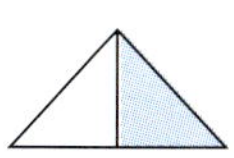

14. 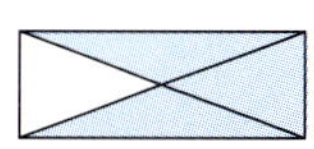

Write as parts out of one hundred.

15. 60% is ______ parts out of 100

16. 20% is ______ parts out of 100

17. 130% is ______ parts out of 100

18. 99% is ______ parts out of 100

19. 0% is ______ parts out of 100

20. 1% is ______ part out of 100

21. 7% is ______ parts out of 100

22. 100% is ______ parts out of 100

23. 25% is ______ parts out of 100

24. 200% is ______ parts out of 100

Write using percent notation.

25. 12 parts out of 100 is ______%

26. 75 parts out of 100 is ______%

27. 40 parts out of 200 is ______%

28. 60 parts out of 300 is ______%

29. 5 parts out of 10 is ______%

30. 0 parts out of 100 is ______%

31. 100 parts out of 100 is ______%

32. 2 parts out of 200 is ______%

33. 150 parts out of 300 is ______%

34. 16 parts out of 10 is ______%

Write each as a fraction with a denominator of 100, then write using decimal notation.

35. 15%

36. 30%

37. 21%

38. 0%

39. 42%

40. 100%

41. 79%

42. 150%

43. 125%

44. 87%

45. 3%

46. 99%

Find each using proportions.

47. 40% of 60

48. 10% of 80

49. 4% of 50

50. 15% of 100

51. 25% of 120

52. 90% of 60

53. 80% of 130

54. 100% of 70

55. 0% of 49

56. 21% of 400

57. 125% of 160

58. 150% of 900

59. 100% of 1

60. 0% of 100

61. 4% of 1,000,000

62. 75% of 44

63. 1% of 360

64. 99% of 1,000

First write as a fraction with a denominator of 100, then as a percent.

65. 11 of the 20 students

66. 1 defect out of 100 items

67. no males in a group of 100

68. 1 lawn mower out of 50

69. four out of five nurses

70. 7 of 50 crimes

71. 1 out of 10 hardware stores

72. all of the workers

73. 300 out of 500 heart attacks

74. 3 out of every 5 car wrecks

75. 9 out of 10 alcoholics

76. 1 out of 2 marriages

Work each problem.

77. Seventy out of every one hundred students at a local college are females. What percent are males? What percent are females?

78. Four out of twenty-five wells drilled produced crude oil. What percent produced oil? What percent did not?

79. Nine out of every ten students passed. What percent passed? What percent did not pass?

80. At one company, three out of every four employees have a personal computer. What percent have personal computers and what percent do not?

81. Upon observing 30 birds, 15 were found to eat crickets. What fraction of the 30 birds ate crickets? What percent of the 30 birds ate crickets?

82. Twelve percent of a company's one hundred employees are in sales. How many employees are *not* in sales?

JOURNAL EXERCISES

1. In your own words, explain the concepts of percent introduced in this section.

2. Can every percent be written using fraction notation? Explain.

3. Can every fraction be written using percent notation? Explain.

4. List several jobs that would require an employee to know how to work with percents.

6.2 PERCENT NOTATION: PERCENTS AND DECIMALS

Every number written in percent notation can be changed to decimal notation, and every number written in decimal notation can be changed to percent notation. In this section we are going to learn how to change a number from one notation to the other.

1 From Percent Notation to Decimal Notation

As you already know, percent means "out of 100." Look again at some of the ways in which 25% can be written:

$$25\% = 25 \text{ out of } 100 = \frac{25}{100} = 0.25 = \frac{1}{4}$$

Actually, the phrase "out of 100" or the "%" symbol can be replaced by either "$\cdot \frac{1}{100}$," or by "÷ 100." For example,

$$25\% = 25 \cdot \frac{1}{100} = \frac{25}{100} = 0.25 \quad \text{or} \quad 25\% = 25 \div 100 = 0.25$$

Either way, we find that 25% is equal to 0.25.

EXAMPLE 1

Write each using decimal notation:

(a) 3% **(b)** 8% **(c)** 125% **(d)** 100%

SOLUTION

(a) $3\% = 3 \cdot \frac{1}{100} = \frac{3}{100} = 0.03$

(b) $8\% = 8 \cdot \frac{1}{100} = \frac{8}{100} = 0.08$

(c) $125\% = 125 \cdot \frac{1}{100} = \frac{125}{100} = 1.25$

(d) $100\% = 100 \cdot \frac{1}{100} = \frac{100}{100} = 1 \text{ or } 1.0$ ■

◇ *Do Practice Exercise 1.*

A percent may also contain a fraction as shown in the next example.

OBJECTIVES

1 Change from percent notation to decimal notation.

2 Change from decimal notation to percent notation.

1. Change each to decimal notation:

(a) 12%

(b) 91%

(c) 101%

(d) 0%

ANSWERS

1. (a) 0.12 (b) 0.91 (c) 1.01 (d) 0

2. Change each to decimal notation:

(a) $4\frac{1}{2}\%$

(b) $3\frac{1}{5}\%$

(c) 12.75%

ANSWERS
2. (a) 0.045 (b) 0.032 (c) 0.1275

EXAMPLE 2

Write 7.5% using decimal notation.

SOLUTION

7.5%

$7.5 \cdot \frac{1}{100}$ Replace % with $\cdot \frac{1}{100}$.

$\frac{7.5}{100}$ Multiply.

$\frac{7.5}{100} \cdot \frac{10}{10}$ Multiply by $\frac{10}{10}$ (= 1) so that the numerator becomes a whole number.

$\frac{75}{1{,}000}$ Multiply.

0.075 Result. ■

In the last example, note that 7.5% can also be written as $7\frac{1}{2}\%$, so that $7.5\% = 7\frac{1}{2}\% = 0.075$.

◇ *Do Practice Exercise 2.*

Multiplying by $\frac{1}{100}$ or equivalently, dividing by 100 is the same as moving the decimal place to the left two places. The rule that follows lists the two steps needed to change from percent notation to decimal notation.

Rule: To Change from Percent Notation to Decimal Notation

Step 1 Move the decimal point to the *left* two places.

Step 2 Drop the % symbol.

The first step in changing from percent notation to decimal notation is to locate the decimal point. Look at the next example.

EXAMPLE 3

Change 15% to a decimal number.

SOLUTION

First, find the decimal point in 15%. No decimal point is to be seen, so the decimal point must be after the 15. In other words, 15% = 15.% = 15.0%

Step 1 .15.0% Move the decimal point to the left two places.

Step 2 0.15 Drop the % symbol and the trailing zero.

15% = 0.15 ■

A percent may be given in mixed number notation in which case it will contain a fraction. In order to change to decimal notation:

1. First change the fraction part of the mixed number to a decimal.
2. Then change the percent to decimal notation.

Look at the next few examples.

EXAMPLE 4

Change $4\frac{1}{2}\%$ to decimal notation.

SOLUTION

First change $4\frac{1}{2}$ to 4.5, then proceed.

$$4\frac{1}{2}\% = 4.5\% = 0.045$$

■

EXAMPLE 5

Change each to decimal notation:

(a) 16% (c) 100% (e) $67\frac{1}{2}\%$

(b) $75\frac{4}{5}\%$ (d) 0% (f) $21\frac{1}{4}\%$

SOLUTION

(a) 0.16 (c) 1.0 (e) 0.675

(b) 0.758 (d) 0.0 (f) 0.2125 ■

◇ *Do Practice Exercises 3 and 4.*

The next example shows how to work with a fraction that has an unlimited number of digits when written using decimal notation.

EXAMPLE 6

Change $25\frac{1}{3}\%$ to decimal notation.

SOLUTION

$$25\frac{1}{3}\% = 25.333\ldots\% = 0.25333\ldots$$

The three dots means that the pattern of digits continues without end. ■

Each percent can be written using decimal notation and using fraction notation as shown in the next example.

3. Change each to decimal notation:

(a) 27%

(b) 0%

(c) 100%

(d) $24\frac{1}{10}\%$

(e) $5\frac{1}{4}\%$

4. Change one hundred twelve percent to decimal notation.

ANSWERS

3. (a) 0.27 (b) 0 (c) 1.00 (d) 0.241 (e) 0.0525

4. 1.12

5. Change 65% to decimal notation then to fraction notation and reduce.

6. Change $15\frac{3}{11}\%$ to decimal notation.

EXAMPLE 7

Change 80% to decimal notation, then to fraction notation.

SOLUTION

$$80\% = 0.80 = 0.8 = \frac{8}{10} = \frac{4}{5} \quad ■$$

◇ *Do Practice Exercises 5 and 6.*

2 From Decimal Notation to Percent Notation

Each decimal number can be written using percent notation by first writing the decimal number as a fraction with a denominator of 100. For example:

$$0.45 = \frac{45}{100} = 45\%$$

EXAMPLE 8

Write each using percent notation:

(a) 0.91 **(b)** 3 **(c)** 0 **(d)** 0.50 **(e)** 1.75

SOLUTION

(a) $0.91 = \frac{91}{100} = 91\%$

(b) $3 = \frac{300}{100} = 300\%$

(c) $0 = \frac{0}{100} = 0\%$

(d) $0.50 = \frac{50}{100} = 50\%$

(e) $1.75 = \frac{175}{100} = 175\%$ ■

A percent may involve a decimal number. In that event, it can also be written as a percent with a fraction as shown in the next example.

EXAMPLE 9

Write 0.705 using percent notation.

SOLUTION

$$0.705 = \frac{705}{1{,}000}$$

We need to write this fraction with a denominator of 100. Using ideas from Chapter 4:

$$\frac{705}{1{,}000} = \frac{x}{100}$$

$$705 \cdot 100 = 1{,}000 \cdot x \qquad \text{Cross multiply.}$$

$$x = \frac{70{,}500}{1{,}000} \qquad \text{Divide by 1,000.}$$

$$x = 70.5$$

ANSWERS

5. $65\% = 0.65 = \frac{13}{20}$

6. 0.152727 . . .

Since 70.5 is the same as $70\frac{1}{2}$, the answer can be written as:

$$70.5\% \quad \text{or} \quad 70\frac{1}{2}\%$$ ■

◇ *Do Practice Exercise 7.*

You may have noticed that in changing from a decimal number to a percent, the decimal place is moved to the *right* two places and a % sign is added. These steps are given in the following rule.

Rule: To Change a Decimal Number to a Percent

Step 1 Move the decimal place to the *right* two places, adding zeros if needed.

Step 2 Place a % sign at the end of the number.

EXAMPLE 10

Change 0.68 to a percent.

SOLUTION

Step 1 0.68. Move decimal point two places to the right.

Step 2 68% place % symbol

→ Decimal point is no longer needed.

0.68 = 68% ■

EXAMPLE 11

Change 3.005 to a percent.

SOLUTION

3.005 Move decimal point two places.

300.5% or $300\frac{1}{2}\%$ place % symbol

→ Decimal point is still needed. ■

◇ *Do Practice Exercises 8 and 9.*

7. Write each using percent notation:

(a) 0.41

(b) 0.01

(c) 1.30

(d) 0.0475

8. Change each to percent notation:

(a) 0.71

(b) 0.01

(c) 0.592

(d) 2.4

(e) 1

9. Change one and seven hundredths to percent notation.

ANSWERS
7. (a) 41% (b) 1% (c) 130%
(d) $4\frac{3}{4}\%$ OR 4.75%
8. (a) 71% (b) 1%
(c) 59.2% OR $59\frac{1}{5}\%$
(d) 240% (e) 100%
9. 107%

name date hour

Change the following to decimal notation.

1. 35% **2.** 40% **3.** 2% **4.** 10%

5. 90% **6.** 17% **7.** 99% **8.** 110%

9. 13% **10.** 150% **11.** 21% **12.** 245%

13. 0% **14.** 100% **15.** 1% **16.** 28%

17. 140% **18.** 225% **19.** 80% **20.** 7%

21. 8.5% **22.** 21.2% **23.** 37.4% **24.** 9.01%

25. 79.8% **26.** $15\frac{1}{2}\%$ **27.** $25\frac{1}{4}\%$ **28.** $30\frac{1}{5}\%$

29. 12.5% **30.** $75\frac{3}{4}\%$ **31.** $102\frac{1}{8}\%$ **32.** $37\frac{2}{3}\%$

Change each to percent notation.

33. 0.51 **34.** 0.04 **35.** 0.25 **36.** 0.60

37. 0.99

38. 1.00

39. 0

40. 0.8

41. 1.25

42. 0.40

43. 0.75

44. 2.35

45. 0.03

46. 2.4

47. 3

48. 1.02

49. 1.2

50. 0.5

51. 0.55

52. 4.5

53. 0.815

54. 0.48

55. 0.6

56. 1.01

Write each as a percent, then write the percent as a mixed number.

57. 0.752

58. 0.806

59. 0.999

60. 0.375

61. 1.255

62. 1.135

63. 0.0325

64. 0.045

Write using both decimal notation and percent notation.

65. 3 out of 4 people

66. 1 out of every 20 children

67. 7 out of every 10 soldiers

68. 1 out of every 100 letters

69. 4 out of 5 college professors

70. 1 out of 8 children

71. 9 out of 10 college graduates

72. 3 out of every 4 men

73. 4 out of every 5 criminals

74. 5 out of 8 automobiles

75. 1 out of 10 high school english teachers

76. 1 out of every 11 individuals

Work each problem.

77. Jess ran six-tenths of a mile. Is six tenths a decimal number, a fraction, or can it be either one? What percent of a mile did he run?

78. Sales tax in one city is 8% of the price. Change 8% to a decimal number.

79. Jan spent 65% of her ten dollars. If all of her change is given in quarters, how many quarters should she receive?

80. Five percent of a class of 80 students are absent. How many students are *not* absent?

81. A bank pays 9.25% interest on a certificate of deposit. Write 9.25% using decimal notation.

82. A sports car costs twice as much as a second automobile. The sports car costs what percent of the cost of the second automobile?

83. A basketball player usually makes 70% of her free throws. How many would you expect her to make out of 20 free throws?

JOURNAL EXERCISES

1. Explain the process behind moving the decimal point two places to the *right* when changing from decimal notation to percent notation.

2. Explain the process behind moving the decimal point two places to the *left* when changing from percent notation to decimal notation.

1 From Percent Notation to Fraction Notation

As we have already seen, there are two different methods to change from percent notation to fraction notation. The first method uses the definition of percent that is "out of one hundred." For example,

$$30\% = 30 \text{ out of } 100 = \frac{30}{100}$$

Then the fraction is simplified or reduced, if possible:

$$30\% = \frac{30}{100} = \frac{3}{10} \quad \text{or} \quad 30\% = \frac{3}{10}$$

EXAMPLE 1

Change each to fraction notation:

(a) 70% **(b)** 0% **(c)** 1% **(d)** 12.5%

SOLUTION

(a) $70\% = \frac{70}{100} = \frac{7}{10}$

(b) $0\% = \frac{0}{100} = 0$

(c) $1\% = \frac{1}{100}$

(d) $12.5\% = \frac{12.5}{100} = \frac{12.5}{100} \cdot \frac{10}{10} = \frac{125}{1{,}000} = \frac{1}{8}$ ■

◇ *Do Practice Exercise 1.*

Alternatively, a number can be changed from percent notation to fraction notation by replacing the % symbol with multiplication by the fraction $\frac{1}{100}$.

$$3.7\% = 3.7 \cdot \frac{1}{100} = \frac{3.7}{100} = \frac{3.7}{100} \cdot \frac{10}{10} = \frac{37}{1{,}000}$$

($\frac{10}{10}$: form of one)

EXAMPLE 2

Change each to fraction notation or mixed number notation as appropriate:

(a) 40% **(b)** 325% **(c)** 99.9% **(d)** 0.03%

SOLUTION

(a) $40\% = 40 \cdot \frac{1}{100} = \frac{40}{1} \cdot \frac{1}{100} = \frac{40}{100} = \frac{2}{5}$

(b) $325\% = 325 \cdot \frac{1}{100} = \frac{325}{1} \cdot \frac{1}{100} = \frac{325}{100} = 3\frac{1}{4}$

(c) $99.9\% = 99.9 \cdot \frac{1}{100} = \frac{99.9}{1} \cdot \frac{1}{100} = \frac{99.9}{100} = \frac{999}{1{,}000}$

(d) $0.03\% = 0.03 \cdot \frac{1}{100} = \frac{0.03}{1} \cdot \frac{1}{100} = \frac{0.03}{100} = \frac{3}{10{,}000}$ ■

OBJECTIVES

1 Change from percent notation to fraction notation.

2 Change from fraction notation to percent notation.

3 Change from mixed number notation to percent notation and vice versa.

1. Change each to fraction notation:

(a) 25%

(b) 40%

(c) 2.5%

ANSWERS

1. (a) $\frac{1}{4}$ (b) $\frac{2}{5}$ (c) $\frac{1}{40}$

2. Change each to fraction notation and reduce if possible:

(a) 60%

(b) 85%

(c) 9.5%

(d) 0.1%

(e) 2%

3. Change ten percent to a fraction.

ANSWERS

2. (a) $\frac{3}{5}$ (b) $\frac{17}{20}$ (c) $\frac{19}{200}$ (d) $\frac{1}{1,000}$ (e) $\frac{1}{50}$

3. $\frac{1}{10}$

Sometimes the percent is a mixed number. In this case, write the mixed number using fraction notation as shown in the next problem.

EXAMPLE 3

Change $5\frac{1}{4}\%$ to fraction notation.

SOLUTION

$5\frac{1}{4}\%$	
$\frac{21}{4}\%$	Change $5\frac{1}{4}$ to $\frac{21}{4}$.
$\frac{21}{4} \cdot \frac{1}{100}$	Replace % with $\cdot \frac{1}{100}$.
$\frac{21}{400}$	Result.

■

◇ *Do Practice Exercises 2 and 3.*

2 From Fraction Notation to Percent Notation

It is possible to change from fraction notation to percent notation by first writing the fraction using decimal notation. Then, the rules of the prior section can be used to change from decimal notation to percent notation. Look at the next examples.

EXAMPLE 4

Write $\frac{3}{4}$ using percent notation.

SOLUTION

First, use the ideas from Chapter 5 to change $\frac{3}{4}$ to a decimal number. In other words divide the denominator into the numerator.

```
    .75
 4)3.00
   2 8
     20
     20
      0
```

Thus $\frac{3}{4} = 0.75$.

Now, change from decimal notation to percent notation by moving the decimal point to the right two places and adding the % sign.

$$\frac{3}{4} = 0.75 = 75\%$$ ■

EXAMPLE 5

Change each to percent notation:

(a) $\frac{1}{4}$ **(b)** $\frac{5}{8}$ **(c)** $\frac{9}{10}$ **(d)** $\frac{11}{10}$

SOLUTION

(a) $\frac{1}{4} = 0.25 = 25\%$

(b) $\frac{5}{8} = 0.625 = 62.5\%$

(c) $\frac{9}{10} = 0.9 = 90\%$

(d) $\frac{11}{10} = 1.1 = 110\%$ ■

◇ *Do Practice Exercises 4 and 5.*

3 From Mixed Number Notation to Percent Notation and Vice Versa

A mixed number has a whole number part and a fraction part. Because a mixed number is greater than one, it will be greater than 100% when written as a percent. In order to write a mixed number as a percent, first write the mixed number as a decimal number. Then change the decimal number to a percent.

EXAMPLE 6

Change $1\frac{2}{5}$ to a percent.

SOLUTION

$$1\frac{2}{5} = 1.4 \quad \text{and} \quad 1.4 = 140\%$$

Thus $1\frac{2}{5} = 1.4 = 140\%$. ■

EXAMPLE 7

Change each to percent notation:

(a) $1\frac{7}{8}$ **(b)** $1\frac{1}{5}$ **(c)** $2\frac{1}{10}$

SOLUTION

(a) $1\frac{7}{8} = 1.875 = 187.5\% = 187\frac{1}{2}\%$

(b) $1\frac{1}{5} = 1.2 = 120\%$

(c) $2\frac{1}{10} = 2.1 = 210\%$ ■

◇ *Do Practice Exercises 6 and 7.*

A percent greater than 100% can be converted to a mixed number notation by first writing it using fraction notation as shown in the next two examples.

4. Change each to decimal notation, then to percent notation:

(a) $\frac{1}{2}$

(b) $\frac{3}{8}$

(c) $\frac{4}{5}$

(d) $\frac{1}{10}$

5. Change seven tenths to decimal notation, then to percent notation.

6. Change $2\frac{1}{2}$ to a percent.

7. Write one and one-fourth as a percent.

ANSWERS

4. (a) $\frac{1}{2} = 0.50 = 50\%$

(b) $\frac{3}{8} = 0.375 = 37.5\%$

(c) $\frac{4}{5} = 0.80 = 80\%$

(d) $\frac{1}{10} = 0.1 = 10\%$

5. $\frac{7}{10} = 0.7 = 70\%$

6. $2\frac{1}{2} = 250\%$

7. $1\frac{1}{4} = 125\%$

8. Change 150% to a mixed number.

9. Write two hundred eighty percent as a mixed number.

EXAMPLE 8

Write 240% as a mixed number.

SOLUTION

$$240\% = 240 \cdot \frac{1}{100} = \frac{240}{100} = 2.4 = 2\frac{2}{5} \quad ■$$

EXAMPLE 9

Change each to mixed number notation:

(a) 125% **(b)** 175% **(c)** 250% **(d)** 201%

SOLUTION

(a) $125\% = \frac{125}{100} = 1.25 = 1\frac{1}{4}$

(b) $175\% = \frac{175}{100} = 1.75 = 1\frac{3}{4}$

(c) $250\% = \frac{250}{100} = 2.50 = 2\frac{1}{2}$

(d) $201\% = \frac{201}{100} = 2.01 = 2\frac{1}{100}$ ■

◇ *Do Practice Exercises 8 and 9.*

The following table shows some common percents and their fraction, decimal, and mixed number equals. Note that mixed number notation applies only if the percent is greater than 100%.

Fraction		*Decimal*		*Percent*		*Mixed Number*
$\frac{1}{5}$	=	0.20	=	20%	=	—
$\frac{1}{4}$	=	0.25	=	25%	=	—
$\frac{1}{3}$	=	$0.3\overline{3}$	=	$33\frac{1}{3}\%$	=	—
$\frac{1}{2}$	=	0.5	=	50%	=	—
$\frac{1}{1}$ (= 1)	=	1.0	=	100%	=	—
$\frac{6}{5}$	=	1.2	=	120%	=	$1\frac{1}{5}$
$\frac{5}{4}$	=	1.25	=	125%	=	$1\frac{1}{4}$
$\frac{3}{2}$	=	1.5	=	150%	=	$1\frac{1}{2}$
$\frac{11}{5}$	=	2.2	=	220%	=	$2\frac{1}{5}$
$\frac{9}{4}$	=	2.25	=	225%	=	$2\frac{1}{4}$

ANSWERS

8. $150\% = 1\frac{1}{2}$

9. $280\% = 2\frac{4}{5}$

Change the following to fraction notation and reduce if possible.

1. 30% **2.** 50% **3.** 5% **4.** 1%

5. 0% **6.** 99% **7.** 100% **8.** 110%

9. 37% **10.** 85% **11.** 74% **12.** 58%

13. 205% **14.** $8\frac{1}{2}\%$ **15.** 320% **16.** 100.1%

17. $16\frac{3}{5}\%$ **18.** 60% **19.** 47.8% **20.** $10\frac{1}{7}\%$

21. $9\frac{1}{4}\%$ **22.** $\frac{1}{8}\%$ **23.** $\frac{3}{4}\%$ **24.** 16.4%

Change each to percent notation.

25. $\frac{1}{4}$ **26.** $\frac{1}{2}$ **27.** $\frac{3}{4}$ **28.** $\frac{3}{5}$

29. $\frac{1}{10}$ **30.** $\frac{3}{8}$ **31.** $\frac{4}{5}$ **32.** $\frac{23}{20}$

33. $\frac{37}{100}$ **34.** $\frac{45}{100}$ **35.** $\frac{90}{200}$ **36.** $\frac{9}{8}$

37. $1\frac{3}{5}$ **38.** $\frac{13}{10}$ **39.** $\frac{300}{1000}$ **40.** $1\frac{7}{8}$

41. $1\frac{1}{2}$ **42.** $2\frac{3}{4}$ **43.** $\frac{19}{20}$ **44.** $\frac{225}{100}$

45. $1\frac{5}{8}$ **46.** $2\frac{3}{5}$ **47.** $1\frac{7}{9}$ **48.** $5\frac{2}{3}$

Complete the following table, reducing fractions if possible.

Percent		*Decimal*		*Fraction*		*Mixed Number*
70%	=	**49.** ______	=	**50.** ______	=	—
51. ______	=	0.20	=	**52.** ______	=	—
53. ______	=	**54.** ______	=	$\frac{5}{8}$	=	—
140%	=	**55.** ______	=	**56.** ______	=	**57.** ______
58. ______	=	1.00	=	**59.** ______	=	**60.** ______
61. ______	=	**62.** ______	=	$\frac{11}{10}$	=	**63.** ______
64. ______	=	**65.** ______	=	**66.** ______	=	$3\frac{1}{2}$
67. ______	=	**68.** ______	=	$\frac{13}{2}$	=	**69.** ______
70. ______	=	**71.** ______	=	**72.** ______	=	$2\frac{2}{3}$

Work each problem.

73. Three-fourths of the students passed an exam. What percent passed the exam? What percent failed the exam?

74. A group of children ate three-fifths of a gallon of ice cream. What percent of a gallon of ice cream did they eat?

75. Sixty percent of the people in one small city own their home. What fraction *do not* own their home?

76. An apartment house with 100 units has a 6% vacancy rate. How many apartments are *occupied*?

77. Twenty percent of the seats in a classroom are filled by females and another one-fourth of the seats are filled by males. What percent of the seats are *empty*?

78. Aunt Hilda passed away and left one-fourth of her estate to Judy and three-fifths of her estate to Becky. The remainder of her estate went to her dog Poochie. What percent of her estate went to Poochie?

JOURNAL EXERCISES

1. Explain the process of changing between percent notation and fraction notation.

2. Explain the process of changing between percent notation and mixed number notation.

3. Can every number that we have studied so far be written using fraction notation? Using decimal notation? Using percent notation? Explain.

4. Why do you think there are several different ways to write the same number? In other words, why have different notations for the number 1.25, such as $1\frac{1}{4}$, $\frac{5}{4}$, or 125%?

OBJECTIVES

1 Use the percent formula to calculate amount.
2 Use the percent formula to calculate rate.
3 Use the percent formula to calculate whole.
4 Choose the correct form of the percent equation.

In the first section of this chapter, we saw that proportions can be used to solve percent problems. In this section, we will see another way to solve percent problems.

The phrase "percent of" or "% of" is important in percent problems. Every percent must be a percent of something, namely a percent **of the whole**. The first step in understanding a percent problem is to determine what value the percent applies to.

Example	*Percent of What*
6% tax	6% of *selling price*
8% interest rate	8% of *amount invested*
3% commission	3% of *selling price*
110% final weight	110% of *original weight*
20% discount	20% of *original price*

The general form of the percent formula is:

PERCENT OF WHOLE IS SOME AMOUNT

In percent problems, the phrase "percent of" is a key phrase for multiplication, and the word "is" means equal. Therefore, the percent formula can be stated as:

$$\text{Percent} \cdot \text{Whole} = \text{Amount}$$

The Whole is the total or complete amount. The Whole is the amount that follows the key phrase "percent of" in a word problem. The Percent multiplied times the Whole is the Amount, or Portion.

Some common examples of usages of percents are written below:

Percent	*of*	*Whole*	*is*	*Some Amount*
6%	of	selling price	is	the sales tax
8%	of	the amount invested	is	interest
4%	of	the work force	is	unemployed
5%	of	the original pay	is	the raise

Sometimes the percent formula is written as:

$$\text{Rate} \cdot \text{Base} = \text{Amount}$$

or as

$$R \cdot B = A$$

where R = Rate = Percent
B = Base = Whole
A = Amount or Portion

Any one variable in the percent formula can be unknown and a value can be found for it as long as the other two values are known. We will work problems solving for each variable, in turn.

1. Find 65% of 300.

2. Find twenty-five percent of one thousand.

3. One percent of 700 is what amount?

ANSWERS
1. 195
2. 250
3. 7

1 Calculate Amount

In the percent formula:

$$\text{Amount} = \text{Rate} \cdot \text{Base}$$

Amount is found by substituting rate and base into the equation, then multiplying. In order to multiply a percent times a number, first change from percent notation to either decimal notation or fraction notation, then multiply.

HELPFUL HINT

You must change percent notation to either decimal notation or fraction notation before multiplying.

EXAMPLE 1

Find 25% of 800.

SOLUTION

The number 800 follows the phrase "percent of," so that 800 must be the Whole or Base. Amount is unknown.

$$\begin{aligned} \text{Percent} &= \text{Rate} = R = 25\% \\ \text{Whole} &= \text{Base} = B = 800 \\ \text{Amount} &= A = \text{unknown} \end{aligned}$$

Substitute values into the percent equation and solve.

Amount = Rate · Base	
Amount = 25% · 800	Substitute values.
Amount = 0.25 · 800	Change % to a decimal.
Amount = 200	25% of 800 is 200 ■

EXAMPLE 2

Find 130% of 20.

SOLUTION

Percent = 130%, Base = 20, Amount is unknown

Amount = Rate · Base	
Amount = 130% · 20	Substitute values.
Amount = 1.3 · 20	Change % to a decimal.
Amount = 26	26 is 130% of 20 ■

◇ *Do Practice Exercises 1 through 3.*

2 Calculate Rate (Percent)

Rate (percent) can be found by division if both Base and Amount are known. To find Rate, divide both sides of the equation by Base.

$$\text{Rate} \cdot \text{Base} = \text{Amount}$$

$$\frac{\text{Rate} \cdot \text{Base}}{\text{Base}} = \frac{\text{Amount}}{\text{Base}} \quad \text{Divide both sides by Base.}$$

$$\frac{\text{Rate} \cdot \cancel{\text{Base}}}{\cancel{\text{Base}}} = \frac{\text{Amount}}{\text{Base}} \quad \text{Divide out forms of one.}$$

$$\text{Rate} = \frac{\text{Amount}}{\text{Base}}$$

The division results in a decimal number that is changed into a percent by moving the decimal place to the right two places and attaching the % symbol.

EXAMPLE 3

60 is what percent of 200?

SOLUTION

Base = 200, Percent = ?, Amount = 60

$$\text{Rate} = \frac{\text{Amount}}{\text{Base}} = \frac{60}{200} = \frac{30}{100} = 0.30 = 30\%$$

Or 60 is 30% of 200 ■

It is not necessary to memorize the equation $\text{Rate} = \frac{\text{Amount}}{\text{Base}}$. Rather, you can always start with the basic form of the percent formula as shown in the next example.

EXAMPLE 4

What percent of 240 is 30?

SOLUTION

Rate = ?, Base = 240, Amount = 30

$$R \cdot B = A \quad \text{percent formula.}$$

$$R \cdot 240 = 30 \quad \text{Substitute values.}$$

$$\frac{R \cdot 240}{240} = \frac{30}{240} \quad \text{Divide both sides by 240.}$$

$$\frac{R \cdot \cancel{240}}{\cancel{240}} = \frac{30}{240} \quad \text{Divide out forms of one.}$$

$$R = 0.125$$

$$R = 12.5\%$$

Note that 12.5% can also be written as $12\frac{1}{2}\%$. ■

◇ *Do Practice Exercises 4 through 6.*

4. 75 is what percent of 150?

5. Twenty is what percent of one thousand?

6. What percent of 100 is 125?

ANSWERS
4. 50%
5. 2%
6. 125%

7. 12 is 10% of what number?

8. Fifty is twenty-five percent of what number?

9. Thirty dollars is forty percent of what value?

3 Calculate Whole (Base)

Base can be found by division if both Rate and Percent are known:

$$\text{Rate} \cdot \text{Base} = \text{Amount}$$

$$\frac{\text{Rate} \cdot \text{Base}}{\text{Rate}} = \frac{\text{Amount}}{\text{Rate}} \qquad \text{Divide both sides by Rate.}$$

$$\frac{\cancel{\text{Rate}} \cdot \text{Base}}{\cancel{\text{Rate}}} = \frac{\text{Amount}}{\text{Rate}} \qquad \text{Divide out forms of one.}$$

$$\text{Base} = \frac{\text{Amount}}{\text{Rate}}$$

EXAMPLE 5

10 is 8% of what number?

SOLUTION

The unknown follows the phrase ''percent of'' and so must be the Base.
Amount = 10, Rate = 8%

$$\text{Base} = \frac{\text{Amount}}{\text{Rate}} = \frac{10}{0.08}$$

$$\begin{array}{r} 125. \\ 0.08\overline{)10.00.} \\ \underline{8} \\ 20 \\ \underline{16} \\ 40 \\ \underline{40} \\ 0 \end{array}$$

10 is 8% of 125 ■

Again, you do not need to memorize Base $= \frac{\text{Amount}}{\text{Rate}}$. You can always start with the basic formula and solve for the unknown as shown in the next example.

EXAMPLE 6

80% of what number equals 40?

SOLUTION

Rate = 80%, Base = ?, Amount = 40

$$R \cdot B = A$$

$$80\% \cdot B = 40 \qquad \text{substitute values}$$

$$0.8 \cdot B = 40 \qquad \text{Change 80\% to 0.8.}$$

$$\frac{0.8 \cdot B}{0.8} = \frac{40}{0.8} \qquad \text{Divide both sides by 0.8.}$$

$$\frac{\cancel{0.8} \cdot B}{\cancel{0.8}} = \frac{40}{0.8} \qquad \text{Divide out forms of one.}$$

$$B = 50$$ ■

◇ *Do Practice Exercises 7 through 9.*

ANSWERS
7. 120
8. 200
9. $75

4 Choose the Correct Form of the Percent Equation

The first step in solving a percent problem is to correctly identify the Base, Rate, and Amount. You must know which one is unknown before substituting values into the percent formula. Look carefully at the following examples.

EXAMPLE 7

18 is 20% of what number?

SOLUTION

18 is part of some number so that 18 must be Amount. Rate = 20% and the Base is unknown.

$$R \cdot B = A$$

$$20\% \cdot B = 18 \quad \text{substitute values}$$

$$0.2 \cdot B = 18 \quad \text{Change 20\% to 0.2.}$$

$$\frac{0.2 \cdot B}{0.2} = \frac{18}{0.2} \quad \text{Divide both sides by 0.2.}$$

$$B = 90 \quad \text{Result.}$$

■

EXAMPLE 8

A recent survey showed that 54 out of 240 college graduates were not able to find jobs in the area they had studied in college. What percent does this correspond to?

SOLUTION

Amount = 54, Base = 240, Rate = ?

$$R \cdot B = A$$

$$R \cdot 240 = 54 \quad \text{substitute values}$$

$$\frac{R \cdot 240}{240} = \frac{54}{240} \quad \text{Divide both sides by 240.}$$

$$R = 0.225 = 22.5\% = 22\frac{1}{2}\%$$

$22\frac{1}{2}\%$ of the 240 college graduates were not able to find jobs in their area of study. ■

◇ *Do Practice Exercises 10 through 12.*

10. 15 is what percent of 60?

11. Eighty is twenty percent of what number?

12. Find 6% of 70.

ANSWERS
10. 25%
11. 400
12. 4.2

name date hour

Change to percent notation. (Review)

1. 0.07 **2.** 0.27 **3.** 0.90 **4.** 0.04

5. 1.12 **6.** 2.5 **7.** $\frac{7}{10}$ **8.** $3\frac{1}{5}$

Change to decimal notation. (Review)

9. 16% **10.** 85% **11.** 27% **12.** 35%

13. 3% **14.** 120% **15.** $4\frac{1}{5}$ **16.** $3\frac{1}{4}$

Solve for Amount.

17. 20% of 70 is ______. **18.** 50% of 40 is ______. **19.** 99% of 2,000 is ______.

20. 1% of 37 is ______. **21.** 7% of 60 is ______.

22. 12% of 40 is ______. **23.** 125% of 80 is ______.

24. What number is forty percent of seven hundred?

25. What quantity equals ninety percent of two hundred ten?

26. What number is twenty percent of ninety?

Solve for Rate.

27. 100 = ______% of 500. **28.** 50 is what percent of 200? **29.** 7 is what percent of 70?

30. 12 is what percent of 4? **31.** 70 is what percent of 35? **32.** ______% of 700 is 210.

33. 28 = ______% of 560. **34.** What percent of 40 is 5?

35. What percent of 300 is 50? **36.** ______% of 90 is 27.

Solve for Base.

37. 5% of what number is 40?

38. 20% of ______ is 70.

39. 1% of ______ is 75.

40. 150% of ______ is 75.

41. 80 is 40% of ______.

42. 25 is 20% of ______.

43. 60% of what number is 240?

44. Eighty percent of what number is two hundred?

45. Sixty is thirty percent of what number?

46. 199.5 = 70% of ______.

Solve for the unknown.

47. 8 is what percent of 32?

48. 100 is 30% of what number?

49. What number is 40% of 280?

50. 20% of 70 is what number?

51. 30% of what number is 390?

52. 25% of what number is 2,000?

53. What percent of 1,000 is 70?

54. What percent of 1,000 is 60?

55. What number is 75% of 500?

56. 700 is what percent of 80?

57. Find 60% of 90.

58. Find three-fourths of 72.

59. What percent is 50 of 10?

60. What percent 80 is 20?

61. 300 is what percent of 60?

62. 30 is what percent of 15?

63. 76 is 40% of what number?

64. 60% of 94 is what number?

65. 5 is what percent of 200?

66. Find 150% of 700.

Convert each percent to a decimal.

	Application	*Percent*
67.	Interest rate	5%
68.	Sales commission	8%
69.	Income tax rate	20%
70.	Sales tax rate	7.5%
71.	Growth rate of a child	10%
72.	Unemployment	$7\frac{1}{2}\%$

Write from smallest to largest by changing each to decimal notation first, then comparing.

73. 25%, 0.27, $\frac{1}{5}$

74. 30%, 0.04, $\frac{1}{2}$

75. $4\frac{1}{4}$, 450%, 6.75

76. 40%, $\frac{3}{5}$, 0.70

77. 1.75, $1\frac{3}{5}$, 165%

78. $\frac{2}{3}$, 6.5%, 1.6

Work each problem.

79. Jesse earns $12,000 per year. His boss told him that everyone at the plant will receive an 8% raise. What will his new salary be?

80. Darla ran three-fourths of a mile and John ran seventy percent of a mile. Who ran farther?

81. Hilda ate twenty of twenty-five raisins in a bag and Jan ate the remainder. What percent of the raisins did Jan eat?

JOURNAL EXERCISES

1. Laurie grabbed thirty percent of the marbles. How many marbles did she leave for her sister? Can this problem be solved? If not, why not?

2. Explain the difference in Amount versus Base as used in this section. How can you tell them apart in a problem?

3. List some everyday examples of the use of percents. Write each in the form shown by:

6% of selling price of home = real estate commission

4. Explain the grading system used in the math course you are currently in. Does it involve percents?

OBJECTIVES

1 Solve problems involving simple interest.

2 Solve problems involving taxes.

Here are some general suggestions that may help you solve percent word problems.

HELPFUL HINT

1. Decide if in fact it is a problem that involves percents.
2. Decide which two of rate, base, or amount are known and which one is unknown.
3. Substitute known values into the percent formula:

$$\text{some amount} = \text{percent of whole}$$
$$\text{Amount} = \text{Rate} \times \text{Base}$$
$$A = R \times B$$

4. Solve for the unknown.
5. Check for the reasonableness of the answer with respect to the original word problem.

This section and the next two show some common applications using percents.

1 Solve Problems Involving Simple Interest

The cost of borrowing money is called **interest**. Interest is the difference between the amount of money borrowed and the amount repaid. The percent formula can be used to calculate simple interest as follows:

Amount	=	Rate	·	Base	·	Time
Interest	=	Interest Rate	·	Investment	·	Time

The length of time over which the loan occurs is also included in the equation. Interest rates are usually given on an annual basis, for example 8% per year. In that event, time should be written as a fraction of a year, such as $\frac{1}{2}$ of a year or as 1 year.

EXAMPLE 1

Vera invested $1,000 in a certificate of deposit at a local bank. By doing so, she promised to leave the money at the bank for one full year. In return, the bank promised to pay her 9% interest for the year. How much interest will she earn?

SOLUTION

She will earn $90 interest for the year. ■

1. Interest of 6% per year is earned on an $8,000 investment. Find the interest for six months and for one year.

2. Jesse loaned his brother $200 for one year and was rewarded with $40 interest. What interest rate was paid?

ANSWERS
1. $240 interest, $480 interest
2. 20%

EXAMPLE 2

What is the interest on a six month loan of $800 at an interest rate of 9% per year?

SOLUTION

$$\begin{aligned} \text{Interest} &= \text{Percent} \cdot \text{Loan Amount} \cdot \text{Time} \\ &= 9\% \cdot \$800 \cdot 6 \text{ months} \\ &= 0.09 \cdot \$800 \cdot \frac{1}{2} \text{ of a year} \\ &= 72 \cdot \frac{1}{2} \\ &= \$36 \text{ Interest} \end{aligned}$$

■

EXAMPLE 3

Tom has a checking account that pays interest on the average balance in the account. During a year in which the average balance was $400, he received $12 in interest. What effective interest rate did he receive?

SOLUTION

Rate = ?, Base = $400, Amount = $12

$$R \cdot B \cdot T = A$$

$$R \cdot 400 \cdot 1 = 12 \qquad \text{Substitute values.}$$

$$\frac{R \cdot 400}{400} = \frac{12}{400} \qquad \text{Divide both sides by 400.}$$

$$R = 0.03 = 3\% \text{ interest rate}$$

■

◇ *Do Practice Exercises 1 and 2.*

Interest rate is a very common application of percent problems. The Application Word Problem at the end of this chapter shows how a loan with monthly payments works. The following calculator application shows how a bank makes money using different interest rates.

Calculator Application

A bank borrows $10,000 from an individual at an interest rate of 7% per year. It then lends the $10,000 at a rate of 9.25% per year. Find the profit.

Cost to the bank:

7% of $10,000 = 0.07 · $10,000 = $700

Interest received by the bank:

9.25% of $10,000 = 0.0925 · $10,000 = $925

Profit to the bank = $925 − $700 = $225

2 Solving Problems Involving Taxes

We live in a society with many different kinds of taxes: property taxes, sales taxes, social security taxes, and income taxes. Many taxes are based on a percentage. For example:

Property tax = Tax Rate × Value of property
Sales tax = Tax Rate × Selling price
Income tax = Tax Rate × Income

The basic percent equation applies to tax problems as follows:

Amount	=	Rate	·	Base
Tax	=	Tax Rate	·	Tax Base

EXAMPLE 4

Items costing \$3.75 and \$5.25 are purchased. What is the total cost if a sales tax of 7% of the total applies?

SOLUTION

The total cost before tax is:

$$\begin{array}{r} \$3.75 \\ +5.25 \\ \hline \$9.00 \end{array}$$

Sales tax must be calculated and added to the cost:

Sales tax = 7% of total cost
Sales tax = 0.07 × \$9.00
Sales tax = \$.63

Thus the total cost is \$9.00 + \$.63 = \$9.63. ■

EXAMPLE 5

Income taxes are withheld from an employee's pay. What is the average tax rate for a year in which \$30 is withheld from each \$200 of gross pay?

SOLUTION

R = ?, Amount = \$30, Base = \$200

$$R \cdot B = A$$
$$R \cdot 200 = 30 \quad \text{substitute values.}$$
$$\frac{R \cdot 200}{200} = \frac{30}{200} \quad \text{Divide both sides by 200.}$$
$$R = 0.15 = 15\% \text{ tax rate}$$ ■

◇ *Do Practice Exercises 3 and 4.*

3. Using the same tax rate as given in example 5, find the tax on \$300.

4. The sales tax on a \$9,000 automobile is \$540 if the car is purchased in one state and \$675 if the car is purchased in a second state. Find the tax rate in both states.

ANSWERS
3. \$45 tax
4. 6%, 7.5%

Calculator Application

One congress person recommended a tax plan with a 10% tax on the first \$20,000 of income per year, and 22% tax on all income above \$20,000. What tax would someone making \$26,000 per year pay under this system?

Tax on first \$20,000:

$$10\% \text{ of } \$20{,}000 = 0.1 \cdot 20{,}000 = \$2{,}000$$

Tax on next \$6,000:

$$22\% \text{ of } \$6{,}000 = 0.22 \cdot 6{,}000 = \underline{\$1{,}320}$$

$$\text{Total Tax} \quad \$3{,}320$$

Work each problem.

1. ABC, Inc. needs to borrow $10,000 for either nine months or for a full year. At 12% simple interest, find the amount of interest associated with each.

2. Find the sales tax on a new car costing $10,500 if the tax rate is 7.5%.

3. Interest of $160 was received on an investment of $1,280. Find the effective interest rate.

4. Jose deposits $2,000 at a bank and receives 6% simple interest. What interest would be earned at the end of nine months? One year?

5. A sales tax on a new $65,000 automobile amounted to $4,875. What is the tax rate?

6. Mr. Tate left an estate valued at $95,000 to his heirs. However, they only received $84,000 after burial and probate expenses. What percent of the total did the heirs receive?

7. Mr. and Mrs. Walker sold their home for $80,000 and invested the money at 9.5% simple interest per year. What yearly income will they derive from this investment?

8. At a 6% simple interest rate, an investment yielded $300 in interest. What amount was invested?

9. Tony is taxed $650 per year on his home which is valued at $81,250. What is the tax rate?

10. The property tax on a house valued at $60,000 is $600. The tax is what percent of the value?

11. An individual pays about 7.5% of her income for social security taxes and about 14% of her income for income taxes. What percent of her income does she take home, after taxes?

12. Jessica pays the following taxes: FICA tax—7.5% of income, federal tax—16.5% of income, and State tax—2% of income. What percent of her income does she take home, after taxes?

13. In one area, gasoline costs $1.20 per gallon which includes a 30 cent tax per gallon. The tax is what percent of the total cost of a gallon of gasoline?

14. Jane made $22 in simple interest one year at a rate of 8%. How much did she have invested?

15. A property tax of $400 was levied on a small commercial building valued at $80,000. What is the tax rate?

16. An individual pays 24.2% of her income in taxes and 4% in interest. What percent of her income remains?

17. One company pays a tax on profits that averages 28%. What is their profit during a year in which their taxes amount to $126,000?

18. A property tax of 0.9% is applied to a house resulting in a tax of $468. What is the value of the house?

19. A large corporation receives $187,000 in annual interest on a $2,200,000 investment. What percent effective interest did they achieve?

20. Martin would like to receive a 10% return on his investment. He invests fifteen thousand dollars and receives a return on investment of $1,800 after one year. Is he satisfied with this investment?

21. A bank borrows *fifty* million dollars from its thousands of customers at an average interest rate of 5% per year. The bank then lends *forty* million of these dollars out to business owners at an average interest rate of 10% per year. Find the difference between interest received and interest paid out.

22. Huge insurance companies have billions of dollars to invest. At 7.5% simple interest, what annual income would be generated from a two and one-half billion dollar investment?

23. Roberta's gross pay is $300 per week. Assume a social security tax of 7.5% and an income tax of 5% are withheld out of each check. What amount does she receive after taxes?

24. A homeowner is required to pay a city tax of $600 per year on his home. Given a tax rate of 0.4%, what value does the county place on the home?

25. The Kings borrowed $8,000 for a medical emergency. They paid it back one year later at 12% simple interest. What interest did they pay?

26. A city levies a 2% tax on all assessed property values within the city. What is the total of assessed values if the city levies forty-two million, five hundred thousand dollars in taxes?

JOURNAL EXERCISES

1. Make a list of all of the different types of taxes you are aware of. Are the tax rates usually given in percent notation? Try to find the tax rates.
2. A King in a make-believe kingdom levied a 99% tax on all of the income of each of his subjects. Explain what this might mean to the people, in terms of daily life.
3. What annual income would you like to make? Just pick a figure! At 6%, how much money would you have to have to invest to earn that amount of interest each year?
4. The national debt of the United States is close to four trillion dollars. At an interest rate of 7%, what is the annual interest payment the United States must pay? Divide the national debt by 250,000,000 to find the average amount of debt per person living in the United States.
5. Why is it important to check to see if your answer to a problem is correct?
6. Explain the steps you use when solving word problems. Be as specific as possible.

1 Solving Problems Involving Commissions

Many sales people are paid based on commissions or a percent of what they sell. Some jobs in which income may be based on sales include:

Real Estate agents	Salespersons in clothing stores
Insurance agents	Travel agents
Stock brokers	Car salespersons

The basic percent equation is used to calculate commissions as follows:

Amount	=	Rate	·	Base
Commission	=	Commission Percent	·	Sales Price

EXAMPLE 1

Chuck and Judy sold $3,200,000 worth of real estate in one year. Find their income if the real estate commission averaged 3%.

SOLUTION

Amount = ?, Rate = 3%, Base = $3,200,000

$$A = R \cdot B = 0.03 \cdot \$3{,}200{,}000 = \$96{,}000 \text{ in commission}$$ ■

EXAMPLE 2

Walter sells furniture manufactured by his company to retail outlets. His commission is 8% of sales made to outlets located in his territory. What were the total sales for a month during which he earned $2,200?

SOLUTION

Rate = 8%, Amount = $2,200, Base = ?

$$\text{Base} = \frac{\text{Amount}}{\text{Rate}} = \frac{\$2{,}200}{0.08} = \$27{,}500 \text{ in sales}$$ ■

◇ ***Do Practice Exercises 1 and 2.***

EXAMPLE 3

Leon receives a fixed amount of $1,000 per month in addition to 3% of monthly sales. Find his income for a month with $30,000 in sales.

SOLUTION

Find 3% of $30,000: R = 3%, Amount = ?, Base = $30,000

$$\text{Commission} = 0.03 \cdot \$30{,}000 = \$900$$

The income is the sum of $1,000 plus the commission:

$$\text{Income} = \$1{,}000 + \$900 = \$1{,}900 \text{ income}$$ ■

OBJECTIVES

1. Solve problems involving commissions.
2. Construct tables using percents.

1. Andrea receives $60 commission for each $400 suit she sells. Find her commission rate.

2. Jamie earns a 20% commission for her Girl Scout troop for every box of $5.00 cookies that she sells. How many boxes must she sell to earn $20 for her troop?

ANSWERS
1. 15% commission
2. 20 boxes

3. The grades in an English class are given below. Complete the table.

Grade	Number of Students	Percent
A	4	
B	8	
C	15	
D	10	
F	3	
Total		

ANSWER
3. A—10%, B—20%, C—37.5%, D—25%, F—7.5%
Totals 40 100%

2 Construct Tables Using Percents

Percents are commonly used to show information in tables and charts. The following examples show you some of the ways in which percents are used.

EXAMPLE 4

The table below shows one family's annual expenditures by category. Complete the table by finding the percent each category is of the total expenses.

Item	Expense	Percent
Housing	$ 6,000	
Food	5,000	
Automobiles	7,000	
Other	7,000	
Total	$25,000	100%

SOLUTION

Housing as a percent of the total is found by:

$$\text{Rate} = \frac{\text{Amount}}{\text{Base}} = \frac{6{,}000}{25{,}000} = 0.24 = 24\%$$

The other three percents are found in the same way. The completed table follows:

Item	Expense	Percent
Housing	$ 6,000	24%
Food	5,000	20%
Automobiles	7,000	28%
Other	7,000	28%
Total	$25,000	100%

◇ *Do Practice Exercise 3.*

Calculator Application

Complete the following table that lists the number of students by age at a small college. Round values to the nearest percent.

Age Group	Number	Percent
17—20	2,105	?
21—23	1,980	?
24—35	1,625	?
Over 35	352	?
Total		

Why do the percents add up to more than 100%?

Answers: 35%, 33%, 27%, 6%; because of rounding errors; columns total 6,062 or 101%

Work each problem.

1. A real estate agent receives one-half of a 6% commission. What would his commission be on a land sale of $400,000?

2. A sales clerk receives a commission of 8% on each sale. What is the commission on a sale of $120?

3. An industrial salesman receives a commission of 2% on each tractor sold. What commission would he receive for the sale of 2 tractors, priced at $45,000 each?

4. A chemical sales person received 2.5% commission on $80,000 in sales one month. What was her commission?

5. A stock broker received $2,800 in commission at a 1% commission rate. What were total sales?

6. A money broker received $4,000 in commissions one month. Given that she makes 0.2% commission of amount of loans placed, how much loan volume did she place?

7. A food broker received $800 commission on $40,000 in sales. What is the commission rate?

8. A forester classified trees by age: 0–20 years—800 trees; 21–40 years of age—300 trees; and 41–100 years of age—3 trees. Place this data in a table which includes a column for percents, and complete the table.

9. A rancher divided her herd into three groups: very young—415 cows; mature—220 cows; and old—35 cows. Present this data in a table, including a column for percents.

10. A real estate agent sold three houses. The sale price and the commission are listed for each: \$45,000–3%; \$60,000–3%; and \$55,000–4%. Find her total commission and her average commission rate.

11. A salesman receives 8% commission on each sale. Find the daily commissions for three days with the following sales: \$2,000, \$800, and \$1,100.

12. A sales clerk receives a base salary of \$800 per month plus a commission of 3% of sales. What is the total income for a month with \$8,000 in sales?

13. A sales person receives a base salary of \$1,200 *per month* plus a commission of 1% of sales. Find the total sales for a month in which she was paid \$1,800.

14. A salesman received a \$1,200 commission on one sale. At a 5% commission rate, what was the size of the sale?

15. A sales person makes a salary of \$500 per month plus a commission of 5% of sales. What is the total monthly income for three months in which sales were \$9,000, \$12,000, and \$15,000?

16. Jesse received a commission of \$900 on sales of \$7,200. What percent commission does Jesse receive?

17. Lupe was paid a \$400 commission for a month with \$12,000 in sales. What commission does she receive?

18. A commission of \$800 was calculated as 6% of total sales. Find the total sales.

19. Three percent of total sales was $1,200. Find the total sales.

20. At a commission rate of 4%, what monthly sales are needed to earn $3,000 per month?

21. Complete the following table by finding the Amounts in the column labeled Income.

Item	*Income*	*Percent of Total Income*
Sales		80%
Investments		12%
Sale of land		8%
Total income	$120,000	100%

Make a table for each, including a column for percent of total.

22. *One Week's Food Requirements for a Large Circus* 7 tons of hay, 900 tons of oats, $\frac{1}{2}$ ton of meat, $\frac{3}{4}$ tons elephant chow, $\frac{1}{2}$ ton of carrots, and $\frac{1}{2}$ ton of apples

23. *Age distribution of Canadians* Age: 0–14 Number: 5,677,000; Age: 15–59 Number: 16,871,000; Age: 60 and up Number: 3,979,000

24. *Student Grades* A's—4, B's—20, C's—35, D's—8, F's—2

25. *Company Sales* First Quarter—$450,000, Second Quarter—$426,000, Third Quarter—$375,000, Fourth Quarter—$580,000

26. *Company Revenues* Sales—$40,000, From Investments—$10,000, From Sales of land—$30,000

JOURNAL EXERCISES

1. List all of the professions you can think of where at least some portion of income is based on a percent of sales. Do you think these sales people need to know how to calculate percents accurately?

2. Look through magazines or science books, and find some tables that use percents. Copy them into your journal and briefly explain their meaning.

3. Create a table using data of your own creation. Be sure to include a column with percents.

4. Assume you are starting up your own company. You will need employees, but you are concerned about whether or not they will work hard for the company. Might there be an advantage to you to pay each person a small monthly salary and a bonus based on a percentage of profits? Explain.

1 Solve Problems Involving Markdown and Markup

The idea of **markdown** corresponds to a discount at a retail store in which the price of an item has been marked down by some percent. Another word sometimes used in place of markdown is **discount**.

The basic percent equation is applied to calculate markdown as follows:

Amount	=	Rate	·	Base
Markdown	=	Discount Rate	·	Original Price

EXAMPLE 1

A coat normally costing \$80 is on sale at a 20% discount. Calculate the sale price of the coat.

SOLUTION

Original Price = Base = \$80, Rate = 20%, Amount = ?

$$\text{Markdown} = \text{Rate} \cdot \text{Base} = 0.2 \cdot 80 = \$16$$

Therefore the sale price of the coat is:

\$80	normal price of coat
− 16	minus the discount or markdown
\$64 = sale price	

■

Sometimes the original price and sale price are given, and the problem is to find the percent of markdown as shown in the next example.

EXAMPLE 2

A stereo system is marked at \$261 rather than its normal price of \$300. By what percent has it been marked down?

SOLUTION

Original Price = \$300, Rate = ?

$$\text{Markdown} = \text{Discount} = \$300 - \$261 = \$39$$

Amount = Rate · Base, or

$$\text{Rate} = \frac{\text{Amount}}{\text{Base}} = \frac{39}{300} = 0.13 = 13\% \text{ markdown}$$ ■

◇ *Do Practice Exercises 1 and 2.*

The idea of **markup** corresponds to a retail business that increases prices above cost before selling them. Look at the next example.

OBJECTIVES

1 Solve problems involving markdown and markup.

2 Solve problems containing more than one percent calculation.

1. A sweater normally costing \$90 was on sale for 10% off. What is the sale price?

2. An automobile normally priced at \$13,500 was listed at \$12,150. What was the percent of markdown?

ANSWERS

1. \$81

2. 10%

3. A used automobile with a sticker price of $4,500 in the window is being advertised at a 10% discount. What is the discounted price?

4. A wine shop uses an 80% markup. What would their selling price be for a bottle costing them $6.50?

EXAMPLE 3

A discount pharmacy marks up specific items by 30%. In other words, they charge the customer 30% more than their cost for each of these items. What is their selling price for an item costing the store $8.20?

SOLUTION

Rate = 30%, Base = $8.20, Amount = ?

Amount = Rate · Base = 30% of $8.20 = $2.46

The final selling price is the cost to the pharmacy plus the markup:

$8.20	cost to pharmacy
+ 2.46	markup
$10.66	final selling price ■

Many times, prices in a store end with a 9. If that were the case in the last example, the price of $10.66 would probably be increased to $10.69, and the markup would be slightly greater than 30%.

Do Practice Exercises 3 and 4.

Calculator Application

An item costs a store $160. They then mark it up by 45%. Finally, they run a sale in which the price is marked down by 15% of the full price. What is the final selling price?

Markup = 45% of $160 = 0.45 · 160 = $72
Full price = $160 + $72 = $232

Markdown = 15% of $232 = 0.15 · 232 = $34.80
Selling price = full price − markdown = $232 − $34.80 = $197.20

2 Solve Problems Containing More than One Percent Calculation

Some problems require that more than one percent be calculated as shown in the following examples. These types of problems are excellent applications for your calculator.

EXAMPLE 4

The cost of a home is $45,000. In addition, a flat fee of $400 and expenses equal to 2% of $45,000 *must also* be paid. Find the down payment given that it is 5% of the combined cost of the house plus all fees.

SOLUTION

First, find the cost of the house, including all fees.

ANSWERS
3. $4,050 after the discount
4. $11.70 selling price

Then, find the down payment, which is 5% of $46,300.

$$\begin{array}{r} 46{,}300 \\ \times \quad .05 \\ \hline \end{array}$$

$2,315.00 is the down payment. ■

The next example may help you figure out the true value of a raise given that taxes take part of any raise.

EXAMPLE 5

Ting's gross pay is $500 per week and her take-home pay after all taxes is $420 per week. She receives a pay raise of 10% of her gross pay. However, taxes take 25% of the increase in pay. What is her additional take-home pay after taxes? What percent increase in take-home pay did Ting actually receive?

SOLUTION

First find Ting's total pay raise.

$$\text{Rate} = 10\%, \quad \text{Base} = \$500, \quad \text{Amount} = ?$$

$$\text{Raise} = \text{Amount} = \text{Rate} \cdot \text{Base} = 10\% \cdot \$500 = \$50$$

But taxes take 25% of the $50 raise:

$$\text{Tax on raise} = 25\% \text{ of } \$50 \text{ raise} = \$12.50$$

The amount of the raise she actually gets to keep is:

$$\text{Amount of raise} - \text{Tax on raise} = \$50 - \$12.50 = \$37.50$$

So that her additional take-home pay after taxes is $37.50.

Her percent increase in take-home pay is calculated by:

$$\text{Rate} = ?, \quad \text{Base} = \text{original take-home pay} = \$420$$

$$\text{Amount} = \text{additional take-home pay} = \$37.50$$

$$\text{Rate} = \frac{\text{Amount}}{\text{Base}} = \frac{\$37.50}{\$420} = 0.08929 \ldots \text{ or about } 8.9\%$$

Ting was given a 10% raise but only receives an 8.9% raise because of additional taxes. ■

Do Practice Exercises 5 and 6.

5. A contractor bought a building for $70,000. She marked it up by 20%, sold it, and paid a tax of 30% on her profit. How much tax must she pay?

6. An individual makes $400 per week. She receives a raise of 8%. What is her new weekly income? Taxes take 20% of her raise. How much of the raise goes to taxes?

ANSWERS

5. tax = $4,200

6. new weekly income = $432 per week; $6.40 of the $32 raise goes to taxes

Work each problem.

1. A car normally costing $18,500 is marked down by 15%. What is the discounted price?

2. A refrigerator normally priced at $850 is on sale for 20% off. What is the sale price?

3. A sweater normally costs $60 but is on sale for 10% off. What is the discounted price?

4. A car dealer marks prices up 20% over her cost. What price would she ask for an automobile costing her $14,000?

5. PC Inc. had sales of $620,000 one year. The manager predicted an increase in sales of 15% for the following year. What sales volume did the manager predict?

6. A fast food restaurant had sales of $800,000 one year but expected sales to *decrease* by 20% next year due to competition. What sales do they expect for next year?

7. A coat normally costing $120 was marked down to $84. What was the percent markdown?

8. A stereo system normally costing $350 was priced at $297.50. What was the percent markdown?

9. An individual had an income based on a salary of $800 per month plus 4% commission on sales. Find the total income for a month in which the individual made two sales, one for $20,000 and one for $28,000.

10. An individual has an income of $32,000 per year before receiving a raise of 2%. What is the new income? If living expenses go up by $800 per year during the same time period, is the individual better off or worse off?

11. Three coyotes were fighting over a dead rabbit. Coyote #1 broke free with 60% of the carcass and ran off. Coyote #2 ended up with three-fourths of the remaining portion of the rabbit. What percent did each coyote get?

12. A home sells for $52,000 plus a finance charge of 3% of the selling price. Find 10% of the total.

13. A local finance company charged $40 interest plus a flat fee of $10 for a one year loan of $200. Find the sum of the interest and the flat fee. This sum is what percent of the loan amount?

14. An employee's income went from $1,500 per month to $1,800 per month. What was the percent increase?

15. A new car depreciates (goes down in value) at the rate of 20% during the first year. After a year, what is the value of an automobile costing $15,000 when new?

16. Find 20% of 75% of 400.

17. What is $12\frac{1}{2}$% of 20% of 100?

18. An insurance agent earns 40% commission on new sales and his manager earns 2.5% on the same sales. Find commissions for both individuals on $4,000 in new sales by the agent.

19. A carpenter makes $22,000 per year yet takes home only $19,500 after taxes. He receives a pay increase of 8% of total income, of which 20% will go for taxes. What is the dollar increase in his take-home pay?

20. Tina makes $1,400 per month before taxes or $1,220 per month after taxes. She receives a pay raise of 5%, of which 15% goes for taxes. What is the dollar increase in her take-home pay?

JOURNAL EXERCISES

1. A human hair grew 400% in one year. Could the length of the hair have been 10 inches at the beginning of the year? Why, or why not?

2. A recent study showed that recycling efforts in one city resulted in: 25% of aluminum being recycled, 30% of newspaper being recycled, 15% of glass being recycled, and 2% of all plastics being recycled. Show this data in a table. Do the percents add up to 100%? If not, why not?

3. Do you find word problems difficult? Explain. Are they easier now than at the beginning of the course?

4. The original price of an item is first marked up 50%. Then, the price is marked down by 50% of the full price. Does the final price equal the original price? Explain. (*Hint:* Let the original price equal $100 and work the problem.)

5. An advertisement offered a 100% discount on all items in stock. What would the price of an item be to a customer? Do you think there might be a mistake? Explain.

CHAPTER 6 REVIEW

SECTION	TOPIC OR TERM	EXAMPLE OR DEFINITION
6.1		
Basics	Percent	Percent is the number out of each 100. 85% of 100 is 85 out of 100
	Find Percents Using Proportions	Find 40% of 320 using proportions. First ratio: $40\% = \frac{40}{100}$ Second ratio: $\frac{x}{320}$ $\frac{40}{100} = \frac{x}{320}$ Set ratios equal. $40 \cdot 320 = 100 \cdot x$ Cross multiply. $12{,}800 = 100x$ Multiply. $\frac{12{,}800}{100} = \frac{100x}{100}$ Divide both sides by 100. $128 = x$ Result.
6.2		
Percent Notation: Percents and Decimals	From Percents to Decimals	Move decimal point to the left 2 places and drop the % sign. $65\% = 0.65, \quad 9\frac{1}{2}\% = 9.5\% = 0.095$
	From Decimals to Percents	Move decimal point to the right 2 places and attach the % sign. $0.01 = 1\%, \quad 3.5 = 350\%, \quad 1 = 100\%$
6.3		
Percent Notation: Percents and Fractions	From Percents to Fractions	Replace % with "$\cdot \frac{1}{100}$" and multiply. $35\% = 35 \cdot \frac{1}{100} = \frac{35}{100} = \frac{7}{20}$
	From Fractions to Percents	Divide denominator into numerator then change to percent notation. $\frac{3}{8} = 0.375 = 37.5\%, 2\frac{3}{4} = 2.75 = 275\%$

SECTION	TOPIC OR TERM	EXAMPLE OR DEFINITION
6.4		
Percent Formula	Percent Formula	Percent of Whole is Amount, or Rate · Base = Amount, or $R \cdot B = A$
	Rate	Rate refers to the percentage rate.
	Base	Base is the whole or total.
	Amount	Amount is a part of the whole.
	Percent Problems	60 is what percent of 300? $R = ?,\quad B = 300,\quad A = 60$ $R \cdot B = A$ $R \cdot 300 = 60$ Substitute. $\frac{R \cdot 300}{300} = \frac{60}{300}$ Divide by 300. 300 300 $R = 0.2 = 20\%$ so that $60 = 20\%$ of 300
6.5		
Word Problems: Interest and Taxes	Simple Interest	Simple interest is the amount of interest calculated by multiplying loan amount times the interest rate times Time as a fraction of a year.
	Calculate Simple Interest	Find 8% simple interest on a one year loan of \$1,200. $R = 8\%,\quad A = ?,\quad B = \$1{,}200$ $A = R \cdot B = 8\% \cdot \$1{,}200 \cdot 1$ $= 0.08 \cdot \$1{,}200 \cdot 1$ $= \$96$ interest
6.6		
Word Problems: Commission and Tables	Commission	Commission is income calculated as a percent of sales.
	Calculate Commission	Find a 3% commission on sales of \$80,000. $R = 3\%,\quad B = \$80{,}000,\quad A = ?$ $A = R \cdot B = 3\% \cdot \$80{,}000$ $= 0.03 \cdot 80{,}000$ $= \$2{,}400$ commission

SECTION	TOPIC OR TERM	EXAMPLE OR DEFINITION
6.7		
Word Problems: Markdown, Markup, and More	Markdown	Markdown is the amount of the original price by which an item has been discounted.
	Calculate Markdown	Find a 25% discount on an item originally priced at \$480. $R = 25\%, \quad B = \$480, \quad A = ?$ $A = R \cdot B = 25\% \cdot \$480$ $= 0.25 \cdot 480$ $= \$120$ discount
	Markup	Markup is the amount by which the cost is increased before an item is sold.

APPLICATION WORD PROBLEM

LOAN REPAYMENT SCHEDULE

Many loans are repaid in equal sized monthly payments called **installments**. Part of each monthly payment is interest on the current amount of the loan. The remainder of the payment goes to reduce the loan. A table that shows the payment number, amount of loan remaining, payment, and amount of payment that goes to pay interest is called a **Payment Schedule**. For example:

Mrs. Smith borrows $9,000 from a bank to purchase a car. The interest rate charged by the bank is 12% per year. The bank tells Mrs. Smith that she must make 36 equal monthly payments of $298.93 to repay the loan.

The interest rate of 12% applies to an entire year, but the payments are monthly. The interest rate must first be broken down into that interest rate that applies to one month, since the payments are monthly.

$$\text{Monthly Interest Rate} = \frac{1}{12} \cdot \text{Annual Interest Rate}$$

$$\text{Monthly Interest Rate} = \frac{1}{12} \cdot (12\%) = 1\%$$

Part of *each payment* is interest and the remainder of the payment reduces the amount of money owed. The first payment is analyzed by first calculating the amount of the first payment that is interest: $R = 1\%$, Base $= \$9{,}000$, $A = ?$

$$\begin{aligned} A &= R \cdot B \\ \text{Interest} &= 1\% \cdot \$9{,}000 \\ \text{Interest} &= .01 \cdot 9{,}000 \\ \text{Interest} &= \$90 \end{aligned}$$

Thus, $90 of the first month's payment of $298.93 is interest and the remaining amount reduces the debt.

$298.93	first payment
− 90.00	interest part of first payment
$208.93	reduces loan amount

The new loan amount is found by subtracting $208.93 from the loan amount of $9,000.

$9,000.00	loan amount before first payment
− 208.93	amount of first payment applied to loan
$8,791.07	loan amount after first payment

The payment schedule for the first payment is:

Payment Number	*Loan Amount*	*Monthly Payment*	*Interest*	*Principal*	*Remaining Loan*
1	$9,000	$298.93	$90.00	$208.93	$8,791.07

The process begins again for the second month except for one important difference. The debt at the beginning of the second month is no longer $9,000, it is now $8,791.07.

$8,791.07	
× .01	monthly interest rate
$ 87.91	interest part of second payment
$ 298.93	payment
− 87.91	interest part of second payment
$ 211.02	amount of payment that reduces loan
$8,791.07	loan amount before second payment
− 211.02	amount of second payment that reduces loan
$8,580.05	loan amount after second payment

The updated payment schedule is now:

Payment Number	*Loan Amount*	*Monthly Payment*	*Interest*	*Principal*	*Remaining Loan*
1	$9,000.00	$298.93	$90.00	$208.93	$8,791.07
2	$8,791.07	$298.93	$87.91	$211.02	$8,580.05

Notice that the monthly payment remains the same even though the loan amount decreases after each payment. As the loan amount decreases, the amount of each payment that goes to interest decreases and the amount of each payment that goes to reduce the loan amount increases. Eventually, after 36 payments of $298.93, the loan amount will be zero.

1. Complete the next five rows of the table for Mrs. Smith's loan (payments 3–7).

2. Complete the first five rows of an installment table for the following home loan:

Loan amount:	$45,000.00
Monthly payment:	$362.08
Annual interest rate:	9% ⟶ first change to monthly rate

CHAPTER 6 REVIEW PROBLEMS

What percent of each geometrical figure is shaded?

1. 

2. 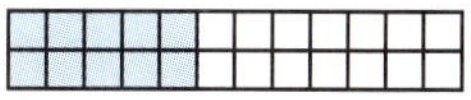

3. 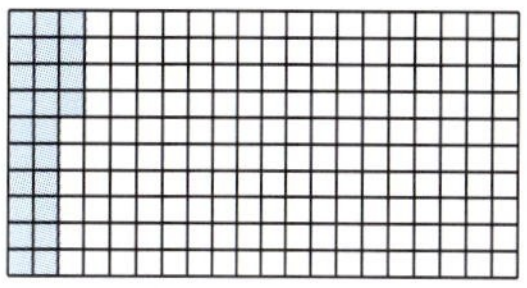

4. 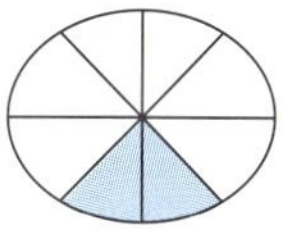

5. 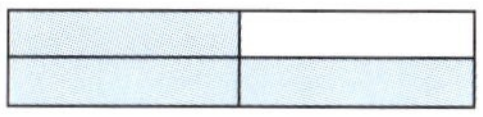

6. 

Change to percent notation.

7. 0.65

8. 0.05

9. 0.12

10. 0.01

11. 1.25

12. 0.001

13. 0.581

14. 1.5

15. $\frac{3}{4}$

16. $\frac{1}{2}$

17. $\frac{7}{8}$

18. $\frac{4}{5}$

19. 0.3

20. $\frac{3}{20}$

21. $\frac{1}{5}$

22. 3

Change to decimal notation.

23. 14%

24. 70%

25. 220%

26. 33%

27. 2.8%

28. 99%

29. 100%

30. 1000%

Change to fraction notation.

31. 40%

32. 15%

33. 95%

34. 21%

35. 100%

36. 180%

37. $4\frac{1}{2}\%$

38. 8.6%

Complete the table. The first column has been done for you.

Fraction	$\frac{3}{4}$	**39.**	**40.**	$\frac{7}{8}$
Decimal	0.75	0.125	**41.**	**42.**
Percent	75%	**43.**	15%	**44.**

Solve for the unknown.

45. 40% of 20 is ______.

46. 60% of ______ is 180.

47. ______% of 40 is 10.

48. 50% of 340 is ______.

49. 2% of ______ is 8.

50. ______% of 60 is 15.

51. 6% of ______ is 48.

52. Find 8% of 94.

53. 300 is 40% of what number?

54. 27 is ______% of 2,700.

Write from smallest to largest.

55. $\frac{1}{2}$, 35%, $\frac{2}{3}$

56. $\frac{1}{4}$, 4%, 0.45

57. 12%, 0.13, $\frac{1}{8}$

58. 75%, 2, 5.14

59. 22, 260%, $\frac{12}{4}$

60. 2.14, 12%, $3\frac{1}{4}$

Solve for the unknown.

61. 6% × 1,000 = Amount

62. 20% × Base = 60

63. Rate × 50 = 15

64. 40% × Base = 32

65. Rate × 100 = 70

66. 4% × 42 = Amount

Work each problem.

67. An item costs seventy-five dollars and has a six percent sales tax added. What is the total cost?

68. Mona deposits two thousand dollars in a certificate of deposit yielding seven percent per year. How much interest does she earn in one year?

69. A factory worker is paid $9.00 per hour. What is her hourly wage after receiving a six percent raise?

70. An item normally costing one hundred fifty dollars is on sale at a discount of 20%. What is the sales price?

71. An automobile originally costing $16,000 is reduced by $960. What percent discount is this?

72. An athlete improved his time for a 100-meter race from eleven seconds to ten and eight-tenths seconds. Find the percent of improvement to the nearest percent.

73. One broker earns 2% of sales. How much must she sell to earn $40,000 per year?

74. A newspaper boy is supposed to deliver 180 papers every Sunday. However, on average, he forgets to make $1\frac{1}{2}\%$ of the deliveries. About how many deliveries does he miss on average? Round your answer off to the nearest whole number.

75. A student took three tests in a class:

Test Number	*Number Correct*	*Number of Problems*
1	60	80
2	70	100
3	30	50

What final grade would the student receive if the scale is:

Range	*Grade*
90–100	A
80–89	B
70–79	C
60–69	D

Assume each test counts equally. In other words, first calculate the percent for each test, then average the percents to find the average test score.

CHAPTER 6 TEST

Change to percent notation.

1. 0.65 **2.** 0.01 **3.** $1\frac{1}{4}$ **4.** $\frac{3}{8}$

Change to decimal notation.

5. 16% **6.** 215% **7.** 100% **8.** $15\frac{1}{2}\%$

Change to fraction notation.

9. 20% **10.** 75% **11.** 62.5% **12.** $3\frac{1}{2}\%$

Complete the table.

Fraction	$\frac{4}{5}$	**13.**	**14.**
Decimal	**15.**	0.22	**16.**
Percent	**17.**	**18.**	5%

Work each problem.

19. Find 75% of 500.

20. Forty is five percent of what number?

21. Twenty-five is what percent of five hundred?

22. The Thompsons moved their life savings of forty thousand dollars into an account paying eight percent per year. What is the interest earned in one year?

23. An item costs eight dollars and has a tax of six percent added to it. What is the change from a twenty dollar bill?

24. Does 50% of 70 equal $\frac{3}{4}$ of 80? If not, which is greater?

25. The United States made the Louisiana purchase of about 800,000 square miles in 1803, and purchased Alaska with an area of about 600,000 square miles in 1867. The size of the Alaska purchase was what percent of the size of the Louisiana purchase?

CUMULATIVE TEST CHAPTERS 5 AND 6

Work each problem.

1. $6.25 + 3$

2. $5.4 \cdot 3.01$

3. $6.75 \div 2.5$

4. $12 - 1.4$

5. $\frac{1}{3} + 0.8$

6. $1.92 \div 3$

List from smallest to largest.

7. $\frac{3}{8}$, 0.275, 40%

8. $\frac{16}{3}$, 3.1, 300%

Evaluate.

9. $0.3 + 4.1 \cdot 5$

10. $(0.27 + 0.01) \div 0.002$

Solve for the unknown.

11. 6% of ______ is 72.

12. ______% of 300 is 30.

13. 20% of 400 is ______.

14. 3 is what percent of 60?

Work each problem.

15. Write 0.28 using both percent notation and fraction notation.

16. Write $2\frac{1}{4}$ using percent notation.

17. In one course, 10% of the students made A's, 15% made B's, and 45% made C's. What percent made grades lower than a C? Present all of this information in the form of a table.

18. Find the sum of seven hundredths with thirty and one tenth.

19. Find forty percent of six hundred.

20. Tom is paid $1,000 per month plus 8% of sales. What is his income for a month during which he sold $4,500.80? What place value do you think the answer should be rounded off to, and why?

21. A carpenter has a board four and eight-tenths feet long, but both ends are bad. Assuming equal amounts are cut off both ends, how much must be cut off to end up with a piece 3.5 feet long?

22. An individual began a diet program weighing one hundred sixty pounds. At the end of the diet the individual weighed one hundred fifty pounds. What percent of his original weight did he lose?

23. An individual pays an average of $15\frac{1}{2}\%$ tax on a salary of twenty-four thousand dollars. What is the amount of tax?

24. Four items were purchased for a total of sixteen dollars and seventeen cents, which included a sales tax of seventy-seven cents. Subtract the tax, then find the average cost of each item.

25. An item normally costs twelve dollars, but is on sale for twenty percent off. What is the sales price of the item?

Measurement

7

■APPLICATION *Sara needs 3 yards of fabric and 180 inches of ribbon for her craft projects. The cost of the fabric is $3.99 per yard and the cost of the ribbon is $0.89 per yard. Find the total cost.*

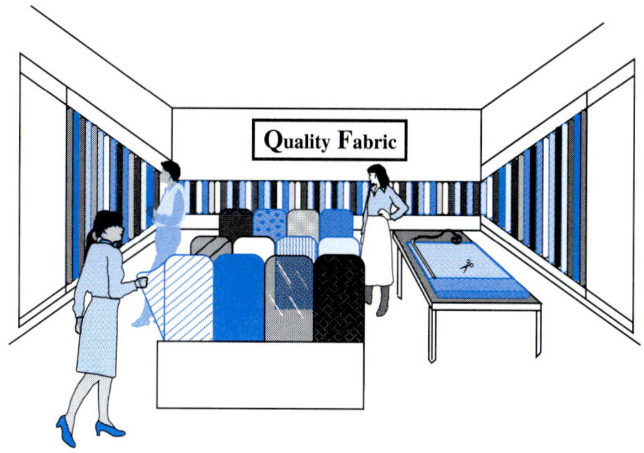

■SOLUTION *Convert 180 inches of ribbon to yards:*

$$\frac{180 \text{ inches}}{1} \cdot \frac{1 \text{ yard}}{36 \text{ inches}} = \frac{180 \text{ yards}}{36} = 5 \text{ yards}$$

$$\begin{aligned}\text{Total cost} &= \text{cost of fabric} + \text{cost of ribbon} \\ &= \frac{3 \text{ yards}}{1} \cdot \frac{\$3.99}{\text{yard}} + \frac{5 \text{ yards}}{1} \cdot \frac{\$0.89}{\text{yard}} \\ &= 3 \cdot (\$3.99) + 5 \cdot (0.89) \\ &= \$11.97 + \$4.45\end{aligned}$$

Total cost = $16.42

7.1 MEASUREMENT—THE U.S. SYSTEM

OBJECTIVES

1. Use the U.S. System of measurement.
2. Convert from one unit to another.
3. Work arithmetic problems using denominate numbers.

1 The U.S. System

The need for a standard system of measurement was recognized even in ancient times. Our early ancestors often used body measurements for these standards. A "cubit" was the length of a person's forearm. Later, a "foot" was set to the length of a human foot. It is said that King Henry I of England decided that the "yard" would be the distance from his nose to the end of his thumb.

A "mile" was the Roman "mille passuum," 1,000 paces, the Roman pace being a double step of about five feet. The foot, yard, and mile became part of a system called the English System. Now, it is often referred to as the **U.S. Customary System**, or just the **U.S. System**, since the U.S. is the only major country where it is still being used. The most common units of length, volume, and weight for the U.S. System are given in Table 1.

Table 1
The U.S. System

Length	*Volume*
12 inches(in) = 1 foot(ft)	8 ounces(oz) = 1 cup(c)
36 inches(in) = 1 yard(yd)	2 cups(c) = 1 pint(pt)
3 feet(ft) = 1 yard(yd)	2 pints(pt) = 1 quart(qt)
5,280 feet(ft) = 1 mile(mi)	4 quarts(qt) = 1 gallon(gal)

Table 1 *(continued)*
The U.S. System

Time	*Weight*
60 seconds(sec) = 1 minute(min)	16 ounces(oz) = 1 pound(lb)
60 minutes(min) = 1 hour(hr)	2,000 pounds(lb) = 1 ton(T)
24 hours(hr) = 1 day(da)	
7 days(da) = 1 week(wk)	
52 weeks(wk) = 1 year(yr)	

2 Convert from One Unit to Another

Sometimes, it is necessary to change a measurement to a unit other than the one given. This is done using a **unit factor** which is a statement of equivalency between two different units of measurement, such as 1 foot = 12 inches. Unit factors equal one when written as fractions, for example:

$$\frac{12 \text{ inches}}{1 \text{ foot}}, \quad \frac{2 \text{ cups}}{1 \text{ pint}}, \quad \frac{1 \text{ week}}{7 \text{ days}}, \quad \frac{1 \text{ ton}}{2{,}000 \text{ pounds}}$$

Note that each fraction equals one so that each is a form of one.

Rule: To Convert Measurements Using Unit Factors

Step 1 Choose a unit factor that contains both the units given in the problem and the desired units.

Step 2 Multiply the quantity given in the problem times the unit factor written as a fraction. The units in the denominator of each fraction should match the units in the numerator of the previous fraction.

Step 3 Divide out matching units in the numerator and denominator, then reduce.

EXAMPLE 1

Change 192 inches to feet.

SOLUTION

Step 1 Choose the unit factor that contains both feet and inches. From Table 1, we see that 12 inches = 1 foot.

Step 2 Multiply 192 inches times the unit factor written as a fraction with inches in the denominator.

unit fraction equals 1

$$\frac{192 \text{ inches}}{1} \cdot \frac{1 \text{ foot}}{12 \text{ inches}}$$

Step 3 Divide out inches, multiply, and reduce.

$$\frac{192 \cancel{\text{inches}}}{1} \cdot \frac{1 \text{ foot}}{12 \cancel{\text{inches}}} = \frac{192 \text{ feet}}{12} = 16 \text{ feet}$$ ■

HELPFUL HINT

Be sure to write the unit factors so that the units in the denominator of one fraction match the units in the numerator of the previous fraction. Those units will divide out leaving the unit required.

◇ *Do Practice Exercise 1.*

Sometimes there is no single statement in Table 1 relating the two units in the problem. For example, there is no statement relating ounces to gallons. Therefore, to change ounces to gallons, several unit factors are needed as shown in the next example.

EXAMPLE 2

The manager of a restaurant estimates she will need 320 glasses of orange juice per week. Each glass holds one cup of juice. How many gallons of orange juice should she order for the week?

SOLUTION

Since each glass holds 1 cup of juice, the manager needs 320 cups of orange juice. The problem is to change 320 cups to gallons. There is no single unit factor in Table 1 involving cups and gallons, so several unit factors are needed. Write each unit factor as a fraction, cancel units, multiply, and reduce.

(changes cups to pints; changes pints to quarts; changes quarts to gallons)

$$\frac{320 \text{ \sout{cups}}}{1} \cdot \frac{1 \text{ \sout{pint}}}{2 \text{ \sout{cups}}} \cdot \frac{1 \text{ \sout{quart}}}{2 \text{ \sout{pints}}} \cdot \frac{1 \text{ gal}}{4 \text{ \sout{quarts}}} = \frac{320 \text{ gal}}{16}$$

$$= 20 \text{ gallons} \quad \blacksquare$$

◇ *Do Practice Exercise 2.*

In some cases two units given in the problem may need to be converted, such as miles per hour to feet per second. Unit factors containing miles and feet are needed, but factors relating hours and seconds are also necessary. See the following example.

EXAMPLE 3

In August, 1991, Al Teague drove his race car 409.986 miles per hour to break a 26 year old land speed record. Find his speed in feet per second to the nearest whole number.

1. An auto shop worker wants to know if 22 quarts of oil will fit into a 5 gallon storage container. Change 22 quarts to gallons. Will the oil fit in the container?

2. Change 198 inches to yards.

ANSWERS
1. 5 gallons 2 quarts; no
2. $5\frac{1}{2}$ yards

3. Movie film moves through a certain projector at a rate of 330 feet per minute. Convert the rate to miles per hour.

SOLUTION

Note that miles per hour can be written as a fraction, $\frac{\text{miles}}{\text{hour}}$, and feet per second can be written "$\frac{\text{feet}}{\text{second}}$."

$$\frac{409.986 \cancel{\text{miles}}}{1 \cancel{\text{hour}}} \cdot \frac{5{,}280 \text{ feet}}{1 \cancel{\text{mile}}} \cdot \frac{1 \cancel{\text{hour}}}{60 \cancel{\text{min}}} \cdot \frac{1 \cancel{\text{min}}}{60 \text{ sec}} = \frac{409.986 \cdot 5{,}280 \text{ feet}}{60 \cdot 60 \text{ second}}$$

$$\approx \frac{601 \text{ feet}}{\text{second}} \quad ■$$

◇ Do Practice Exercise 3.

Calculator Application

The OH-6A helicopter has a rotor diameter of 26.33 ft. Use your calculator to find the diameter in yards rounded to the nearest hundredth yard.

Answer: 8.78 yd

3 Work Arithmetic Problems Using Denominate Numbers

A **denominate number** is a number followed by a unit of measure, for example:

7 ounces   10 seconds   6 feet 4 inches

When adding or multiplying denominate numbers, the answer may need to be simplified. That is, it may need to be rewritten in a different but equivalent form. This is shown in the next example.

EXAMPLE 4

Anna worked 6 hours 45 minutes on Monday, 3 hours 20 minutes on Tuesday, and 2 hours 40 minutes on Wednesday. Find the total amount of time worked.

SOLUTION

$$\begin{array}{rr} & 6 \text{ hours} \quad 45 \text{ minutes} \\ & 3 \text{ hours} \quad 20 \text{ minutes} \\ + & 2 \text{ hours} \quad 40 \text{ minutes} \\ \hline & 11 \text{ hours} \; 105 \text{ minutes} \end{array}$$

105 minutes is more than one hour, so divide 105 minutes by 60. The quotient is the number of hours and the remainder is the number of minutes left.

$$\begin{array}{r} 1 \\ 60\overline{)105} \\ \underline{60} \\ 45 \end{array}$$

105 minutes = 1 hour 45 minutes

$$11 \text{ hours } 105 \text{ minutes} = 11 \text{ hours} + 1 \text{ hour } 45 \text{ minutes}$$
$$= 12 \text{ hours } 45 \text{ minutes} \quad ■$$

ANSWER
3. 3.75 mph

EXAMPLE 5

It took a runner 1 minute 37 seconds to run one-fourth of a mile. At the same rate, how long would it take him to run 1 mile?

SOLUTION
It will take four times as long. Multiply 4 times the minutes and 4 times the seconds, then simplify.

4 · 1 minute = 4 minutes	=	4 minutes
4 · 37 seconds = 148 seconds	=	+ 2 minutes 28 seconds
Total time	=	6 minutes 28 seconds ■

This problem can also be solved using proportions.

◇ *Do Practice Exercises 4 and 5.*

Subtraction of denominate numbers may involve borrowing from the next larger unit. In that case, write the unit borrowed in its equivalent form before adding it to the smaller unit. In the next example, it is necessary to change one gallon to four quarts before subtracting.

EXAMPLE 6

A photographer had 3 gallons 1 quart of developing fluid. She used 3 quarts for the day's work. How much developing fluid does she have left?

SOLUTION
It is necessary to borrow since 3 quarts cannot be subtracted from 1 quart.

2 4 qt	
~~3~~ gal 1 qt	Borrow 1 gallon and change it to 4 quarts.
− 3 qt	
2 gal 5 qt	Add 4 qt to 1 qt to get 5 qt.
− 3 qt	Subtract.
2 gal 2 qt	Result. ■

◇ *Do Practice Exercise 6.*

4. Add 6 feet 8 inches to 5 feet 7 inches and simplify.

5. Multiply 5 times 4 hours 20 minutes and simplify.

6. A baby weighed 7 pounds 2 ounces at birth. He lost 5 ounces in the next few days. What was the new weight?

ANSWERS
4. 12 feet 3 inches
5. 21 hours 40 minutes
6. 6 pounds 13 ounces

7. Divide: $3\overline{)4 \text{ pounds } 5 \text{ ounces}}$

Division may also require that a measure be rewritten in its equivalent units. Look at the next example.

EXAMPLE 7

Divide 48 feet 4 inches by 5.

SOLUTION

```
      9 feet
  5)48 feet   4 inches        Divide 5 into 48.
    45 feet
  ┌── 3 feet
  │                           Change 3 feet to 36 inches.
  │   9 feet   8 inches
  │5)48 feet   4 inches
  │  45 feet
  └──────→ 36 inches
           40 inches          Add 4 inches to 36 inches.
           40 inches          Divide 5 into 40 inches.
            0                 Subtract.
```

48 feet 4 inches can be divided into 5 equal parts of equal lengths of:

9 feet 8 inches each ■

◇ *Do Practice Exercise 7.*

ANSWER
7. 1 pound 7 ounces

name date hour

Convert each given measurement to the indicated equivalent form.

1. 3 feet = _____ inches
2. 5 gallons = _____ quarts
3. 1 mile = _____ yards
4. 1 ton = _____ ounces
5. 1 gallon = _____ pints
6. 10 minutes = _____ seconds
7. 63 inches = _____ feet
8. 3 quarts = _____ cups
9. $\frac{1}{2}$ gallon = _____ cups
10. 57 feet = _____ yards
11. 2 years = _____ weeks
12. 100 days = _____ weeks
13. 1,000 seconds = _____ minutes
14. 2.4 tons = _____ pounds
15. 252 inches = _____ yards
16. 128 ounces = _____ pounds
17. 3.5 miles = _____ feet
18. 3 years = _____ weeks
19. 96 cups = _____ gallons
20. 330 seconds = _____ minutes

Simplify each.

21. 9 ft 20 in
22. 6 hr 100 min
23. 6 gal 10 qt
24. 5 lb 64 oz
25. 7 yd 17 ft
26. 3 hr 90 min 150 sec

27. 3 wk 18 da

28. 3 yd 13 ft 28 in

29. 7 da 180 min

30. 1 mi 3,000 yd

Do the indicated operation and simplify your answer.

31.
```
  6 ft 9 in
+ 5 ft 8 in
```

32.
```
  7 hr 48 min
+ 4 hr 35 min
```

33.
```
  5 gal 3 qt
+ 2 gal 2 qt
```

34.
```
  17 qt 1 pt
+  2 qt 1 pt
```

35.
```
  8 cups 5 oz
- 2 cups 7 oz
```

36.
```
  50 yd 1 ft
-  5 yd 2 ft
```

37.
```
  4 wk 3 days
- 1 wk 6 days
```

38.
```
  8 lb  8 oz
- 3 lb 10 oz
```

39.
```
5 gal 3 qt
       × 3
```

40.
```
50 yd 2 ft
       × 8
```

41.
```
2 wk 5 days
        × 3
```

42.
```
6 hr 30 min
        × 5
```

43. $7\overline{)38 \text{ gal } 2 \text{ qt}}$

44. $9\overline{)42 \text{ min } 18 \text{ sec}}$

45. $6\overline{)19 \text{ hr } 12 \text{ min}}$

46. $10\overline{)52 \text{ cups } 4 \text{ oz}}$

Solve the following problems, converting units as necessary.

47. A passenger jet flew at an altitude of 6.5 miles. What was the altitude in feet?

48. The average person uses a total of about 90 gallons of water a day. How many quarts per day does a person use?

49. Nylon manufacturers worldwide release about 60,000 tons of nitrous oxide into the atmosphere each month. How many pounds of nitrous oxide is released in a month?

50. The total weight of an aircraft and the items it carries on one trip is given below. Find the total weight in tons with partial tons being expressed in pounds.

Empty aircraft	65,000 lb	Passengers	8,000 lb
Crew	770 lb	Baggage	1,000 lb
Oil	900 lb	Fuel	6,000 lb

51. About $\frac{1}{4}$ of a young adult's nights sleep is made up of what is called dream sleep. Dream sleep is also called rapid eye movement sleep. If a person slept 6 hours 40 minutes, how long was the period of rapid eye movement sleep?

52. The weather forecaster predicted a high tide of 10.2 feet for Boston Harbor. How many inches is the tide predicted to be?

53. During the Ice Age so much water was frozen in the glaciers that sea level was 520 feet lower than it is today. How many yards lower was sea level in the Ice Age than it is now?

54. Measurements taken in one part of Louisiana over the past 50 years show that the land has subsided nearly 4 inches a year. At that rate, how many years would it take for the area to be a foot lower than it is now?

55. A person weighing 150 pounds will use approximately 300 calories for every hour of bicycling. How many calories will be used in 2 hours and 15 minutes of bike riding?

56. A person weighing 125 pounds will use approximately 650 calories for every hour of aerobic exercise. How many calories will be used in 1.75 hours of aerobic exercise?

57. In the ionosphere, which is the area 40 to 400 miles above the earth, the speed of the wind can be 400 inches per second. Find the wind speed in feet per second.

58. The Apennines is a 600 mile chain of mountains on the moon. The highest peak in the range is Mount Huygens at nearly 20,000 feet above the surface. Find the height of Mt. Huygens in miles to the nearest tenth of a mile.

59. A large turkey should be cooked about 20 minutes for each pound of weight. What should the total cooking time be for a 20 pound turkey? Write the answer in simplified form using hours and minutes.

60. Mike works as a lifeguard for 6 hours 15 minutes each day. How many hours does he work in 6 days? Express partial hours as minutes.

61. Theresa's child was at the day-care center 4 hours 20 minutes on Monday, 5 hours 25 minutes on Wednesday, and 3 hours 15 minutes on Friday. Find the total time the child was at the day-care center.

62. A farmer needs 240 feet of barbed wire. It costs \$6.80 for a 20 yard spool of wire. How much will the 240 feet of wire cost?

63. 1,760 pounds is what part of a ton? Write your answer as a fraction and as a decimal.

64. A car has a gas tank capacity of 16 gallons, a transmission fluid capacity of 6 pints, and a cooling system capacity of 12 quarts. What is the total amount of fluid in these three systems?

JOURNAL EXERCISES

1. In 1870, Jules Verne wrote a book titled *20,000 Leagues Under the Sea.* A league is equal to 3 nautical miles or 18,241 feet. State 20,000 leagues in feet and in nautical miles. Is it possible to travel 20,000 leagues under the sea, or was Jules Verne exaggerating?

2. Write about a situation in which you needed to make a measurement conversion. Discuss your confidence in your ability to change the measurement from one unit to another. Would you be more confident in doing this type of problem now?

3. Explain what a unit factor is and how it is used in converting from one unit of measurement to another.

4. Imagine you live in a country with no measurement system. What would it be like? What type of activities would be difficult to do with no measurement system?

1 The Metric System

The metric system came into existence about 200 years ago when France decided to make an entirely new system of measurement. They chose the **meter** for the basic unit of length, and defined it to be one ten millionth of the distance from the equator to the North Pole. Most countries in the world now use the **metric system**.

You can visualize a meter as being about the length of a guitar. The names of the other units of length are formed by attaching a prefix to the word "meter." The most common prefixes and their meanings are shown in Table 2.

Table 2
Metric Prefixes

kilo—thousand	deci—one-tenth
hecto—hundred	centi—one-hundredth
deka—ten	milli—one-thousandth

EXAMPLE 1

How many meters are in 1 kilometer?
How many meters are in 1 centimeter?

SOLUTION
"Kilo" means a thousand, so there are 1,000 meters in 1 kilometer. "Centi" means one hundredth, so there is one hundredth of a meter in one centimeter. ■

◇ *Do Practice Exercise 1.*

A centimeter is one hundredth of a meter and is about the thickness of an audiocassette tape. A millimeter is one thousandth of a meter and is about the thickness of a credit card. Some common metric units of length and their abbreviations are shown in Table 3.

Table 3
The Metric System

Length

1 kilometer(km)	=	1,000	meters(m)	
1 hectometer(hm)	=	100	meters(m)	
1 dekameter(dam)	=	10	meters(m)	
1 decimeter(dm)	=	.1	meters(m)	1 meter = 10 dm
1 centimeter(cm)	=	.01	meters(m)	1 meter = 100 cm
1 millimeter(mm)	=	.001	meters(m)	1 meter = 1,000 mm

The **mass** of an object is defined as the amount of matter in the object. The terms mass and weight are often used interchangeably when referring to objects on the earth. The basic unit of mass is the **gram**. A gram is the mass of water contained in a box that is 1 centimeter long on each side.

OBJECTIVES

1 Use the metric system.
2 Convert from one unit to another.
3 Work application problems.

1. How many meters are in a millimeter?

ANSWER
1. One thousandth of a meter is in one millimeter.

The chief unit for measuring **capacity** or volume is the **liter**. A liter is the amount of liquid that will fill a box that is 10 centimeters long on each side.

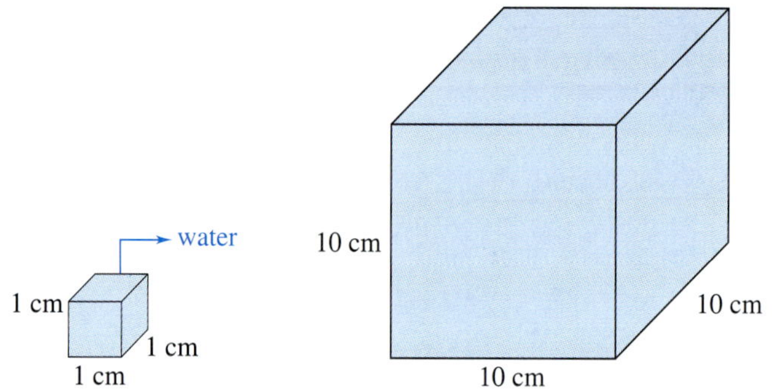

Mass of One Gram — Capacity of One Liter

Other units of mass (weight) and capacity (volume) are formed by adding the prefixes listed in Table 2 to the words ''gram'' and ''liter.'' These measurements are shown in Table 4.

Table 4
The Metric System

Capacity			*Mass*		
1 kiloliter(kL)	=	1,000 liters(L)	1 metric ton(t)	=	1,000 kilograms
1 hectoliter(hL)	=	100 liters	1 kilogram(kg)	=	1,000 grams(g)
1 dekaliter(daL)	=	10 liters	1 decigram(dg)	=	0.1 grams(g)
1 deciliter(dL)	=	0.1 liters	1 centigram(cg)	=	0.01 grams(g)
1 centiliter(cL)	=	0.01 liters	1 milligram(mg)	=	0.001 grams(g)
1 milliliter(mL)	=	0.001 liters	10 decigrams	=	1 gram
10 deciliters(dL)	=	1 liter	100 centigrams	=	1 gram
100 centiliters(cL)	=	1 liter	1,000 milligrams	=	1 gram
1,000 milliliters(mL)	=	1 liter			

2 Convert from One Unit to Another

The metric system has an advantage over the U.S. System in that each unit is 10 times the unit before it. Using meters as the unit of measurement, we have:

1 kilometer = 10 hectometers
1 hectometer = 10 dekameters
1 dekameter = 10 meters
1 meter = 10 decimeters
1 decimeter = 10 centimeters
1 centimeter = 10 millimeters

The following diagram shows the metric prefixes in order from largest to smallest.

kilo hecto deka units deci centi milli
(grams)
(meters)
(liters)

In order to change from a unit of measurement to the next *smaller* unit of measurement on the right, multiply by 10. This is accomplished by moving the decimal point one place to the *right*. If changing more than one unit, move the decimal point more than one place.

EXAMPLE 2

Change 4.5 centigrams to milligrams.

SOLUTION

We are changing from **centi**grams to the next smaller unit, **milli**grams:

kilo hecto deka grams deci centi milli

Since moving one unit to the right, multiply by 10. This is done by moving the decimal point one place to the right.

4.5 cg = 45. mg = 45 mg ■

In order to change from a unit of measurement to the next *larger* unit of measurement on the left, divide by 10. This is accomplished by moving the decimal point one place to the *left*. If changing more than one unit, move the decimal point more than one place.

EXAMPLE 3

Change 125 decimeters to meters.

SOLUTION

The problem is to change **deci**meters to **meters**, which is one unit to the left. Divide by 10 by moving the decimal one place to the left.

kilo hecto deka meters deci centi milli

125. dm = 12.5 m = 12.5 m

■

Rule: To Convert from One Metric Unit to Another

Step 1 Write the metric prefixes from largest to smallest.

Step 2 Count the number of moves to go from the given unit to the desired unit.

Step 3 Move the decimal point in the same direction and the same number of places counted in Step 2.

◇ *Do Practice Exercises 2 and 3.*

2. Change 3.8 grams to decigrams.

3. Change 50 millimeters to centimeters.

ANSWERS
2. 38 decigrams
3. 5.0 centimeters

4. Change 3.5 liters to milliliters.

5. Change 40.8 centigrams to grams.

6. Neil ran in a 10K (10 kilometers) fun run on Labor Day. How many meters did he run?

EXAMPLE 4

Convert 6.5 kilometers to meters.

SOLUTION

Step 1 kilo hecto deka meter deci centi milli

Step 2 1 2 3 — 3 moves are needed.

Step 3 6.5 km = 6 5 0 0. m = 6,500 m — Move decimal 3 places.

6.5 kilometers = 6,500 meters ■

EXAMPLE 5

A swimmer won the 800 meter freestyle event by 215 centimeters. By how many meters did the swimmer win?

SOLUTION

The problem is to change centimeters to meters.

kilo hecto deka meter deci centi milli

2 1

Since the problem involves moving two units to the left, move the decimal point 2 places to the left.

215 cm = 2.15 m = 2.15 m ■

◇ *Do Practice Exercises 4 and 5.*

EXAMPLE 6

In an undercover operation, agents confiscated 2.5 kilos (kilograms) of cocaine. How many milligrams of cocaine was this?

SOLUTION

In order to change from kilograms to milligrams count 6 units to the right. Move the decimal point 6 places to the right.

kilo hecto deka grams deci centi milli

1 2 3 4 5 6

Since we are moving 6 units, move the decimal point 6 places to the right and add commas as needed:

2.50000000 kg = 2,500,000 milligrams ■

◇ *Do Practice Exercise 6.*

3 Application Problems

Unit factors can be used to solve problems involving metric measurements, using the same methods explained in Section 7.1. Look at the next examples.

ANSWERS
4. 3,500 milliliters
5. 0.408 grams
6. 10,000 meters

EXAMPLE 7

If the measurements of the thickness of the braking disc on a car vary more than 13 micrometers, the discs should be resurfaced or replaced. State 13 micrometers in millimeters if 1 meter(m) = 1,000,000 micrometers(μm).

SOLUTION

The unit factors needed are 1 m = 1,000,000 μm, and 1 meter = 1,000 millimeters.

$$\frac{13\ \cancel{\mu m}}{1} \cdot \frac{1\ \cancel{\text{meter}}}{1{,}000{,}000\ \cancel{\mu m}} \cdot \frac{1{,}000 \text{ millimeters}}{1\ \cancel{\text{meter}}} = \frac{13 \text{ mm}}{1{,}000} = 0.013 \text{ mm} \quad \blacksquare$$

◇ *Do Practice Exercise 7.*

EXAMPLE 8

Dr. Kemper ordered 35 mg of Vistaril for a patient who is nauseated. The label on the ampule of Vistaril reads "Vistaril 50 mg/ml." How many milliliters of Vistaril should the patient receive?

SOLUTION

The label on the ampule means that there are 50 mg of Vistaril in each milliliter of fluid. In other words, 50 mg = 1 ml is the unit conversion factor.

$$\frac{35\ \cancel{\text{mg}}}{1} \cdot \frac{1 \text{ ml}}{50\ \cancel{\text{mg}}} = \frac{35 \text{ ml}}{50} = 0.7 \text{ ml}$$

Note that this problem could also be solved using the following proportion:

$$\frac{50 \text{ mg}}{1 \text{ ml}} = \frac{35 \text{ mg}}{x \text{ ml}} \quad \blacksquare$$

◇ *Do Practice Exercise 8.*

EXAMPLE 9

Dr. Yuan ordered 250 mg of Doriden for a patient. The label on the bottle of Doriden reads "0.5 gm/tablet." How many tablets should the patient take?

SOLUTION

The conversion factors needed are 0.5 gm = 1 tablet and 1 gm = 1,000 mg.

$$\frac{250 \text{ mg}}{1} \cdot \frac{1 \text{ gm}}{1{,}000 \text{ mg}} \cdot \frac{1 \text{ tablet}}{0.5 \text{ gm}} = \frac{250 \text{ tablets}}{(1{,}000)(0.5)} = \frac{250 \text{ tablets}}{500}$$

$$= \frac{1}{2} \text{ tablet} \quad \blacksquare$$

◇ *Do Practice Exercise 9.*

7. Change 817 μm to centimeters.

8. Dr. Jackson ordered 500 mg of Keflin. The bottle of Keflin reads "1,000 mg/3ml." How many milliliters of Keflin should the patient receive?

9. The doctor ordered 0.2 grams Equanil. The bottle reads "400 mg/1 tablet." How many tablets should be given?

ANSWERS
7. 0.0817 cm
8. 1.5 milliliters
9. $\frac{1}{2}$ tablet

name date hour

Give the prefix that represents each.

1. 1,000

2. 0.01

3. 100

4. 10

5. 0.1

6. 0.001

Convert each measurement.

7. 1 kilogram = _______ grams

8. 1 liter = _______ milliliters

9. 1 meter = _______ centimeters

10. 2 kilometers = _______ meters

11. 550 milliliters = _______ liters

12. 1 meter = _______ millimeters

13. 6 grams = _______ milligrams

14. 7.6 liters = _______ milliliters

15. 2.5 milligrams = _______ grams

16. 80 grams = _______ kilograms

17. 200 millimeters = _______ meters

18. 6.2 decimeters = _______ centimeters

19. 0.2 kiloliters = _______ milliliters

20. 50 m = ______ mm

21. 52 mm = ______ cm

22. 320 ml = ______ L

23. 0.4 m = ______ mm

24. 0.5 t = ______ g

25. 6000 kg = ______ t

26. 25 cm = ______ mm

27. 743 g = ______ mg

28. 820 cg = ______ g

29. 480 L = ______ kL

30. 6 km = ______ m

31. 50 cm = ______ m

32. 0.25 m = ______ cm

33. 7 cg = ______ mg

34. 3.5 L = ______ ml

35. 0.4 g = ______ kg

36. 100 g = ______ mg

37. 1500 ml = ______ L

38. 20 g = ______ kg

39. 290 cm = ______ mm

Work the following problems.

40. Gene takes 50 mg of Pamelor each day to prevent migraine headaches. How many grams does he take in a week?

41. Grace had soup and crackers for lunch. The soup contained 1,100 mg sodium and the crackers had 190 mg sodium. How many grams of sodium did her lunch contain?

42. A plumber buys a pipe that is 10 meters long. How much is left after he uses a piece measuring 240 centimeters?

43. How many glasses containing 250 ml each can be filled from a 3 liter drink bottle?

44. Laura's doctor instructed her to take 250 mg of ampicillin 3 times a day for 10 days. How many grams of ampicillin will she have taken at the end of the 10 days?

45. A mechanic needs 4.5 liters of oil to change the oil in a truck. How many milliliters of oil are needed?

46. An athlete trains in Toronto by running 105 kilometes per week. How many meters does she average per day?

47. The doctor ordered 0.5 grams of Chloromycetin. The bottle reads ''250 mg/tablet.'' How many tablets should be given?

48. Blood pressure is measured in millimeters of mercury. A blood pressure reading of 160/80 means that a column of mercury would rise 160 millimeters if subjected to the same force as that exerted on the blood in the artery. Change 160 millimeters to centimeters.

49. The normal adult daily protein requirement is about 65 grams. By mid-afternoon a patient at a hospital has consumed 27,000 mg of protein. How much is remaining of the person's daily requirement of protein?

50. A study showed that drinking two cups of coffee (300 mg of caffeine) resulted in 2 hours less total sleep. This included up to 66 minutes more time required to fall asleep and twice as much wakefulness afterward. If Phyllis had one cup of coffee, how many grams of caffeine did she have?

51. Foods high in protein reduce the time it takes to fall asleep because they contain a chemical called L-tryptophan. Milk is high in protein, containing 8 grams per 8 ounce serving. How many centigrams are in one 8 ounce serving of milk?

JOURNAL EXERCISES

1. If 1 ounce of a certain brand of cereal contains 150 milligrams of sodium, how many grams of sodium are in an 8 ounce serving? Can this problem be solved using both the techniques of this section and using proportions? If so, solve it both ways.

2. In creating a system of measurement, a basic unit must be chosen. The ancients used parts of the body such as a foot. The French used $\frac{1}{10{,}000{,}000}$ of the distance from the equator to the north pole (the meter). You are chosen to create an entirely new system of measurement. What would you use for the basic unit and why?

3. Explain what you are doing mathematically when you move the decimal point three places to the right in order to change from meters to millimeters. What mathematical operation are you using?

4. Copy the common measurements of both the U.S. and Metric Systems into your journal. Which system is easier to use and why?

1 Convert Between U.S. and Metric Measurements

In today's world of international travel and trade, it is common to have the need to change from one system of measurement to another. For example, highway signs change from miles per hour in the United States to kilometers per hour in Canada and Mexico. Gasoline in the United States is sold in gallons whereas it is sold in liters in Europe. The weight limits of bridges in the United States are given in pounds but in kilograms in most other countries of the world.

Table 5 includes statements that involve measurements from both the U.S. System and the Metric System. Most of these numbers are approximations since there is not an exact metric measurement equal to each U.S. measurement.

Table 5
U.S.—Metric Conversions

Length
1 inch = 2.54 centimeters*
1 foot ≈ 0.305 meters
1 yard ≈ 0.914 meters
1 mile ≈ 1.61 kilometers

Capacity
1 quart ≈ 0.946 liters
1 gallon ≈ 3.785 liters

Weight
1 ounce ≈ 28.35 grams
1 pound ≈ 0.4545 kilograms
1 ton ≈ 907 kilograms

*1 inch = 2.54 centimeters exactly
The other conversions shown in the table are approximate.

The measurements below are given in both the U.S. and metric systems for comparison.

Height

Weight

Capacity

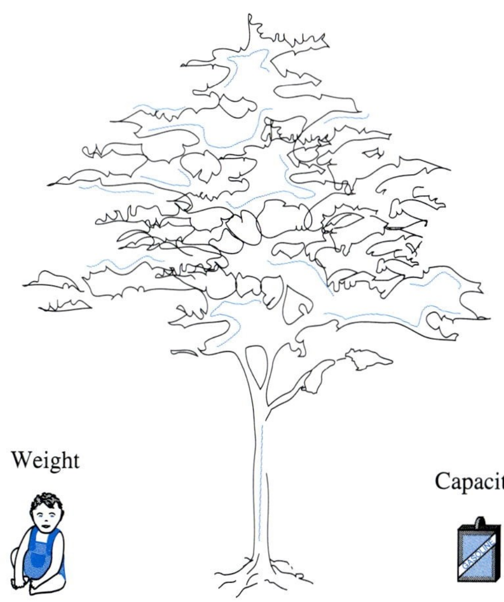

7 lb ≈ 3.2 kg

31 ft ≈ 9.5 m

1 gal ≈ 3.8 L

OBJECTIVES

1 Convert between U.S. and metric measurements.
2 Convert between Fahrenheit and Celsius Temperatures.

1. Angela put 12 gallons of gas in her car. How many liters of gas, rounded to the nearest tenth, was this?

2. How many quarts are in 4 liters? Round your answer to the nearest tenth.

ANSWERS
1. 45.4 liters
2. 4.2 quarts

The method of multiplication by a unit factor can be used to change from one system to the other. The unit factors needed can be found in Table 5. The next few examples show the process.

EXAMPLE 1

The Boston Marathon is about 26 miles long. State this distance in kilometers.

SOLUTION
From Table 5, we see that 1 mile = 1.61 kilometers. This information is used to form the unit factor, with miles in the denominator.

$$\frac{26 \cancel{\text{miles}}}{1} \cdot \underbrace{\frac{1.61 \text{ km}}{1 \cancel{\text{mile}}}}_{\text{unit factor}} = (26) \cdot (1.61) \text{ km} = 41.86 \text{ km} \quad \blacksquare$$

◇ *Do Practice Exercise 1.*

EXAMPLE 2

How many yards, to the nearest yard, are in 1,000 meters (1 kilometer)?

SOLUTION
Choose the statement "1 yd = 0.914 meters" from Table 5, then proceed:

$$\frac{1000 \cancel{\text{meters}}}{1} \cdot \underbrace{\frac{1 \text{ yard}}{0.914 \cancel{\text{meters}}}}_{\text{unit factor}} = \frac{1{,}000 \text{ yd}}{0.914} \approx 1{,}094 \text{ yd} \quad \blacksquare$$

In order to get a mental picture of the length of 1 kilometer, consider the last example. We found that 1 kilometer is equal to 1,094 yards. This can be pictured as approximately the length of 11 football fields, without the end zones.

◇ *Do Practice Exercise 2.*

Sometimes more than one unit factor is needed to make a change from one measurement to another. In this case, it may be necessary to form unit factors using information from more than one table. Look carefully at the next example.

EXAMPLE 3

A Canadian basketball player is 208 cm tall. How tall is he in feet and inches, rounded to the nearest inch?

SOLUTION

208 cm can be changed to inches using the unit factor 2.54 cm = 1 in found in Table 5. In turn, inches can be changed to feet using 12 in = 1 ft from Table 1 in the first section of this chapter.

from Table 5 (applies to $\frac{1 \text{ in}}{2.54 \text{ cm}}$); from Table 1 (applies to $\frac{1 \text{ ft}}{12 \text{ in}}$)

$$\frac{208 \cancel{\text{cm}}}{1} \cdot \frac{1 \cancel{\text{in}}}{2.54 \cancel{\text{cm}}} \cdot \frac{1 \text{ ft}}{12 \cancel{\text{in}}} = 6.824 \text{ ft}$$

The problem asks for feet and inches so that the decimal portion, .824 feet from above must be changed to inches. This can be done by multiplying .824 feet times the unit factor, 1 foot = 12 inches.

$$\frac{0.824 \cancel{\text{foot}}}{1} \cdot \frac{12 \text{ inches}}{1 \cancel{\text{foot}}} = 9.888 \text{ inches} \approx 10 \text{ inches}$$

The basketball player is 208 cm or 6 ft 10 in tall. ■

◇ *Do Practice Exercise 3.*

Table 6 shows equivalents that are useful in household and medical applications.

Memorize

Table 6
Volume Equivalents for Household and Medical Applications

English to English	*Metric to Metric*	*English to Metric*
1 oz ≈ 2 Tbs	1000 ml = 1 Liter	1 tsp ≈ 5 cc
1 Tbs = 3 tsp	1 ml = 1 cc	1 Tbs ≈ 15 cc
		1 oz ≈ 30 cc

tsp = teaspoon cc = cubic centimeters
Tbs = tablespoon oz = ounces

EXAMPLE 4

A premature baby receives 10 ml of formula at each feeding. How many teaspoons of formula is the baby given?

SOLUTION

$$\frac{10 \cancel{\text{ml}}}{1} \cdot \frac{1 \cancel{\text{cc}}}{1 \cancel{\text{ml}}} \cdot \frac{1 \text{ tsp}}{5 \cancel{\text{cc}}} = \frac{10 \text{ tsp}}{5} = 2 \text{ tsp}$$ ■

◇ *Do Practice Exercise 4.*

3. Convert 1,984 grams to pounds and ounces. Round the answer to the nearest ounce.

4. A child is to receive 1 tsp of decongestant cough medicine every four hours during the day (4 tsp each day). The bottle contains 8 oz. How many days will the bottle last?

ANSWERS
3. 4 pounds 6 ounces
4. 12 days

2 Convert Between Fahrenheit and Celsius

There are two common scales for measuring temperature, the Fahrenheit scale (F) and the Celsius scale (C). The temperature at which water freezes is 32° Fahrenheit or 0° Celsius so that 32°F = 0°C. The boiling point of water is 212° Fahrenheit or 100° Celsius so that 212°F = 100°C. A room temperature of 72° Fahrenheit would be approximately equal to 22° Celsius. These temperatures are shown on the following thermometers.

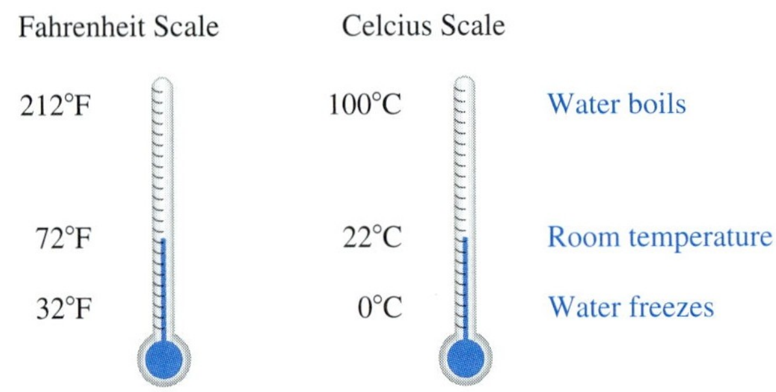

If you wish to change a temperature given in degrees Celsius to degrees Fahrenheit, use either of the following formulas:

$$F = \frac{9}{5}C + 32 \qquad \text{or} \qquad F = (1.8)C + 32$$

In the following example 45°C is changed to degrees Fahrenheit:

$$\begin{aligned} F &= (1.8)C + 32 && \text{formula} \\ F &= (1.8)(45) + 32 && \text{Substitute 45 for C.} \\ F &= 81 + 32 && \text{Multiply.} \\ F &= 113 && \text{Result.} \end{aligned}$$

so that 45°C = 113°F.

If you wish to change a temperature given in degrees Fahrenheit to degrees Celsius, use either of the following formulas:

$$C = \frac{5}{9}(F - 32) \qquad \text{or} \qquad C = \frac{(F - 32)}{1.8}$$

In the following example, 86°F is changed to degrees Celsius:

$$\begin{aligned} C &= \frac{5}{9}(F - 32) && \text{formula} \\ C &= \frac{5}{9}(86 - 32) && \text{Substitute 86 in place of F.} \\ C &= \frac{5}{9} \cdot 54 && \text{contents of parentheses} \\ C &= \frac{5}{9} \cdot \frac{54}{1} && \text{Write 54 as } \frac{54}{1}. \\ C &= \frac{5}{1} \cdot \frac{6}{1} && \text{Cross reduce.} \\ C &= 30 && \text{Result.} \end{aligned}$$

so that 86°F = 30°C.

EXAMPLE 5

Melinda's 2 year old son has a fever of 103°F. What is his temperature on the Celsius scale to the nearest tenth of a degree?

SOLUTION

First, choose the form of the equation that results in degrees Celsius, given degrees Fahrenheit. Using the decimal version of the formula gives the following:

$$C = \frac{(F - 32)}{1.8} \quad \text{formula}$$

$$C = \frac{(103 - 32)}{1.8} \quad \text{Substitute.}$$

$$C = \frac{71}{1.8} \quad \text{Subtract.}$$

$$C \approx 39.4°C \quad \text{Result.}$$

◇ *Do Practice Exercises 5 and 6.*

5. Change 12° Celsius to Fahrenheit. Round the answer to the nearest tenth degree.

6. Change 100° Fahrenheit to Celsius. Round the answer to the nearest tenth degree.

ANSWERS
5. 53.6° Fahrenheit
6. 37.8° Celsius

name date hour

EXERCISE SET 7.3

Convert the given measurement to its indicated equivalent form and round to the nearest tenth.

1. 1 kilogram = _______ pounds

2. 1 meter = _______ feet

3. 1 kilometer = _______ miles

4. 1 liter = _______ quarts

5. 1 centimeter = _______ inches

6. 1 meter = _______ yards

7. 6 inches = _______ centimeters

8. 5,000 pounds = _______ metric tons

9. 5 feet = _______ meters

10. 30 cc = _______ ounces

11. 10 quarts = _______ liters

12. 100 liters = _______ quarts

13. 5 gallons = _______ liters

14. 1,400 grams = _______ pounds

15. 10 tons = _______ kilograms

16. 2 ounces = _______ teaspoons

17. 1 meter = _______ inches

18. 3 tablespoons = _______ cc

19. 3 yards = _______ meters

20. 6.5 kilometers = _______ miles

21. 3 liters = _______ quarts

22. 5 kilograms = _______ pounds

23. 1 inch = _______ millimeters

24. 5 liters = _______ quarts

25. 1 ton = _______ milligrams

26. 10 centimeters = _______ inches

27. 30 cc = _______ teaspoons

28. 125 pounds = _______ kilograms

29. 8 ounces = _______ milliliters

30. 5.5 feet = _______ meters

31. 300 miles = _______ kilometers

32. 5 pounds = _______ grams

33. 20 gallons = _______ kiloliters

34. 90 kilometers = _______ miles

35. 2 millimeters = _______ inches

36. 100 yards = _______ meters

37. 6 meters = _______ feet

38. 1,000 tons = _______ kilograms

39. 12 quarts = _______ milliliters

40. 80 kilometers = _______ feet

41. 18°C = _______ °F

42. 25°C = _______ °F

43. 85°F = _______ °C

44. 20°F = _______ °C

Solve the following problems, converting units as necessary. Round your answers to the nearest tenth.

45. Richard lost 41 pounds while on a diet. How many kilograms of weight did he lose?

46. The speed limit on a highway in Ontario is 90 kph (kilometers per hour). What is the speed limit in miles per hour?

47. The clearance height of a highway overpass near Mexico City is 4 meters. State this height in feet.

48. The distance from Chicago to New York is approximately 714 miles. Express this distance in kilometers.

49. A common track event is the 50 yard dash. Fifty yards is equivalent to how many meters?

50. A diabetic receives 3 cc of insulin daily. How many milliliters of insulin does he receive? How many ounces of insulin does he receive?

51. The cable car over the Royal Gorge in Colorado is 1,200 feet above the Arkansas River at the bottom of the gorge. State this height in kilometers.

52. It takes about 1,000,000 gallons of water to fill an Olympic size swimming pool. How many liters of water is this?

53. A driver purchases 42.5 liters of gasoline for 42,000 pesos. How many gallons did she purchase? What was the cost in dollars, if one dollar is equal to 3,200 pesos?

54. At an exchange rate of 3,200 pesos to one dollar, how many pesos would it take to purchase a bicycle costing one hundred sixty-nine dollars?

55. A mechanic needs an eight millimeter wrench but only has a one-half inch wrench. Will the one-half inch wrench work? Why, or why not?

56. The San Andreas fault runs about 600 miles along the coast of California. How many kilometers long is the fault?

57. Horses are often measured in "hands." A "hand," originally the width of a palm, is 4 inches. If a horse is 15 hands high, what is its height in inches? In meters?

58. A recipe calls for 4 tablespoons of milk. How many ounces of milk does it require? How many cups is that equivalent to?

59. Which package of cookies contains more, one that weighs 12 ounces or one that contains 400 grams? How much more does it contain?

60. Which gas tank is larger, one that holds 16 gallons, or one that holds 60 liters? How much more does it hold?

61. The scale on a city map shows 1 inch being equivalent to 0.3 kilometers. The distance from the airport to the convention center is 10 inches on the map. How many miles is it from the airport to the convention center?

62. The chart on the back of a dress pattern package says it will take 2.8 meters of material to make the dress in a size 12. How many yards of material does the dress require?

63. The pasteurization process destroys any bacteria that may be in milk. Milk is pasteurized by being heated to 60° Celsius for 20 to 30 minutes. State this temperature in degrees Fahrenheit.

64. The label on a roll of photographic film says it may be damaged by temperatures higher than 120°F. Convert 120°F to degrees Celsius.

65. Major league baseball catchers must wear about 10 pounds of protective gear. How many kilograms of gear do they wear?

JOURNAL EXERCISES

1. Measure your height, the distance from your home to the college, and the number of gallons of gas it takes to fill up your car. Write your height in centimeters, the distance to the college in kilometers, and the amount of gas in liters.

2. State some differences between the U.S. and Metric Systems of measurement. Is one system easier than the other? Which?

3. List some possible advantages if every country in the world used the same measurement system.

4. Make a list of things that would have to be changed in order for the United States to convert totally to the Metric System. (*Hint:* Gas pumps would have to be changed to measure liters instead of gallons.)

CHAPTER 7 REVIEW

SECTION	TOPIC OR TERM	EXAMPLE OR DEFINITION
7.1		
Measurement—The U.S. System	U.S. System	The U.S. system is a system of measurement used in the United States.
	Unit Factor	A unit factor is a fraction in which the numerator is equal to the denominator, for example: $\frac{36 \text{ inches}}{1 \text{ yard}} = 1 \qquad \frac{1 \text{ pound}}{16 \text{ ounces}} = 1$
	Denominate Number	A denominate number is a number followed by a unit of measure.
	Conversions	Convert 64 ounces to pounds: $\frac{64 \text{ ounces}}{1} \cdot \frac{1 \text{ lb}}{16 \text{ oz}} = 4 \text{ lb}$ ($\frac{1 \text{ lb}}{16 \text{ oz}}$ → unit factor)
	Simplify Denominate Numbers	4 hrs 100 min = 5 hrs 40 min
	Add Denominate Numbers	4 days 10 hours + 20 hours 4 days 30 hours = 5 days 6 hrs
	Multiply Denominate Numbers	5 · (10 min 45 sec) = 50 min 225 sec = 53 min 45 sec
	Subtract Denominate Numbers	2 gal 1 qt = 1 gal 5 qts − 3 qts = − 3 qts 1 gal 2 qts
	Divide Denominate Numbers	5 ft 10 in 3)17 ft 6 in 15 ft 2 ft = 24 in — Add 6 to 24. 30 in — Divide 3 into 30. 30 in 0
7.2		
Measurement—The Metric System	Metric System	The metric system is a system of measurement based on the meter, liter, and gram.
	Meter	The meter is the basic unit of length in the metric system.
	Gram	The gram is the basic unit of mass or weight in the metric system.
	Liter	The liter is the basic unit of capacity or volume in the metric system.
	Conversions	4.5 meters = 4 5 0. cm = 450 cm 55 grams = 0.0 5 5 kg = 0.055 kg

SECTION	TOPIC OR TERM	EXAMPLE OR DEFINITION
7.3		
Conversions Between U.S. and Metric Systems	Fahrenheit	The Fahrenheit scale is the temperature scale used in the U.S.
	Celsius	The Celsius temperature scale is used in most countries other than the United States.
	Conversions	Change 115 pounds to kilograms: $\frac{115\text{ lbs}}{1} \cdot \frac{0.4545\text{ kg}}{1\text{ lb}} \approx 52.3\text{ kg}$ Change 8°C to degrees Fahrenheit: $F = 1.8C + 32$ formula $= 1.8(8) + 32$ Substitute. $= 14.4 + 32$ Multiply. $= 46.4°C$ Result.

NURSING ACROSS SEVERAL CONTINENTS

After graduation from a nursing school in the United States, Anna decided to join the Peace Corps and work overseas for two years. She was sent to a remote part of Africa where she worked with a German doctor named Dr. Schneider.

Dr. Schneider ordered 2 milligrams of a particular medication per kilogram of body weight for a patient. Anna began to work on the problem.

The only weight scale available was based on pounds. Anna weighed the patient and found that the patient weighed 108 pounds. The medicine came only in liquid form with a concentration of 100mg/1mL. In other words, there is 100 milligrams of medicine in each milliliter of volume. Help Anna find the correct dosage. Round all answers to the nearest tenth.

1. Convert the patient's weight from pounds to kilograms.

2. Find the number of milligrams of medication the patient will need for her body weight.

3. Use the unit factor from the label of the medication bottle to find out the number of milliliters needed.

CHAPTER 7 REVIEW PROBLEMS

Convert each using data from the tables given in this chapter and round to the nearest hundredth.

1. 1 meter = _______ centimeters

2. 1 foot = _______ inches

3. 1 gram = _______ decigrams

4. 1 pound = _______ kilograms

5. 1 yard = _______ inches

6. 1 metric ton = _______ kilograms

7. 1 kilogram = _______ grams

8. 1 mile = _______ kilometers

9. 1 minute = _______ seconds

10. 1 liter = _______ centiliters

11. 1 centimeter = _______ millimeters

12. 1 ton = _______ pounds

13. 1 inch = _______ centimeters

14. 1 quart = _______ liters

15. 1 mile = _______ feet

16. 1 gallon = _______ quarts

17. 1 yard = _______ meters

18. 1 quart = _______ pints

19. 40°C = _______ °F

20. 38°F = _______ °C

Convert each, rounding to the nearest tenth.

21. 20 feet = _______ inches

22. 8 inches = _______ centimeters

23. 8 miles = _______ kilometers

24. 4 quarts = _______ pints

25. 25 yards = _______ meters

26. 7 years = _______ weeks

27. 7 quarts = _______ liters

28. 180 pounds = _______ kilograms

29. 8 tons = _______ kilograms

30. 12 hectograms = _______ grams

Calculate each and simplify, if possible.

31. 6 · (10 hours 3 minutes)

32. 7 · (12 pounds 8 ounces)

33.

```
   5 feet 10 inches
+          5 inches
```

34.

```
   3 hours 50 minutes
+  5 hours 25 minutes
```

35.

```
   7 gallons 1 quart
−  1 gallon  3 quarts
```

36.

```
   1 mile 70 feet
−        240 feet
```

37. $6\overline{)41 \text{ feet } 6 \text{ inches}}$

38. $3\overline{)7 \text{ quarts } 1 \text{ pint}}$

Follow the directions for each problem. Round your answers to the nearest tenth.

39. Change 4 feet 3 inches to centimeters.

40. Change 1 mile 2 yards to meters.

41. Change 2 gallons 1 quart to liters.

42. Change 7 pounds 2 ounces to grams.

43. A professional football player is 6 feet 4 inches tall and weighs 240 pounds. Write his height in centimeters and weight in kilograms.

44. A professional basketball player is 7 feet tall and weighs 280 pounds. Write his height in centimeters and his weight in kilograms.

45. A piece of mining equipment weighs 40 tons. Write the weight in kilograms.

46. An automobile weighs 2,000 kilograms. Write the weight in pounds.

47. An individual weighs 64 kilograms before, and 61 kilograms after a diet. How many *pounds* were lost?

48. A person's blood pressure is reduced 20% during the stages of deep sleep. If normal systolic pressure is 120, what would it be during sleep?

49. At age 14, Tony jogged 1 mile 800 yards without stopping. At age 16 he jogged 3 times as far. Write the distance jogged at age 16 in a simplified form using miles and yards.

50. The average weight of the players on a professional football team is 228 pounds each. What would you expect the combined weight of all eleven starting players to be in pounds and in kilograms?

51. The height of three basketball players is 6 feet 4 inches, 6 feet 8 inches, and 6 feet 7 inches. Find the average height to the nearest inch.

52. The distance from Montreal, Canada, to San Francisco is approximately 2,900 miles. How many kilometers is this?

53. Jerry ran 6 miles in 42 minutes. At the same pace how long will it take him to run 8 miles?

54. A car travels at 100 kilometers per hour. How many miles per hour is that?

55. Two percent of a metric ton of ore is gold. How many pounds of gold are in one metric ton of this ore?

56. A dump truck weighs 6 metric tons when empty and 9 metric tons when full of garbage. How many *pounds* of garbage does the truck hold?

CHAPTER 7 TEST

Convert each, rounding to the nearest tenth.

1. 6 inches = _______ centimeters

2. 21 pounds = _______ ounces

3. 3 kilometers = _______ meters

4. 22 kilograms = _______ pounds

5. 4 meters = _______ centimeters

6. 60 minutes = _______ seconds

7. 3 tons = _______ pounds

8. 5 tons = _______ kilograms

9. 30 centimeters = _______ inches

10. 12 weeks = _______ days

11. 3 hours = _______ seconds

12. 7 yards = _______ inches

13. 75°F = _______ °C

14. 12 liters = _______ gallons

Solve each and simplify, if possible.

15. 6 · (25 pounds 8 ounces)

16. $5\overline{)16 \text{ feet } 3 \text{ inches}}$

17.

$$\begin{array}{r} 61 \text{ minutes } 44 \text{ seconds} \\ +\ \ 3 \text{ minutes } 21 \text{ seconds} \\ \hline \end{array}$$

18.

$$\begin{array}{r} 6 \text{ gallons } 1 \text{ quart} \\ -\ 5 \text{ gallons } 2 \text{ quarts} \\ \hline \end{array}$$

Solve each.

19. An automobile weighing 1,800 kilograms is brought into the United States. A tax must be paid on the weight, in pounds, of the automobile. How many pounds does the car weigh?

20. The heights of the five starting members of a high school basketball team are: 6 feet, 5 feet 10 inches, 6 feet 3 inches, 6 feet 1 inch, and 6 feet 6 inches. Find the average height to the nearest inch.

21. An insect colony contains about 100 insects, each weighing about two milligrams. Find their total weight in ounces.

22. One individual weighs 135 pounds and another weighs 60 kilograms. Find their average weight in pounds and in kilograms.

23. A tank holds 75 liters of gasoline. How many gallons of gasoline does this correspond to to the nearest tenth of a gallon?

24. A motorcycle is traveling at 80 miles per hour. How many kilometers per hour does this correspond to to the nearest tenth of a km?

25. A particular German shepherd dog weighs 80 pounds. State this weight in kilograms.

Geometry

8

■**APPLICATION** *Mr. Jackson plans to install central air conditioning in his home. The recommended size of the air conditioner, in tons, is found by dividing the area of the home by 700. What size air conditioner should Mr. Jackson purchase?*

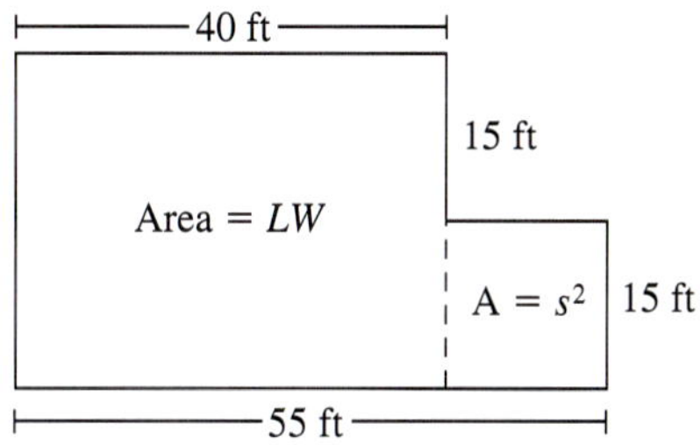

■**SOLUTION** *The area of the house is the sum of the area of a rectangle 40 feet by 30 feet and the area of a square with dimensions 15 feet on each side.*

$$\begin{aligned}\text{Area of house} &= LW + s^2\\ &= (40)(30) + 15^2\\ &= 1200 + 225\\ &= 1425 \text{ square feet}\end{aligned}$$

Recommended air conditioner size:

$$1425 \div 700 \approx 2.036$$

Therefore, Mr. Jackson should buy a 2 ton unit.

8.1 BASICS

OBJECTIVES

1 Understand basic geometric concepts.
2 Define lines and congruent line segments.
3 Define several types of angles.

1 Basic Concepts

The study of geometrical concepts goes back far in recorded history. For example, the master craftsmen who built the ancient pyramids of Egypt used geometry. The pyramid builders were concerned with the geometrical properties of length, width, and height of each stone.

Length, width, and height provide information used to describe an object. If something has only one of these properties, it is said to be **one-dimensional**. If it has both length and width, it is **two-dimensional**. **Three-dimensional** objects have length, width, and height.

A **point** is a geometric concept that has no dimension. It has neither width, length, nor height. A point is a location or position. For example, the location of a hurricane is shown by a point representing the center, or eye, of the hurricane.

2 Lines

A tightly stretched string can be thought of as one-dimensional, even though the string has a small thickness. A **line** is a one-dimensional figure that has length, but neither width nor height. One example of a line of infinite length is the number line:

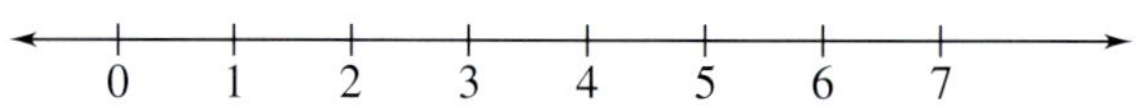

A line of a given length with **endpoints**, or points on each end, is called a **line segment**. The notation used for a line segment connecting point A to point B is $\overline{AB}$.

Line segment $\overline{AB}$

Two line segments are **congruent line segments** if they have the same length. The symbol for ''congruent'' is $\cong$. For example:

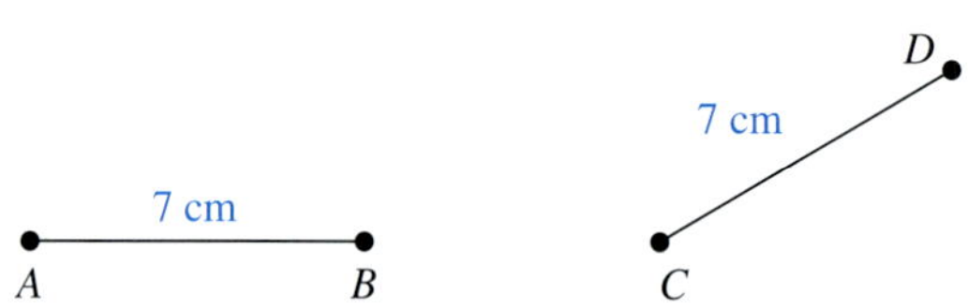

$\overline{AB} \cong \overline{CD}$

Line segment $\overline{AB}$ is congruent to line segment $\overline{CD}$ since they both have the same length.

EXAMPLE 1

Name the congruent line segments in the following diagram.

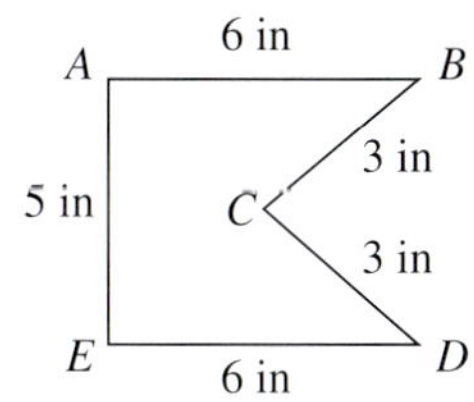

SOLUTION

$\overline{AB} \cong \overline{ED}$ both have length 6 in

$\overline{CB} \cong \overline{CD}$ both have length 3 in ■

◇ *Do Practice Exercise 1.*

A line with an endpoint on only one end is called a **ray**:

3 Angles

Two rays or line segments from the same endpoint form an **angle**. In each case, the common endpoint is called the **vertex**. The symbol for angle is $\angle$.

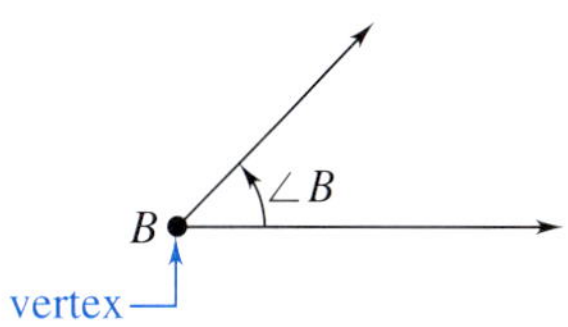

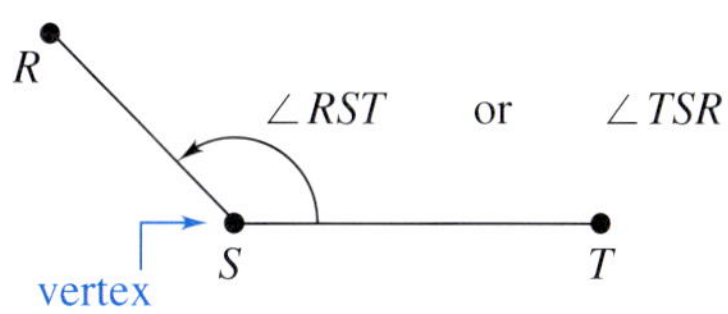

EXAMPLE 2

The path taken by a pool ball is shown below. Name the angle in the diagram.

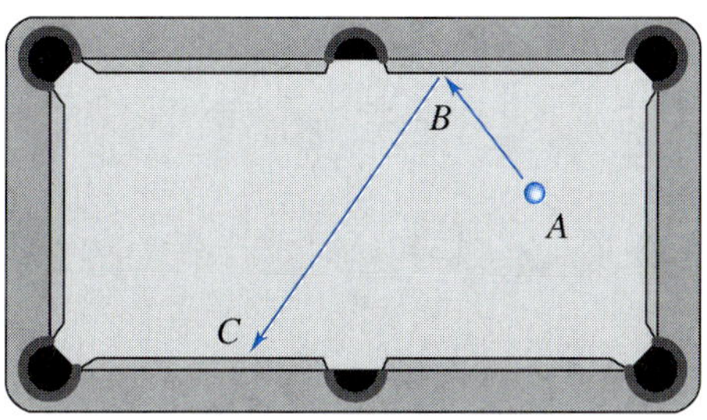

SOLUTION

$\angle ABC$ or $\angle CBA$ ■

◇ *Do Practice Exercise 2.*

1. Name all line segments congruent to $\overline{AB}$.

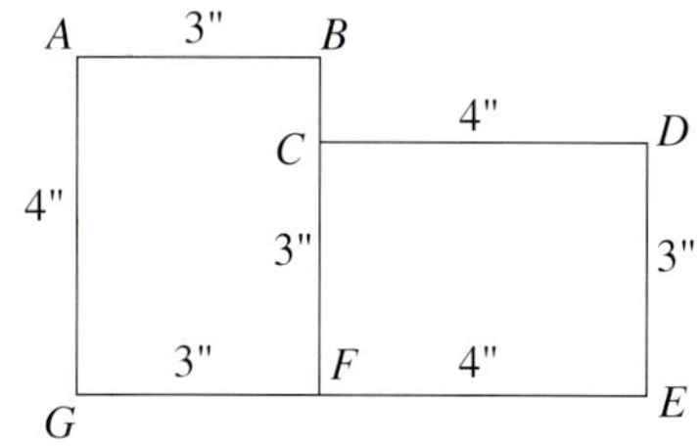

2. Name two angles in the following diagram.

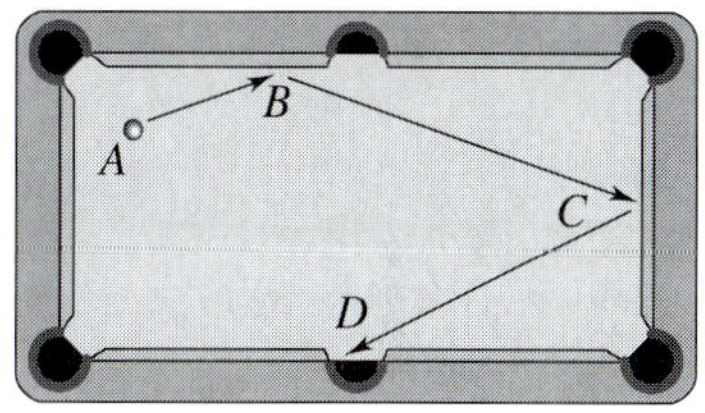

ANSWERS

1. $\overline{AB} \cong \overline{GF} \cong \overline{DE} \cong \overline{CF}$
2. $\angle ABC$, $\angle BCD$

3. One-tenth of a circle is how many degrees?

An angle is measured in **degrees**. One degree, written 1°, is $\frac{1}{360}$ of a circle. In other words, there are 360 degrees in one complete circle. The minute hand of a clock travels 360° as the time goes from 4:30 P.M. to 5:30 P.M. In going from 5:30 to 5:45, the minute hand travels 90°. Here are some common angles.

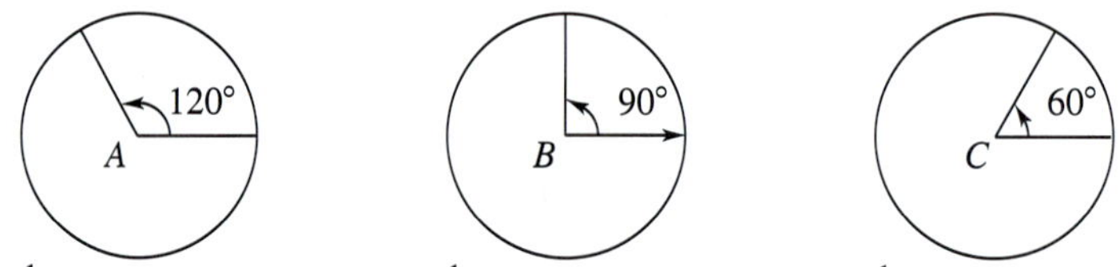

$\frac{1}{3}$ of a revolution	$\frac{1}{4}$ of a revolution	$\frac{1}{6}$ of a revolution
$\frac{1}{3}$ of 360° = 120°	$\frac{1}{4}$ of 360° = 90°	$\frac{1}{6}$ of 360° = 60°

The **measure of an angle**, written **m∠**, is the number of degrees between the two rays of an angle. Sometimes for convenience the angle is said to be equal to the number of degrees without using the word measure.

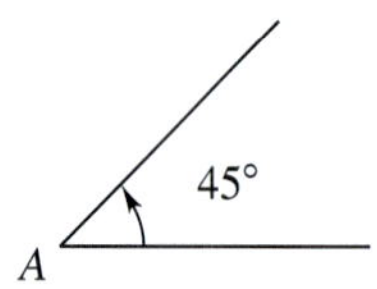

m $\angle A = 45°$

Measure of $\angle A$ = 45 degrees   or $\angle A = 45°$

EXAMPLE 3

Find the measure of each angle in degrees:

(a) $\frac{1}{5}$ of a circle   **(b)** $\frac{1}{2}$ of a circle

SOLUTION

(a) $\frac{1}{5}$ of a circle $= \frac{1}{5} \cdot 360° = 72°$

(b) $\frac{1}{2}$ of a circle $= \frac{1}{2} \cdot 360° = 180°$ ■

◇ *Do Practice Exercise 3.*

Two angles are **congruent angles** if they have the same number of degrees. The line segments making up the angles *do not* have to be the same length, for example:

$\angle A \cong \angle C$

Angle A is congruent to Angle C.

In other words, the two angles have the same number of degrees.

ANSWER

3. 36°

An angle in which the two rays form a straight line has a measure of 180°. This is called a **straight angle**.

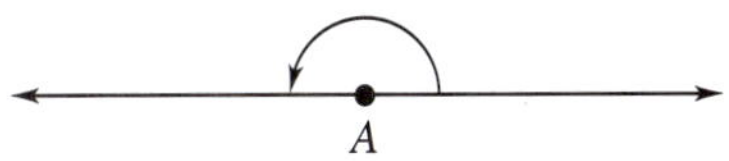

$\angle A = 180°$

In the military, a Sergeant orders his platoon to do an about face or to turn around and go in the exact opposite direction. He is asking them to make a 180 degree turn.

Two angles that add up to 180° are called **supplementary angles**. For example:

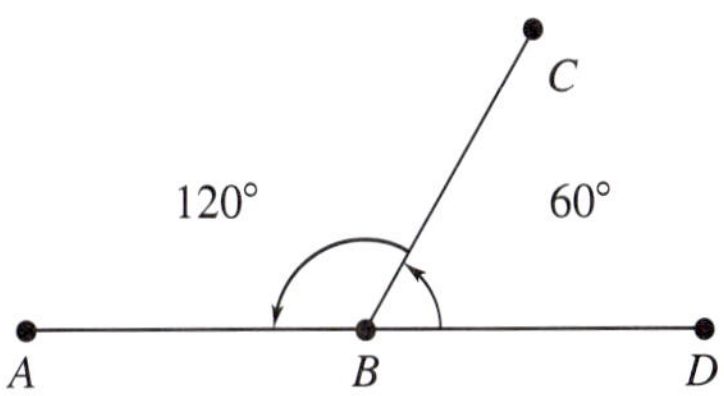

$\angle ABC$ and $\angle CBD$ are supplementary angles since the sum of the two angles equals 180°.

EXAMPLE 4

Find $\angle CBD$ given that $\angle ABC$ and $\angle CBD$ are supplementary angles.

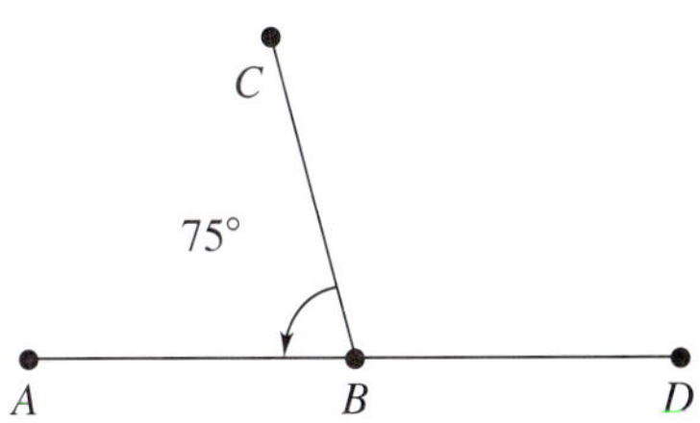

SOLUTION

$\angle ABC$ and $\angle CBD$ must total 180° because they are supplementary angles. The unknown angle can be found by subtracting 75° from the total of 180°.

$\angle CBD = 180° - 75°$	Subtract 75 from the total.
$\angle CBD = 105°$	Result. ■

◇ *Do Practice Exercise 4.*

A **right angle** is an angle measuring 90°. The symbol for a right angle is a small square drawn at the vertex of the angle.

$\angle A = 90°$

4. $\angle ABC$ and $\angle CBD$ are supplementary angles. Find $\angle CBD$.

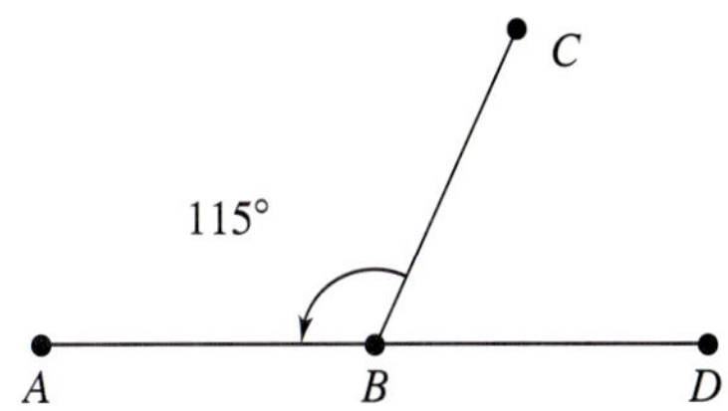

ANSWER

4. $\angle CBD = 65°$

5. $\angle A$ and $\angle B$ are complementary angles. Find $\angle A$ given that $\angle B$ is 40°.

Some examples of everyday things that are 90° angles are the corners of doors, the corners of a page of a book, and the directions of north and east, or east and south.

Two angles that add up to 90°, or to right angles, are called **complementary angles**. For example:

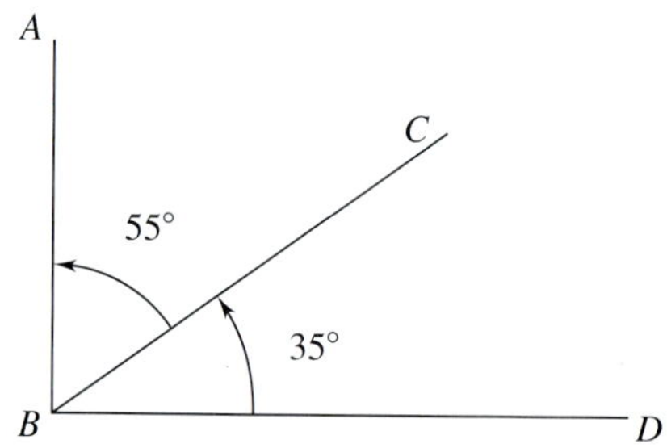

$\angle ABC$ and $\angle CBD$ are complementary angles since $55° + 35° = 90°$.

EXAMPLE 5

$\angle B$ and $\angle C$ are complementary angles. Find $\angle C$ if $\angle B$ is 20°.

SOLUTION

$\angle B$ plus $\angle C$ must total 90° since they are complementary angles. The unknown angle can be found by subtracting 20° from the total of 90°.

$\angle C = 90° - 20°$ Subtract 20° from the total.

$\angle C = 70°$ Result. ■

◇ ***Do Practice Exercise 5.***

ANSWER

5. $\angle A = 50°$

Sketch an example of each.

1. A line segment

2. A 90° angle

3. A right angle

4. Complementary angles

5. A ray

6. Congruent line segments

7. A straight angle

8. Supplementary angles

9. A 270° angle

10. Congruent angles

Express each angle in degrees.

11. $\frac{1}{4}$ of a circle

12. $\frac{1}{2}$ of a circle

13. $\frac{1}{5}$ of a circle

14. 1 complete circle

15. $\frac{3}{4}$ of a circle

16. $\frac{2}{3}$ of a circle

Name a line segment congruent to AB in each figure.

17.

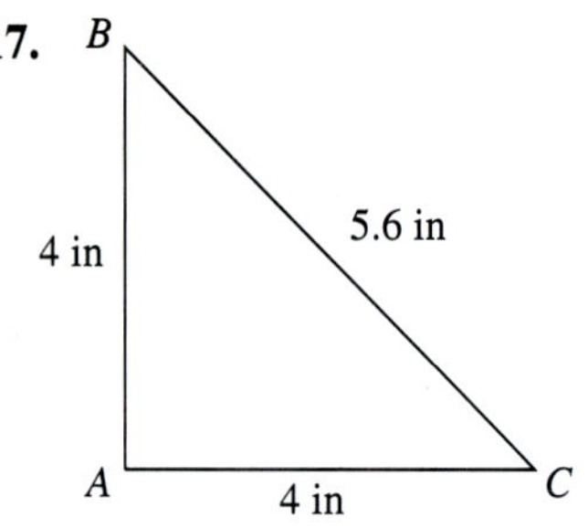

18.

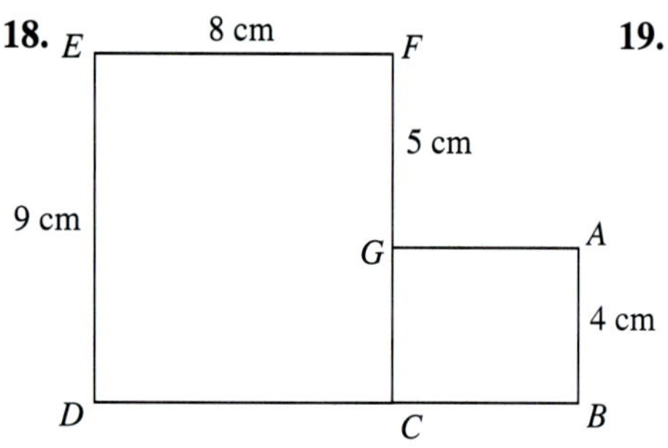

19.

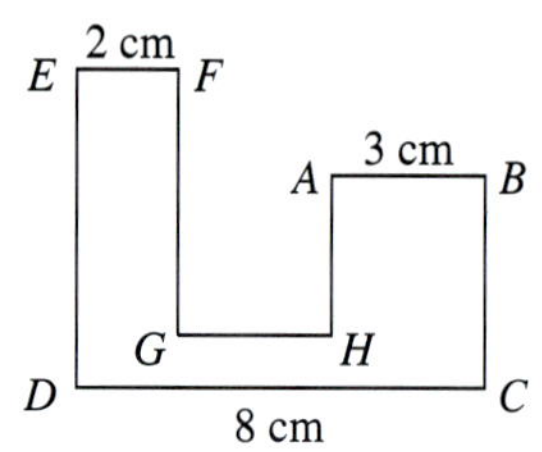

Find the measure of the indicated angle.

20. $\angle A$ and $\angle B$ are supplementary angles. Find $\angle B$ given $\angle A = 85°$.

21. $\angle A$ and $\angle B$ are supplementary angles. Find $\angle B$ given $\angle A = 179°$.

22. $\angle A$ and $\angle B$ are supplementary angles with $\angle B$ equal to 95°. Is $\angle A$ larger than $\angle B$?

23. $\angle C = 75°$. Find $\angle D$ if it is supplementary to $\angle C$.

24. $\angle A$ and $\angle B$ are complementary angles. Find $\angle B$ if $\angle A$ is 79°.

25. $\angle D$ and $\angle F$ are complementary angles. Find $\angle F$ given $\angle D = 34°$.

26. $\angle D$ and $\angle F$ are complementary angles. Find $\angle D$ given $\angle F = 45°$.

27. $\angle x = 1°$. Find $\angle y$ if it is complementary to $\angle x$.

28. $\angle A$ is 50°. Find $\angle B$ if it is supplementary to $\angle A$. Find $\angle B$ if it is complementary to $\angle A$.

29. $\angle B$ measures 65°. Find $\angle A$ if it is complementary to $\angle B$. Find $\angle A$ if it is supplementary to $\angle B$.

30. Name an angle that is complementary to itself.

Answer the following with true *or* false.

31. $\angle D = 57°$ and $\angle F = 33°$. $\angle D$ and $\angle F$ are complementary.

32. $\angle A = 128°$ and $\angle B = 62°$. $\angle A$ and $\angle B$ are supplementary.

33. A right angle has more than 100°.

34. Right angles and complementary angles are the same thing.

35. $\angle A = 180°$, so $\angle A$ is a right angle.

36. A straight angle can be made up of two right angles.

37. If two adjacent angles are supplementary, they form a straight angle.

38. Two right angles are supplementary.

39. There are six right angles in one complete circle

40. If two angles are complementary, together they form a right angle.

41. $\angle A$ and $\angle B$ are supplementary. $\angle A$ and $\angle C$ are complementary. $\angle A = 40°$. Then $\angle B$ and $\angle C$ are supplementary.

42. $\angle A$ and $\angle B$ are complementary. $\angle B$ and $\angle C$ are complementary. Therefore $\angle A$ and $\angle C$ are also complementary.

Find the measure of angle x.

43.

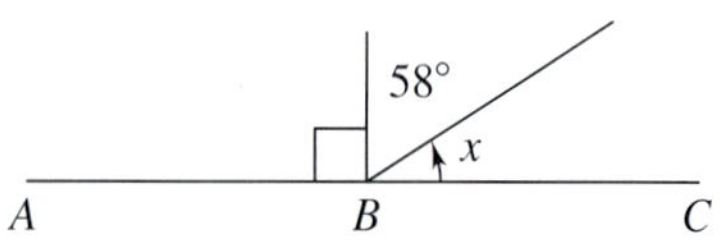

44.

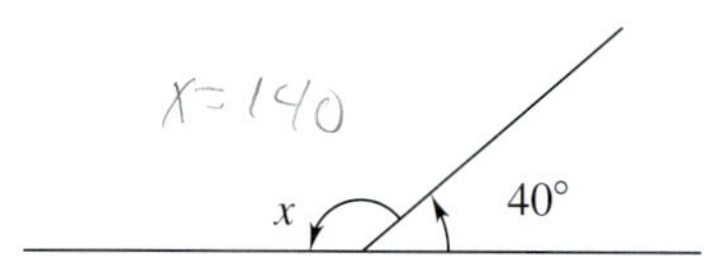

45.

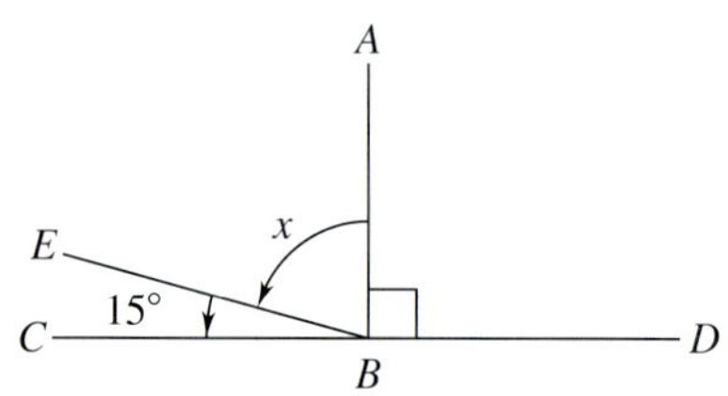

46.

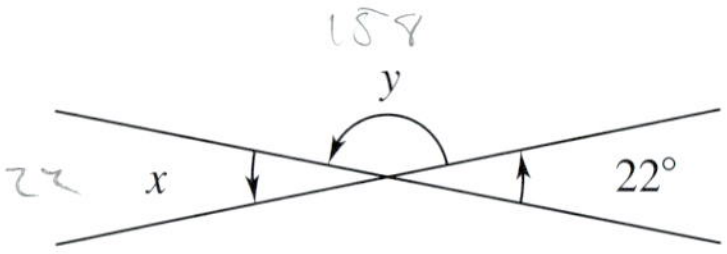

47.

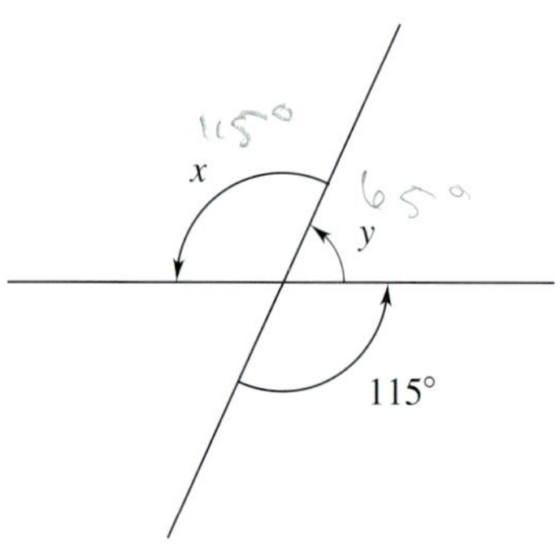

48.

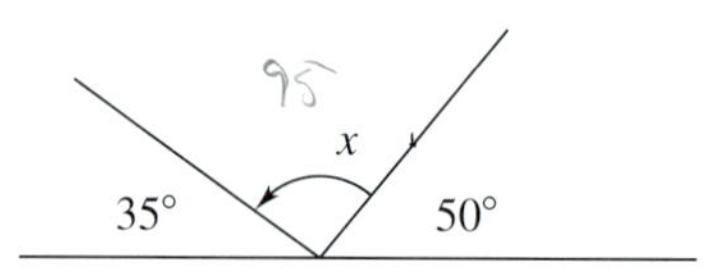

49. Jan is driving due north and turns to the west. What degree turn did she make?

50. A straight angle is divided into three congruent angles. What is the measure of each angle?

51. A right angle is added to a straight angle. Find the measure of the resulting angle.

52. How many degrees does the minute hand of a clock travel in one minute?

6 degrees

53. A passenger jet makes a 12° climb as it takes off from the runway. Find the angle the path of the jet makes with a vertical line drawn to the ground.

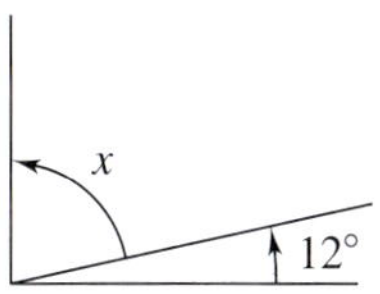

54. A roller coaster makes a 70° drop. Find the measure of the angle between the tracks and the vertical supports.

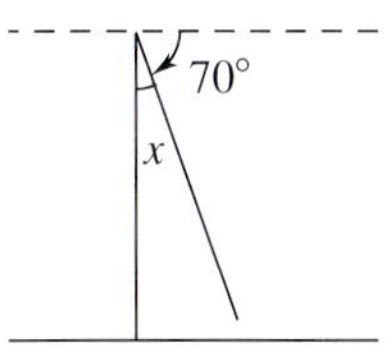

JOURNAL EXERCISES

1. In your own words, what does the word "angle" mean?
2. What is the difference between a line, a line segment, and a ray?
3. Explain complementary angles and supplementary angles in your own words.
4. Use approximate angles and distances to give directions from your house or apartment to the nearest hospital.
5. Is a line segment of one inch in length made up of points? If so, how many points are needed?
6. $\overline{AB}$ and $\overline{BC}$ represent telephone lines along the normal route of a long distance phone call from Los Angeles to Chicago. If the computer systems show that these lines are full, the call is rerouted through point D. Which is the larger angle, $\angle ABC$ or $\angle ADC$? What happens to the length of the route between A and C as the angle of the path gets smaller?

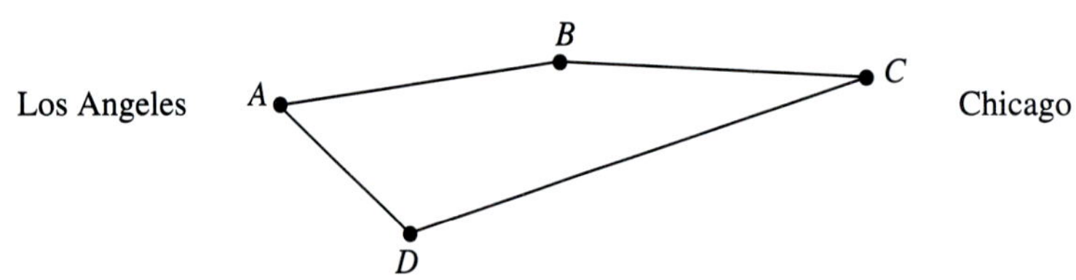

OBJECTIVES

1 Define polygons.
2 Identify types of triangles.
3 Define parallel and perpendicular lines.
4 Identify special quadrilaterals.

1 Polygons

Two-dimensional figures have both length and width, but no height or thickness. A **polygon** is a two-dimensional figure having three or more sides. Polygons are named according to the number of sides they have. For example, the shape of home plate on the baseball field is called a **pentagon** because it has five sides (''penta'' means five).

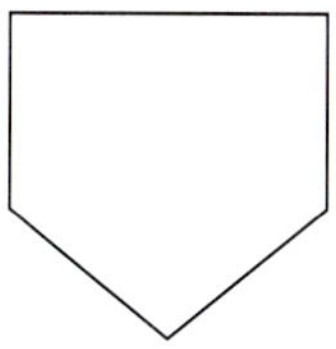

Home plate is a pentagon.

Other two-dimensional figures include the **triangle** with three sides, the **quadrilateral** with four sides, the **hexagon** with six sides, and the **octagon** with eight sides.

EXAMPLE 1

Name the following polygons:

(a) 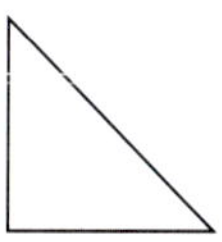 **(b)** 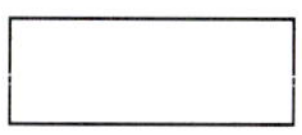 **(c)** 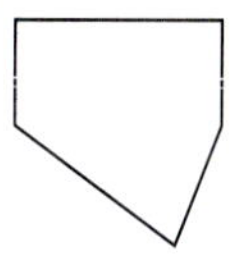 **(d)** 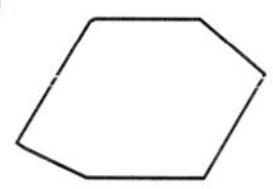

SOLUTION

(a) triangle (3 sides) **(b)** quadrilateral (4 sides) **(c)** pentagon (5 sides) **(d)** hexagon (6 sides) ■

◇ *Do Practice Exercise 1.*

2 Triangles

A triangle is a polygon with three sides. Each point where two sides meet is called a **vertex**. A triangle has three vertices (plural of vertex). The notation for triangle is $\triangle$, so that $\triangle ABC$ means the triangle with vertices at A, B, and C.

The three angles formed at the vertices of a triangle always add to 180°. To see that this is true, cut a triangle out of a piece of paper. Tear the three corners off and fit them together to form a straight line.

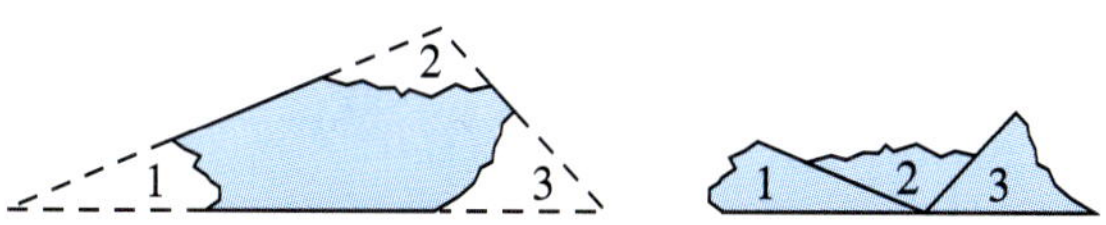

HELPFUL HINT

The angles of every triangle add to 180°.

1. Name each figure according to its number of sides:

(a) 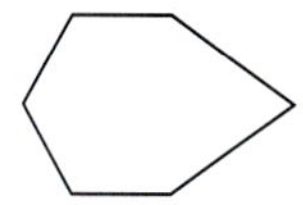

(b) 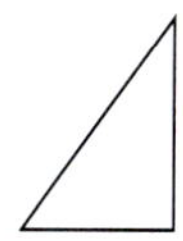

(c) 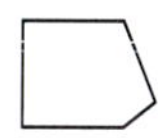

ANSWERS

1. (a) hexagon (b) triangle (c) pentagon

2. Find the third angle of a triangle if the first angle is 20° and the second angle is 75°.

EXAMPLE 2

In the triangle below, find $\angle A$ if $\angle B$ if 60° and $\angle C$ is 80°.

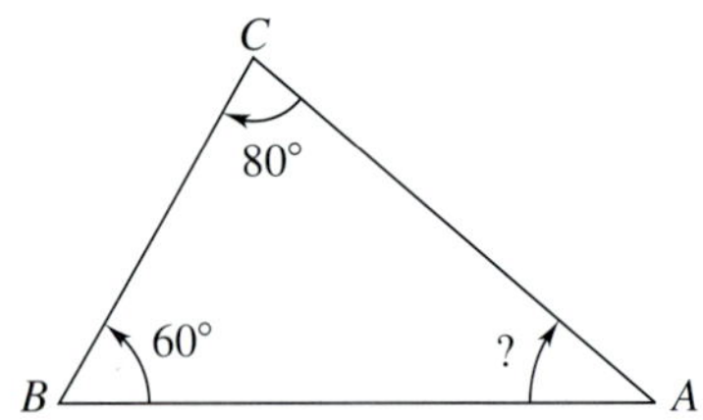

SOLUTION

The unknown angle can be found by subtracting the total of the other two angles from 180°.

$$\begin{aligned}\angle A &= 180° - (80° + 60°)\\ &= 180° - 140°\\ &= 40°\end{aligned}$$

◇ *Do Practice Exercise 2.*

An **isosceles triangle** is a triangle with two sides of equal length. That is, two sides are congruent. The two angles formed by each congruent side with the third side are called **base angles**. These two angles are also congruent.

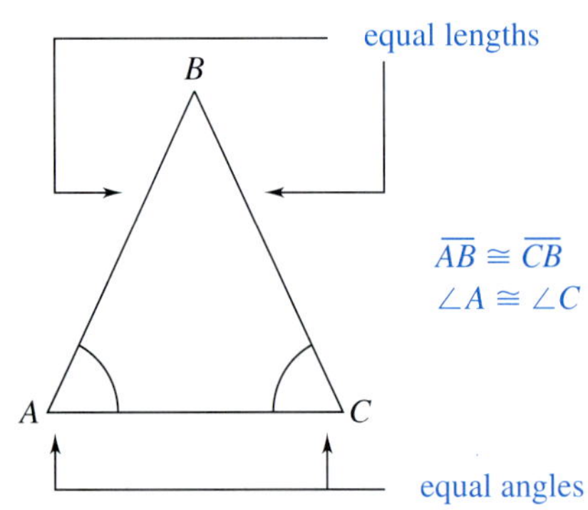

Isosceles triangle

EXAMPLE 3

$\triangle ABC$ is an isosceles triangle. The base angles equal 70° each, and $\overline{AC}$ = 7 in. Find $\angle A$ and side $\overline{AB}$.

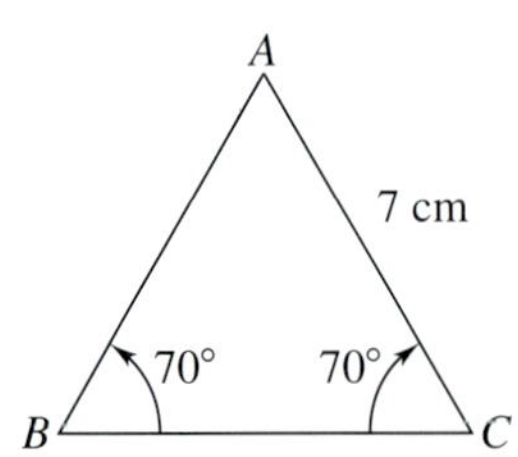

ANSWER
2. 85°

SOLUTION

$\overline{AB}$ is the side congruent to side $\overline{AC}$ since $\angle B$ and $\angle C$ are the congruent base angles.

$$\overline{AB} \cong \overline{AC} = 7 \text{ cm}$$
$$\angle A = 180^\circ - (70^\circ + 70^\circ)$$
$$\angle A = 180^\circ - 140^\circ$$
$$\angle A = 40^\circ$$

■

An **equilateral triangle** is a triangle in which all three sides are the same length. All three of its angles are also the same size. For example:

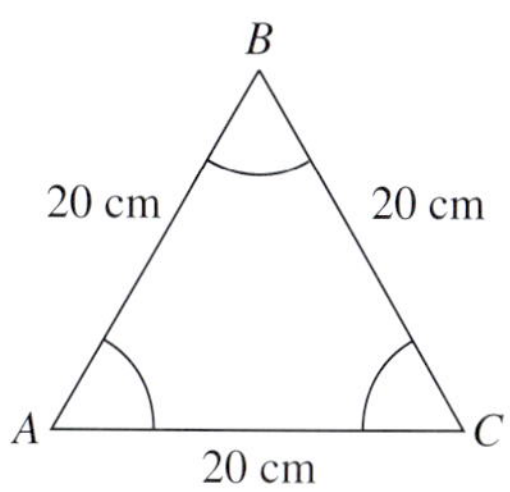

Equilateral triangle

$\overline{AB} \cong \overline{BC} \cong \overline{AC}$ equal lengths
$\angle A \cong \angle B \cong \angle C$ equal angles

◇ *Do Practice Exercises 3 and 4.*

3 Parallel and Perpendicular Lines

Two lines that never cross and are always the same distance apart are said to be **parallel**. The symbol for parallel lines is two vertical bars: ||

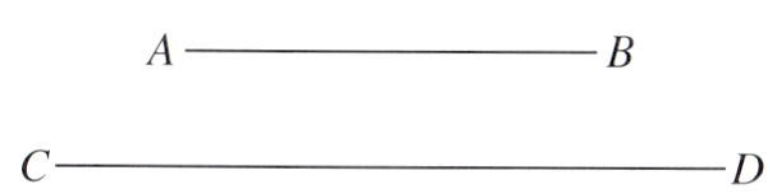

$\overline{AB} \parallel \overline{CD}$
Line segment $\overline{AB}$ is parallel to line segment $\overline{CD}$.

The main rails of railroad tracks are an example of parallel lines. They are always the same distance apart. The sides of most ladders are also parallel.

Perpendicular lines are lines that form right angles when they meet. The symbol for perpendicular is $\perp$.

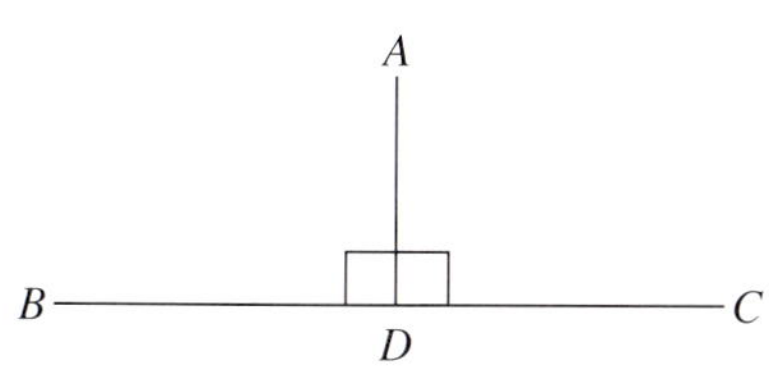

$\overline{AD} \perp \overline{BC}$
Line segment $\overline{AD}$ is perpendicular to line segment $\overline{BC}$.
$\angle ADB = \angle ADC = 90^\circ$

Most city streets cross at 90° angles and are perpendicular to each other. Corners of rooms meet at 90° angles making the walls perpendicular to each other.

3. $\triangle ABC$ is an isosceles triangle. Its base angles equal 65°. Find the third angle.

4. One side of an equilateral triangle is 14 inches. Find the distance around the edge of the triangle.

ANSWERS
3. 50°
4. 42 inches

5. AB is perpendicular to CD and $\angle ABE = 60°$. What is the measure of $\angle EBD$?

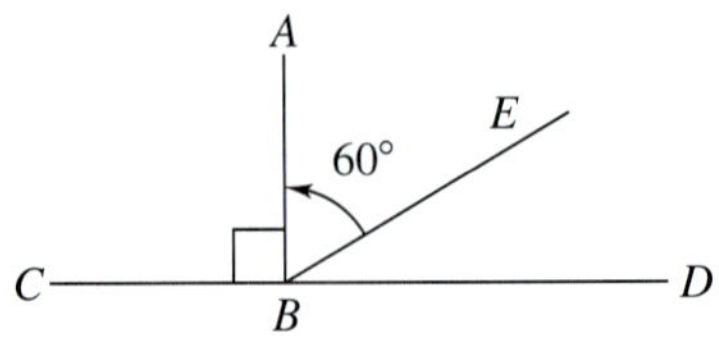

6. Name all of the rectangles in the following figure. Give the length and width of each.

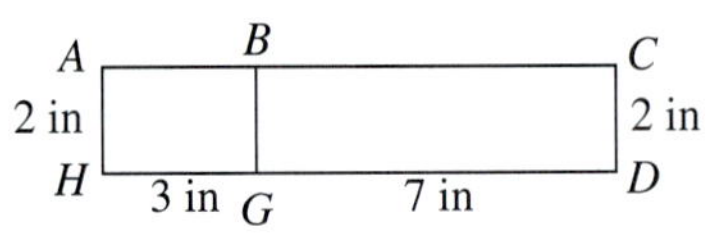

ANSWERS

5. $\angle EBD = 30°$

6.

Rectangle	*Length*	*Width*
ACDH	3 in + 7 in = 10 in	2 in
ABGH	3 in	2 in
BCDG	7 in	2 in

EXAMPLE 4

$\overline{BD}$ is perpendicular to $\overline{AC}$. $\angle EDB = 15°$. What is the measure of $\angle EDA$?

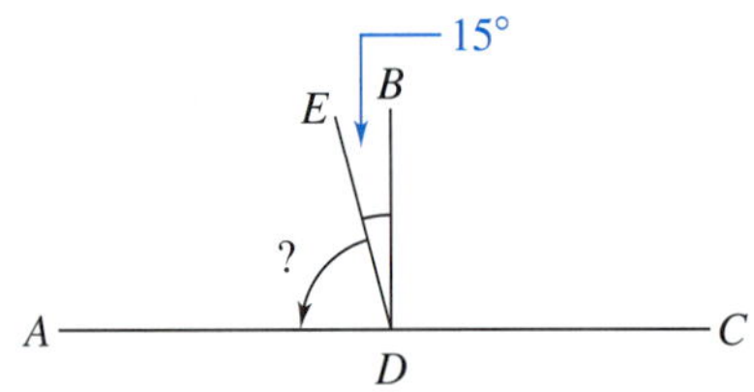

SOLUTION

Since $\overline{BD}$ is perpendicular to $\overline{AC}$, they must meet at a 90° angle, implying that $\angle ADB$ is 90°. $\angle ADE$ and $\angle EDB$ add to 90° since together they form a right angle.

$\angle ADE = 90° - 15°$ Subtract 15° from the total.

$\angle ADE = 75°$ Result. ■

◇ *Do Practice Exercise 5.*

4 Quadrilaterals

There are several special types of quadrilaterals: a parallelogram, a rhombus, a rectangle, a square, and a trapezoid. Definitions of each follow.

A **parallelogram** is a quadrilateral in which opposite sides are parallel and equal.

A **rhombus** is a parallelogram in which all sides are the same length.

A **rectangle** is a parallelogram in which all four angles are right angles.

A **square** is a rectangle in which all sides are equal.

A **trapezoid** is a quadrilateral in which only two sides are parallel.

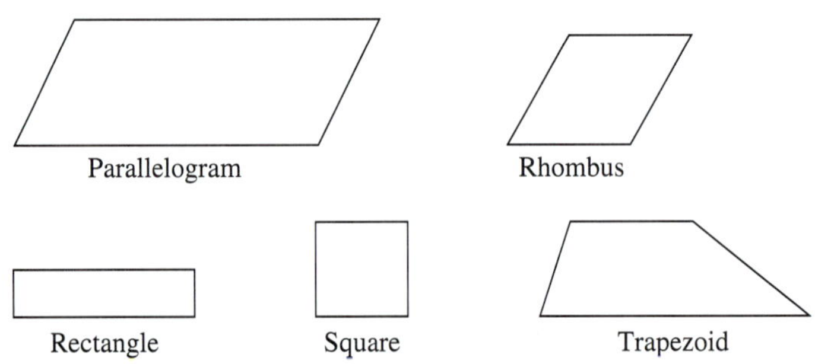

Several types of figures may be combined to make a **composite figure**. For example, rectangles of different sizes may be combined, or a triangle and a rectangle may be combined. The figures may overlap or share common parts.

◇ *Do Practice Exercise 6.*

Sketch an example each of the following figures.

1. Pentagon
2. Rhombus
3. Hexagon
4. Isosceles triangle
5. Rectangle
6. Quadrilateral
7. Parallelogram
8. Trapezoid
9. Equilateral triangle
10. Square

Find the measure of the indicated angle.

11. Two angles of a triangle are 60° each. Find the third angle.
12. Two angles of a triangle are 55° and 72°. What is the measure of the third angle?
13. Two angles of a triangle are 85° and 45°. What is the measure of the third angle?
14. Two angles of a triangle are 90° and 45°. What is the measure of the third angle?

15. What is the measure of each of the three congruent angles of an equilateral triangle?

16. One of the base angles of an isosceles triangle is 35°. What are the measures of the other two angles?

17. What is the sum of the four angles of a rectangle?

18. One side of an equilateral triangle is 13 cm. What is the sum of the other two sides?

19. The total length of the three sides of an equilateral triangle is 213 cm. What is the length of one side?

20. The total length of the three sides of an isosceles triangle is 14 inches. The base of the triangle is 6 inches. What is the length of each of the congruent sides?

Answer the following with true *or* false.

21. A rectangle is a type of parallelogram.

22. A trapezoid is a type of parallelogram.

23. Opposite sides of a parallelogram are equal in length.

24. A rectangle has four right angles.

25. The adjacent sides of a square are perpendicular to each other.

26. A square is a type of rhombus.

27. A trapezoid has two pairs of parallel sides.

28. Opposite sides of a trapezoid are congruent.

29. A parallelogram always has at least one right angle.

30. A hexagon has five sides.

31. Two perpendicular lines are also parallel.

32. A rectangle can be built out of squares.

33. A pentagon can be divided into two triangles.

34. Two lines that meet at right angles are perpendicular.

Find the indicated angle.

35. In the rectangle, find $\angle ADB$.

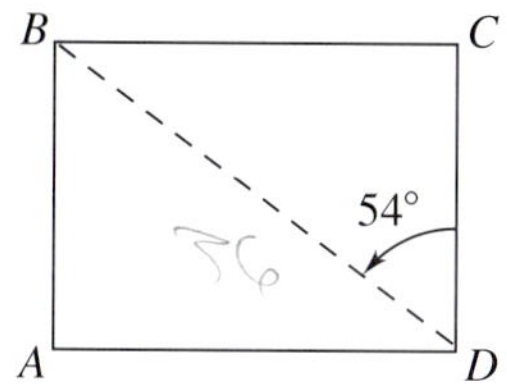

36. Find $\angle A$.

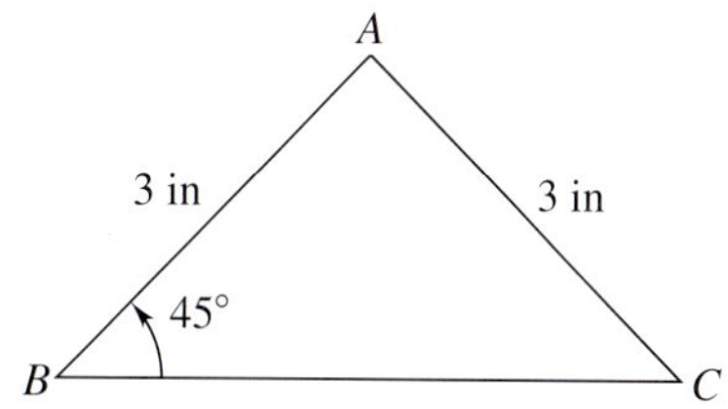

37. Find $\angle A$.

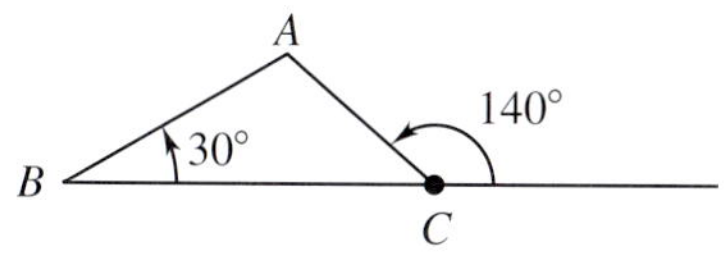

38. Find $\angle C$.

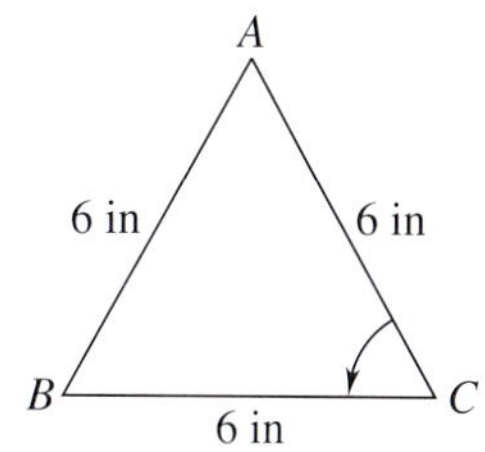

39. $AB \perp CD$. Find $\angle ABE$.

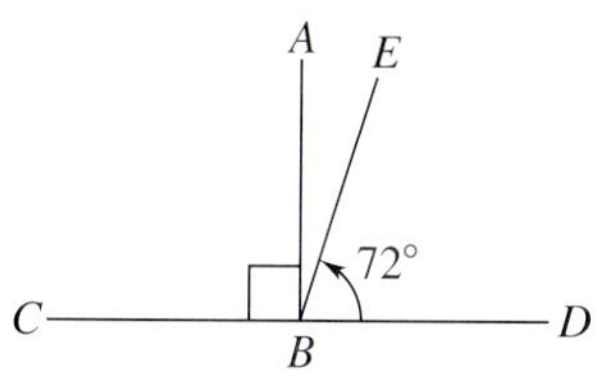

40. $\angle DCE = 33°$. Find $\angle ACB$.

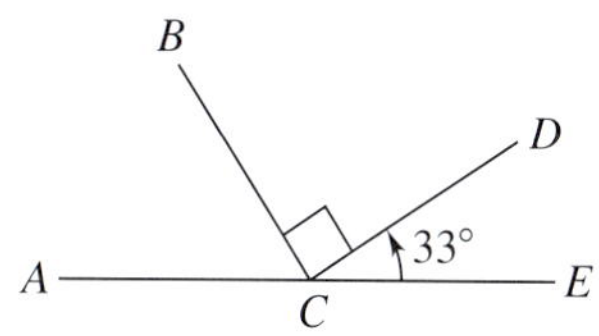

Answer the following questions.

41. Name all the rectangles in the given floor plan of a house. Give the length and width of each.

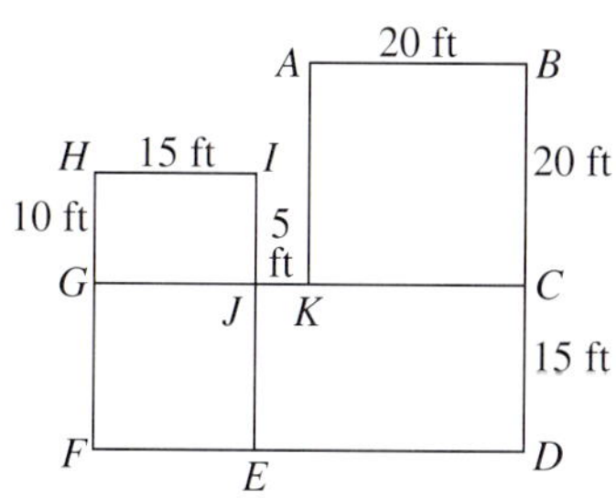

42. In the drawing of a solar eclipse below, $\angle ABC = 140°$ and $\angle BCA = 35°$. Find $\angle A$.

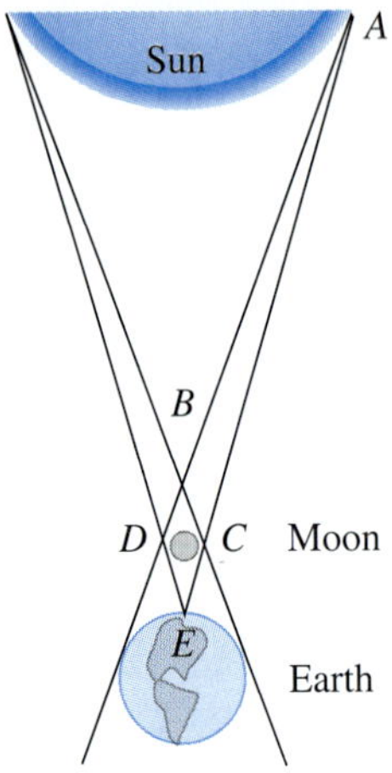

43. In the diagram below of a refracting telescope, $\angle ABC = 14°$. $\triangle ABC$ is isosceles. Find the measure of the base angles A and C.

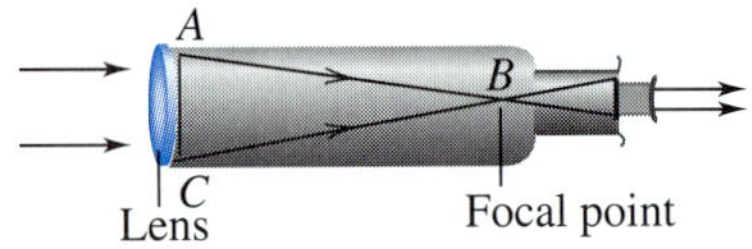

44. A contractor plans to cut cabinet doors from a 3 foot by 72 inch piece of wood. Each cabinet door is to be 3 feet by $1\frac{1}{2}$ feet. How many rectangular doors can be cut from the piece of wood?

JOURNAL EXERCISES

1. What is geometry? Explain the concept of dimension. Can you name a real item that has two dimensions?

2. Look up the definition of symmetry in the dictionary and write it down. Which of the geometric figures described in this section are symmetric? Explain. Why is symmetry important to an architect?

3. The angles in a triangle add up to 180°. What are the sum of the angles in a rectangle? The angles of a pentagon total 540°, and the angles of a hexagon add up to 720°. Try to find a pattern or relationship between the total number of degress in a polygon and its number of sides. What would the total degrees be for the angles of a 7-sided polygon?

1 Perimeter

Perimeter is the distance around the outer edge of a polygon. It is the sum of the distances along each of the edges of the geometric figure. For example, the perimeter of a swimming pool is the distance around the outer edges of the pool.

The **perimeter of a triangle** is the sum of the lengths of the three sides of the triangle.

P = 1st side + 2nd side + 3rd side

$P = a + b + c$

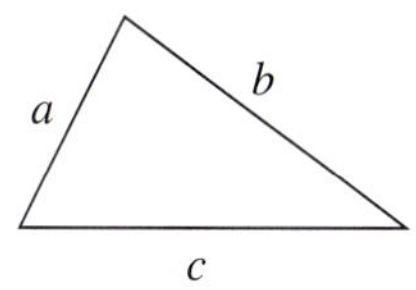

EXAMPLE 1

Find the perimeter of the triangle whose sides are 6 in, 7 in, and 10 in.

SOLUTION

P = 6 in + 7 in + 10 in = 23 inches ■

◇ *Do Practice Exercise 1.*

The **perimeter of a parallelogram** is the sum of the lengths of all four sides. In a parallelogram, however, the opposite sides are equal. The perimeter formula can be written as follows:

$P = a + a + b + b$

$P = 2a + 2b$

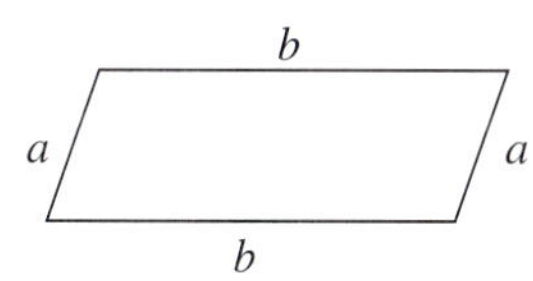

The meaning of $2a$ is two times a, where a is the variable that represents the length of one side of the parallelogram. $2b$ means two times b, where b is the length of the other side.

The **perimeter of a rectangle** is very similar to that of a parallelogram. The only difference is that L is used for length instead of "b", and W is used for width instead of "a".

$P = L + L + W + W$

$P = 2L + 2W$

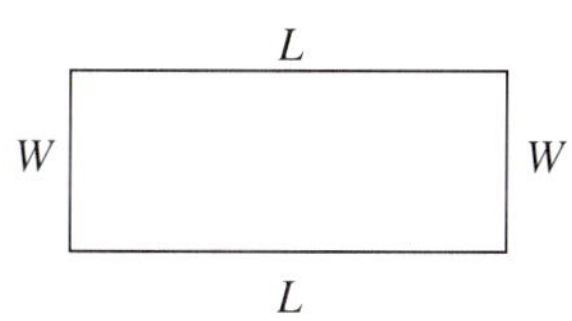

EXAMPLE 2

Find the perimeter of a rectangle whose length is 7.5 cm and whose width is 4.3 cm.

SOLUTION

$P = 2L + 2W$	formula
P = 2(7.5 cm) + 2(4.3 cm)	Substitute.
P = 15 cm + 8.6 cm	Multiply.
P = 23.6 cm	Result. ■

OBJECTIVES

1. Know how to find the perimeter of polygons.
2. Know how to find the circumference of circles.
3. Find perimeter and circumference of simple and composite figures.

1. Find the perimeter of a triangle whose sides are 12 cm, 5 cm, and 15 cm.

ANSWER
1. 32 cm

2. Find the perimeter of a parallelogram with adjacent sides 14 in and 12 in.

3. Find the perimeter of a rectangle with length 60 cm and width 38 cm.

Notice that the perimeter has the same units as those given by the problem. For example, the perimeter will be in centimeters if the lengths are given in centimeters.

HELPFUL HINT

When finding perimeter, the units of the answer are the same as the units given in the problem.

◇ *Do Practice Exercises 2 and 3.*

The lengths of the sides of a square are equal to one another, as are those of a rhombus. To find the **perimeter of a square** or the **perimeter of a rhombus**, multiply four times the length of one side.

$$P = 4 \cdot s$$
$$P = 4s$$

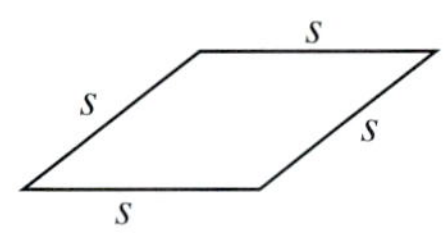

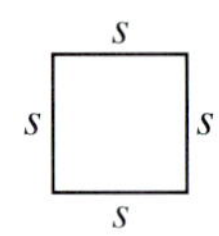

The sides of a trapezoid are not necessarily equal. The formula for finding the **perimeter of a trapezoid**, then is the sum of the four sides.

$$P = a + b + c + d$$

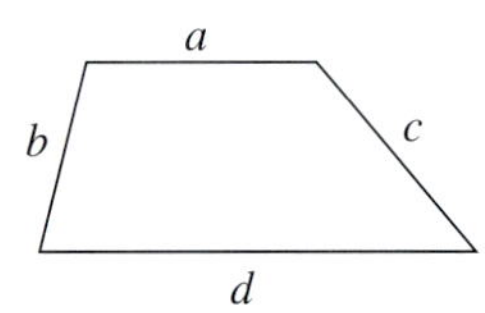

2 Circles and Circumference

A **circle** is a figure formed by drawing a line through all points equally distant from a point called the **center** of the circle. For example, the outer edge of a coin is in the shape of a circle.

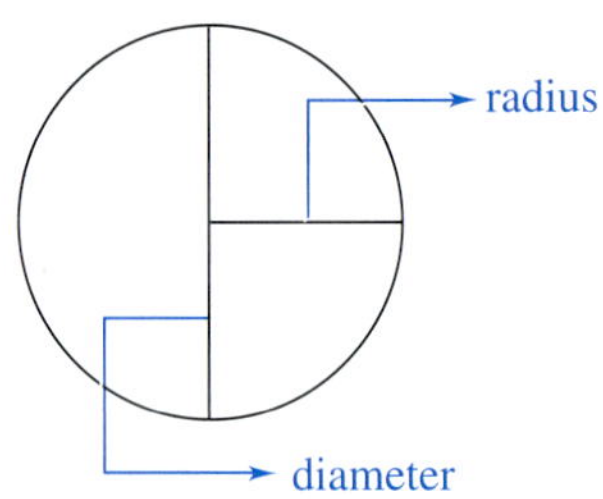

ANSWERS
2. 52 in
3. 196 cm

A line drawn from any point on the circle to the center is called the **radius**. The length of the radius of a circle is always the same no matter which point on the edge one starts from. A line drawn from a point on one side of the circle to a point on the other side passing through the center is the **diameter**. The diameter is exactly twice the radius.

HELPFUL HINT

The diameter is two times the radius. The radius is one-half of the diameter.

Circumference is the distance around the outer edge of a curved geometric shape such as a circle. The formula for the circumference of a circle involves a number π, called **pi**. The following examples help explain the value of pi.

The distance around the outside of a tire of a compact car is about 69 inches. The diameter of the tire is approximately 22 inches. The distance around the edge of a nickel is approximately 6.6 cm. The diameter of a nickel is about 2.1 cm. The ratio of the circumference to the diameter of both is shown below:

measurements for the tire

$$\frac{\text{circumference}}{\text{diameter}} = \frac{69 \text{ in}}{22 \text{ in}} \approx 3.14$$

measurements for the nickel

$$\frac{\text{circumference}}{\text{diameter}} = \frac{6.6 \text{ cm}}{2.1 \text{ cm}} \approx 3.14$$

Notice that both ratios are approximately equal to 3.14. In fact, this ratio is the same value for all circles. The value for the ratio is commonly referred to as π and is usually rounded off to 3.14.

The formula for the circumference of a circle is:

$$C = \pi d \quad \text{or} \quad C = 2\pi r \quad (\text{since } 2r = d)$$

where r = radius and d = diameter.

EXAMPLE 3

Find the circumference of a circle with radius 5 in.

SOLUTION

$C = \pi \cdot 2 \cdot r$	formula for circumference
$C = (3.14)(2)(5)$	Substitute values.
$C = 31.4$ in.	Result. ■

◇ *Do Practice Exercise 4.*

4. Find the circumference of a circle with radius 3 cm.

ANSWER
4. 18.84 cm

The formulas for perimeter and circumference are summarized in the following table.

Formulas for Perimeter and Circumference

Triangle	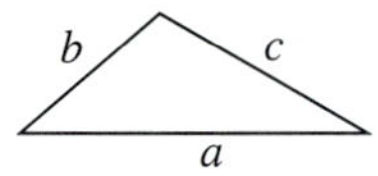	$P = a + b + c$
Parallelogram	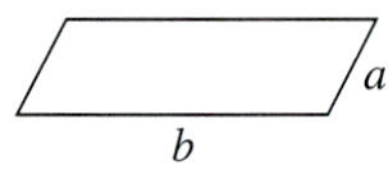	$P = 2a + 2b$
Rectangle	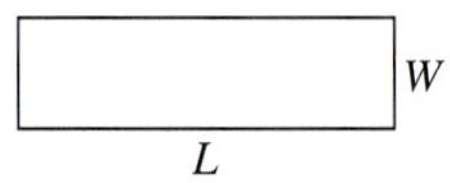	$P = 2L + 2W$
Square	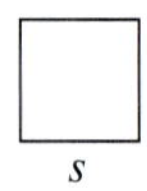	$P = 4s$
Rhombus	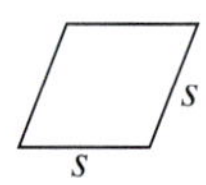	$P = 4s$
Trapezoid	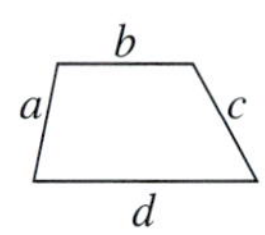	$P = a + b + c + d$
Circle	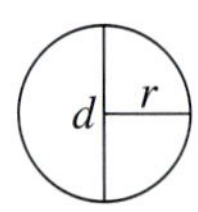	$C = 2\pi r$ $C = \pi d$

3 Calculate Perimeters and Circumferences

The perimeter of any polygon can be found by adding the lengths of all of the sides together. When adding lengths, be sure to add only the lengths of the outside edges of the figure.

EXAMPLE 4

Find the perimeter of each geometric figure:

(a)

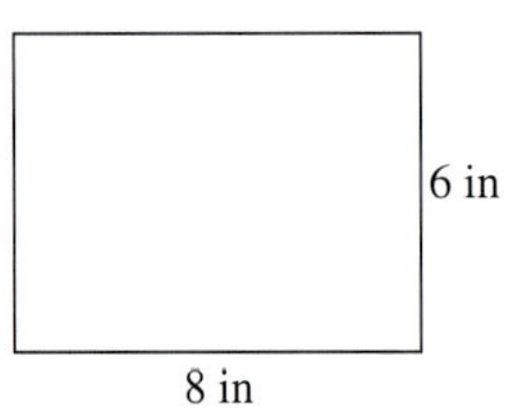

(b)

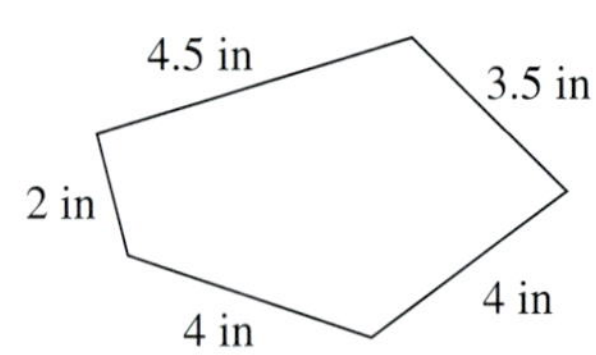

SOLUTION

(a) $P = 2L + 2W$

$P = 2(8'') + 2(6'')$

$P = 16'' + 12''$

$P = 28$ in.

(b) $P =$ sum of all sides

$P = 2'' + 4'' + 4'' + 3.5'' + 4.5''$

$P = 18$ in. ■

◇ Do Practice Exercise 5.

5. Find the perimeter of a rectangle with length 14 cm and width 7 cm.

6. David is putting a circular flower bed with a diameter of five feet, around a new tree. How much edging material will be needed to go around the flower bed? Use $\pi = 3.14$.

EXAMPLE 5

An individual jogs once around a lake that is shaped like a circle with a radius of one half of a mile. How far does the individual jog? Use 3.14 for the value of π.

SOLUTION

The answer is given by the circumference of the circle:

$C = 2\pi r$	formula for circumference
$C = 2\,(3.14)\left(\frac{1}{2}\text{ mile}\right)$	Substitute values.
$C = 3.14(2)\left(\frac{1}{2}\text{ mile}\right)$	Commutative rule
$C = 3.14(1\text{ mile})$	Multiply.
$C = 3.14$ miles	Result.

Thus the individual jogs a little over three miles. ■

◇ Do Practice Exercise 6.

The perimeter of a composite figure can be found by breaking it into basic shapes such as rectangles or circles or even semicircles. This method is shown in the following example.

EXAMPLE 6

Find the length of the track around a football field. The football field including end zones is 120 yards long and the width including the track is 63.7 yards. Use $\pi = 3.14$ and round off the answer to the nearest yard.

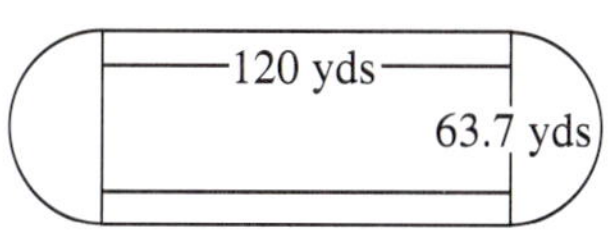

SOLUTION

First, recognize that the shape can be broken into three parts, two semicircles and a rectangle:

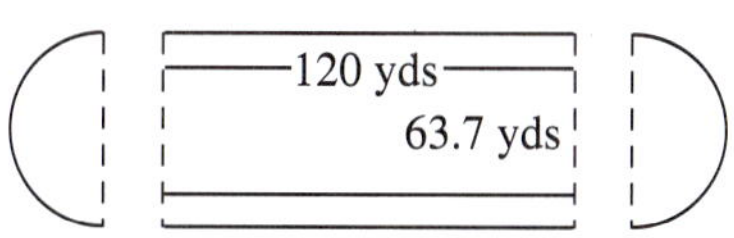

ANSWERS
5. 42 cm
6. 15.7 feet

7. The owner of an apartment complex installed a swimming pool in the shape shown in the figure. What is the perimeter of the pool? Use $\pi = 3.14$.

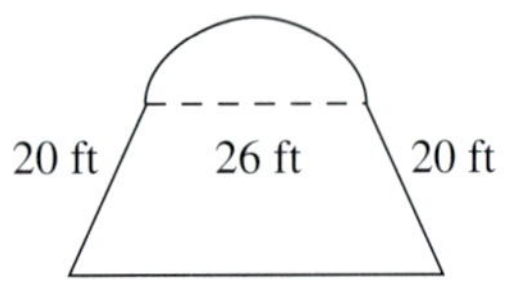

The total length of the track is the sum of the lengths of the two semicircles plus twice the length of the rectangle. The length of the two semicircles equals the circumference of a circle with the same diameter.

$$\begin{aligned}\text{Length} &= \begin{bmatrix}\text{Circumference}\\ \text{of circle}\end{bmatrix} + 2\cdot\begin{bmatrix}\text{Length of}\\ \text{rectangle}\end{bmatrix} \\ &= \pi d + 2L && \text{formula} \\ &\approx (3.14)(63.7\text{ yd}) + (2)(120\text{ yd}) && \text{Substitute.} \\ &\approx 200.018\text{ yd} + 240\text{ yd} && \text{Multiply.} \\ &\approx 440\text{ yards} && \text{Result.}\end{aligned}$$

◇ *Do Practice Exercise 7.*

ANSWER
7. 122.82 feet

Name each geometric figure and find the perimeter or circumference. Use $\pi = 3.14$.

1. 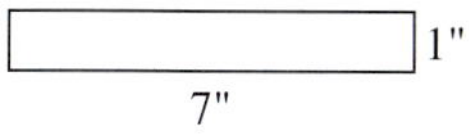

2. 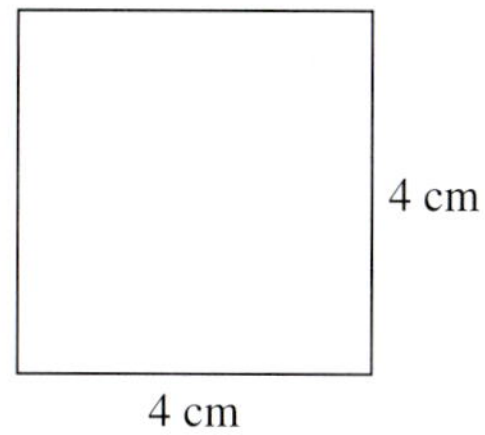

3. 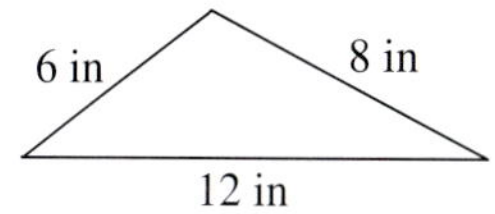

4. 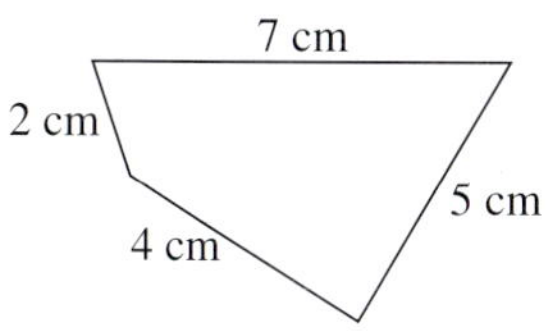

5. 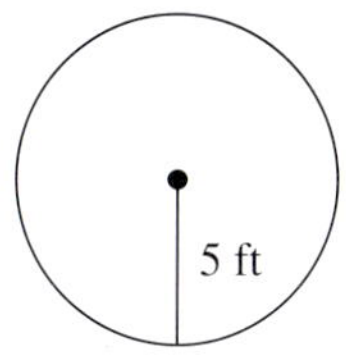

6. 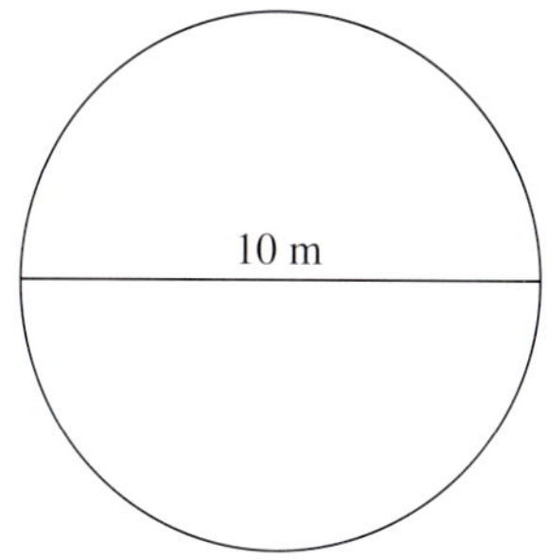

Find the perimeter or circumference of each geometric figure. Use $\pi = 3.14$.

7. A circle of radius 1 in

8. A circle of radius 8 in

9. A circle with diameter 5.5 cm

10. A circle with diameter 1.4 ft

11. A square with side 3 feet

12. A square with side $5\frac{1}{4}$ cm

13. A rhombus with side 2.3 cm

14. A rhombus with side 16 in

15. A parallelogram with adjacent sides 12 cm and 22 cm

16. A parallelogram with adjacent sides 5 in and 8.7 in

17. A rectangle with width of $4\frac{1}{3}$ m and length of $9\frac{2}{3}$ m

18. A rectangle with a length of 7 meters and a width of a 4 meters

19. A triangle of sides 2.5 cm, 4.2 cm, and 3.0 cm

20. A triangle with sides 4.3 in, 2.5 in, and 6.58 in

21. An isosceles triangle with congruent sides of 45 mm and base 20 mm

22. An isosceles triangle with base 16 m and sides of 18 m each

23. An equilateral triangle with sides of 19 in

24. An equilateral triangle with sides of length 75 ft

25. A trapezoid with sides of 10.2 cm, 4.5 cm, 12 cm, and 3.8 cm

26. A quadrilateral of sides $6\frac{1}{2}$ in, 3 in, 4 in, and $2\frac{1}{2}$ in

27. A semicircle with a diameter of 10 mm

28. A semicircle with radius 12 in

Find the perimeter of each composite figure.

29.

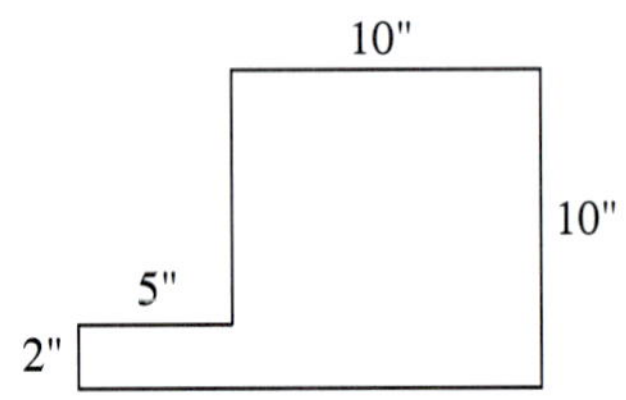

30.

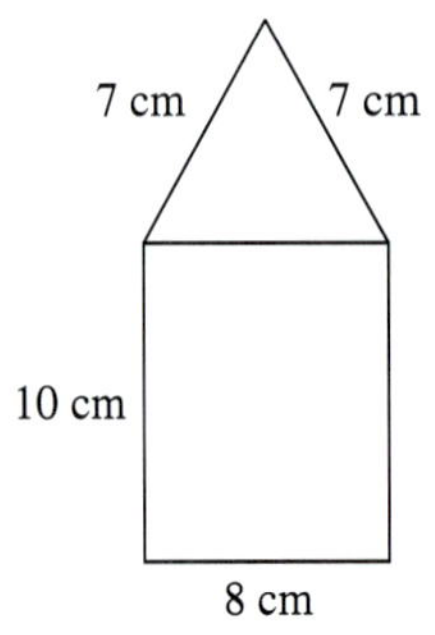

31.

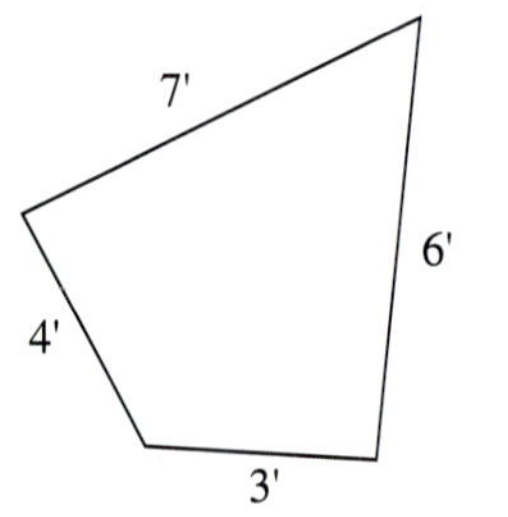

32.

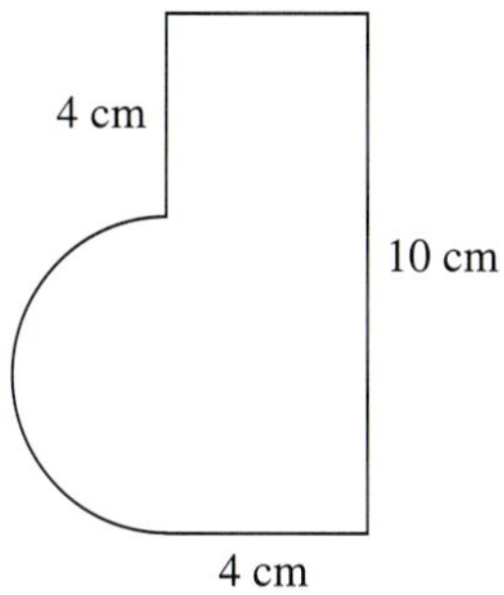

33.

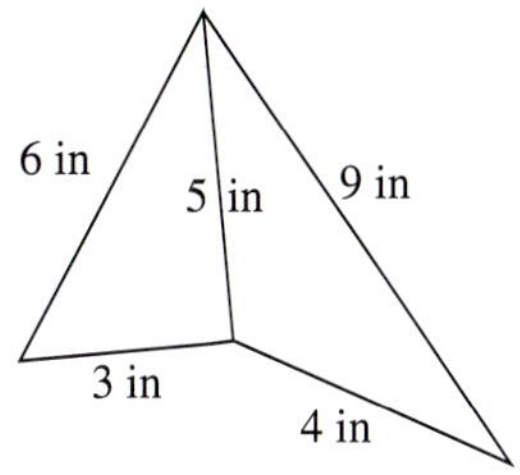

34.

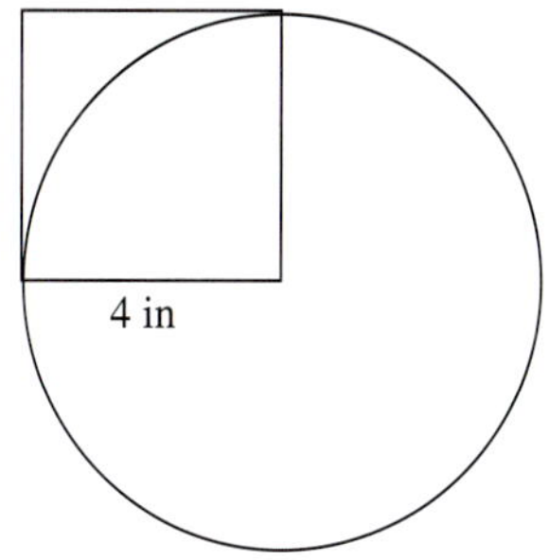

35.

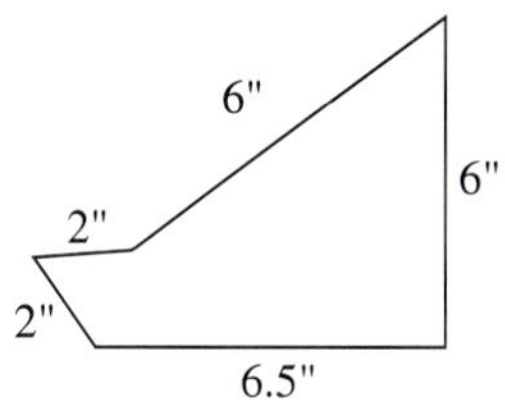

36.

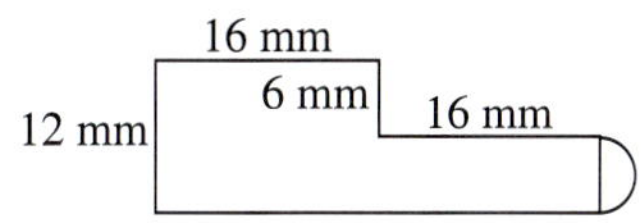

37.

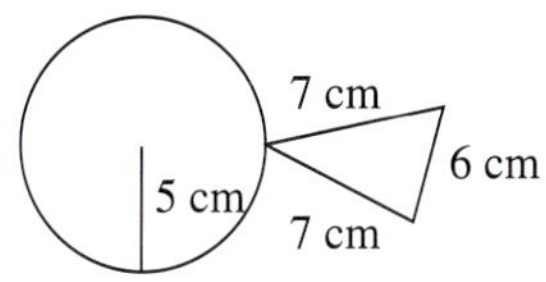

Sketch and find the perimeter or circumference of each. Use 3.14 *for the value of* π.

38. A triangle with sides 8 ft, 8 ft, and 10 ft on top of a square with side 10 ft (like a roof on a house).

39. A square with a side of length 5 cm centered on top of a rectangle of width 4 cm and length 10 cm.

40. A semicircle of radius 4 in on top of a rectangle of length 14 in and width 8 in.

41. A rectangle with length $4\frac{2}{3}$ in and width $3\frac{1}{2}$ in.

Solve the following problems.

42. Lucille has a garden plot shaped like a rectangle with a width of 12 feet and a length of 40 feet. How many feet of fencing material is needed to keep stray dogs out of the garden?

43. A Boy Scout troop is installing one campsite marker at each of 40 sites. The signs are hexagons with 5″ sides. The boys are outlining the signs with reflective tape. How many feet of tape will they need?

44. A physical therapist is helping a stroke patient in a walking program. The patient's stride is 1.7 feet. About how many steps must the patient take to walk around a 12 foot by 13 foot room? Round your answer to the nearest unit. A month later, the patient's stride is 2.2 feet. How many steps does the walk now require? Round to the nearest unit.

45. A racetrack is shaped like a rectangle with a semicircle on each end. Find the distance around the track if the rectangle is 1,000 meters wide and 2,000 meters long.

46. Name the type of polygon that approximates the closed part of the outline of the constellation, Leo. Find the perimeter of the figure in inches.

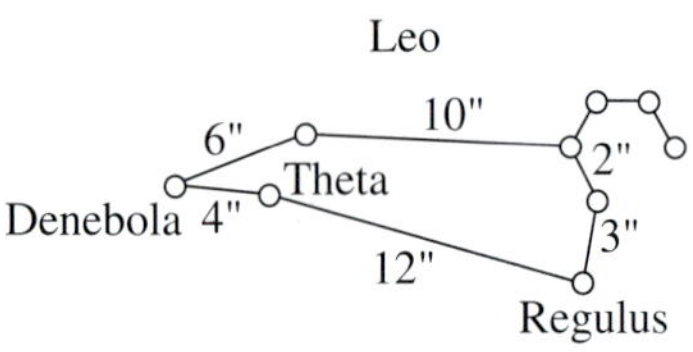

47. Use a triangle, a trapezoid, and a parallelogram to estimate the length of the 1986 Tour de France route. The lengths are given in inches. Each inch on the map represents about 80 miles.

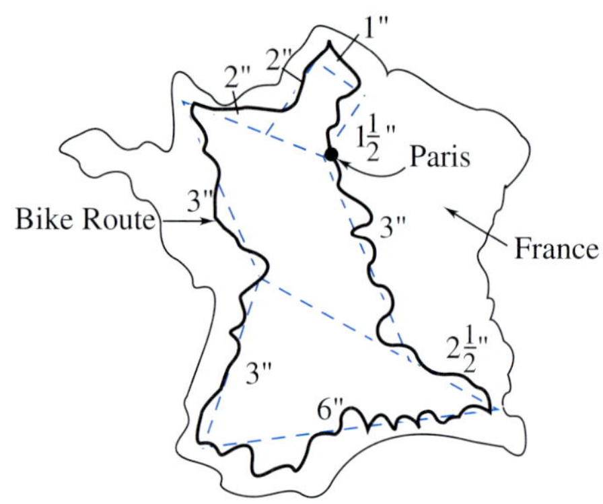

Route Map of the 1986 Tour de France

48. A certain car tire has a diameter of 26 inches. Find the circumference of the tire. How many revolutions will the wheel make in 1 mile (5,280 feet)?

49. Approximate the circumference of each, assuming $\pi \approx 3.14$.

Body	*Approximate Radius*
Sun	430,000 miles
Earth	4,000 miles
Jupiter	45,000 miles
Pluto	1,000 miles

JOURNAL EXERCISES

1. Measure the circumference of a basketball by wrapping a string around the ball. Lay the string out flat and divide it into three equal pieces with a small piece left over as shown. Explain what the length of one of these pieces represents. (*Hint:* Use $C = 2\pi r = \pi d$ where d is the diameter.)

2. Sketch each figure in this section. Turn your paper upside down and draw them again. Which figures appear different when upside down? Which ones are the same either way?

OBJECTIVES

1 Learn the formulas for area of basic figures.

2 Find area of composite figures.

1 Formulas for Area

Area is a measure of the space enclosed by a geometric figure. The measure of area is the number of small squares measuring 1 unit on each side that will fit into the geometric figure. Each square represents **one square unit**. For example, the area of a storeroom floor measuring 4 feet by 3 feet is 12 square feet. In other words, it would take 12 tiles measuring 1 foot on each side to tile the floor of the storeroom.

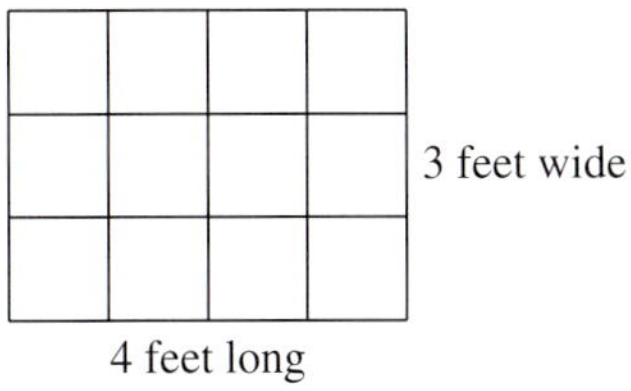

Area is 12 square feet.

The **area of a rectangle** can be found by multiplying the length times the width. $\boldsymbol{A = L \cdot W}$ is the formula for the area of a rectangle. For the storeroom above:

$A = L \cdot W$		formula
$A = (4 \text{ ft})(3 \text{ ft})$		Substitute values.
$A = 12 \text{ sq ft}$ or	12 ft^2	Result.

Multiplying feet times feet always gives square feet and may be written as sq ft or as ft^2.

HELPFUL HINT

When finding area, the units in the answer will always be square units.

A special case of a rectangle is one in which the length and width are equal to one another. The figure described is a square. The formula for the **area of a square** with side s is $A = s \cdot s$, or $A = s^2$.

The diagonal of a rectangle divides the rectangle into two equal parts. The area of each triangle formed is one-half of the area of the entire rectangle.

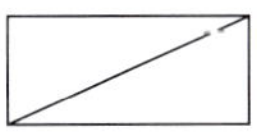

area of each triangle $= \frac{1}{2}$ of area of rectangle

$A = \frac{1}{2} L \cdot W$

The **height of a triangle** is the length of the perpendicular line from a vertex to the opposite side, the **base**. In the rectangle above, the height of each triangle is the same as the width of the rectangle. The base of the triangle is the length of the rectangle. Using the words "base" and "height" instead of "length" and "width" gives the following formula for the **area of a triangle**.

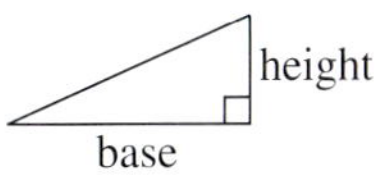

$$A = \frac{1}{2} \text{base} \cdot \text{height}$$
$$= \frac{1}{2} bh$$

1. Find the area of the triangle with height 5 cm and base 8 cm.

2. Find the area of a parallelogram with a height of 20 mm and a base of 25 mm.

This formula works for all types of triangles as seen in Example 1b below.

EXAMPLE 1

Find the area of the given triangles:

(a)

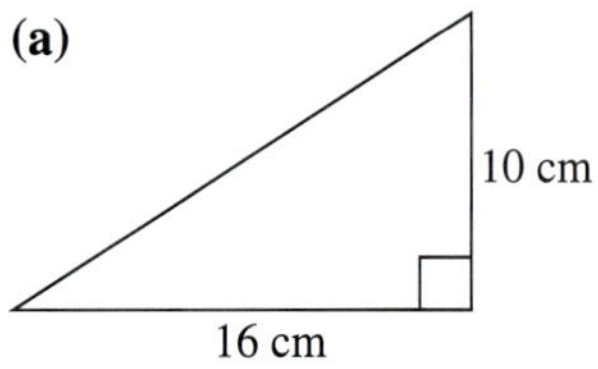

(b)

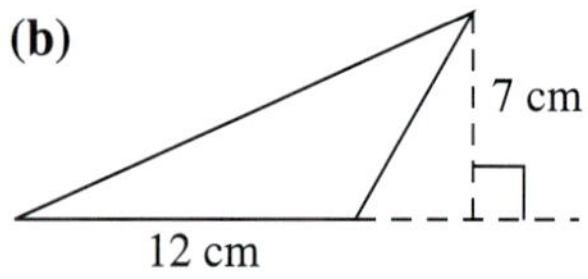

SOLUTION

(a) $A = \frac{1}{2}bh$

$A = \frac{1}{2}(16 \text{ cm})(10 \text{ cm})$

$A = (8 \text{ cm})(10 \text{ cm})$

$A = 80 \text{ cm}^2$

(b) $A = \frac{1}{2}bh$

$A = \frac{1}{2}(12 \text{ cm})(7 \text{ cm})$

$A = (6 \text{ cm})(7 \text{ cm})$

$A = 42 \text{ cm}^2$ ■

◇ ***Do Practice Exercise 1.***

The diagonal of a parallelogram also divides it into two identical triangles. So the **area of a parallelogram** is twice the area of one triangle.

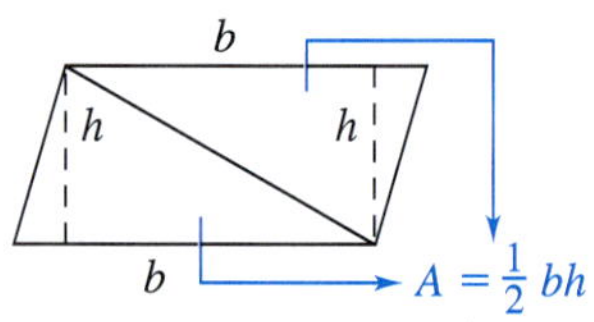

$A \text{ parallelogram} = 2(\frac{1}{2}bh)$

$A \text{ parallelogram} = bh$

EXAMPLE 2

Find the area of the parallelogram shown.

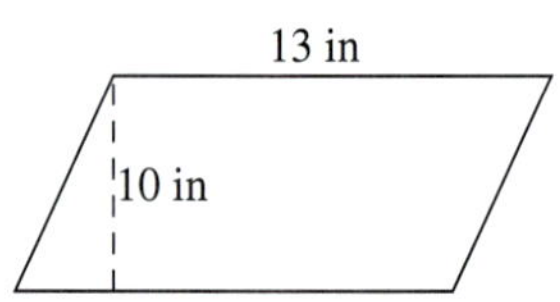

SOLUTION

$A = bh$	formula
$A = (13 \text{ in})(10 \text{ in})$	Substitute values.
$A = 130 \text{ in}^2$	Result. ■

◇ ***Do Practice Exercise 2.***

ANSWERS
1. 20 cm^2
2. 500 mm^2

Solve the following problems. Use $\pi \approx 3.14$.

39. A contractor put an arched window in a new home. The rectangular area is 4 ft by 5.5 ft. Find the area enclosed by the window.

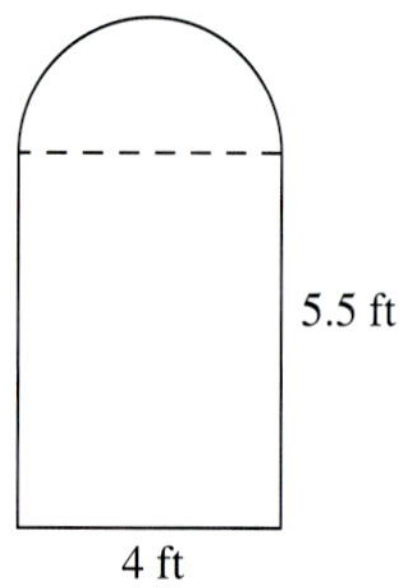

40. The kitchen area of a house is shown. A counter measuring 12 feet wide by $2\frac{1}{2}$ feet deep is marked with x's.

Tile is to be placed everywhere in the kitchen except under the counter. How many square feet of tile is needed?

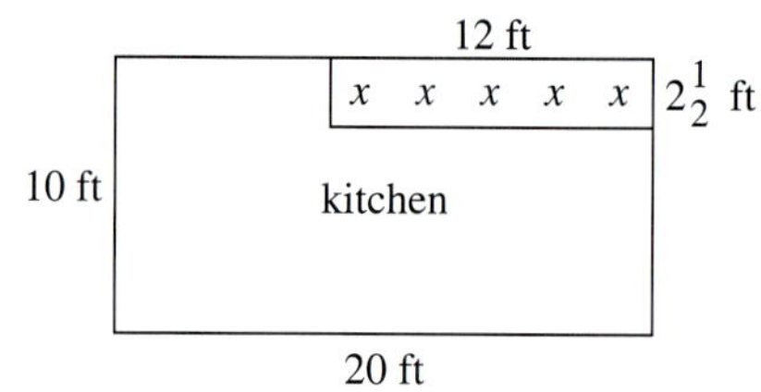

41. The prices of two homes are \$60,000 and \$75,000. Using the floor plans shown, find the price per square foot of each home, rounded to the nearest cent. Which house costs less per square foot?

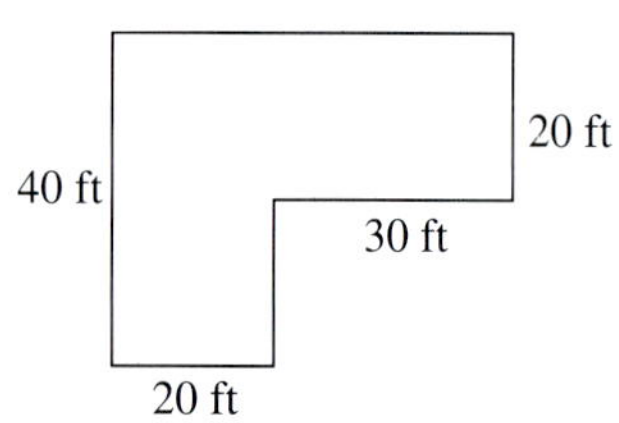

\$60,000

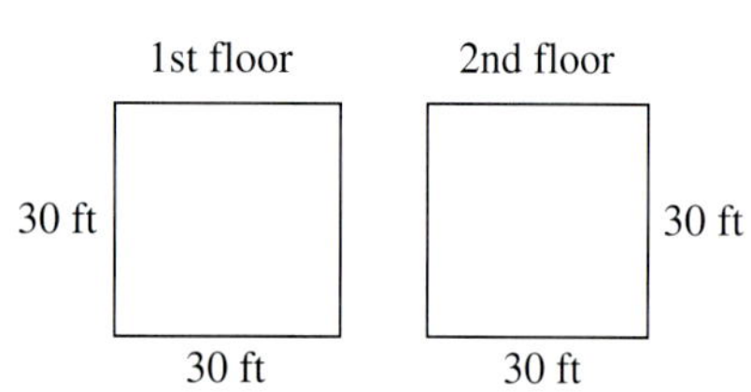

\$75,000

42. An unfolded sheet of newspaper is a rectangle of size 2 ft by 2.25 ft. If a roll of paper contains 9,000 square feet, how many sheets can be printed from one roll?

The diagonal of a trapezoid divides it into two triangles that are not identical. Therefore, the formula for the **area of a trapezoid** must include both of the bases of the trapezoid. To find the area, multiply $\frac{1}{2}$ times the height times the sum of the two parallel sides.

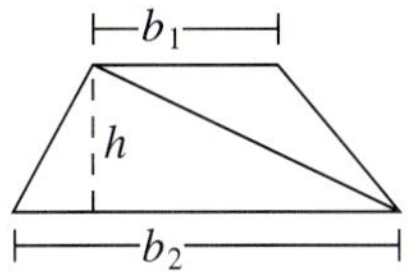

$$A = \frac{1}{2}h(b_1 + b_2)$$

Like the formula for the circumference of a circle, the formula for the **area of a circle** involves the number pi, π. The area of a circle is the product of π and the square of the radius. The square of the radius is the product r times r. Thus the formula is:

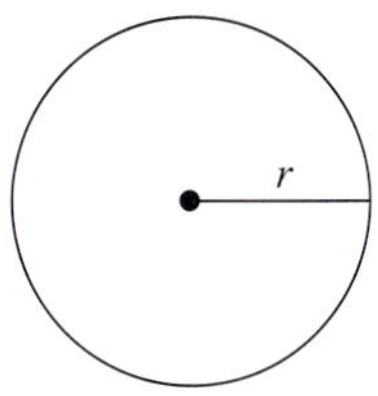

$$A = \pi r^2$$

EXAMPLE 3

Find the area of a circle with a radius of 6 cm. Use $\pi = 3.14$.

SOLUTION

$A = \pi r^2$	formula
$A = (3.14)(6 \text{ cm})^2$	Substitute values.
$A = (3.14)(36 \text{ cm}^2)$	Evaluate the exponent.
$A = 113.04 \text{ cm}^2$	Result. ■

◇ *Do Practice Exercise 3.*

3. Find the area of a circle with radius 5 ft. Use $\pi = 3.14$.

ANSWER

3. 78.5 sq ft

A summary of the formulas for area are given in the following table.

Formulas for Area

Rectangle	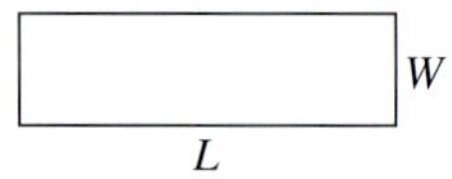	$A = L \cdot W$
Square	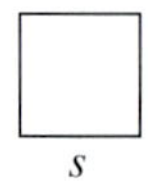	$A = s^2$
Triangle	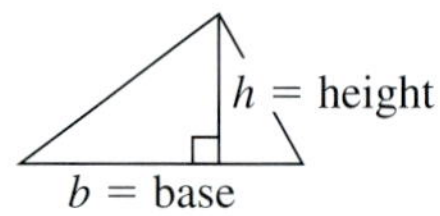	$A = \frac{1}{2}b \cdot h$
Parallelogram	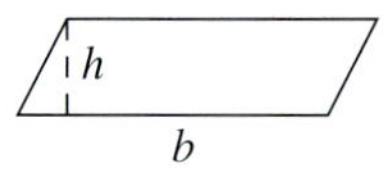	$A = bh$
Trapezoid	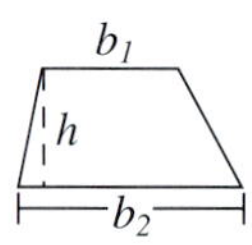	$A = \frac{1}{2}h(b_1 + b_2)$
Circle	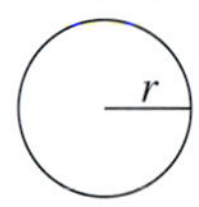	$A = \pi r^2$

For practice in using the formulas, work the following example.

EXAMPLE 4

Find the figure with the larger area.

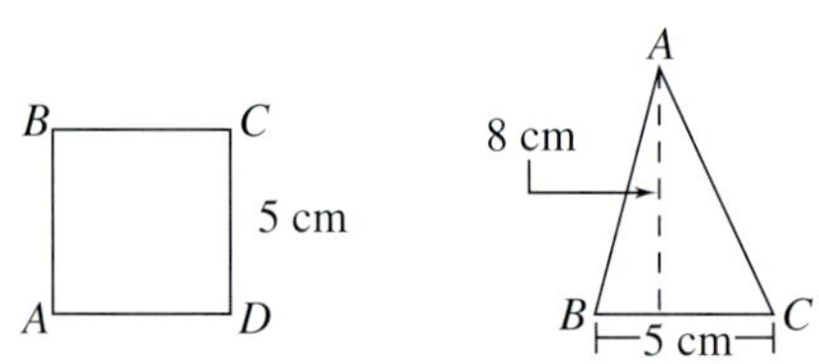

Square *ABCD* Triangle *ABC*

SOLUTION

Area of a square $= s^2 = 5 \cdot 5 = 25 \text{ cm}^2$

Area of a triangle $= \left(\frac{1}{2}\right)bh = \left(\frac{1}{2}\right)(5)(8) = (4)(5) = 20 \text{ cm}^2$

The square has the larger area. ■

The following example is an application involving both perimeter and area.

EXAMPLE 5

A farmer wishes to enclose a rectangular pasture 1,000 meters by 200 meters with a fence and sod the area with a special type of grass. How much fence and how much sod will be needed?

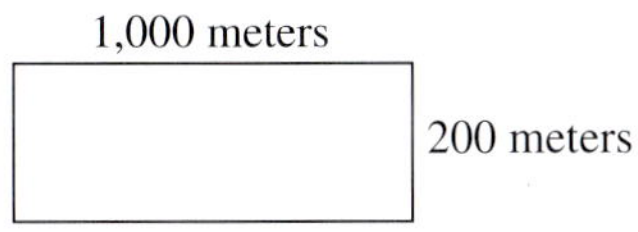

SOLUTION

Find the perimeter to determine the amount of fence needed and find the area to determine the amount of sod needed.

Perimeter	$= 2L + 2W$	formula
	$= 2(1{,}000\text{ m}) + 2(200\text{ m})$	Substitute.
	$= 2{,}000\text{ m} + 400\text{ m}$	Multiply.
	$= 2{,}400$ meters	Result.
Area	$= L \cdot W$	formula
	$= (1{,}000\text{ m})(200\text{ m})$	Substitute.
	$= 200{,}000$ square meters	Result.

The farmer needs 2,400 meters of fencing material and 200,000 square meters of sod. ■

◇ *Do Practice Exercise 4.*

2 Area of Composite Figures

The area of composite figures can be found by breaking the figure into individual shapes, then finding the area of each shape and adding or subtracting as needed.

EXAMPLE 6

Find the shaded area.

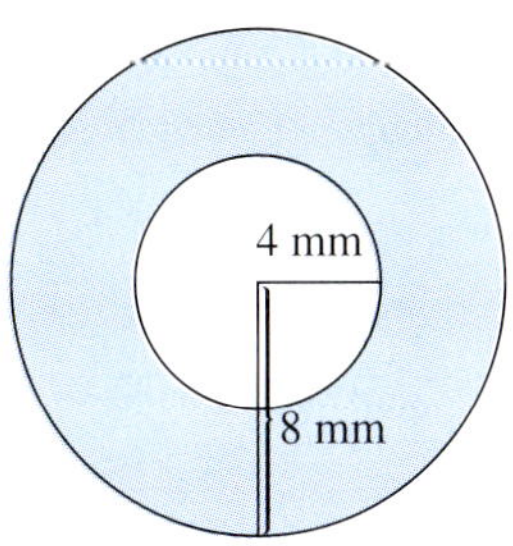

4. A day-care center is installing new carpet in a room measuring 20 feet by 18 feet. How many square feet of carpet is needed? How many feet of molding for the edges is needed?

ANSWER

4. 360 square feet of carpet and 76 feet of molding.

5. Find the area of the metal plate shown.

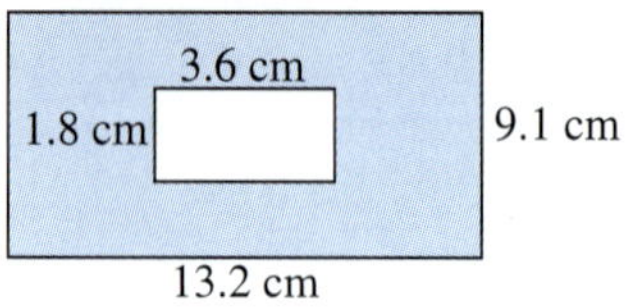

6. Calculate the area.

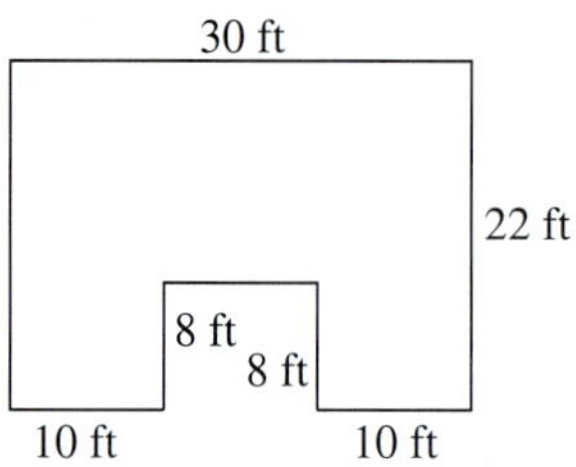

ANSWERS
5. 113.64 cm^2
6. 580 square feet

SOLUTION

The area shaded equals the area of the large circle minus the area of the small circle.

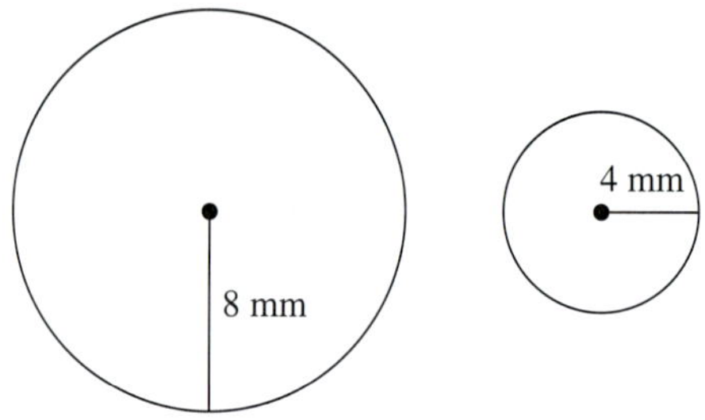

$A = \pi r^2$ large circle $\qquad$ $A = \pi r^2$ small circle

$$
\begin{aligned}
A_{\text{shaded}} &= A_{\text{large circle}} - A_{\text{small circle}} \\
&= \pi r^2_{\text{large circle}} - \pi r^2_{\text{small circle}} && \text{formula} \\
&= (3.14)(8 \text{ mm})^2 - (3.14)(4 \text{ mm})^2 && \text{Substitute.} \\
&= (3.14)(64 \text{ mm}^2) - (3.14)(16 \text{ mm}^2) && \text{Evaluate exponents.} \\
&= 200.96 \text{ mm}^2 - 50.24 \text{ mm}^2 && \text{Multiply.} \\
&= 150.72 \text{ square millimeters} && \text{Result.}
\end{aligned}
$$

◇ **Do Practice Exercise 5.**

EXAMPLE 7

Sean plans to sell his home. He is told by a realtor that a reasonable selling price is \$40 per square foot. Find the square footage of Sean's house and his approximate selling price.

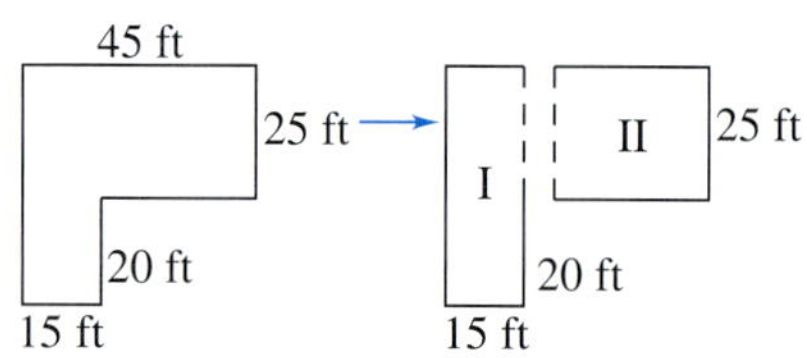

SOLUTION

Split the figure into two rectangles as shown.

$$
\begin{aligned}
\text{Area} &= \text{Area}_{\text{rec I}} + \text{Area}_{\text{rec II}} \\
&= L \cdot W + L \cdot W && \text{formula} \\
&= (25 \text{ ft} + 20 \text{ ft})(15 \text{ ft}) + (45 \text{ ft} - 15 \text{ ft})(25 \text{ ft}) && \text{Substitute.} \\
&= (45 \text{ ft})(15 \text{ ft}) + (30 \text{ ft})(25 \text{ ft}) && \text{Simplify.} \\
&= 675 \text{ sq ft} + 750 \text{ sq ft} && \text{Multiply.} \\
&= 1{,}425 \text{ square feet} && \text{size of house}
\end{aligned}
$$

$$
\begin{aligned}
\text{Selling price} &= \$40 \cdot 1{,}425 \text{ sq ft} \\
&= \$57{,}000 && \text{price of house}
\end{aligned}
$$

◇ **Do Practice Exercise 6.**

name date hour

Find the area of each. Use π = 3.14.

1.

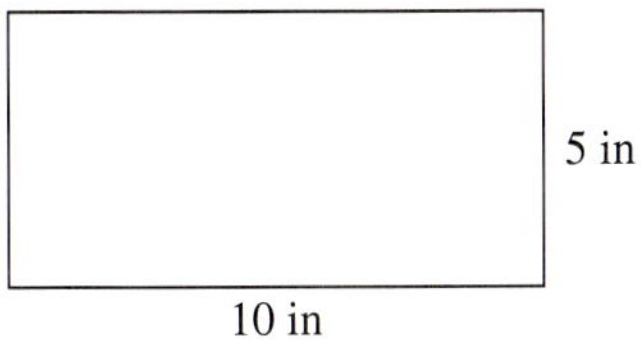

2.

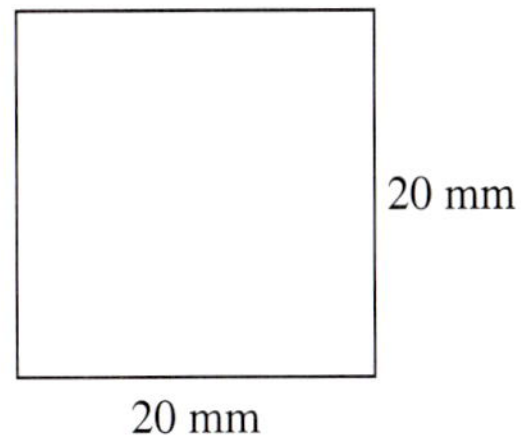

3.

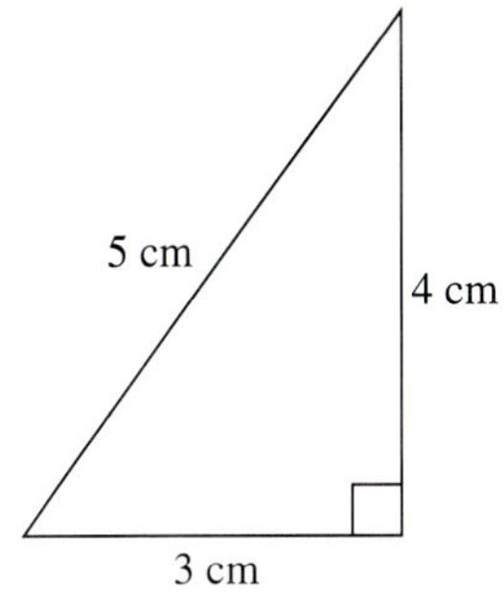

4.

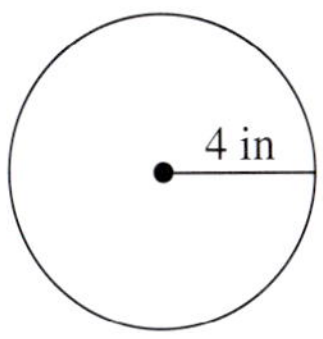

5.

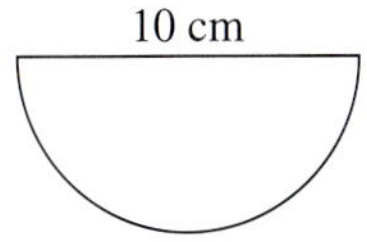

6.

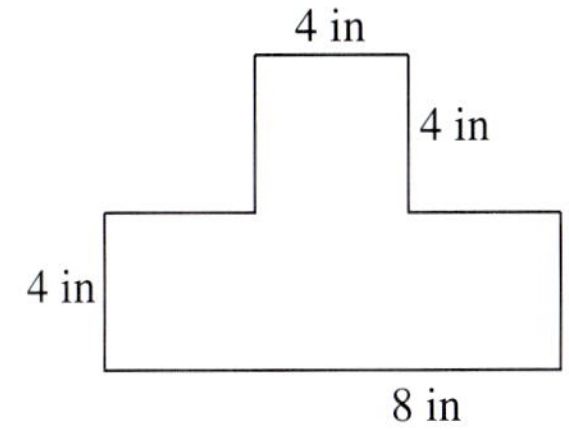

7.

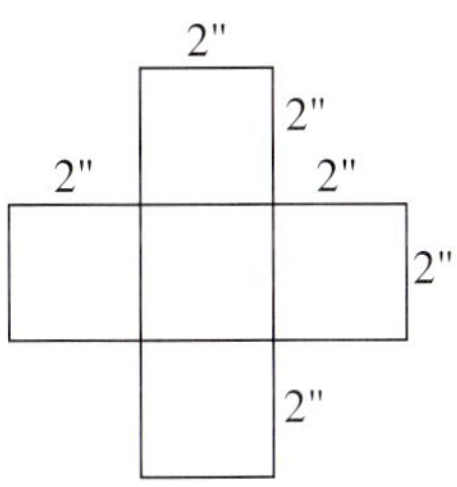

8.

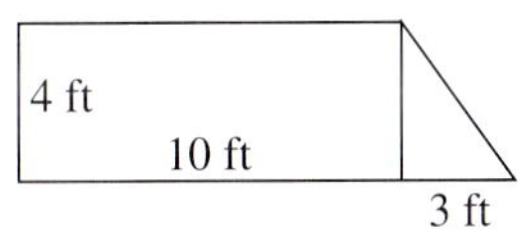

9.

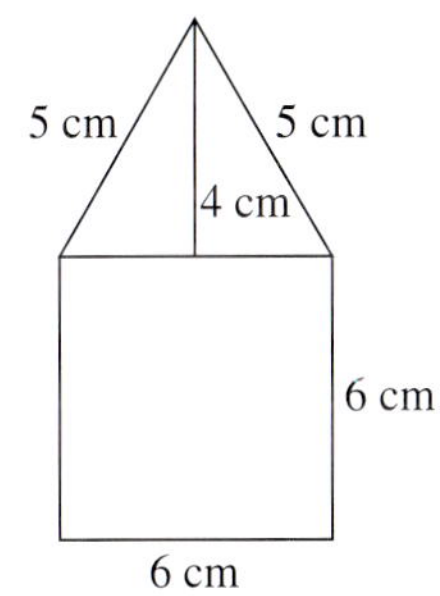

10.

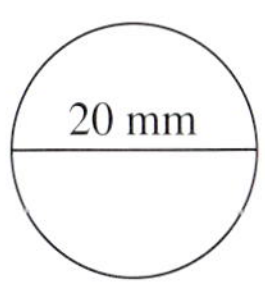

11.

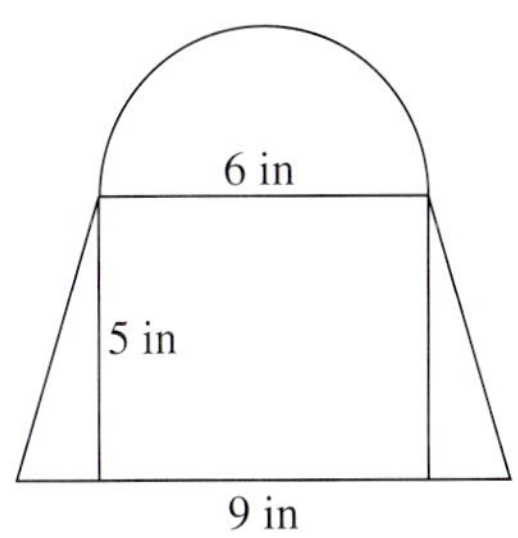

12.

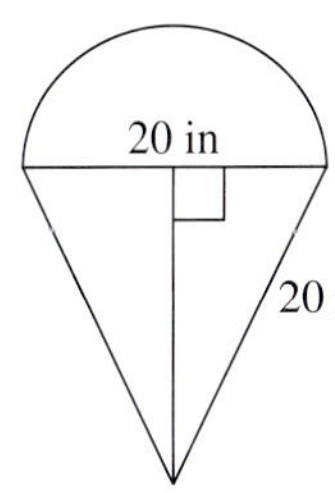

43. Melissa has designed two different plans for a deck. Find the area of each deck. She buys 60 two-by-fours, each 8 feet long and 4 inches wide. How many square feet will the boards cover? For which plan does she have enough lumber?

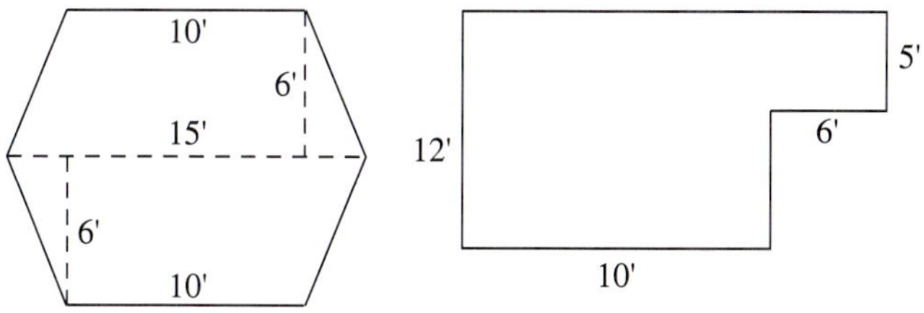

44. Find the area of the deck built around this house rounded to the nearest square foot.

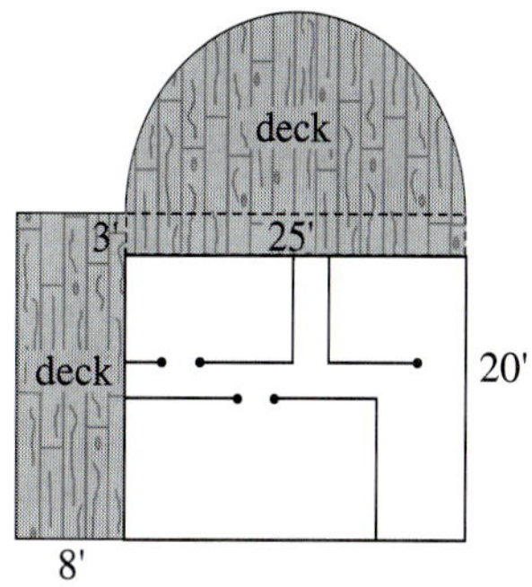

45. A dart game is made up of a round target placed on top of a square backboard:

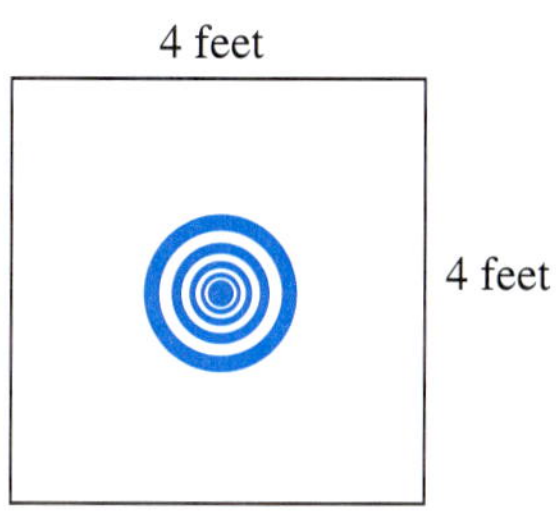

circle has radius of 1 foot

Find the chances of hitting the circular target if the probability of hitting the circle $= \frac{\text{area of circle}}{\text{area of square}}$. (Round to the nearest hundredth.) Then express the answer as a percent.

46. A manager plans to place a small restaurant in a local shopping mall. She needs 1,200 square feet. If the space is shaped like a rectangle with a front of 30 feet, what will be its length?

47. A city plans to build a park on one of two different tracts of land. The first is a rectangular piece measuring 220 feet by 175 feet. The second is a square block with a length of 195 feet. Which piece of land would provide the largest area for the park?

JOURNAL EXERCISES

1. Find the approximate area and perimeter of your home or apartment.

2. Explain the differences between perimeter and area. Draw the following four rectangles: 1 x 7, 2 x 6, 3 x 5, 4 x 4. Find the perimeter and area of each. Explain how the shape of a figure affects its perimeter and area.

3. How would you find the surface area of a three dimensional figure shaped like a drink can? What three two-dimensional (flat) surfaces would you have to add together? (*Hint:* Take the can apart in a way so that it can be laid flat on a table.)

1 Volume

Each two-dimensional figure has length and width and also has a perimeter and an area. Three-dimensional objects have three measurements, length, width, and height. Volume can be found using these three measurements. **Volume** is the amount of space inside a three-dimensional figure. It can be measured by counting the number of cubes measuring one unit along each side that will fit into the object.

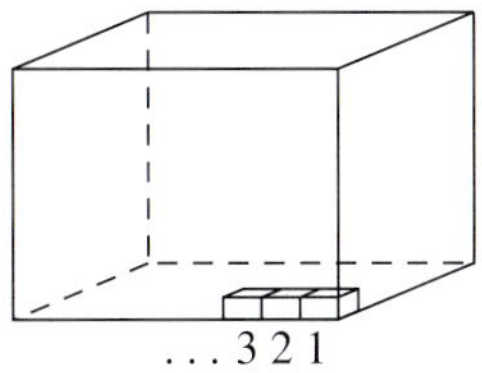

Volume = Number of cubes measuring 1 unit on each side which will fit into the large geometric figure.

EXAMPLE 1

Find the volume of a book with measurements as shown.

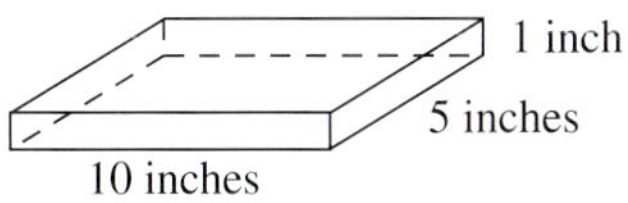

SOLUTION

The volume of this book is the number of cubes measuring 1 inch by 1 inch by 1 inch that can be placed inside the book. The book is 5 inches by 10 inches by 1 inch. The number of cubes in the book is:

$$5 \times 10 \times 1 = 50 \text{ cubes measuring } 1'' \text{ on each side}$$

Thus volume = 50 cubic inches = 50 in^3 ■

HELPFUL HINT

When finding volume, the units in the answer are always cubic units.

The book in the last example is a **rectangular solid**. All six of its surfaces are rectangles. The formula for finding the **volume of a rectangular solid** is as follows.

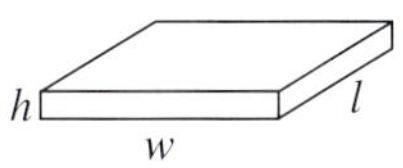

$V = l \cdot h \cdot w$

◇ *Do Practice Exercise 1.*

OBJECTIVES

1 Learn the formulas for volume.
2 Calculate volumes of composite figures.
3 Find the surface area of basic figures (Optional).

1. Jim has 1,200 cubic feet of boxes to move. He plans to rent a moving truck that measures 22 feet by a 8 feet by 9 feet. What is the volume of the truck? Will Jim have enough space?

ANSWER

1. volume = 1,584 cubic feet so he will have enough space

2. Find the volume of a rectangular pyramid with a base measuring 12 m × 15 m and a height of 9 m.

A **rectangular pyramid** is a three-dimensional figure whose base is a rectangle. Its sides are triangles that share a common vertex at the top of the figure.

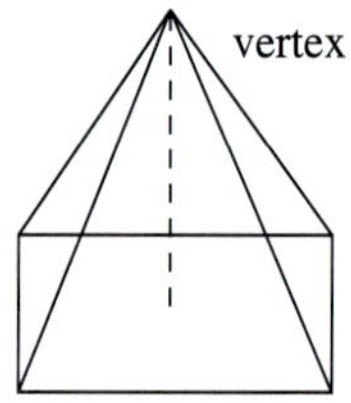

$$V = \frac{1}{3} lwh$$

The formula for the volume of a rectangular pyramid is closely related to that of a rectangular solid. The **volume of a rectangular pyramid** is exactly one third the volume of a rectangular solid having the same base and height. The height in the formula is the height from the common vertex at the top of the pyramid straight down, perpendicular to the rectangle at the base of the pyramid.

EXAMPLE 2

Find the volume of the rectangular pyramid shown.

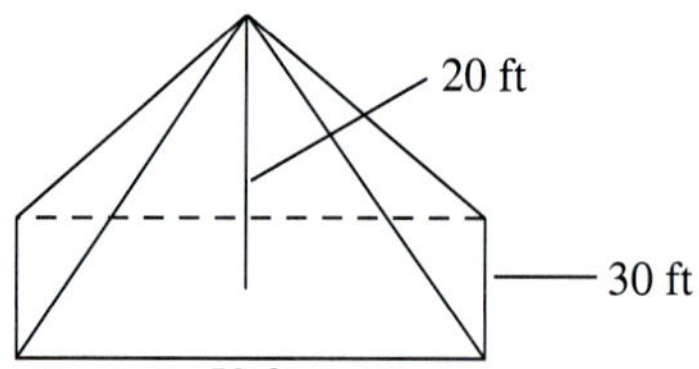

SOLUTION

$$V = \frac{1}{3} l \cdot w \cdot h \qquad \text{formula}$$

$$V = \frac{1}{3}(50 \text{ ft} \cdot 30 \text{ ft} \cdot 20 \text{ ft}) \qquad \text{Substitute.}$$

$$V = \frac{1}{3}(30{,}000 \text{ ft}^3) \qquad \text{Multiply.}$$

$$V = 10{,}000 \text{ ft}^3 \qquad \text{Result.}$$

■

◇ *Do Practice Exercise 2.*

ANSWER
2. 540 m^3

A figure the shape of a can of vegetables or a drink can is called a **right circular cylinder**. The top and bottom bases are both circles and the sides form right angles with the bases. The formula for the **volume of a right circular cylinder** is the area of the circle on the bottom ($A = \pi r^2$) times the height (h).

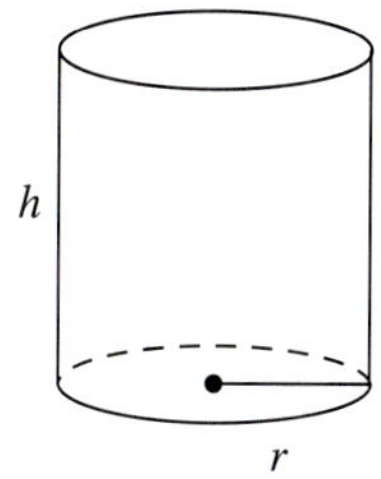

$$V = \pi r^2 h$$

The formula for the **volume of a right circular cone** is closely related to the volume of the right circular cylinder. The volume of the right circular cone is exactly one third the volume of the right circular cylinder with the same height.

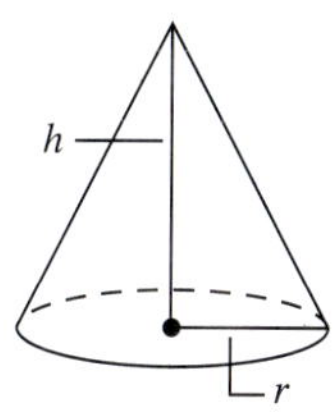

$$V = \frac{1}{3}\pi r^2 h$$

The geometrical figure that is the shape of a basketball is called a **sphere**. Its volume involves the radius cubed, that is the product $r \cdot r \cdot r$.

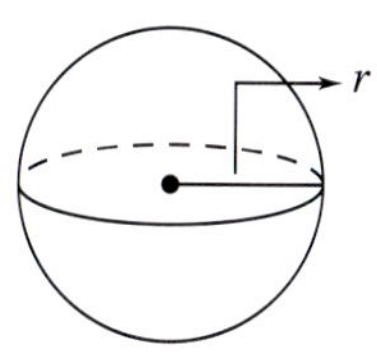

$$V = \frac{4}{3}\pi r^3$$

EXAMPLE 3

Which of the two geometrical shapes will hold more fluid?

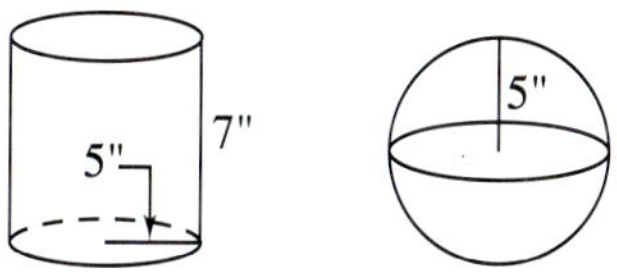

right circular cylinder sphere

3. Which of the two figures has the larger volume, the right circular cylinder or the rectangular solid?

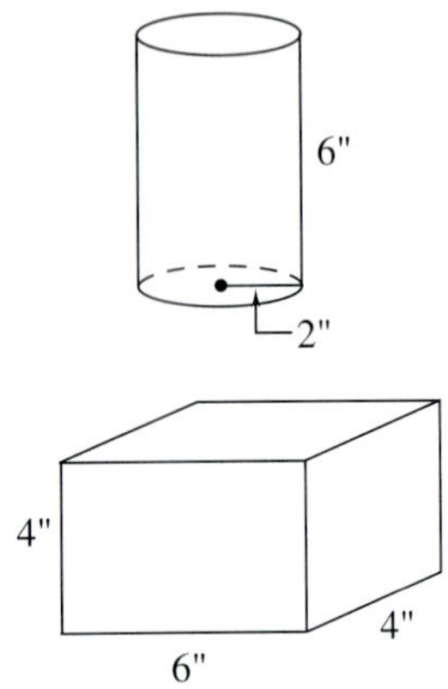

SOLUTION

The shape with the larger volume will hold more fluid. The first step is to calculate both volumes, then compare them.

Volume of cylinder	$= \pi r^2 h$	Volume of sphere	$= \dfrac{4}{3}\pi r^3$
	$= (3.14)(5\text{ in})^2(7\text{ in})$		$= \dfrac{4}{3}(3.14)(5\text{ in})^3$
	$= (3.14)(25\text{ in}^2)(7\text{ in})$		$= \dfrac{4}{3}(3.14)(125\text{ in}^3)$
	$= 549.5\text{ in}^3$		$= \dfrac{1570\text{ in}^3}{3}$
			$= 523.3\text{ in}^3$

Thus the right circular cylinder will hold more fluid. ■

◇ *Do Practice Exercise 3.*

A summary of the formulas for the volume of three-dimensional figures is given below.

Formulas for Volume

Rectangular Solid	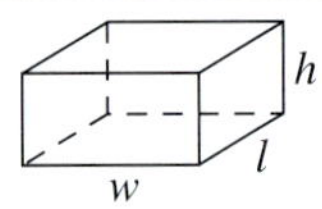	$V = lwh$
Rectangular Pyramid	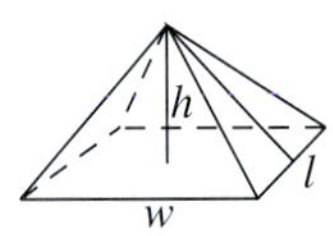	$V = \dfrac{1}{3}lwh$
Right Circular Cylinder	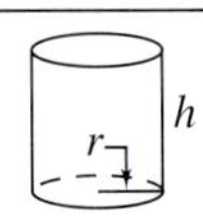	$V = \pi r^2 h$
Right Circular Cone	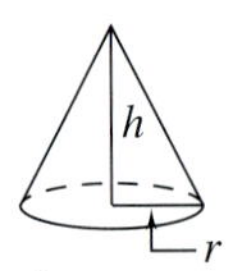	$V = \dfrac{1}{3}\pi r^2 h$
Sphere	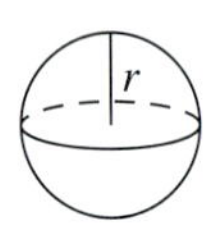	$V = \dfrac{4}{3}\pi r^3$

ANSWER

3. volume of cylinder = 75.36 cu in
volume of rectangular solid = 96 cu in
volume cylinder < volume rectangular solid

2 Volume of Composite Three-Dimensional Figures

Three-dimensional composite figures may be separated into two or more basic solids in order to find their volume. The tent-shaped figure in the next example can be separated into a rectangular solid and a rectangular pyramid.

Both the formula for the volume of the rectangular solid and the formula for the volume of the rectangular pyramid contain height:

$$V_{\text{rec.solid}} = lwh \qquad V_{\text{pyramid}} = \left(\frac{1}{3}\right) lwh$$

However, the two heights are different. The h in the volume formula for the rectangular solid is the height of the "box" (5 feet). The h in the volume formula for the rectangular pyramid is the height of the pyramid (3 feet).

EXAMPLE 4

A tent has the shape of a rectangular pyramid mounted on top of a rectangular solid, as shown:

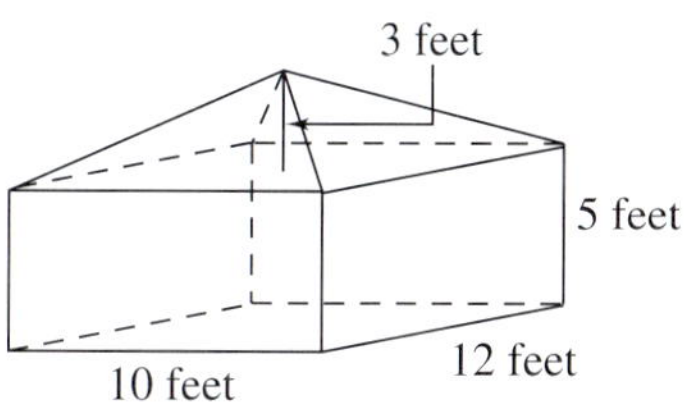

What is the volume of the tent?

SOLUTION

The total volume is the sum of the volume of the rectangular solid and the volume of the rectangular pyramid.

$$\text{Vol}_{\text{total}} = \text{Vol}_{\text{rec.sol.}} + \text{Vol}_{\text{rec.pyr.}}$$

$$\text{Vol}_{\text{total}} = lwh + \left(\frac{1}{3}\right) lwh \quad \text{formulas}$$

$$\text{Vol}_{\text{total}} = (10 \text{ ft})(12 \text{ ft})(5 \text{ ft}) + \left(\frac{1}{3}\right)(10 \text{ ft})(12 \text{ ft})(3 \text{ ft}) \quad \text{Substitute.}$$

$$\text{Vol}_{\text{total}} = 600 \text{ ft}^3 + 120 \text{ ft}^3 \quad \text{Multiply.}$$

$$\text{Vol}_{\text{total}} = 720 \text{ cubic feet} \quad \text{Result.}$$

◇ *Do Practice Exercise 4.*

4. Find the volume of the following figure:

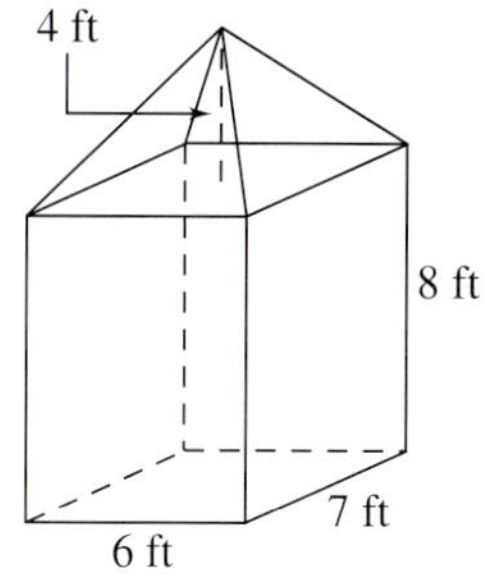

ANSWER
4. 392 ft^3

5. Find the volume of the right circular cylinder topped by half of a sphere.

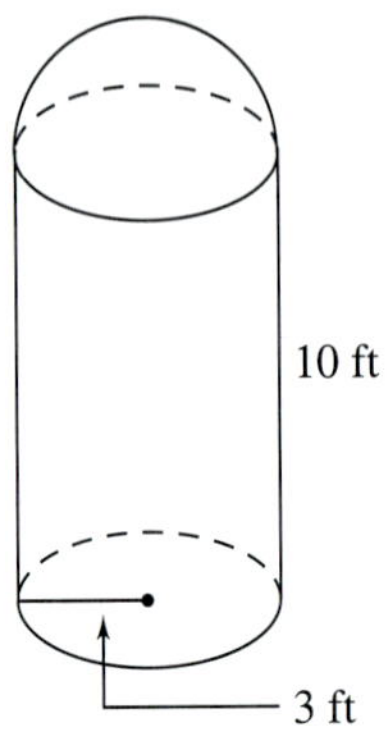

6. Find the surface area of the rectangular solid shown.

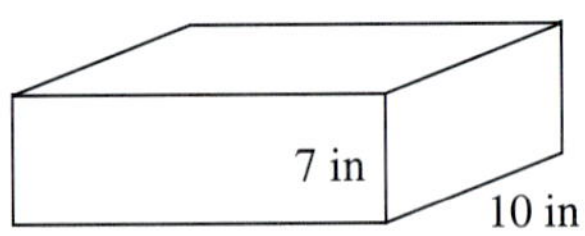

ANSWERS
5. 339.12 cu ft
6. 650 in^2

EXAMPLE 5

Find the volume of the figure.

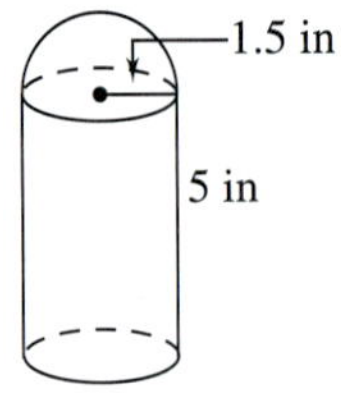

SOLUTION

The figure can be split into a right circular cylinder and half of a sphere.

$$\text{Volume} = \text{Volume}_{\text{cylinder}} + \frac{1}{2}\text{Volume}_{\text{sphere}}$$

$$\text{Volume} = \pi r^2 h + \left(\frac{1}{2}\right)\left(\frac{4}{3}\right)\pi r^3$$

$$\text{Volume} \approx (3.14)(1.5 \text{ in})^2(5 \text{ in}) + \left(\frac{2}{3}\right)(3.14)(1.5 \text{ in})^3$$

$$\text{Volume} \approx (3.14)(11.25 \text{ in}^3) + \frac{(2)(3.14)(3.375 \text{ in}^3)}{3}$$

$$\text{Volume} \approx 35.325 \text{ in}^3 + \frac{21.195 \text{ in}^3}{3}$$

$$\text{Volume} \approx 35.325 \text{ in}^3 + 7.065 \text{ in}^3$$

$$\text{Volume} \approx 42.39 \text{ in}^3$$

◇ *Do Practice Exercise 5.*

3 Surface Area (Optional)

The **surface area** of a three-dimensional figure is the total area of all the surfaces of the figure. The **surface area of a rectangular solid** is the total area of the six rectangles it is made of: the top and bottom, the front and back, and the left and right sides.

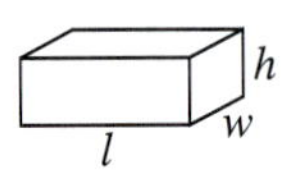

Surface Area = $2lw + 2wh + 2lh$

EXAMPLE 6

Find the surface area of the rectangular solid shown.

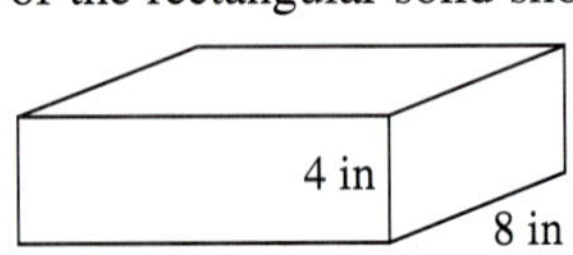

SOLUTION

$$\begin{aligned}\text{Surface Area} &= 2lw + 2wh + 2lh \\ &= 2 \cdot 12 \text{ in} \cdot 8 \text{ in} + 2 \cdot 8 \text{ in} \cdot 4 \text{ in} + 2 \cdot 12 \text{ in} \cdot 4 \text{ in} \\ &= 192 \text{ in}^2 + 64 \text{ in}^2 + 96 \text{ in}^2 \\ &= 352 \text{ sq in}\end{aligned}$$

◇ *Do Practice Exercise 6.*

The formula for the **surface area of a cylinder** can be illustrated by removing the circular top and bottom of the cylinder. The remaining part of the cylinder is cut and opened flat, making a rectangle.

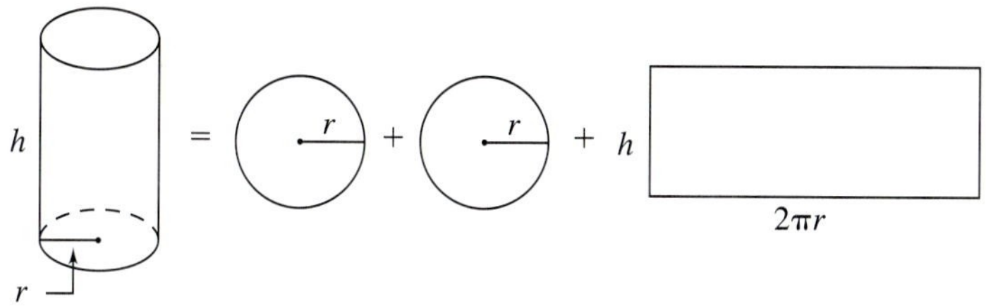

The circumference of the cylinder, which is $2\pi r$, becomes the length of the rectangle. The width of the rectangle is the height of the cylinder. Thus, the surface area of the side of the cylinder is the area of the rectangle, $(2\pi r)(h)$. The area of both the top and bottom of the cylinder is πr^2. Adding the area of the side to the area of the top and bottom gives the following formula:

$$\text{Surface area of a cylinder} = 2\pi rh + 2\pi r^2$$

EXAMPLE 7

Find the surface area of a tank in the shape of a cylinder with radius 5 meters, and height 15 meters.

SOLUTION

$$\begin{aligned}\text{Area} &= 2\pi rh + 2\pi r^2\\ &\approx 2(3.14)(5\text{ m})(15\text{ m}) + 2(3.14)(5\text{ m})^2\\ &\approx 471\text{ m}^2 + 157\text{ m}^2\\ &\approx 628\text{ m}^2\end{aligned}$$

■

◇ *Do Practice Exercise 7.*

The **surface area of a sphere** is given by the formula:

$$A = 4\pi r^2$$

EXAMPLE 8

Find the surface area of a sphere with radius 4 cm.

SOLUTION

$$\begin{aligned}A &= 4\pi r^2\\ &\approx 4(3.14)(4\text{ cm})^2\\ &\approx 200.96\text{ cm}^2\end{aligned}$$

■

◇ *Do Practice Exercise 8.*

7. Find the surface area of a cylinder with height 12 ft and radius 3 ft.

8. Find the surface area of a sphere with radius 10 in.

ANSWERS
7. 282.6 ft^2
8. 1,256 in^2

Find the perimeter (or circumference) and the area of each. Use $\pi \approx 3.14$. (Review)

1.

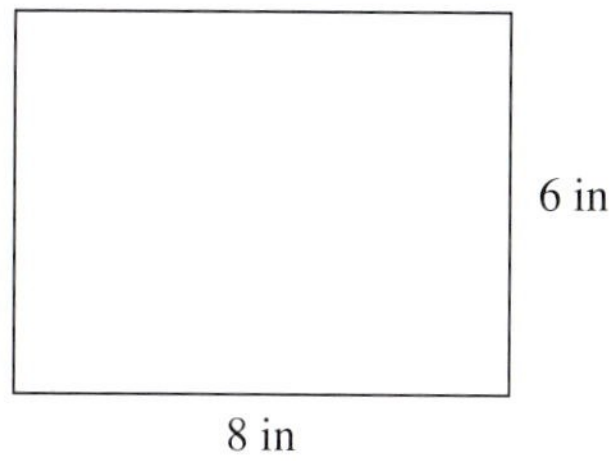

2.

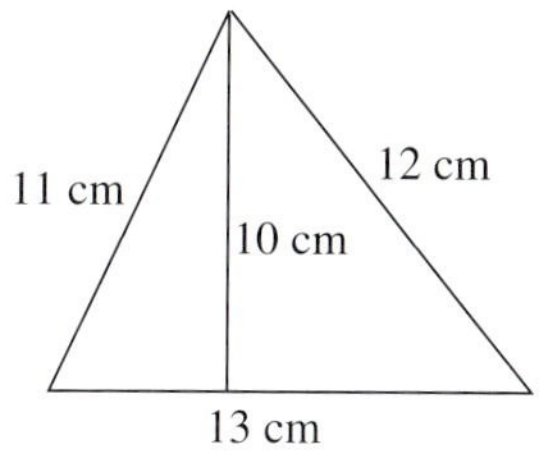

3.

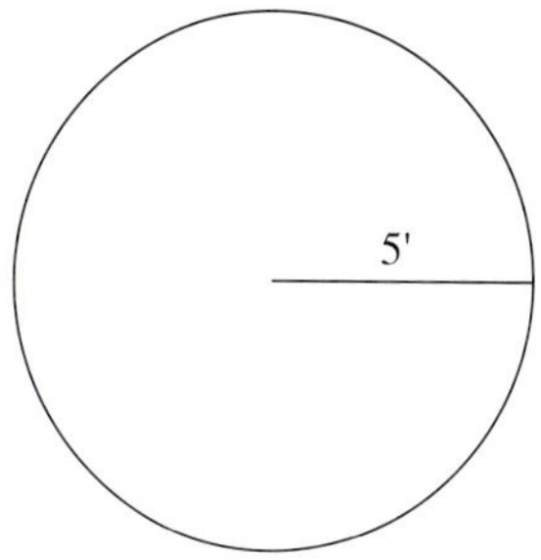

4.

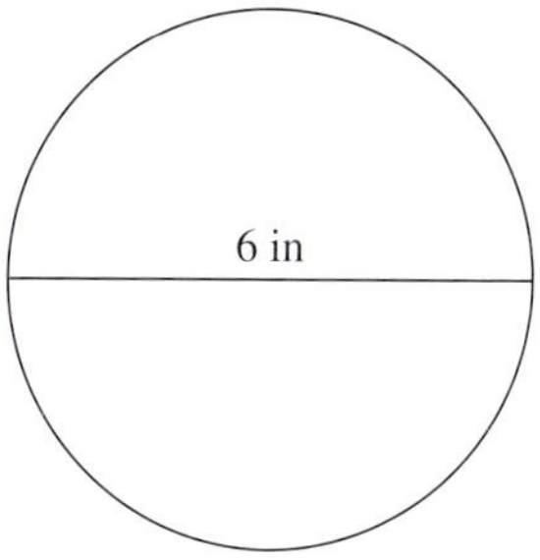

5.

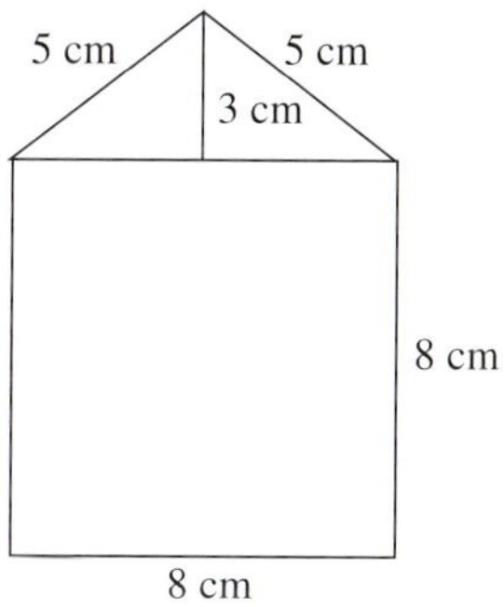

6.

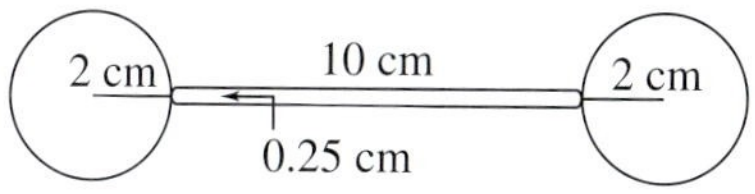

Find the volume of each. Round your answers to the nearest hundredth. Use $\pi \approx 3.14$.

7.

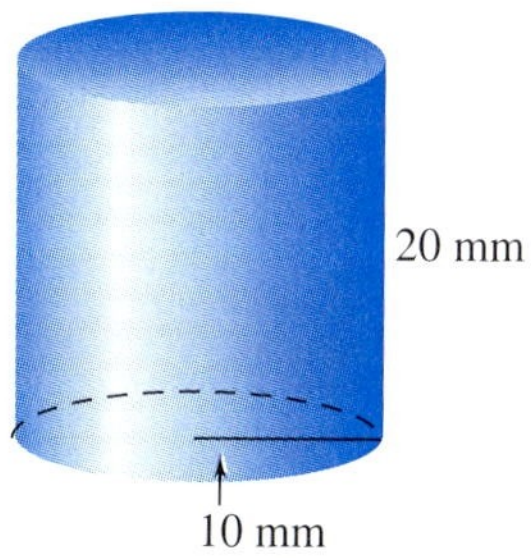

8.

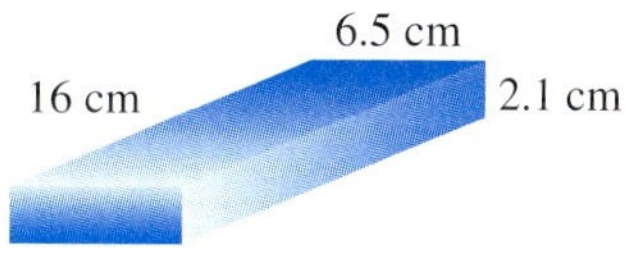

9.

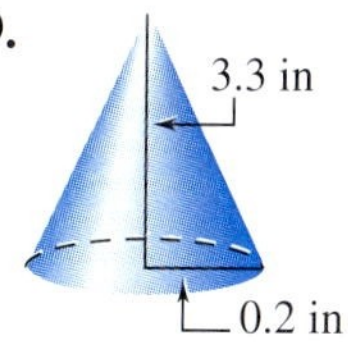

10.

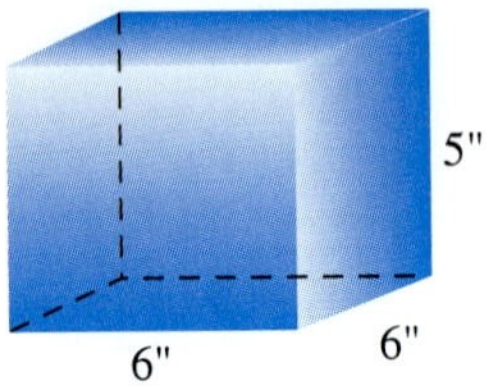

11.

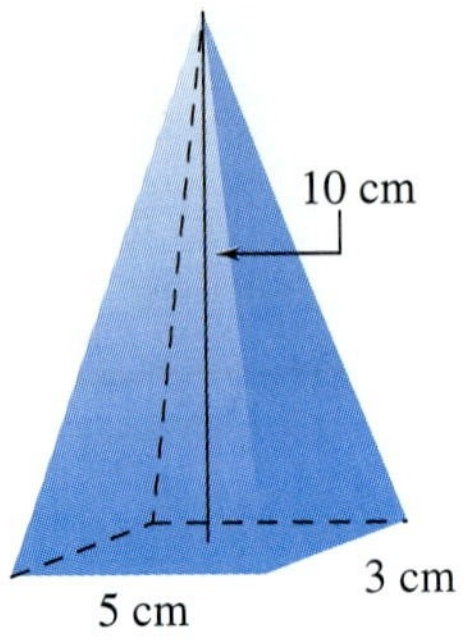

12.

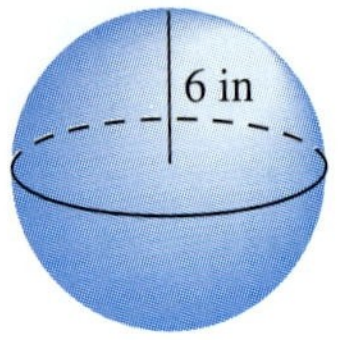

13.

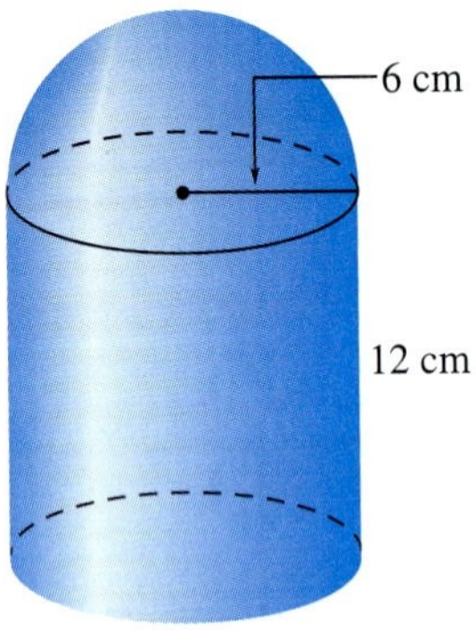

14.

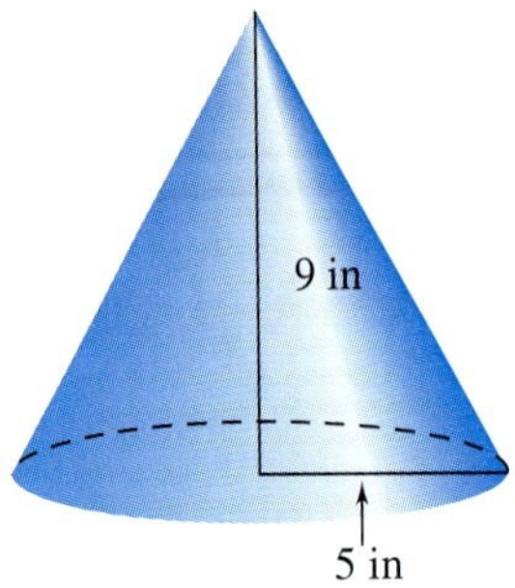

15.

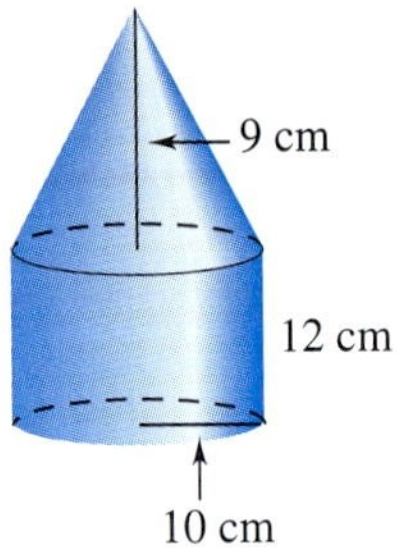

16.

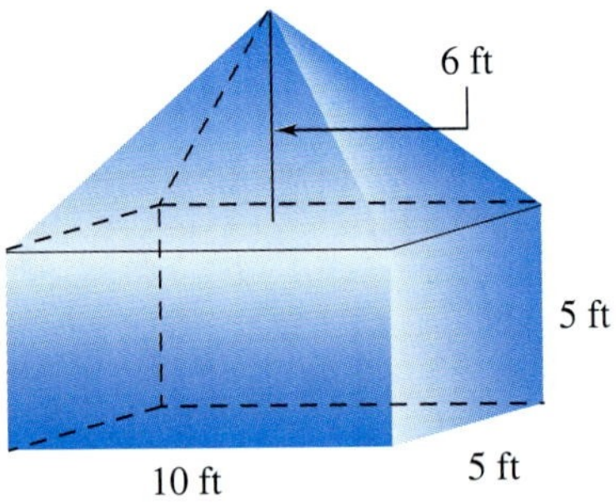

17.

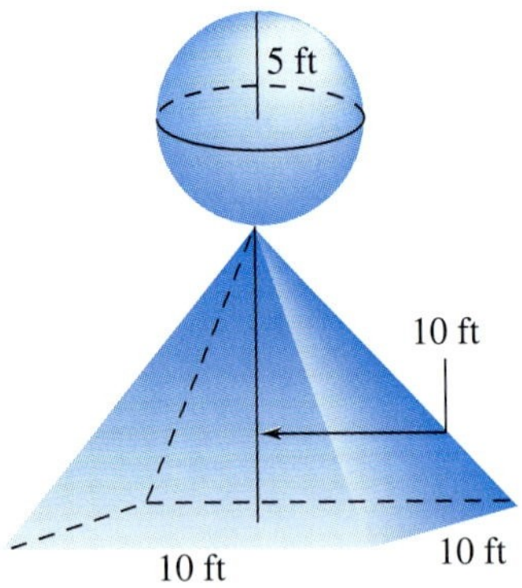

18.

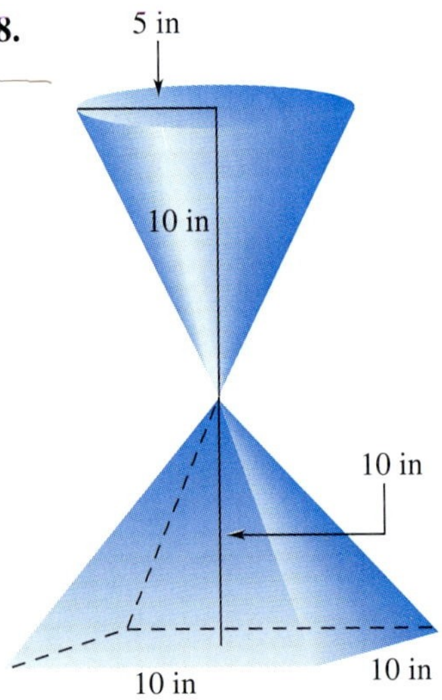

19.

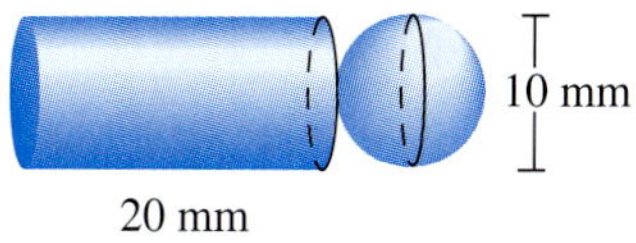

20.

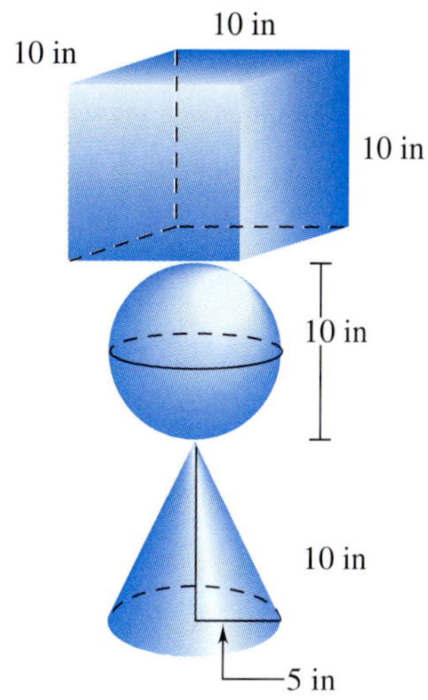

Sketch each and find the volume. Round answers to the nearest hundredth. Use $\pi \approx 3.14$.

21. A sphere of radius 12 in

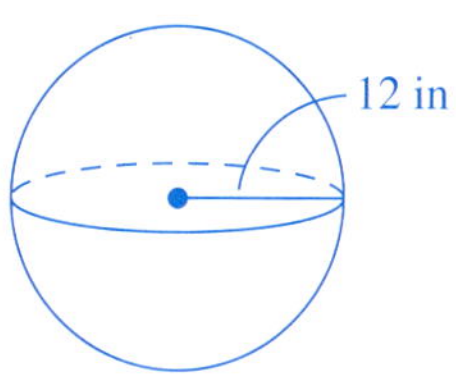

22. A rectangular solid of dimensions 3 in by 7 in by 4 in

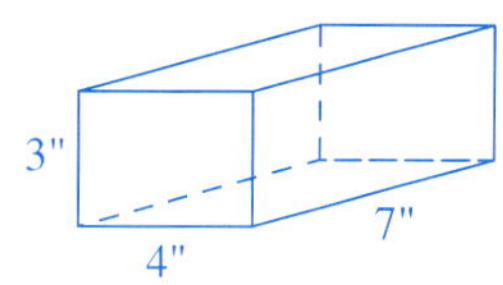

23. A rectangular pyramid with a length of 15 cm, a width of 12 cm, and a height of 9 cm

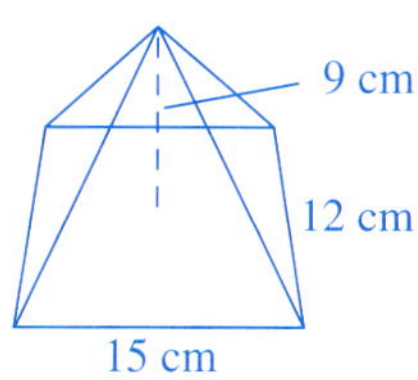

24. A right circular cylinder with a height of 9 in and a radius of 6 in

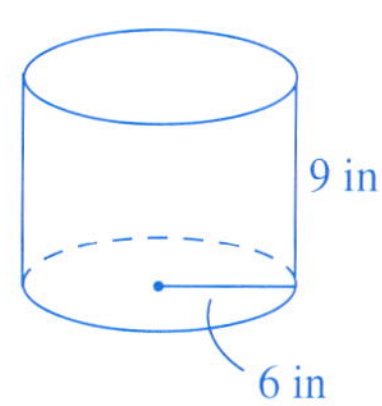

25. A right circular cone with a height of 18 cm and a radius of 8 cm

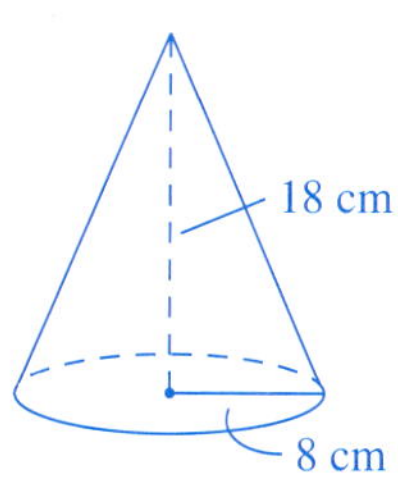

26. A rectangular solid with a length of 3.5 in, a width of 1.75 in, and a height of 2.2 in

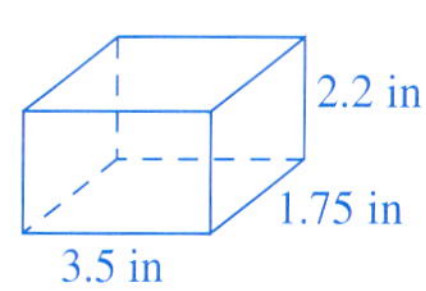

27. A sphere with a radius of 1.4 cm

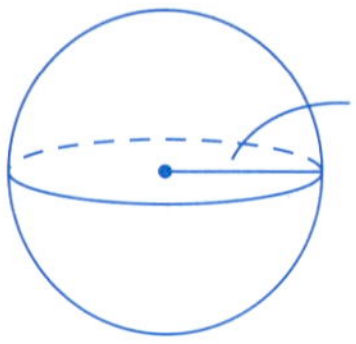

28. A right circular cylinder with a height of 9.25 cm and a radius of 0.8 cm

29. A right circular cone with a height of 5 in and a radius of 2.6 in

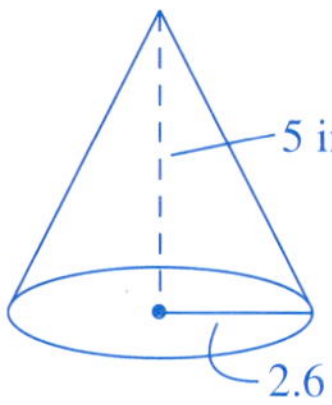

Match the formula to the correct figure.

30. Triangle

31. Sphere

32. Circle

33. Rectangle

34. Right circular cylinder

35. Right circular cone

(a) $A = \pi r^2$

(b) $V = \frac{1}{3}\pi r^2 h$

(c) $P = 2L + 2W$

(d) $V = \pi r^2 h$

(e) $A = \frac{1}{2}bh$

(f) $V = \frac{4}{3}\pi r^3$

Solve the following problems.

36. The earth is shaped like a sphere with a radius of four thousand miles.
   (a) Find the volume of the earth.
   (b) Find the distance from the North Pole to the South Pole along the surface of the earth.
   (c) Find the distance around the equator.

37. Find the volume of the first Wonder of the World, the Khufu pyramid in Egypt. Its base is a square of width 762.8 feet. The height is 485 feet. Of the Seven Wonders of the World, this pyramid is the oldest and only surviving "wonder."

38. The largest pyramid in the world is the pyramid of Quetyalcoatl at Cholula near Mexico City. Its base covers an area of 45 acres. Find the volume of the pyramid if its height is 177 feet and the base is 1,500 feet by 1,500 feet.

39. How much fluid will a rectangular tank hold if the dimensions are 60 cm × 40 cm × 15 cm?

40. A small trailer has a length of 10 feet, a width of 6 feet, and a height of 4 feet. How many cubic feet of material will the trailer hold?

41. Find the volume of cement needed to make the two steps shown.

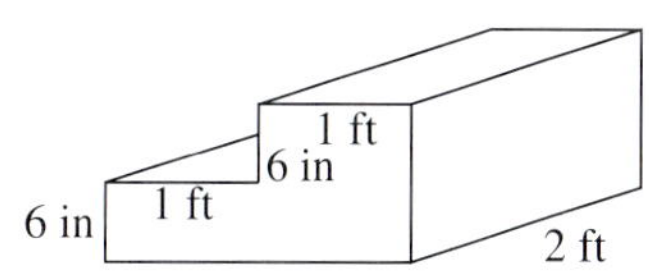

42. Find the volume of the barrel-shaped barbecue grill shown. What volume of charcoal would be needed to fill the barrel $\frac{1}{8}$ full of charcoal? Round your answer to the nearest hundredth.

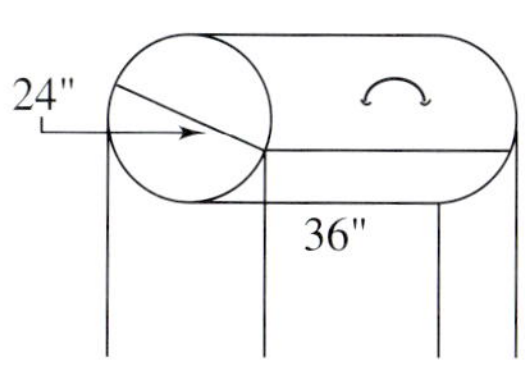

43. How does the volume of a right circular cylinder change if the radius of the base doubles in length from 3 to 6? Does the volume also double?

44. Recycling a stack of newspapers measuring 4 ft by 1.25 ft by 1 ft will save a 40 foot pine tree. A student government sponsored a newspaper drive and collected a stack of newspapers measuring 7 ft $\times$ 5 ft $\times$ 2 ft. How many 40 foot trees did they save?

45. A builder is preparing the front and back yard of a new home. The yard measures 30 ft by 40 ft. If she wants to spread the topsoil 6 in thick, how many cubic feet of topsoil will she need?

46. Find the volume of an ice cave or igloo in the shape of a dome with diameter 10 feet.

Solve the following problems involving surface area. Use $\pi = 3.14$. *(Optional)*

47. How much cardboard is required to make a box with dimensions 12 inches by 16 inches by 2 inches?

48. A fast food restaurant is interested in saving money in its packaging. The restaurant presently uses boxes that are 4 in by 4 in by 3 in for its hamburgers. How much less cardboard would be required to make a box with dimensions 3.8 in by 3.8 in by 2.8 in? If this change saves \$0.003 per box, how much money would be saved on 500,000 boxes?

49. How many square feet of sheet metal will it take to make a cylindrical tank with radius 4 feet and height 9 feet?

50. The material to cover a ball costs \$2.50 per square foot. How much would it cost to cover a ball with radius 1 foot?

JOURNAL EXERCISES

1. Create your own problem using volumes and bring it to class. Exchange your problem with a problem written by another student and solve.

2. Explain the difference between volume and surface area.

3. Explain the differences between a meter, a square meter, and a cubic meter. Sketch each.

1 Similar Figures

Similar figures are figures that have the same shape but are different sizes. For example, a photo and its enlargement are similar figures. Everything in the enlargement is the same shape as in the original, only larger.

Similar geometric figures have identical angles, but the length of the sides are not necessarily the same. The symbol for "is similar to" is $\sim$.

Triangle ABC below is similar to triangle DEF. The three angles of $\triangle ABC$ are the same size as the three angles of $\triangle DEF$, and the lengths of the sides of $\triangle ABC$ are exactly three times the lengths of the sides of $\triangle DEF$.

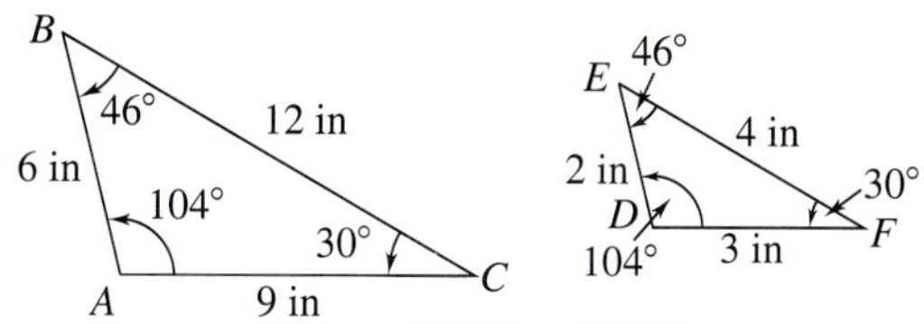

$\triangle ABC \sim \triangle DEF$

In similar triangles, each side of the first triangle corresponds to a specific side of the second triangle. Use the following hint to help identify corresponding sides of similar triangles.

HELPFUL HINT

The sides of similar triangles opposite congruent angles are corresponding sides.

EXAMPLE 1

Name a pair of corresponding sides for the following similar triangles.

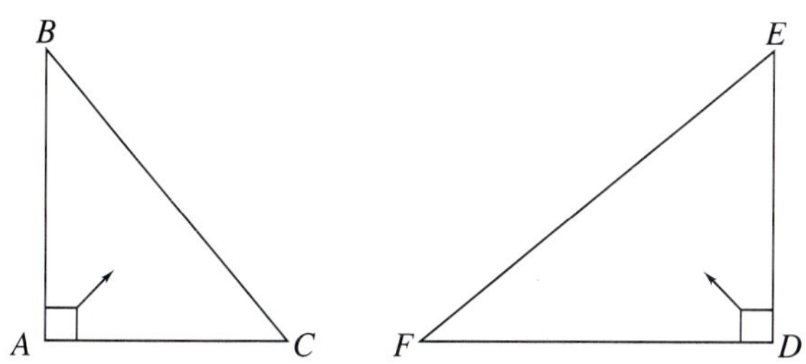

SOLUTION

Sides $\overline{BC}$ and $\overline{EF}$ are opposite congruent right angles A and D. Therefore, $\overline{BC} \leftrightarrow \overline{EF}$. ■

◇ *Do Practice Exercise 1.*

Angles formed directly across from one another when two lines cross are called **vertical angles**. Vertical angles are always congruent. Therefore, the sides of triangles opposite vertical angles are corresponding sides. In Practice Exercise 1, $\angle ABC$ and $\angle FBE$ are vertical angles. So $\overline{AC}$ and $\overline{FE}$ are corresponding sides.

OBJECTIVES

1. Know the basics of similar figures and corresponding parts.
2. Show that two triangles are similar.
3. Find the missing side of similar triangles.
4. Solve application problems using similar triangles.

1. Name a pair of corresponding sides in the two similar triangles shown below.

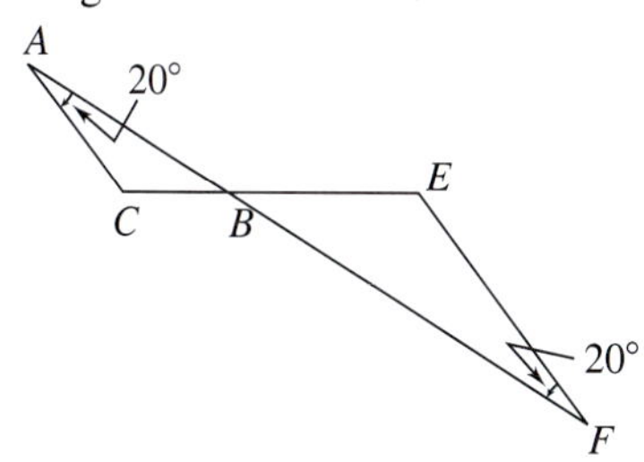

ANSWER

1. $\overline{BC} \leftrightarrow \overline{BE}$

2. $\triangle ABC \sim \triangle DEC$. Name the corresponding sides of the triangles.

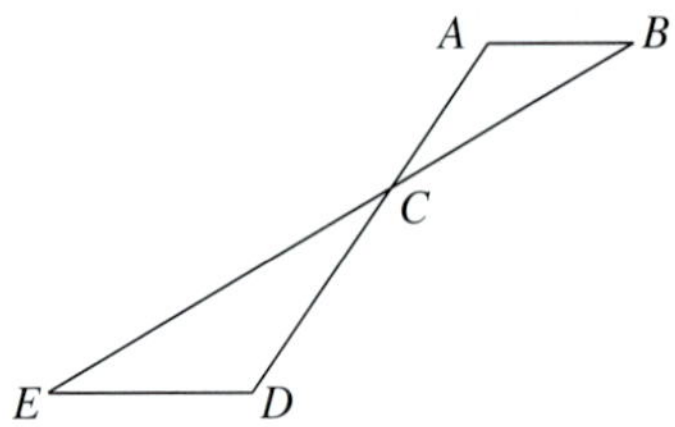

Another method for identifying corresponding sides is noting the order in which the vertices of the two similar triangles are named. The angles at the two vertices named first are corresponding, congruent angles. Therefore, the sides opposite these angles are corresponding sides. The same is true of the vertices named second and those named last.

EXAMPLE 2

$\triangle ABC \sim \triangle DEF$. Name the corresponding sides of the similar triangles.

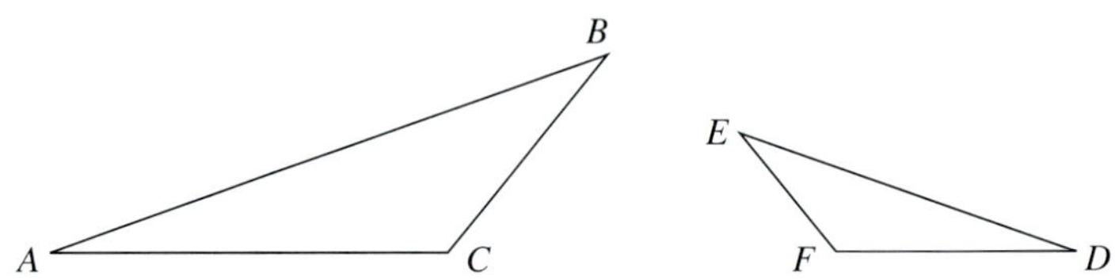

SOLUTION

$\angle A \cong \angle D$	These vertices are named first.
$\overline{BC} \leftrightarrow \overline{EF}$	sides opposite congruent angles *A* and *D*
$\angle B \cong \angle E$	These vertices are named second.
$\overline{AC} \leftrightarrow \overline{DF}$	sides opposite second vertices
$\angle C \cong \angle F$	These vertices are named last.
$\overline{AB} \leftrightarrow \overline{DE}$	sides opposite last vertices ■

◇ *Do Practice Exercise 2.*

2 Show that Two Triangles Are Similar

A triangle's shape is determined by the size of its angles. In order to determine whether two triangles are similar, show that the angles of one have the same measure as the angles of the other. If two angles of the two triangles are congruent, the third angle will also be congruent. This is true since the sum of the angles of a triangle equal 180°. The following rule can be used to determine whether two triangles are similar.

Angle-Angle Rule

Two triangles are similar if at least two pairs of corresponding angles are congruent.

ANSWER

2. $\overline{EC} \leftrightarrow \overline{BC}$, $\overline{AB} \leftrightarrow \overline{DE}$, $\overline{AC} \leftrightarrow \overline{DC}$

EXAMPLE 3

Show that $\triangle ABC \sim \triangle EDC$.

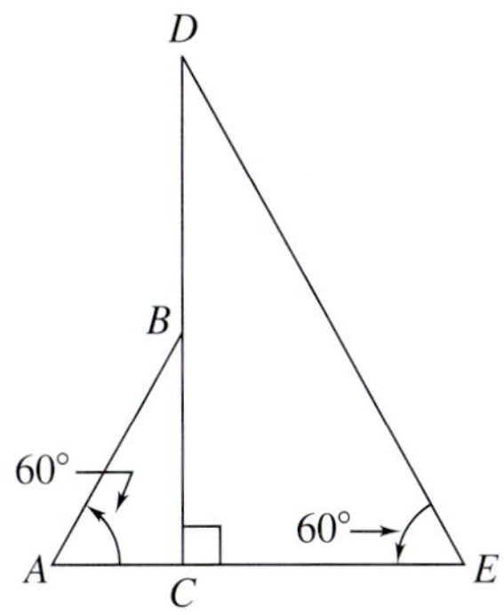

SOLUTION

	$\angle BAC \cong \angle DEC$	both equal 60°
	$\angle ACB \cong \angle ECD$	both equal 90°
Therefore	$\triangle ABC \sim \triangle EDC$	Angle-Angle rule ■

◇ *Do Practice Exercise 3.*

3 Find the Missing Side of Similar Triangles

The lengths of the sides of similar triangles are proportional to one another. For example, the sides of one triangle may be four times as long as the sides of the other. Or, perhaps the sides of one triangle are one-half as long as the sides of a similar triangle. The lengths of corresponding sides of similar triangles can be written as ratios, and these ratios are all equal.

HELPFUL HINT

Ratios of corresponding sides of similar triangles are equal.

Proportions can be used to find the length of a missing side in a pair of similar triangles. A proportion can be written using a known ratio on one side of the proportion and a ratio involving the unknown on the other side of the proportion.

3. Show that $\triangle ABC \sim \triangle DBE$.

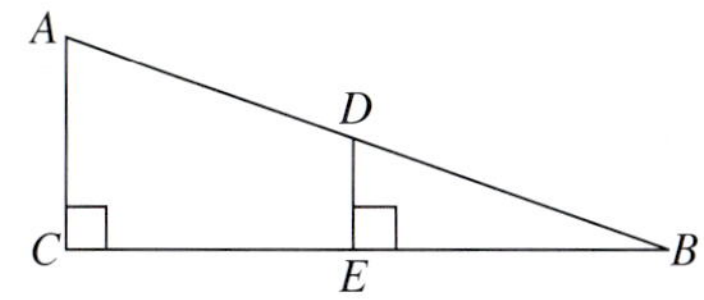

ANSWER

3.	$\angle B \cong \angle B$	same angle
	$\angle C \cong \angle DEB$	both right angles
	$\triangle ABC \sim \triangle DBE$	Angle-Angle rule

4. $\triangle ABC \sim \triangle DEF$. Use proportions to find the length of $\overline{EF}$.

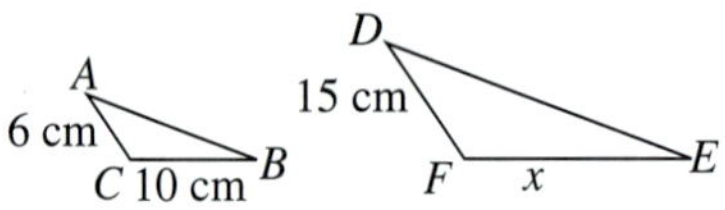

EXAMPLE 4

Given that $\triangle ABC \sim \triangle DEF$, use proportions to find the length of side DE.

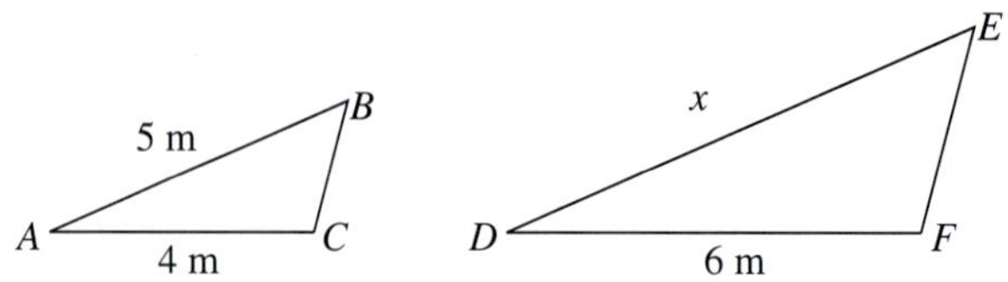

SOLUTION

$$\frac{AC}{DF} = \frac{AB}{DE} \quad \text{Ratios of corresponding parts are equal.}$$

$$\frac{4}{6} = \frac{5}{x} \quad \text{Substitute values.}$$

$$4x = 5 \cdot 6 \quad \text{Cross multiply.}$$

$$4x = 30 \quad \text{Simplify.}$$

$$\frac{4x}{4} = \frac{30}{4} \quad \text{Divide both sides by 4.}$$

$$x = 7\frac{1}{2}\text{ m} \quad \text{Result.}$$

■

◇ *Do Practice Exercise 4.*

4 Applications Involving Similar Triangles

There are many problems, such as finding the height of an object, or the distance between two points, that can be solved using similar triangles.

EXAMPLE 5

A 6 foot tall man casts a 4 foot shadow. At the same time a tree casts a 16 foot shadow. Find the height of the tree.

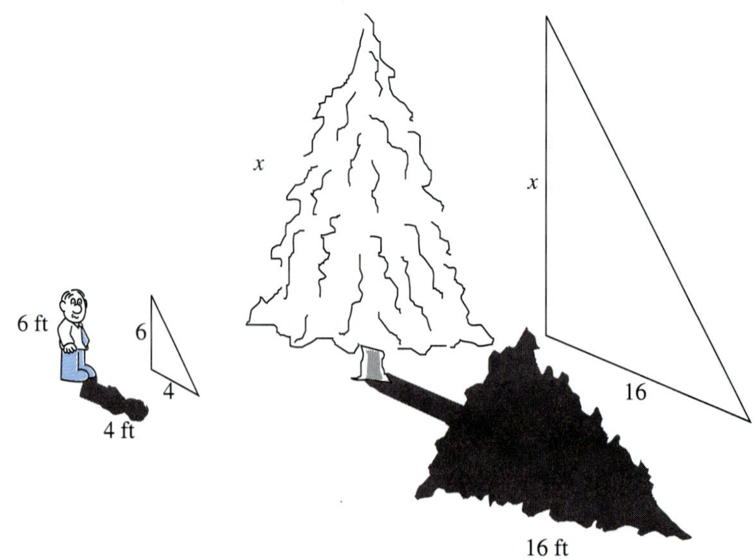

ANSWER

4. $\overline{EF} = 25$ cm.

SOLUTION

$$\frac{4}{16} = \frac{6}{x}$$ Ratios of corresponding sides are equal.

$$4x = 96$$ Cross multiply.

$$\frac{4x}{4} = \frac{96}{4}$$ Divide both sides by 4.

$$x = 24 \text{ feet}$$ Result. ■

◇ *Do Practice Exercise 5.*

EXAMPLE 6

To find the distance across the river to the camp on the other side, Jeff measured the distance along the edge of the river as shown in the diagram. Find $\overline{AB}$, rounding the answer to the nearest foot.

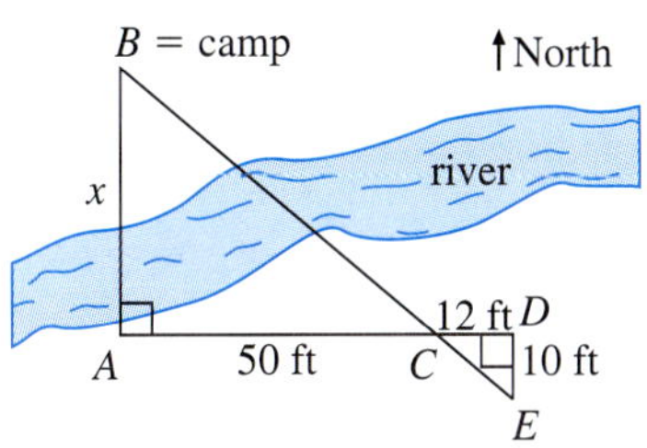

SOLUTION

Using right angles and vertical angles, $\triangle ABC \sim \triangle DEC$ by the Angle-Angle rule. The distance across the river, AB, can be found using proportions.

$$\frac{AB}{DE} = \frac{AC}{DC}$$ Ratios of corresponding sides are equal.

$$\frac{x}{10} = \frac{50}{12}$$ Substitute values.

$$12x = 500$$ Cross multiply.

$$\frac{12}{12}x = \frac{500}{12}$$ Divide both sides by 12.

$$x \approx 42 \text{ ft}$$ Result. ■

◇ *Do Practice Exercise 6.*

5. Rudy is 5 feet tall and casts a 3 foot shadow. At the same time, the shadow of a building is 12 feet. How tall is the building?

6. A park ranger sees a boat out on the lake and wants to know how far it is from the shore. Using the measurements in the diagram, find the distance from A to B.

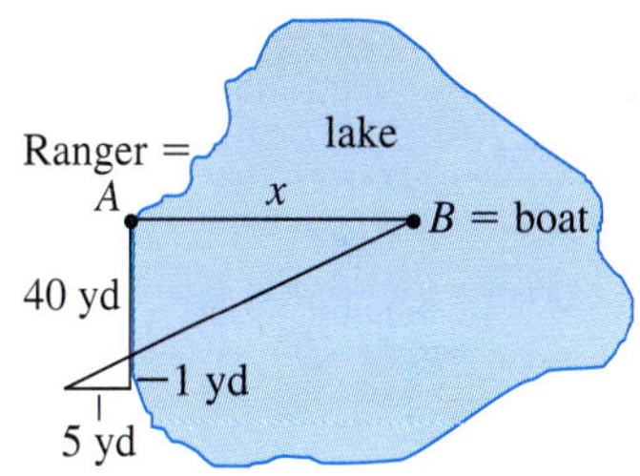

ANSWERS
5. 20 feet
6. 200 yd

Name the corresponding sides of the following pairs of similar triangles.

1.

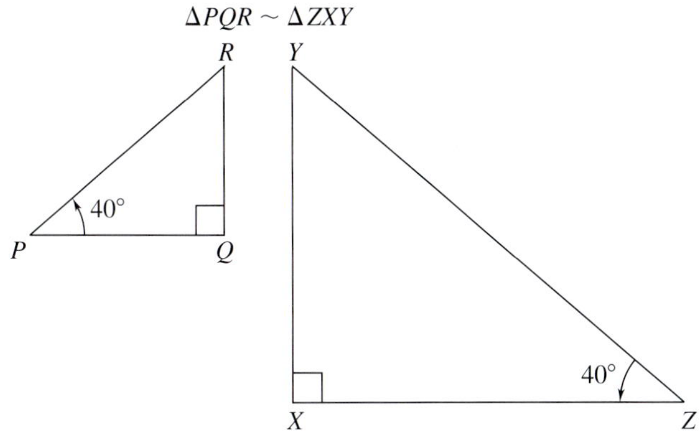

2.

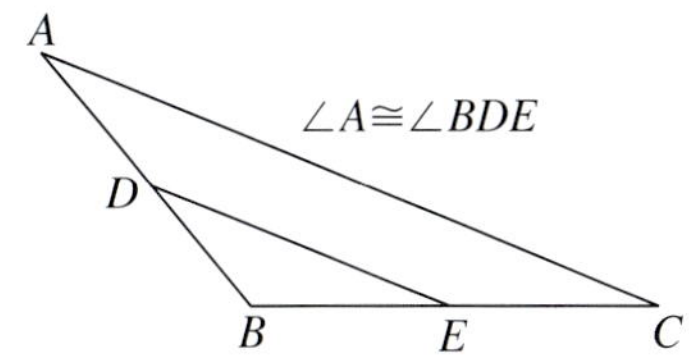

3.

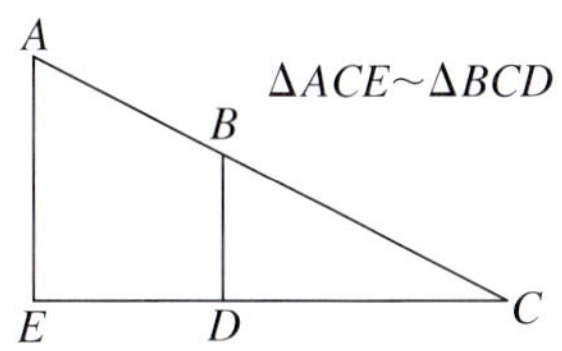

4.

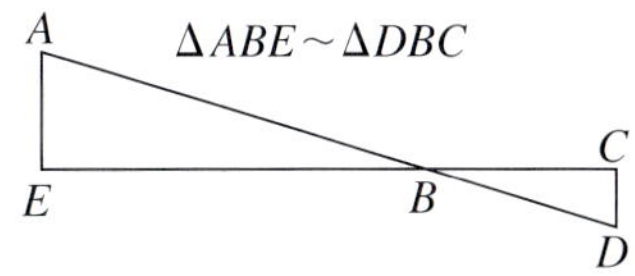

5.

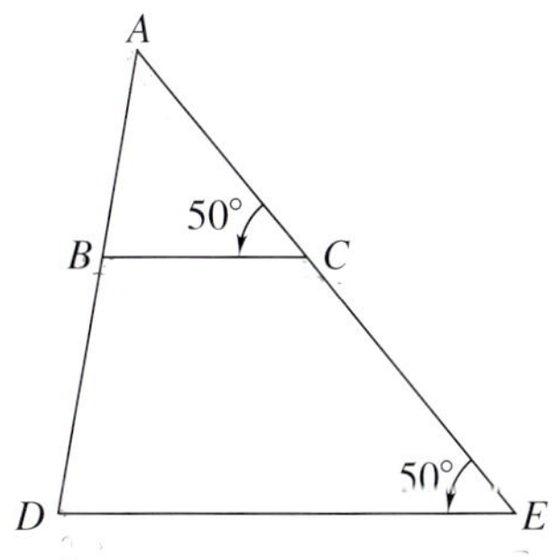

6.

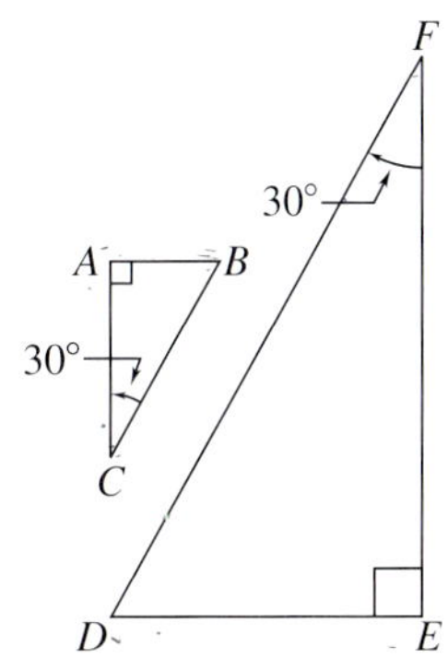

Write the ratios of corresponding sides for the pairs of similar triangles. Reduce the ratio to lowest terms.

7.

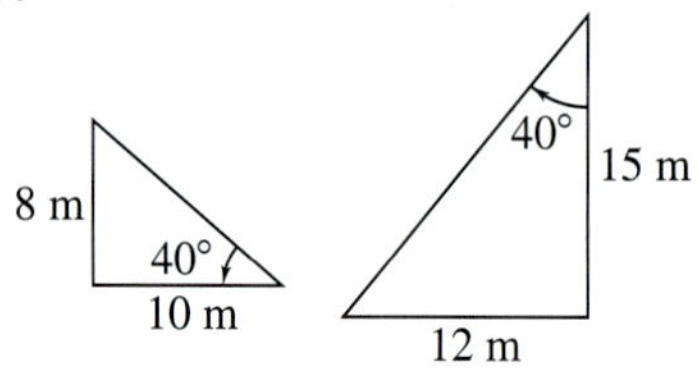

8.

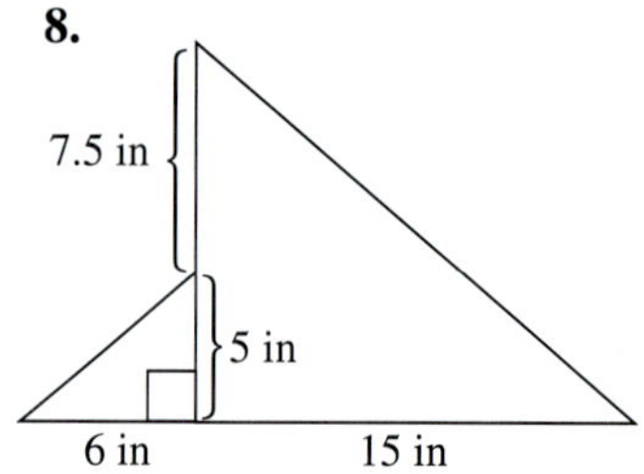

9.

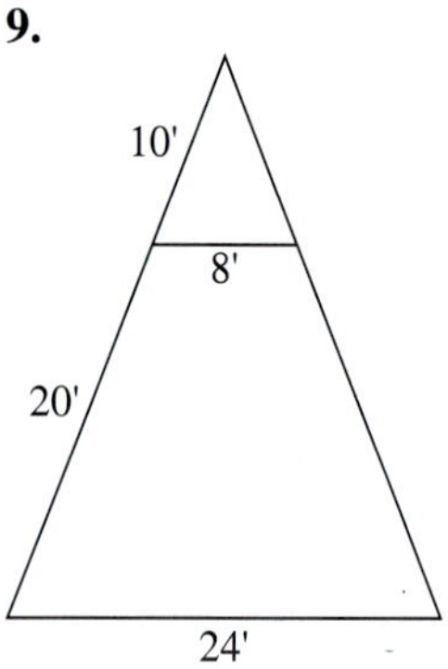

10.

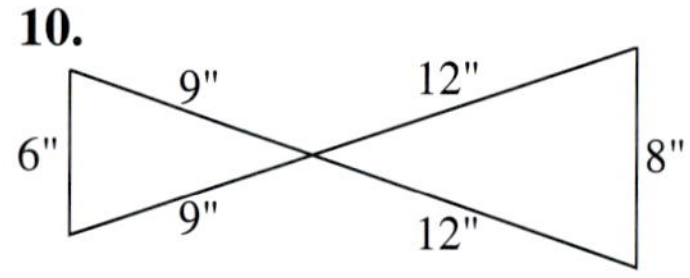

Use proportions to find the length of the side marked with an x in the following similar triangles.

11.

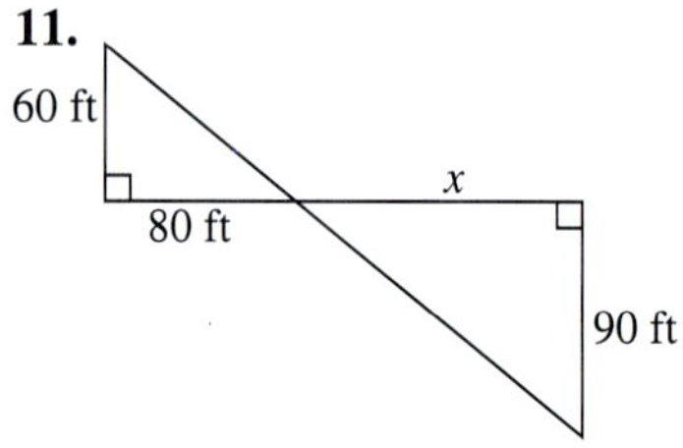

12.

$\angle A \cong \angle D$, $\angle B \cong \angle E$

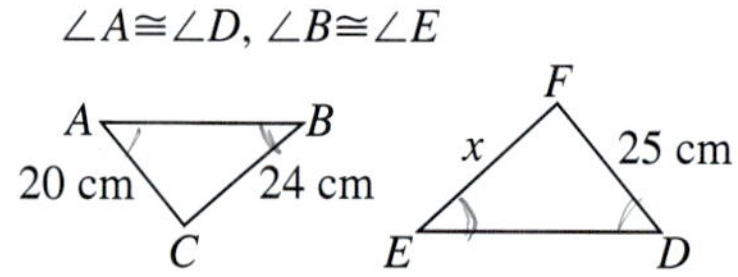

13.

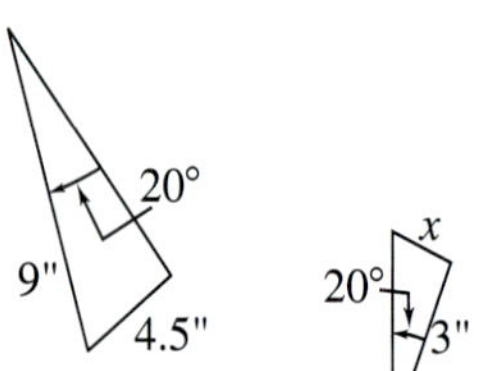

14.

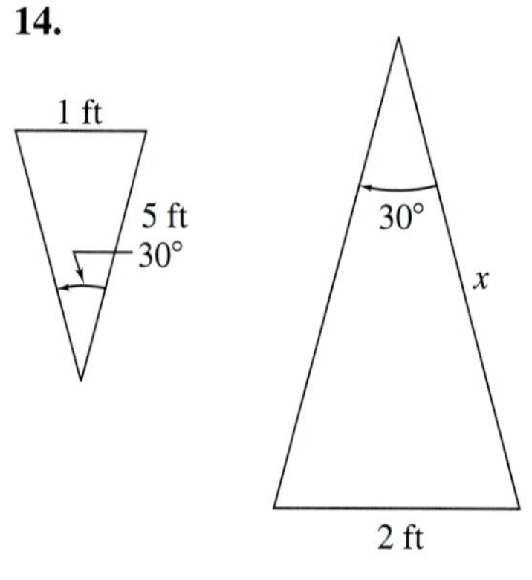

15.

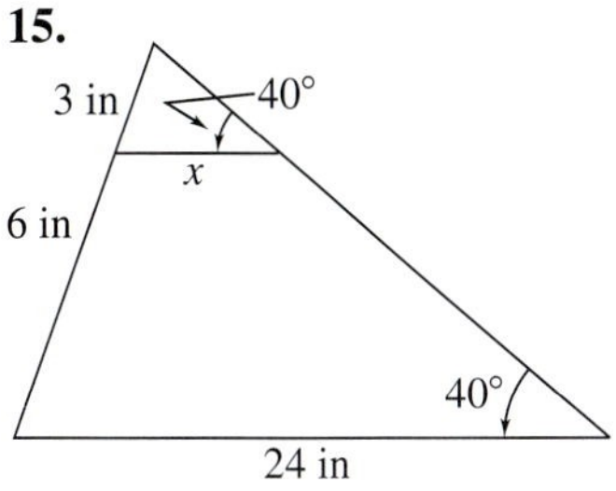

16.

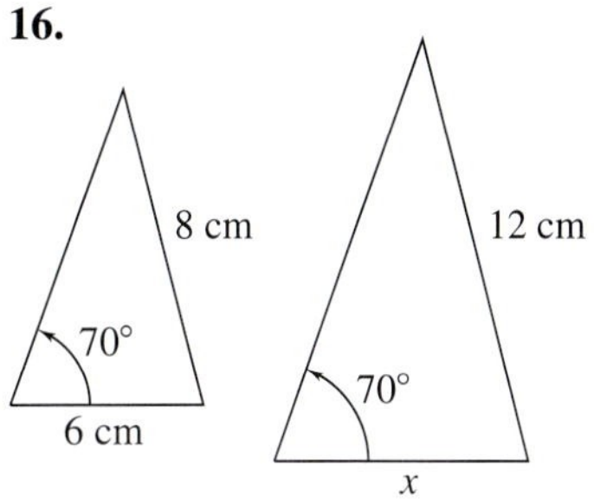

Solve the following problems.

17. The drawing of a triangular support on the blueprint of a bridge has three equal sides measuring 2.5 inches. If the ratio of sides of the triangle on the blueprint to the sides of the triangle on the bridge is 1 to 40, how long is the side of the triangle on the bridge?

18. A 35 millimeter slide film of Belinda in her graduation gown is to be projected onto a screen. The projection bulb is 300 mm from the film and 15 m from the screen as shown. Find the height of the screen.

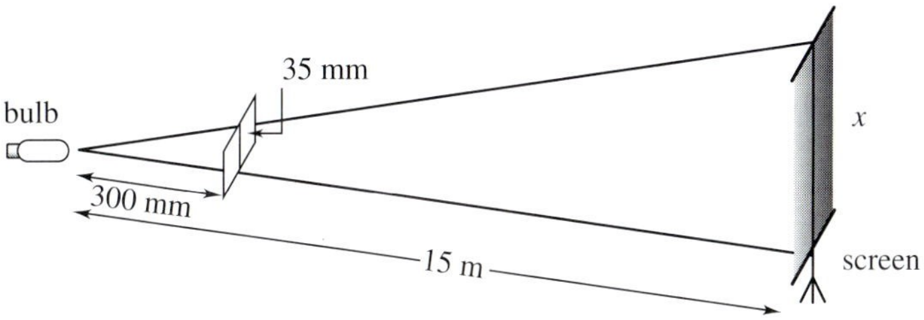

19. Mr. Rodriguez is installing a TV antenna on the roof of his house. He hooks a four foot guy wire two feet from the bottom of the antenna to a spot on the roof nearby. How much wire would Mr. Rodriguez need to install a parallel wire from the top of the eight foot antenna to the edge of the roof?

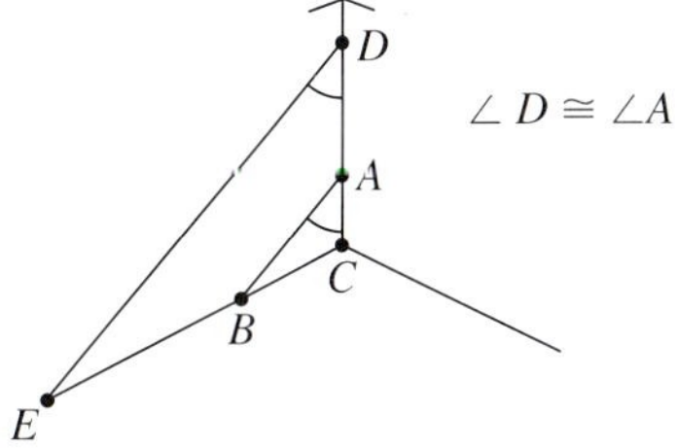

20. A high adventure club uses a rapelling tower for practice. The height of the tower can be found by lining up the top of the tower with the top of a 5 foot stick. Use similar triangles to find the height of the tower.

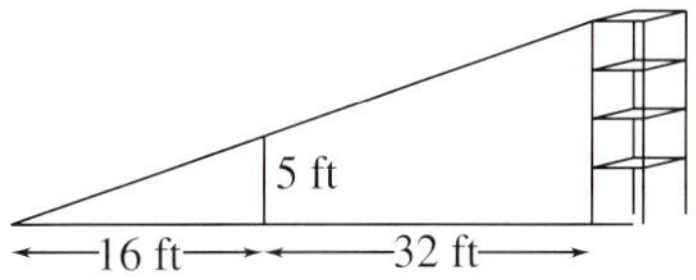

21. The diagram shows a plan similar to one used in 600 B.C. for digging a tunnel through a mountain. Using this plan, two digging teams starting at opposite ends A and B meet at the center with only a small error. Using similar triangles PBO and ABC, find the length of the tunnel (AB).

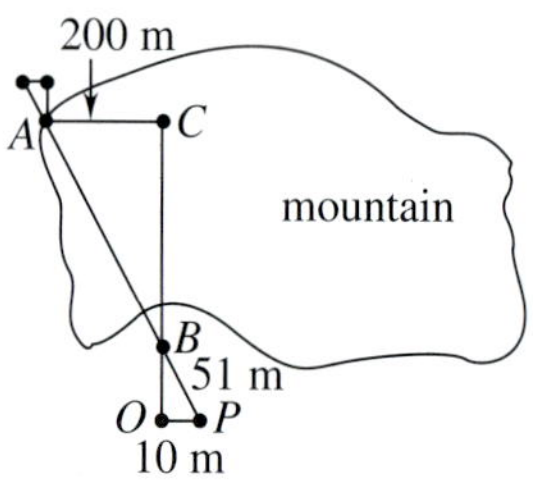

22. Backpackers on a mountain trail want to find the altitude change over the next $\frac{1}{8}$ mile. At one point they measure a 5 foot rise over a 15 foot distance. Find the rise in altitude over the next $\frac{1}{8}$ mile assuming the trail is equally steep. (1 mile = 5,280 feet).

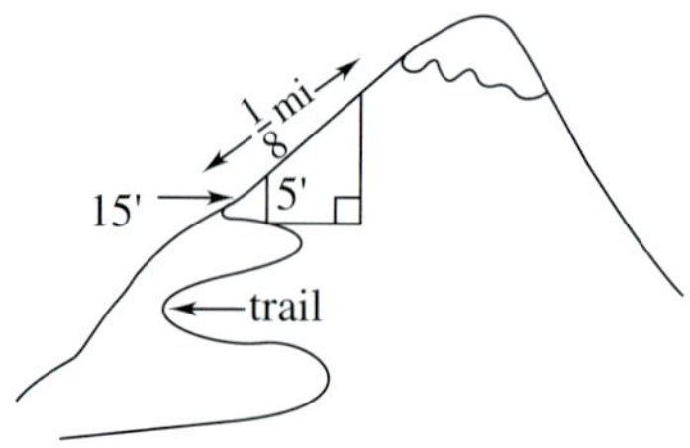

23. Gerry is 6 feet tall and casts a shadow 4 feet long. At the same time of day, a monument casts a shadow 10 feet long. How tall is the monument?

24. Sandra found the distance from a swimmer to the beach by putting a stick in the sand at point P and measuring the distances shown. The triangles formed are similar. How far out in the water is the swimmer?

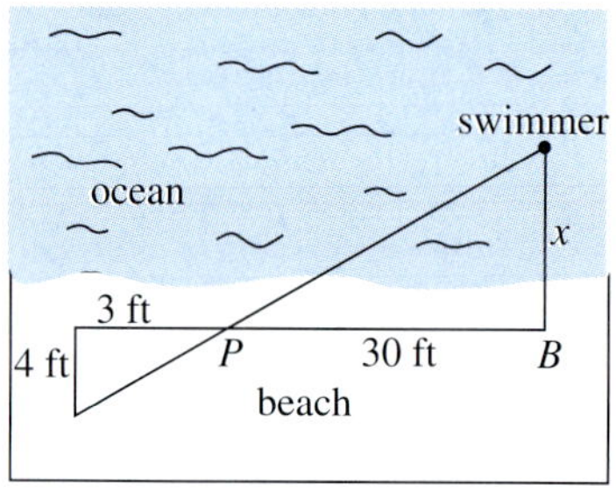

JOURNAL EXERCISES

1. Explain the concept of similar triangles.

2. Explain what it means to say the sides of similar triangles are proportional.

3. Explain how you could measure the height of a building using similar triangles and shadows. Draw a diagram to support your explanation.

1 Pythagorean Theorem

Right triangles are triangles that have one 90° angle. The other two angles are complementary, that is, they add to 90°.

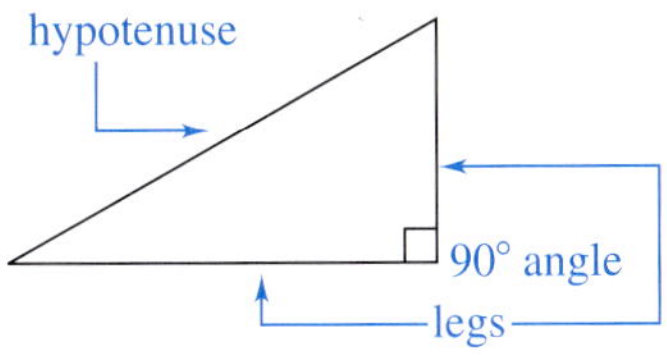

The **hypotenuse** of a right triangle is the side opposite the 90° angle. It is also the longest side of a right triangle. The other two sides are called **legs** of the triangle.

Examine the following right triangle with squares drawn on each side. The area of the three squares is shown on the right.

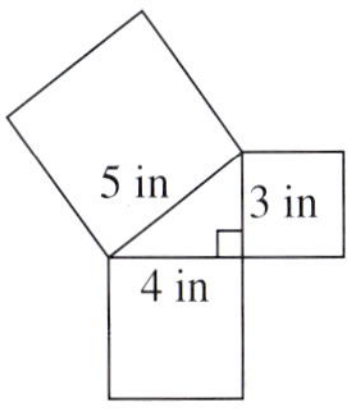

Area of square on hypotenuse = 5 in × 5 in = 25 in^2

Area of square on first leg = 3 in × 3 in = 9 in^2

Area of square on second leg = 4 in × 4 in = 16 in^2

Notice that the areas of the squares have the following relationship.

$$\begin{aligned} A_{\text{1st leg}} + A_{\text{2nd leg}} &= A_{\text{hypotenuse}} \\ 9 \text{ in}^2 + 16 \text{ in}^2 &= 25 \text{ in}^2 \\ 25 \text{ in}^2 &= 25 \text{ in}^2 \end{aligned}$$

Since the days of the Babylonians, the sides of right triangles have been known to have this relationship. Then, the Greek mathematician, Pythagoras, formally wrote it as a theorem, or rule.

The Pythagorean Theorem:

The square of the hypotenuse of a right triangle is equal to the sum of the squares of the other two sides of the right triangle.

If c is the length of the hypotenuse and a and b are the lengths of the two legs, the Pythagorean theorem gives the following equation:

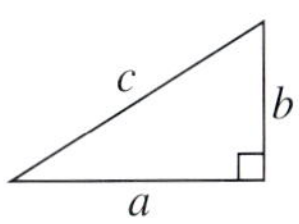

$$(\text{hypotenuse})^2 = (\text{1st leg})^2 + (\text{2nd leg})^2$$
$$c^2 = a^2 + b^2$$

OBJECTIVES

1 State and use the Pythagorean theorem.
2 Find the missing side of a right triangle.
3 Solve application problems with right triangles.

1. Show that the triangle with sides of lengths 5, 12, and 13 satisfies the Pythagorean theorem.

2. Find the square root of 81.

EXAMPLE 1

Show that the sides of the following triangle satisfy the Pythagorean theorem.

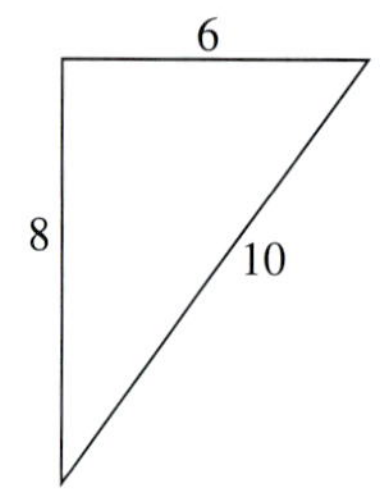

SOLUTION

From the Pythagorean theorem:

$$(\text{hypotenuse})^2 = (\text{length of 1st leg})^2 + (\text{length of 2nd leg})^2$$

$$c^2 = a^2 + b^2 \quad \text{Pythagorean theorem}$$

$$10^2 = 6^2 + 8^2 \quad \text{Substitute values.}$$

$$100 = 36 + 64 \quad \text{Evaluate exponents.}$$

$$100 = 100 \quad \text{Result.}$$

■

Therefore, the triangle is a right triangle, since it satisfies the Pythagorean theorem.

◇ *Do Practice Exercise 1.*

2 Find the Missing Side of a Right Triangle

Using the Pythagorean theorem to find the hypotenuse or leg of a right triangle involves finding the square root of a number.

The **square root** of a quantity is that number such that when multiplied times itself gives the original quantity. For example, the square root of 16 is 4 since the product of 4 times itself is 16. The symbol for square root is $\sqrt{\ }$, called a **radical**.

EXAMPLE 2

Find the square root of each:

(a) 49 **(b)** 1 **(c)** 100 **(d)** 64

SOLUTION

(a) $\sqrt{49} = 7$ since $7 \cdot 7 = 49$

(b) $\sqrt{1} = 1$ since $1 \cdot 1 = 1$

(c) $\sqrt{100} = 10$ since $10 \cdot 10 = 100$

(d) $\sqrt{64} = 8$ since $8 \cdot 8 = 64$ ■

◇ *Do Practice Exercise 2.*

ANSWERS

1. $5^2 + 12^2 = 13^2$

$25 + 144 = 169$

$169 = 169$

2. 9

A **perfect square** is a number whose square root is a whole number. Some perfect squares are 16 with a square root of 4, 25 with a square root of 5, and 36 with square root of 6. For work in this course, numbers that are not perfect squares can be left under the radical sign. Their square roots are decimal values that could be found with a calculator using a square root key.

$\sqrt{16} = 4$ — 16 is a perfect square.

$\sqrt{36} = 6$ — 36 is a perfect square.

$\sqrt{29}$ — 29 is not a perfect square. Leave in radical form.

Calculator Application

Use your calculator to find the square root of 29. Round your answer off to the nearest thousandth.

Answer: 5.385

EXAMPLE 3

Use the Pythagorean theorem to find the hypotenuse of the following right triangle.

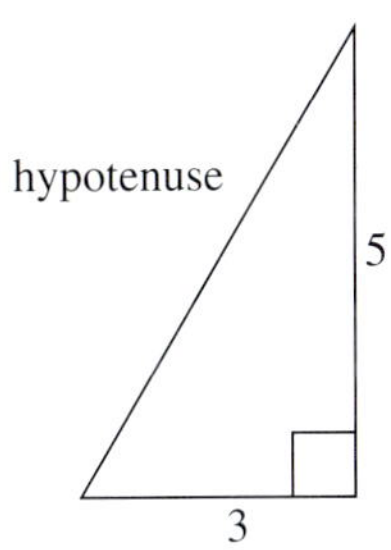

SOLUTION

$c^2 = a^2 + b^2$	formula
$c^2 = 3^2 + 5^2$	Substitute values.
$c^2 = 9 + 25$	Evaluate exponents.
$c^2 = 34$	Add, then find the square root.
$c = \sqrt{34}$	Result. ■

◇ *Do Practice Exercise 3.*

3. Find the hypotenuse of a right triangle with legs of lengths 2 and 7.

ANSWER

3. $\sqrt{53}$

4. Find the second leg of the right triangle in which one leg is 12, and the hypotenuse is 15.

EXAMPLE 4

Find the second leg of the right triangle in which one leg is 24 and the hypotenuse is 26.

SOLUTION

$a^2 + b^2 = c^2$	Pythagorean theorem
$24^2 + b^2 = 26^2$	Substitute values.
$576 + b^2 = 676$	Evaluate exponents.
$576 - 576 + b^2 = 676 - 576$	Subtract 576.
$b^2 = 100$	Simplify.
$b = \sqrt{100}$	Find the square root.
$b = 10$	Result. ■

◇ *Do Practice Exercise 4.*

The lengths of the sides of right triangles are sometimes called Pythagorean Triples. Pythagorean Triples are whole numbers that satisfy the equation $c^2 = a^2 + b^2$. Some of the more common ones are shown below.

3, 4, 5 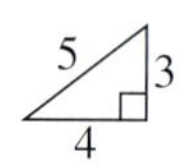

5, 12, 13 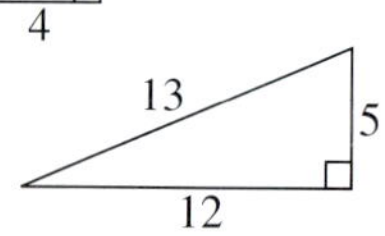

Multiples of the Pythagorean Triples, such as 6, 8, and 10 also satisfy the Pythagorean theorem. Recognizing Pythagorean Triples can be a quick way to find the missing side of a right triangle, as seen in Example 5.

3 Application Problems

Problems involving directions that are at right angles to one another, such as north and west, can be solved using right triangles.

EXAMPLE 5

David drove north 12 miles, and then west 5 miles. How far was David from his original position?

SOLUTION

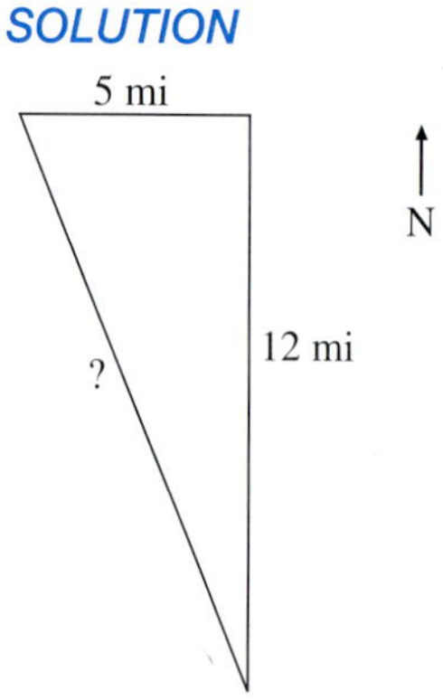

$a^2 + b^2 = c^2$	formula
$5^2 + 12^2 = c^2$	Substitute values.
$25 + 144 = c^2$	Evaluate exponents.
$169 = c^2$	Add, then find the square root.
$13 = c$	Result.

ANSWER
4. 9

Note that by recognizing this as a Pythagorean Triple, you could have immediately known that the answer was 13 miles. ■

◇ *Do Practice Exercise 5.*

EXAMPLE 6

Find the perimeter and area of a right triangle with one leg of 1.5 cm and a hypotenuse of 2.5 cm.

SOLUTION

First, use the Pythagorean theorem to find the length of the second leg, then add the lengths of the three sides.

$$c^2 = a^2 + b^2 \quad \text{Pythagorean theorem}$$
$$(2.5)^2 = (1.5)^2 + b^2 \quad \text{Substitute values.}$$
$$6.25 = 2.25 + b^2 \quad \text{Evaluate exponents.}$$
$$6.25 - 2.25 = 2.25 - 2.25 + b^2 \quad \text{Subtract 2.25.}$$
$$4 = b^2 \quad \text{Simplify.}$$
$$\sqrt{4} = b \quad \text{Find the square root.}$$
$$2 = b \quad \text{Result.}$$
$$\text{Perimeter} = 2 \text{ cm} + 1.5 \text{ cm} + 2.5 \text{ cm}$$
$$\text{Perimeter} = 6 \text{ cm}$$

The two legs of the right triangle form the base and height of the triangle.

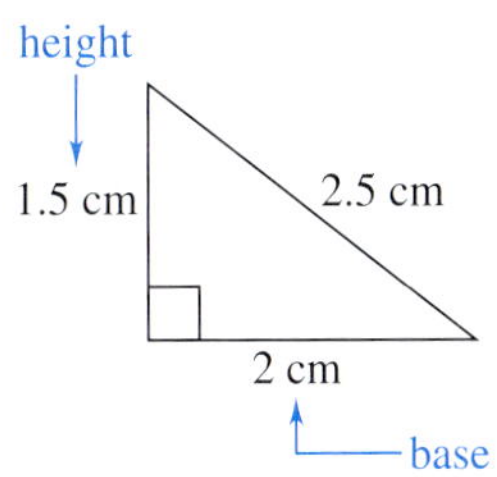

$$A = \frac{1}{2} b \cdot h \quad \text{formula for area}$$
$$A = \frac{1}{2}(2 \text{ cm})(1.5 \text{ cm}) \quad \text{Substitute values.}$$
$$A = (1)(1.5 \text{ cm}^2) \quad \text{Multiply.}$$
$$A = 1.5 \text{ cm}^2 \quad \text{Result.}$$ ■

◇ *Do Practice Exercise 6.*

5. How long is the front edge of a corner garden with sides 3 feet and 4 feet?

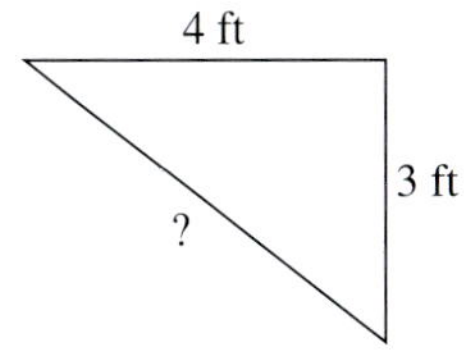

6. Find the perimeter and area of the right triangle having perpendicular sides of lengths 9 ft and 12 ft.

ANSWERS
5. 5 ft
6. area = 54 ft^2, perimeter = 36 ft

name date hour

The following numbers represent the three sides of a triangle. Determine whether or not they satisfy the Pythagorean theorem $c^2 = a^2 + b^2$.

1. 9, 12, 15

2. 2.5, 6, 6.5

3. 2, 3, 5

4. 10, 24, 26

5. 5, 11, 13

6. 4, 7.5, 8.5

In the following right triangles, use the Pythagorean theorem to find the length of the side marked with an x. Answers that are not perfect squares may be left in radical form.

7.

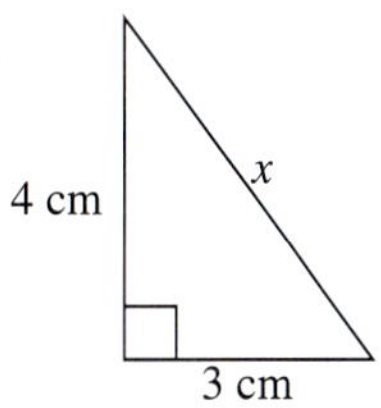

8.

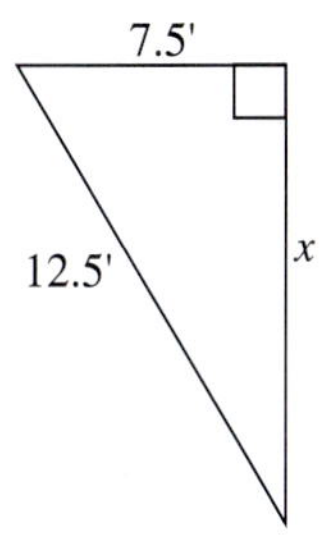

9.

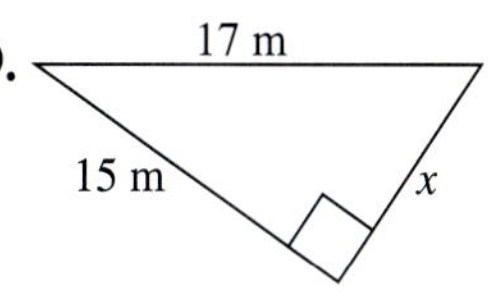

10.

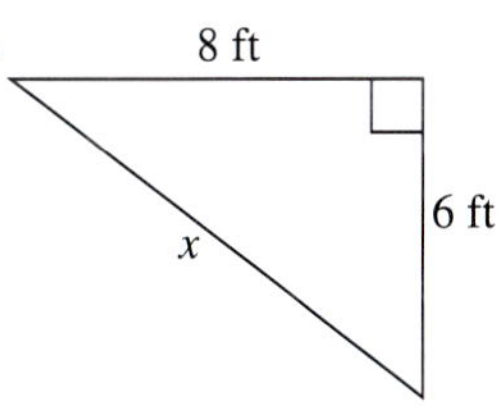

11.

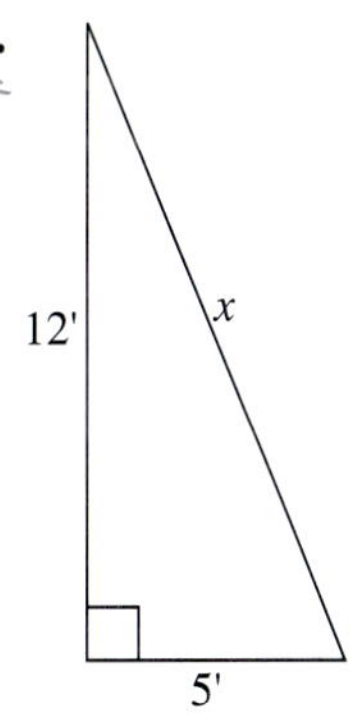

12.

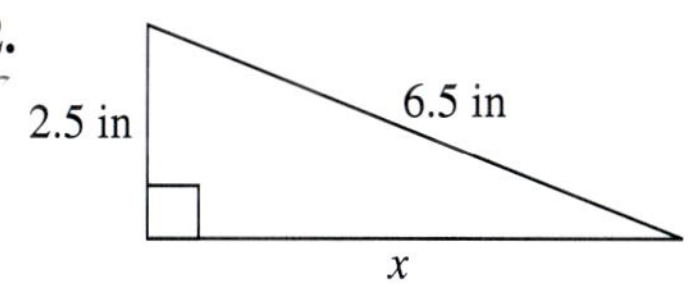

13.

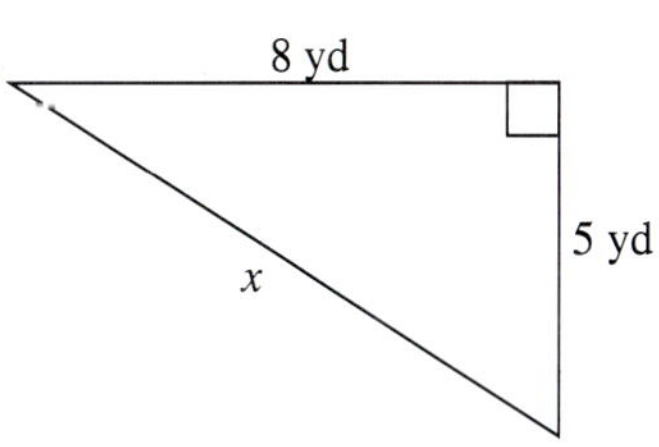

14.

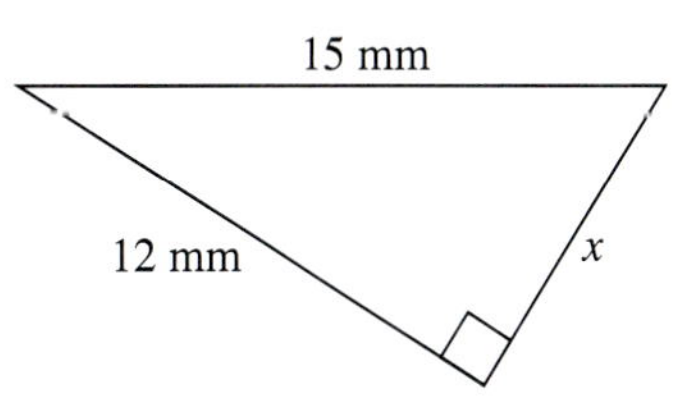

15.

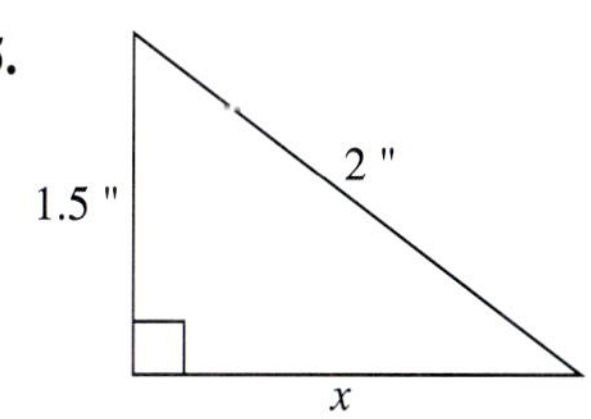

16.

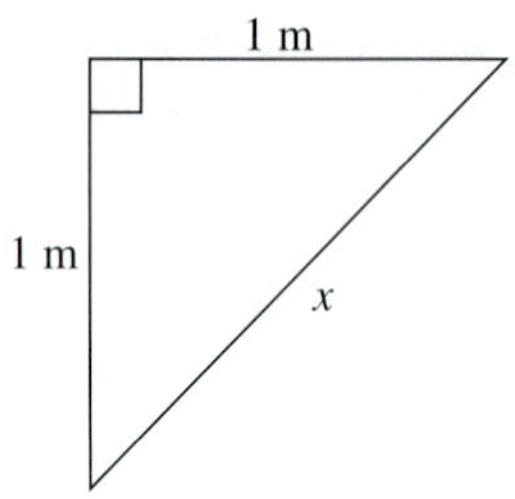

17.

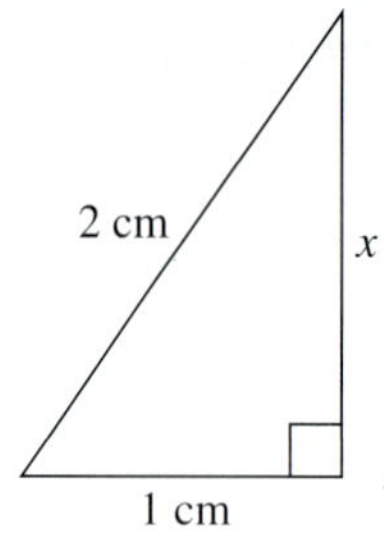

The following problems give two of the sides of a right triangle. Find the third side. Answers that are not perfect squares may be left in radical form.

18. $\text{leg}_1 = 4, \quad \text{hypotenuse} = 9$

19. $\text{leg}_1 = 0.4, \quad \text{leg}_2 = 1.2$

20. $\text{hypotenuse} = 18, \quad \text{leg}_1 = 9$

21. $\text{leg}_1 = 7.5, \quad \text{hypotenuse} = 8.5$

22. $\text{leg}_1 = 24, \quad \text{hypotenuse} = 26$

23. $\text{hypotenuse} = 15, \quad \text{leg}_1 = 12$

Work each problem. Leave answers that are not perfect squares in radical form, or, if a calculator is available, find the value of the radical to the nearest unit.

24. Mr. and Mrs. Flores have a small garden shaped like a right triangle with hypotenuse of unknown length. The other two sides of the triangle are of length eight feet and six feet, respectively. Find the length of the hypotenuse and the perimeter of the garden.

25. A rectangle has a width of five meters and height of twelve meters. Find the length of the diagonal.

26. Find the length of the diagonal of the end of the box shown.

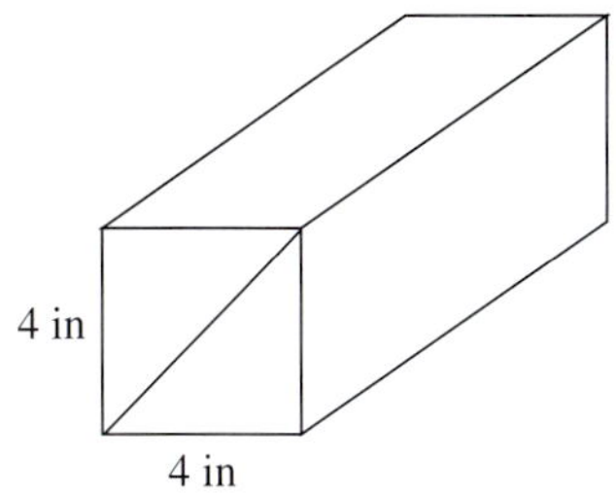

27. A right triangle with hypotenuse of 5″ is located inside of a circle of radius 4″. Find the shaded area.

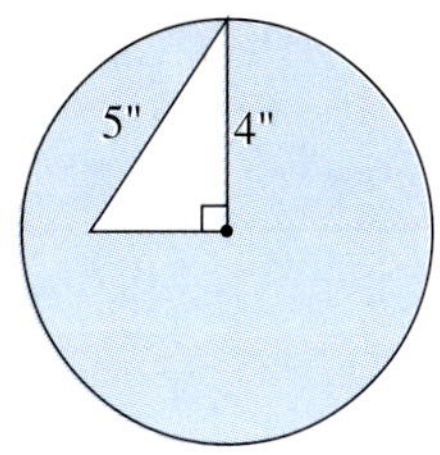

28. A group of hikers walk 10 miles south, then turn west and walk another 8 miles. How far are they from their starting point?

29. A plumber needs to run a pipeline from A to B, but there is a tree obstructing the path. How much more pipe will be needed to go around the tree in a right angle as shown rather than in a straight path?

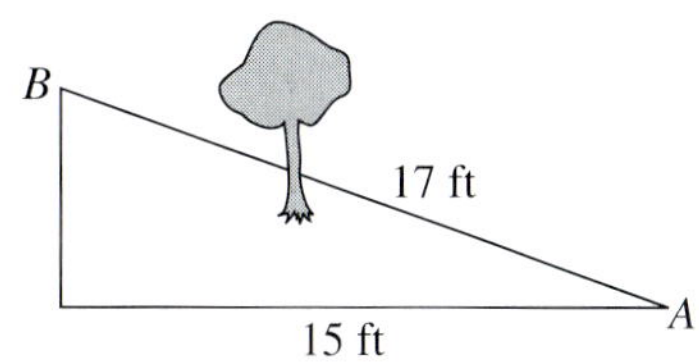

30. An airplane takes off and flies at an angle of 9° to the ground. After flying through the air a distance of 2,557 feet, he has covered a distance of 2,525.5 feet on the ground. Find the altitude of the plane at that time.

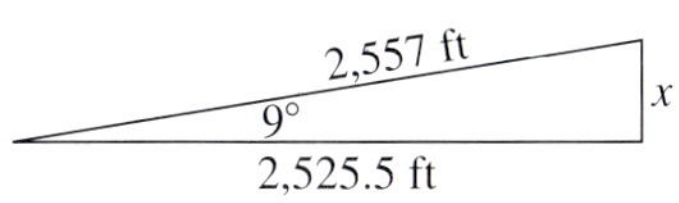

31. An airplane is flying north at 100 knots per hour when it encounters a strong cross wind from the west at 30 knots per hour. Find the resulting ground speed of the plane (AC).

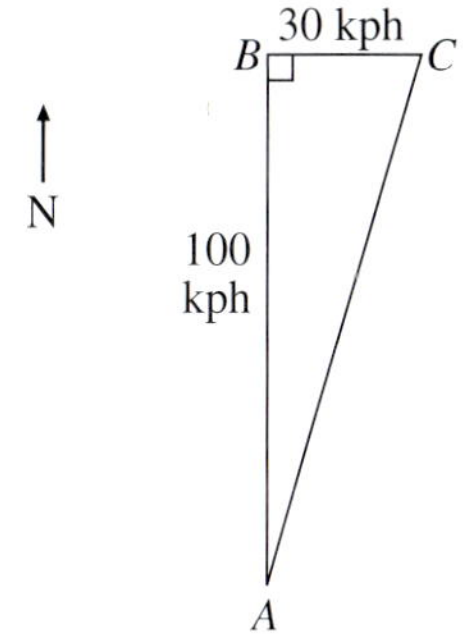

32. An airplane is towing a glider. The tow rope has a tension force of 300 pounds exerted on it by the airplane. If the amount of lift (*BC*) is 102.6 pounds, find the horizontal drag the glider is creating (*AB*).

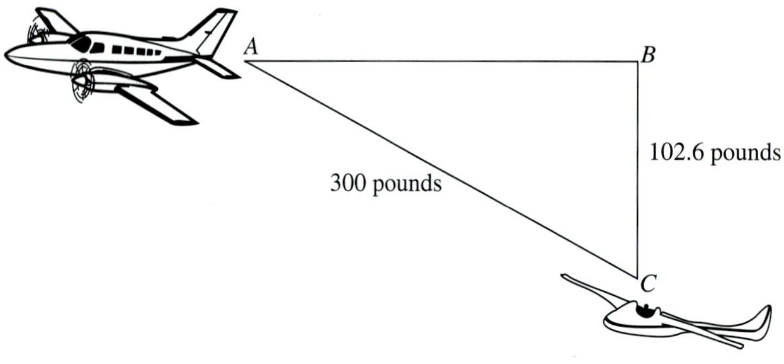

JOURNAL EXERCISES

1. Now that you have finished the chapter on geometry, explain what you think is the most useful concept presented.

2. Explain the Pythagorean theorem in your own words.

3. In what ways is the geometry you studied here different from any you may have previously studied? In what ways is it the same?

4. How do you feel about geometry at this time? Express your opinion regarding its difficulty level, its practicality, and its application to your life and work.

CHAPTER 8 REVIEW

SECTION	TOPIC OR TERM	EXAMPLE OR DEFINITION
8.1		
Basics	Point	A point is a location or position.
	Line	A line is a one-dimensional figure of infinite length.
	Line Segment	A line segment is a line with two endpoints.
	Congruent Line Segments	Congruent line segments are line segments of equal length.
	Ray	A ray is a line with one endpoint.
	Angle	An angle is formed when two lines meet at a point.
	Vertex	The vertex is the point where the two sides of an angle meet.
	Degree	Degree is a unit of measure for angles.
	Congruent Angles	Congruent angles are angles with the same number of degrees.
	Straight Angle	A straight angle measures 180°.
	Supplementary Angles	Supplementary angles are two angles that add to 180°, such as 60° and 120°.
	Right Angle	A right angle measures 90°.
	Complementary Angles	Complementary angles are two angles that add up to 90°, such as 30° and 60°.
8.2		
Polygons, Parallel, and Perpendicular Lines	Polygon	A polygon is a two-dimensional figure with three or more sides.
	Triangle	A triangle has three sides.
	Quadrilateral	A quadrilateral has four sides.
	Pentagon	A pentagon has five sides.
	Hexagon	A hexagon has six sides.
	Octagon	An octagon has eight sides.
	Isosceles Triangle	An isosceles triangle is a triangle with two equal sides.
	Equilateral Triangle	An equilateral triangle has three sides of equal length.
	Parallel Lines	Parallel lines are lines that never meet.
	Perpendicular Lines	Perpendicular lines meet at a 90° angle.
	Parallelogram	A parallelogram is a quadrilateral with opposite sides parallel.
	Rhombus	A rhombus is a parallelogram with equal sides.
	Rectangle	A rectangle is a parallelogram with four right angles.
	Square	A square is a rectangle with four equal sides.
	Trapezoid	A trapezoid is a quadrilateral with two parallel sides.

SECTION	TOPIC OR TERM	EXAMPLE OR DEFINITION
8.3		
Perimeter and Circumference	Perimeter	The perimeter is the distance around the outer edge of a figure. It is the sum of the lengths of all the sides of the figure. Perimeter is measured in units.
	Circle	A circle is all points equidistant from a point called the center.
	Radius	The radius of a circle is the line drawn from the center to any point on the circle.
	Diameter	The diameter of a circle is the line drawn across the circle through the center. The diameter is two times the radius.
	Circumference	The circumference is the distance around the outer edge of a circle. $C = 2\pi r$ or $C = \pi d$.
	Pi, π	π is a number approximately equal to 3.14.
8.4		
Area	Area	Area is the amount of space covered by a two-dimensional figure. Area is measured in square units.
	Area of Polygons	rectangle $A = lw$ square $A = s^2$ triangle $A = \frac{1}{2}bh$ parallelogram $A = bh$ trapezoid $A = \frac{1}{2}h(b_1 + b_2)$ circle $A = \pi r^2$
8.5		
Volume	Volume	Volume is the amount of space inside a three-dimensional figure. Volume is measured in cubic units.
	Rectangular Solid	A rectangular solid is a three-dimensional figure shaped like a box. All sides are rectangles.
	Rectangular Pyramid	A rectangular pyramid is a three-dimensional figure whose base is a rectangle and whose other four sides are triangles that share a common vertex at the top.
	Right Circular Cylinder	A right circular cylinder is a three-dimensional figure shaped like a can.

SECTION	TOPIC OR TERM	EXAMPLE OR DEFINITION
8.5		
Volume	Right Circular Cone	A right circular cone is a three-dimensional funnel-shaped figure whose base is a circle.
	Sphere	A sphere is a three-dimensional figure shaped like a ball.
	Volume of Three-dimensional Solids	rectangular solid $V = lwh$ rectangular pyramid $V = \frac{1}{3}lwh$ right circular cylinder $V = \pi r^2 h$ right circular cone $V = \frac{1}{3}\pi r^2 h$ sphere $V = \frac{4}{3}\pi r^3$
	Surface Area	The surface area of a three-dimensional figure is the total area of all the surfaces that make up the figure.
	Surface Area of Solids	rectangular solid $SA = 2wl + 2wh + 2hl$ right circular cylinder $SA = 2\pi rh + 2\pi r^2$
8.6		
Similar Triangles	Similar Figures	Similar figures are figures with the same shape but different sizes.
	Angle-Angle Rule	If two angles of one triangle are congruent to two angles of a second triangle, then the triangles are similar.
8.7		
Right Triangles	Right Triangle	A right triangle is one with a 90° angle.
	Hypotenuse	The hypotenuse of a right triangle is the side opposite the right angle. It is the longest side of the triangle.
	Legs	The legs of a right triangle are the two sides that form the 90° angle.
	Pythagorean Theorem	The Pythagorean theorem states that the square of the length of the hypotenuse equals the sum of the squares of the lengths of the other two sides.
	Square Root	The square root of a quantity is that number whose product with itself is the original quantity. For example, $\sqrt{36} = 6$, $\sqrt{81} = 9$.
	Perfect Square	A perfect square is a number whose square root is a whole number. 25 and 49 are perfect squares.
	Pythagorean Triples	Three whole numbers that satisfy the Pythagorean theorem, such as 3, 4, 5 or 5, 12, 13.

APPLICATION WORD PROBLEM

SELLING THE FAMILY LAND

The Schmitz family has owned a large piece of land near Chicago for many years. Most of the family members have moved away and they have decided to sell the land. They divided it into plots as shown.

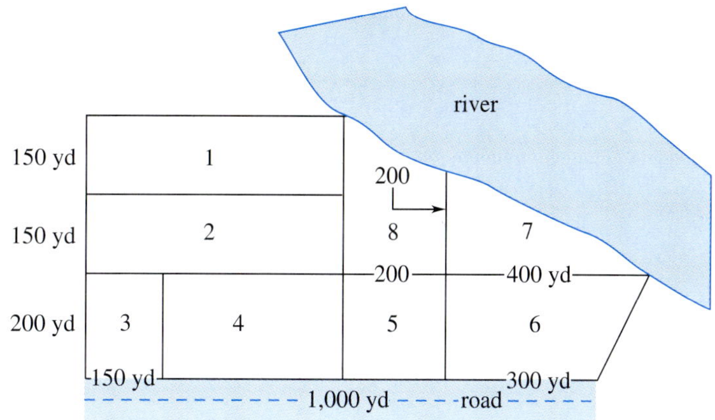

Road front property will be sold for $1,200 an acre. Riverfront property will be sold for $1,500 an acre. Property that is neither on the road nor on the river will be sold for $900 an acre. One acre is equal to 4,840 square yards.

1. Find the area of each plot, first in square yards and then in acres rounded to the nearest tenth.

2. Find the value of each plot of land using the selling prices given in the problem.

3. Find the total value of the eight pieces of land.

4. If there are four heirs what amount will each inherit?

5. Maureen, one of the daughters, decides that she would rather have land than the money. She prefers plots 6 and 7. Does the value of these two plots exceed her share of the inheritance? How much difference is there?

6. Maureen will have to pay a property tax of 2% of the value of the property. How much tax will she owe?

CHAPTER 8 REVIEW PROBLEMS

Solve the following problems.

1. Plot the following points on a number line: 1.5, 8, $3\frac{3}{4}$, and $\frac{21}{4}$

2. Sketch a hexagon, a rectangular solid, a pentagon, and a right circular cylinder.

3. Name all line segments congruent to $\overline{AB}$.

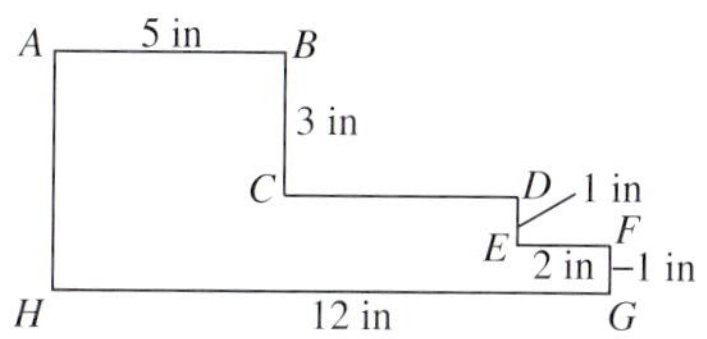

4. Find the number of degrees in $\frac{2}{3}$ of a circle.

5. Find the number of degrees in 25% of a circle.

6. A base angle of an isosceles triangle is 35°. Find the other two angles.

7. Two angles of a triangle measure 80° and 30°. Find the third angle.

8. Find the sum of a straight angle and a right angle.

9. Find the number of degrees in each angle of an equilateral triangle.

10. $\angle A$ is 50°. $\angle B$ is the complement of $\angle A$. Find $\angle B$.

11. $\angle C$ is 98°. $\angle D$ is the supplement of $\angle C$. Find $\angle D$.

Find the perimeter of each.

12.

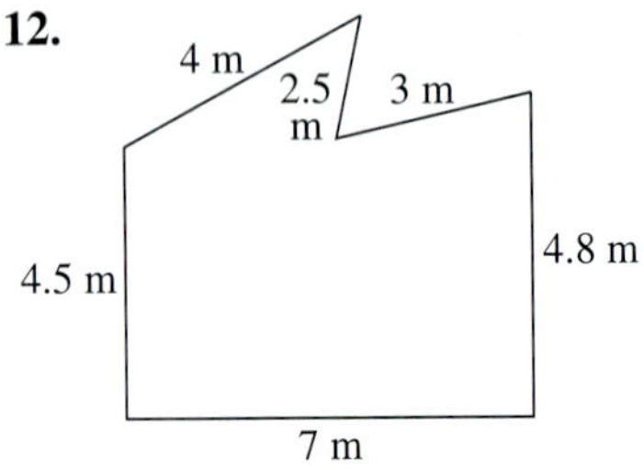

13.

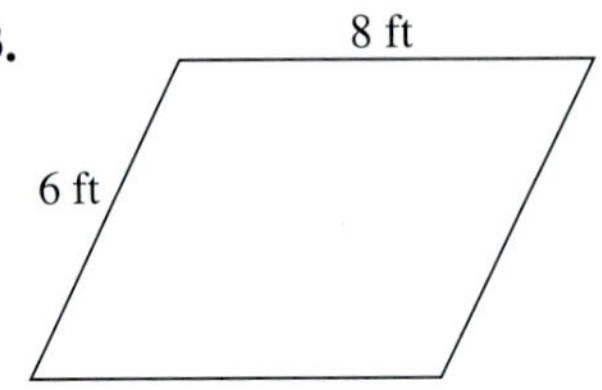

14.

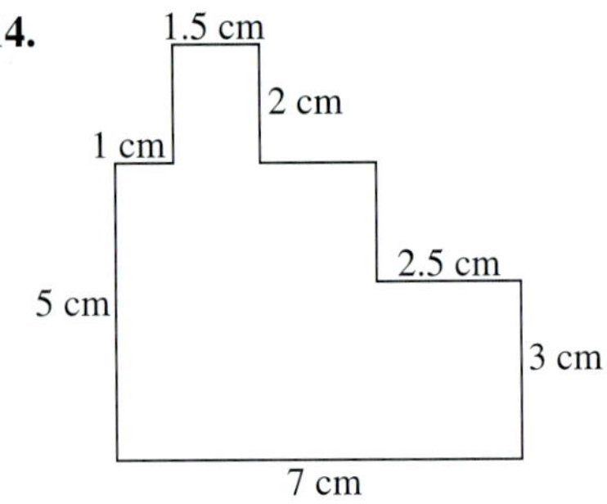

Find the perimeter (or circumference) of each. Round your answer to hundredths. Use $\pi \approx 3.14$.

15. Rectangle of length 8 meters and width 5.5 meters

16. Circle with radius 11 inches

17. Square with side of 7 millimeters

18. Semicircle with diameter 16 centimeters

19. Triangle with sides $3\frac{1}{2}$ cm, 5 cm, and $7\frac{1}{2}$ cm

20. Pentagon with each side equal to 3.2 inches

Find the area of each.

21.

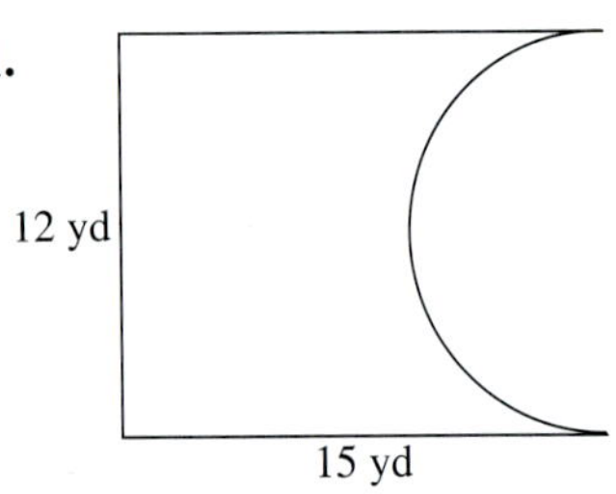

22.

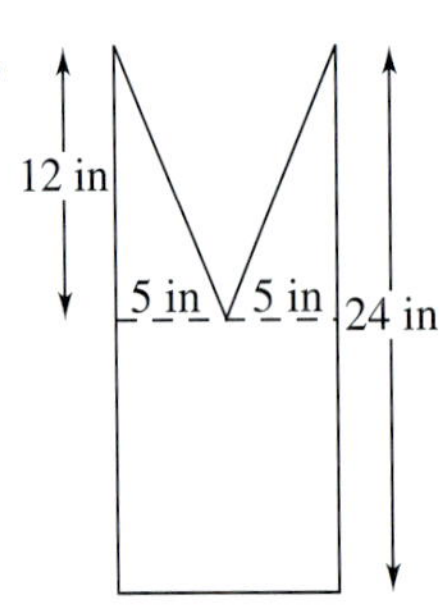

23.

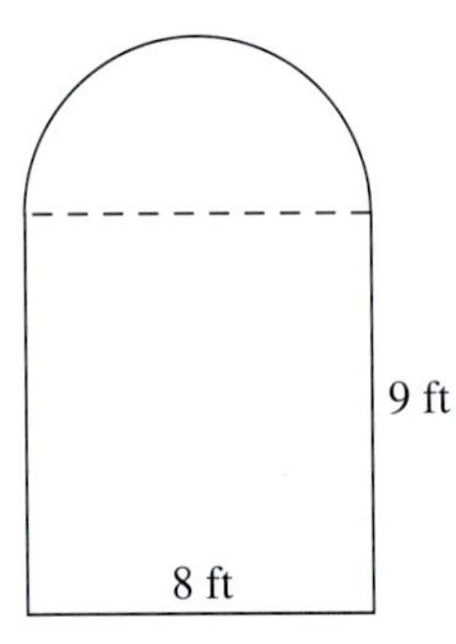

Find the area of each. Round your answers to hundredths. Use $\pi \approx 3.14$.

24. Triangle with a base of 14 m and a height of 6 m

25. Circle with diameter 16 centimeters

26. Rectangle with length 13.25 ft and width 5.5 ft

27. Square with side $2\frac{3}{4}$ inches

28. Parallelogram with base of 17 in and height of 6 in

29. Trapezoid with a small base of 9 cm, a large base of 15 cm, and a height of 8 cm

Find the volume of each. Round your answers to hundredths. Use $\pi \approx 3.14$.

30.

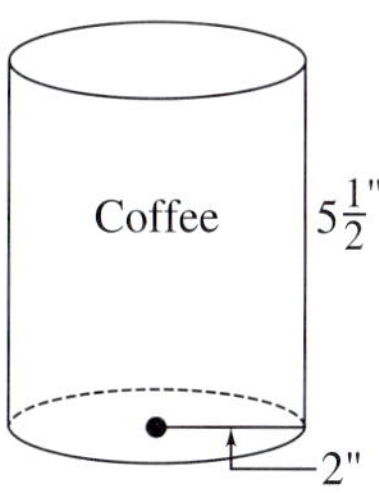

31.

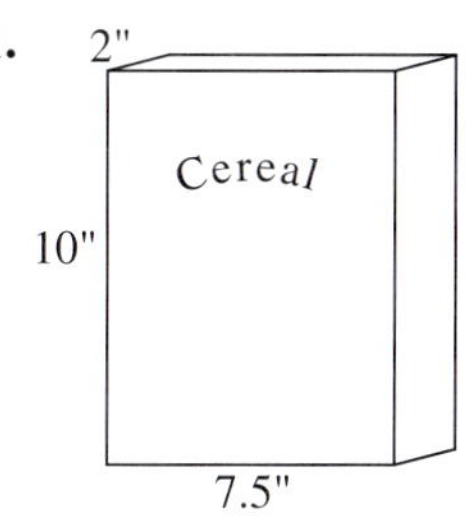

32.

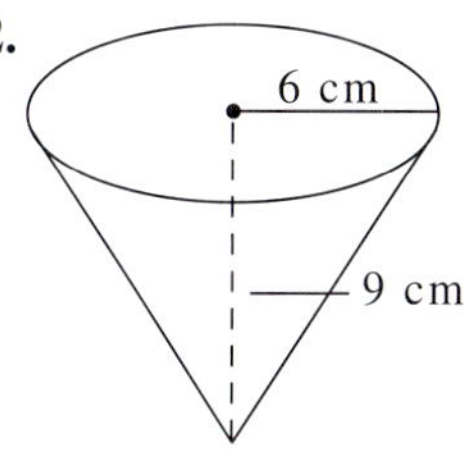

Find the volume of each. Round your answers to hundredths. Use $\pi \approx 3.14$.

33. Box of width 6 in, height 9 in, and depth 3 in

34. Right circular cone with radius 2 ft and height 3.6 ft

35. Sphere with radius 6 ft

36. Rectangular pyramid with height 5.2 in, length of base 3.7 in, and width 2 in

Work each problem.

37. Find the surface area of a box of dimensions 6 in by 9 in by 3 in.

38. $\angle A$ and $\angle B$ are supplementary angles. Find $\angle B$ if $\angle A = 58°$.

39. $\angle A$ and $\angle B$ are complementary angles. Find $\angle B$ if $\angle A = 74°$.

40. $\angle A$ and $\angle B$ are supplementary angles. Find $\angle B$ if $\angle A = 15°$.

Find $\angle x$ in the following.

41.

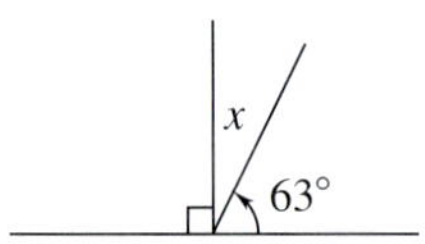

42.

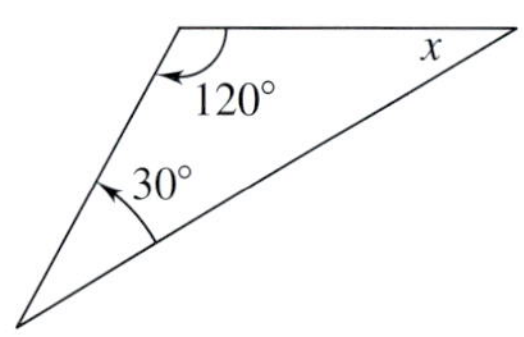

43.

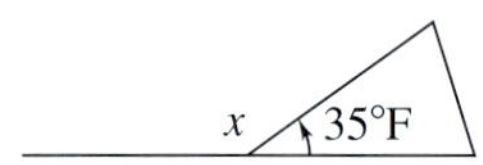

Find $\angle x$ in each figure.

44.

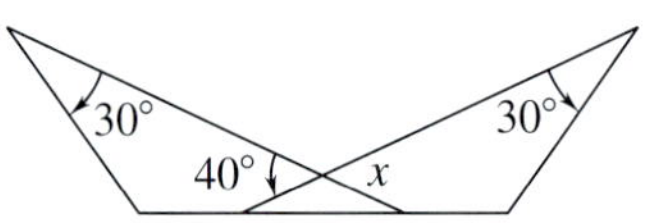

45.

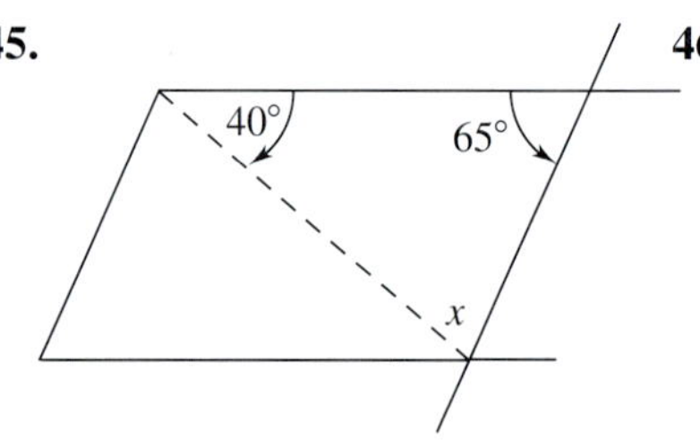

46.

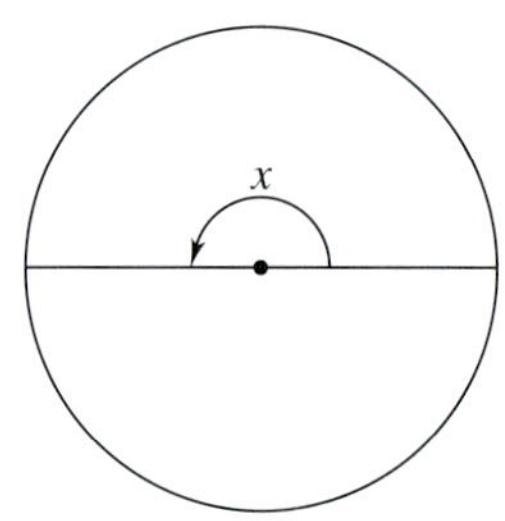

Write the ratios of corresponding sides for the similar triangles. If possible, reduce each ratio.

47.

$\Delta ACD \sim \Delta ABE$

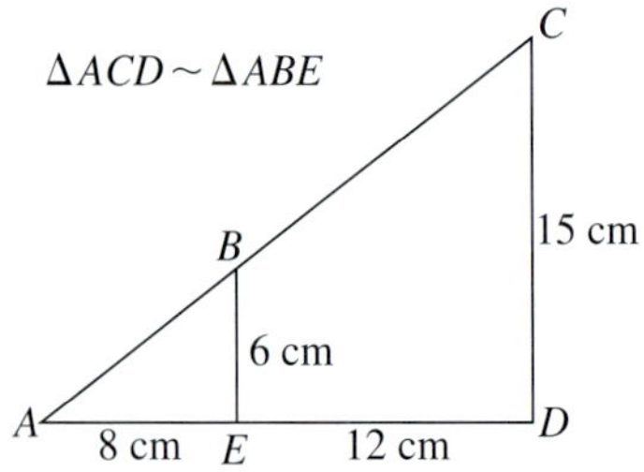

48.

$\Delta ABE \sim \Delta DCE$

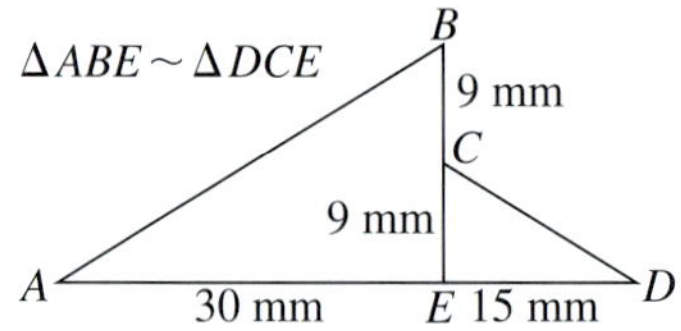

Use proportions to find the missing sides.

49.

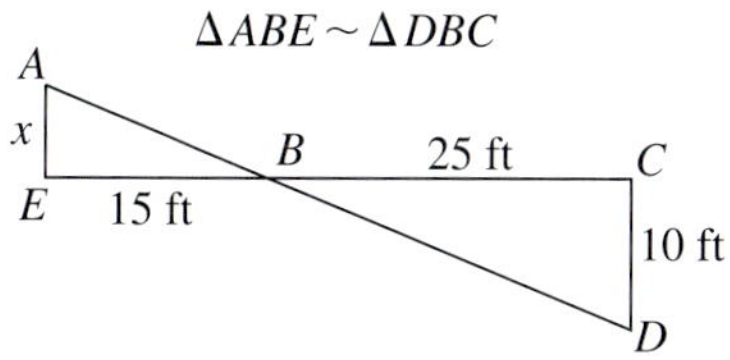

50.

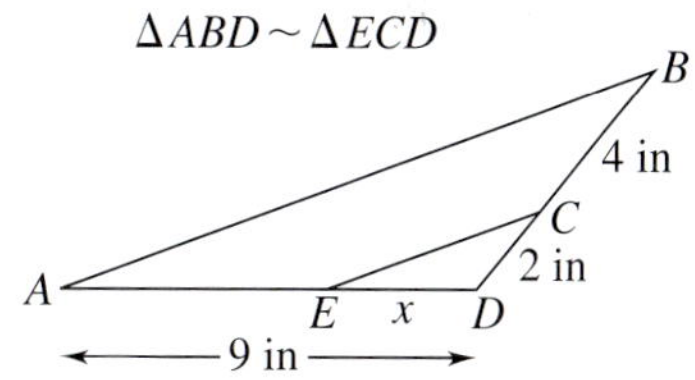

Find the length of the missing side of each figure.

51.

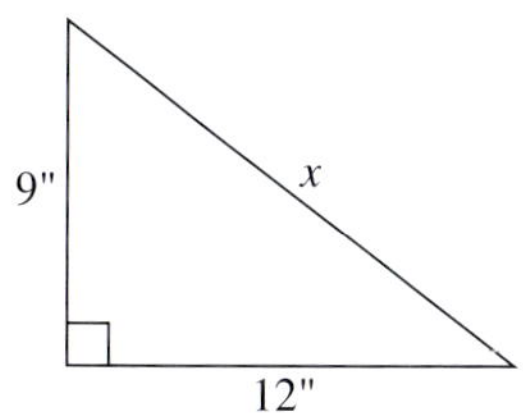

52.

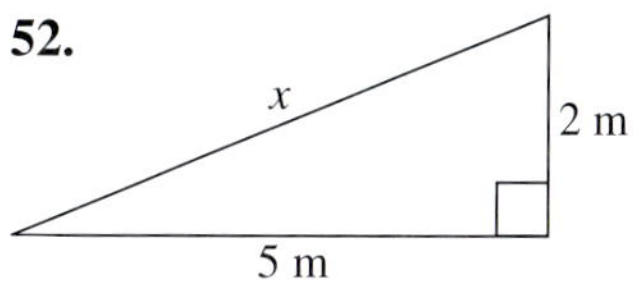

53.

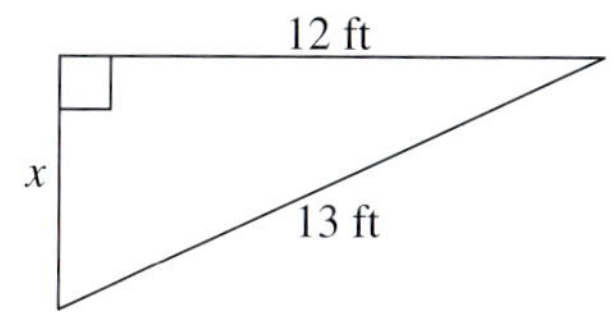

Solve each of the following. Round your answers to the nearest hundredth. Use $\pi = 3.14$.

54. Last year Jerry grew 3.5 inches. How many centimeters did he grow?

55. David bought 38 liters of gas in Vancouver, Canada. How many gallons did he buy to the nearest gallon?

56. Find the circumference of a pipe with a radius 3.8 centimeters.

57. There are 52 white keys on a piano keyboard. Each key is $\frac{7}{8}$ inches wide with an exposed length of $5\frac{1}{2}$ inches. What is the perimeter of the piano keyboard?

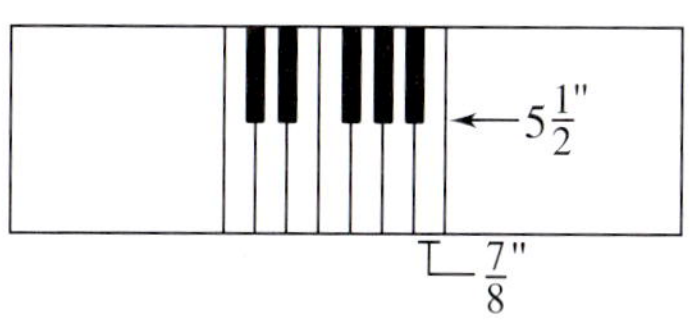

58. A group of people walk along the inside walls of the shopping mall pictured. What distance is one lap?

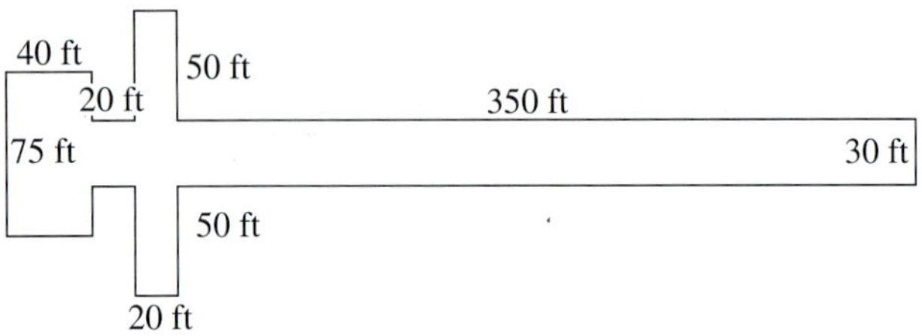

59. A contractor plans to frame an arched window with brick. Each brick has a thickness of $2\frac{1}{2}$ inches and the mortar between bricks is about $\frac{1}{2}$ inch thick. How many bricks will it take to line the window? Round your answer to the nearest unit.

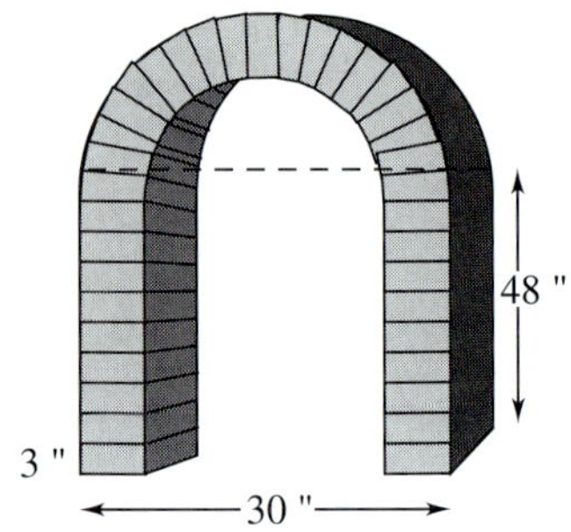

CHAPTER 8 TEST

Sketch and find the perimeter (or circumference) of each. Round off answers to tenths. Use $\pi = 3.14$.

1. Square with sides 8 meters

2. Right triangle with base 4 feet and height 3 feet

3. Circle with diameter 4 inches

4. Parellelogram with base 6 cm and sides 4 cm

5. Trapezoid with two equal sides with lengths of 6 in each and two parallel bases with lengths of 4 in and 8 in

Sketch and find the volume of each. Round your answers to tenths. Use $\pi = 3.14$.

6. Rectangular solid with dimensions $10' \times 10' \times 10'$

7. Sphere with radius of 5 meters

8. Right circular cylinder with diameter $2''$ and height $5''$

9. Right circular cone with radius 4 feet and height 8 feet

10. Rectangular pyramid with length 90 ft, width 80 ft, and height 100 ft

Answer the following problems.

11. A triangle has angles 35° and 66°. Find the third angle

12. One angle of an isosceles triangle measures 48°. Find the two base angles.

13. Find the number of degrees in $\frac{4}{5}$ of a circle.

Solve the following problems.

14. Find the area of the strike zone for a baseball batter if the width of home plate is eighteen inches and the height of the zone is two feet.

15. A can of vegetables has a height of 4 inches and a radius of 1.8 inches. Find the volume of the can. Round your answer to hundredths.

16. Find the square footage of the home shown.

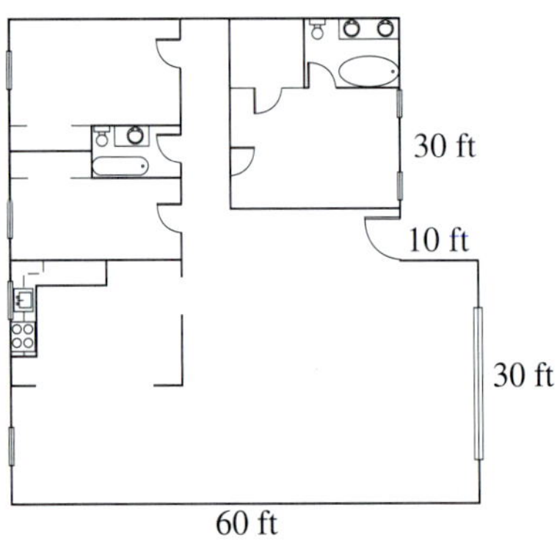

17. $\triangle ABC$ and $\triangle DEF$ are similar. $AB = 2$ cm, $BC = 3$ cm, and $DE = 16$ cm. Find the length of EF.

18. $\angle A$ and $\angle C$ are supplementary angles. Find $\angle C$ if $\angle A$ measures 60°.

19. Find $\angle A$, given that $\angle C = 105°$.

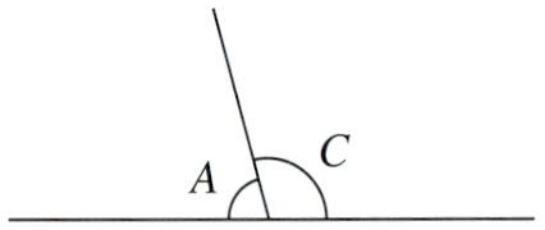

20. Are the two triangles similar? Why or why not?

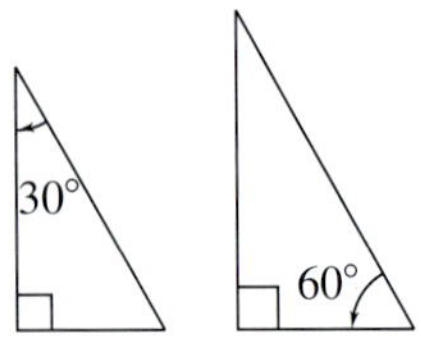

CUMULATIVE TEST CHAPTERS 7 AND 8

Convert the following measurements to their indicated equivalent form.

1. 1,000 yards = _______ feet

2. 20 cups = _______ quarts

3. 5 pounds = _______ ounces

4. 6.5 feet = _______ inches

5. 4.8 meters = _______ cm

6. 750 ml = _______ liters

7. 0.4 kg = _______ grams

8. 80 mg = _______ cg

9. 8 pounds = _______ kilograms

10. 18 inches = _______ centimeters

11. 12.4 gallons = _______ liters

12. 200 miles = _______ km

Perform the following calculations with denominate numbers.

13. Add 14 pounds 12 ounces to 6 pounds 8 ounces.

14. Subtract 5 hours 50 minutes from 8 hours 15 minutes.

Solve the following problems.

15. $\angle A$ and $\angle B$ are complementary angles. Find $\angle A$ if $\angle B = 78°$.

16. One of the base angles of an isosceles triangle is 35°. What are the measures of the other two angles of the triangle?

17. Find the perimeter of a parallelogram with adjacent sides of 3.4 in and 6 in.

18. Find the perimeter and area of a right triangle with legs of 9 ft and 12 ft.

19. The diameter of a circle is 14 cm. Find the circumference to the nearest tenth. Use $\pi = 3.14$.

20. Find the area of the figure below.

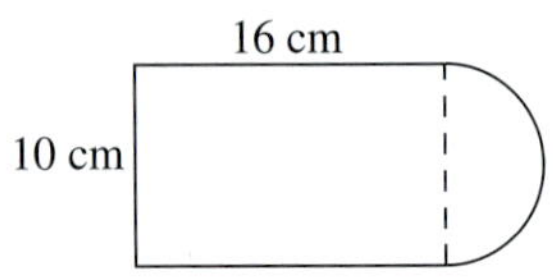

21. Find the volume of a right circular cylinder with height 8 ft and radius 6 ft.

22. Which has a larger volume, a sphere with a radius of 10 in, or a rectangular solid with length 6 in, width 5 in, and height 3 in?

23. Find the missing side of the following similar triangles.

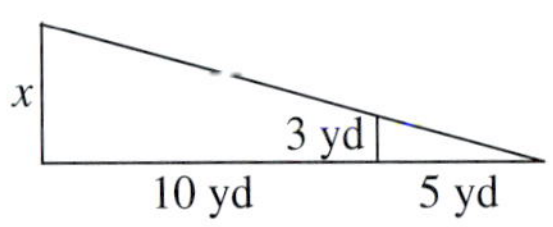

24. Find the hypotenuse of the right triangle with legs 5 and 12.

25. Find the second leg of the right triangle with a hypotenuse of length 2.5 and first leg of length 1.5.

Signed Numbers

9

APPLICATION *The temperature was −8°F at the time a drilling crew started their shift. It then fell six degrees during the next four hours. What is the new temperature?*

SOLUTION *The solution is found by subtracting signed numbers:*

$$-8°F - 6°F = (-8) - (+6) = -14°F$$

The new temperature is −14°F.

9.1 INTRODUCTION TO NEGATIVE NUMBERS

OBJECTIVES

1 Identify signed numbers.

2 Know the meaning of signed numbers.

1 Signed Numbers

All of the numbers that we have studied so far, other than zero, have been *positive numbers*. Every positive number can be written either with or without a positive sign. For example, the numbers 5 and +5 are the exact same number. The number +5 is read as "positive five." Positive decimal numbers, mixed numbers, and fractions can also be written with or without a positive sign as shown below:

$$10.17 = +10.17$$

$$1{,}000\frac{5}{6} = +1{,}000\frac{5}{6}$$

$$\frac{1}{4} = +\frac{1}{4}$$

Not all numbers are greater than zero; some are less than zero. Every number less than zero is a **negative number**, and every negative number is written using a negative sign. Thus, −1 is read as "negative one," $-\frac{1}{2}$ is "negative one-half," and $-4\frac{1}{5}$ is "negative four and one-fifth." Every negative number is less than zero.

1. Read each:

(a) -7

(b) $+8.5$

(c) $-12\frac{4}{5}$

(d) -6.02

(e) $\frac{2}{3}$

EXAMPLE 1

Read each:

(a) $-\frac{4}{5}$ **(b)** -7.8 **(c)** $-2\frac{1}{4}$ **(d)** 8

SOLUTION

(a) $-\frac{4}{5}$ = negative four-fifths

(b) -7.8 = negative seven and eight tenths

(c) $-2\frac{1}{4}$ = negative two and one-fourth

(d) 8 = positive eight ■

◇ *Do Practice Exercise 1.*

Each positive number has an **opposite** that is a negative number. For example, the opposite of $+5$ is -5 and the opposite of $+\frac{1}{4}$ is $-\frac{1}{4}$. The $+$ and $-$ signs in front of numbers give rise to the name signed numbers. Therefore, a **signed number** is any number with either a $+$ or a $-$ sign in front of it.

HELPFUL HINT

Every negative number must have a negative sign in front of it.

The **integers** are made up of the negative whole numbers, the number zero, and the positive whole numbers:

$$\ldots\ -4,\ -3,\ -2,\ -1,\quad 0,\quad 1,\ 2,\ 3,\ 4,\ \ldots$$

negative numbers 0 positive numbers

Numbers are placed in order of size on the number line, with smaller numbers to the left of greater numbers. For example, the integers -3, 1, and $+2$ are shown on the number line:

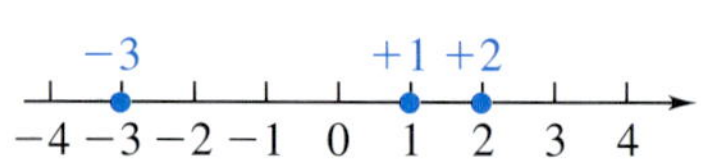

Notice that $1 = +1$ and that $+2 = 2$. Of the three numbers, -3 is the smallest so it is farthest to the left on the number line. Using the symbol $<$ that was introduced in Chapter 1, we have:

$$-3 < 1 \quad \text{and} \quad -1 < 2$$

(< : is less than)

ANSWERS
1. (a) negative seven
(b) positive eight and five tenths
(c) negative twelve and four-fifths
(d) negative six and two-hundredths
(e) positive two-thirds

EXAMPLE 2

Draw a number line and plot -2, 0, and 4 as points. Is $-2 < 0$?

SOLUTION

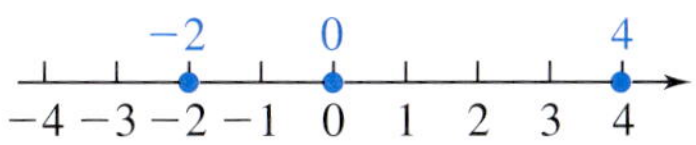

Yes, $-2 < 0$. ■

◇ *Do Practice Exercise 2.*

The number -1 is read as negative one. Negative one is to the left of *every* positive number on the number line, therefore -1 is less than *any* positive number. For example:

$$-1 < +10 \qquad -1 < +\frac{1}{2} \qquad -1 < 0.00001$$

The number 0 is greater than -1 since 0 is to the right of -1 on the number line. The number -2 is greater than -3 since it is to the right of -3 on the number line. But, -2 is less than -1 since -2 is to the left of -1 on the number line.

EXAMPLE 3

State whether each is true or false:

(a) $-6 < 9$ **(b)** $-8 < -9$ **(c)** $+4 < -8$ **(d)** $10 > -4$

SOLUTION

(a) true **(b)** false **(c)** false **(d)** true ■

◇ *Do Practice Exercise 3.*

A negative sign in front of a signed number can be thought of as "the opposite of." Thus -4 can be interpreted as "the opposite of 4." Similarly, the number $-(-4)$ is the opposite of (-4), which is $+4$.

$$-(-4) = \text{opposite of } (-4) = +4$$

EXAMPLE 4

Simplify each:

(a) $-(-10)$ **(b)** $-(+3)$ **(c)** $-(-700)$ **(d)** $-(+9)$

SOLUTION

(a) $-(-10)$ = opposite of -10 = $+10$
(b) $-(+3)$ = opposite of $+3$ = -3
(c) $-(-700)$ = opposite of -700 = $+700$
(d) $-(+9)$ = opposite of $+9$ = -9 ■

2. Plot -5 and $+1$ on a number line.

3. State whether each is true or false:

(a) $-3 < +5$

(b) $-7 < -2$

(c) $-3 > -7$

(d) $0 > +2$

(e) $-4 < -2$

(f) $-1 < -3$

ANSWERS

2.

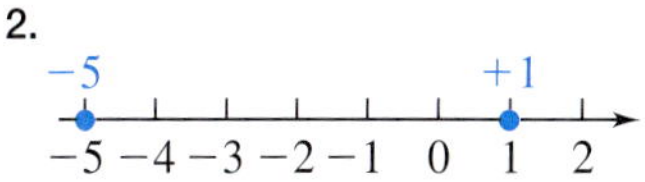

3. (a) true (b) true (c) true (d) false (e) true (f) false

4. Simplify $-(-8)$.

5. Simplify $+(-9)$.

In contrast, a plus in front of a number does not change the sign of a number. For example:

$$+(-2) = -2$$
$$+(-5) = -5$$
$$+(+7) = +7$$

◇ *Do Practice Exercises 4 and 5.*

Fractions and decimals can also be negative as well as positive. For example, the opposite of $+\frac{1}{2}$ is $-\frac{1}{2}$. The fraction $+\frac{1}{2}$ is between 0 and $+1$ so the fraction $-\frac{1}{2}$ must be between 0 and -1.

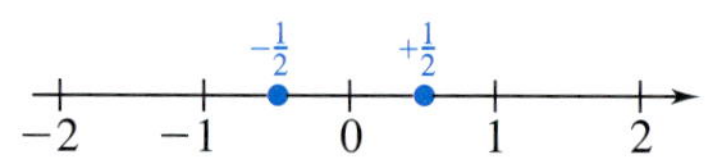

The fraction $+\frac{1}{2}$ is greater than the fraction $+\frac{1}{4}$ but the fraction $-\frac{1}{2}$ is *less than* $-\frac{1}{4}$. The reason for this can be seen on the number line.

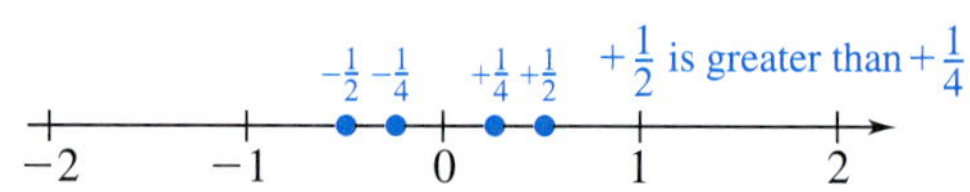

From left to right, the order in which these numbers occur is $-\frac{1}{2}$, $-\frac{1}{4}$, $+\frac{1}{4}$, and $+\frac{1}{2}$. In other words:

is less than

$$-\frac{1}{2} < -\frac{1}{4} < +\frac{1}{4} < +\frac{1}{2}$$

EXAMPLE 5

Graph the mixed number $-1\frac{3}{5}$ on the number line.

SOLUTION

The number $-1\frac{3}{5} = -\left(1\frac{3}{5}\right)$ is between -2 and -1 on the number line. It is closer to -2 than it is to -1 since the fraction three-fifths is more than one-half. One could also plot $+1\frac{3}{5}$ on the number line, then find its opposite.

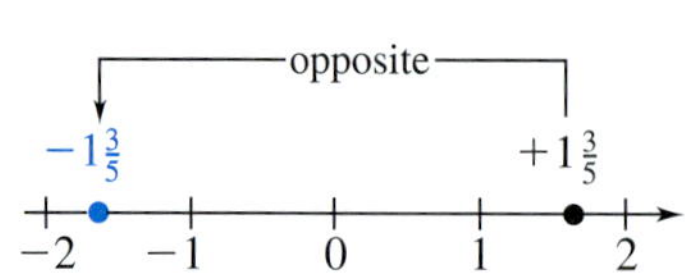

■

ANSWERS
4. +8
5. −9

The mixed number $-1\frac{1}{2}$ and the fraction $-\frac{3}{4}$ are shown on the following number line. Note that $-1\frac{1}{2} < -\frac{3}{4}$ since $-1\frac{1}{2}$ is to the left of $-\frac{3}{4}$.

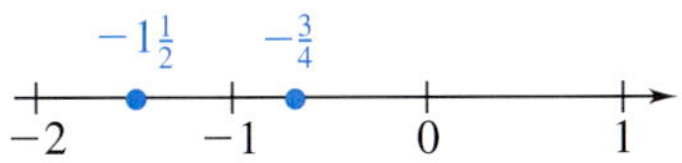

EXAMPLE 6

State which of the following are true or false:

(a) $-\frac{3}{4} > -\frac{1}{2}$ **(b)** $-5\frac{1}{2} < -2$

(c) $-3\frac{1}{4} > -8$ **(d)** $-2 < -2\frac{1}{4}$

SOLUTION

(a) false **(b)** true **(c)** true **(d)** false ■

◇ *Do Practice Exercises 6 and 7.*

2 Meaning of Signed Numbers

It is important to be able to solve problems containing signed numbers. However, it is equally important to have a feeling for the meaning of signed numbers. You have been working with positive numbers throughout the course. The following examples give you some different ways to think about negative numbers.

EXAMPLE 7

The freezing point of water is 0° Celsius. Sketch a thermometer and show the temperature $-5°$ Celsius on it.

SOLUTION

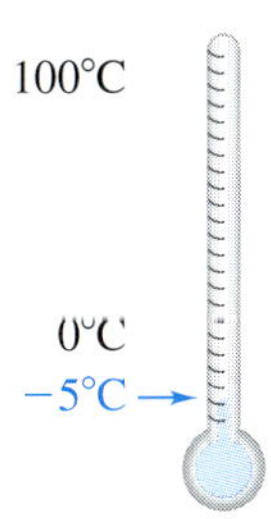

■

EXAMPLE 8

An individual has \$100 in a checking account and writes a check for \$120. How much money would the individual have in the account if the bank honored the check (by mistake)?

SOLUTION

The individual would have $-\$20$ in the account. This means that the individual would owe the bank \$20. ■

6. Is $-\frac{3}{4}$ greater than or less than -1?

7. Is $-1\frac{1}{2}$ greater than or less than -1?

ANSWERS

6. $-\frac{3}{4} > -1$

7. $-1\frac{1}{2} < -1$

EXAMPLE 9

The Dow Jones Average is one measure of stock prices on the New York Stock Exchange. What is the meaning of a sign reading:

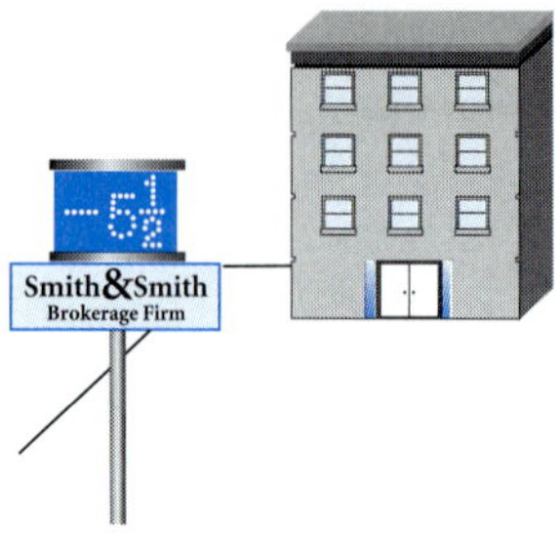

SOLUTION

The $-5\frac{1}{2}$ means that the Dow Jones Average went down that day by $5\frac{1}{2}$ points. ■

EXAMPLE 10

A military submarine dove to 500 feet below sea level. Write this as a negative number.

SOLUTION

The submarine dove to -500 feet. ■

name date hour

Draw a number line and graph each.

1. $+3, 0$

2. $-1, 4$

3. $-5, -4$

4. $-10, +10$

5. $-5, +\frac{1}{2}$

6. $3.5, -\frac{1}{4}$

7. $5\frac{1}{4}, -5\frac{1}{4}$

8. $-3, -3.9$

9. $-1, -1\frac{1}{2}$

10. $5.3, -1$

11. $-3.5, 0$

12. $-\frac{1}{2}, +2\frac{1}{2}$

13. $1\frac{1}{2}, \frac{11}{5}$

14. $-2, -1\frac{1}{2}$

Find the opposite of each.

15. $+7$

16. -4

17. $\frac{21}{2}$

18. $-\frac{1}{3}$

19. 2.7

20. -6.4

21. $5{,}000$

22. -2.7651

23. $+7.5$

24. $+\frac{4}{5}$

Find the greater of the two.

25. $6, \quad -4$

26. $3, \quad 3.2$

27. $0, \quad 4$

28. $1\frac{1}{4}, \quad -3$

29. $-7.0, \quad -17$

30. $-5.2, \quad -5.3$

31. $3, \quad -2$

32. $6, \quad -8$

33. $5.5, \quad -2$

34. $-7\frac{1}{2}, \quad \frac{3}{4}$

State which are true *and which are* false.

35. $6 < 19$

36. $5 > 12$

37. $3 < 4$

38. $7 > 65$

39. $-1 > 0$

40. $-2 < -4$

41. $-\frac{1}{2} < \frac{1}{2}$

42. $-4 > -5$

43. $-5 > -10$

44. $-3.5 < 3$

45. $0 > 8$

46. $-5 < -1$

47. $-2 > -1.5$

48. $2.5 > \frac{1}{2}$

49. $1.7 < -7.5$

50. $-2 < -7$

51. $-3 > -4.5$

52. $-1 < -5$

53. $+17 > -12.2$

54. $-4 > -8$

55. $-75 < -10$

56. $+21 < +2\frac{1}{3}$

57. $16\frac{1}{4} > -21$

58. $-80 < 0$

59. $1{,}200 > -1{,}000\frac{3}{8}$

60. $-3\frac{1}{2} > -2\frac{1}{4}$

61. $-40 < -41$

62. $-1 > -\frac{3}{4}$

Solve the following.

63. It is five degrees above zero and the temperature falls seven degrees. What is the new temperature?

64. At dawn it is two degrees below zero. The sun comes out and warms the area by nine degrees. What is the final temperature?

65. An arctic temperature is measured at 0°F before falling eight degrees. What is the new temperature?

JOURNAL EXERCISES

Follow the instructions given for each.

1. Draw a number line and plot -4 on it. Add 8 by moving eight units on the number line. Did you move to the right or to the left? Why?

2. Is the number 0 positive, negative, or neither? Explain.

3. Evaluate $-(-7)$. Then evaluate $-[-(-7)]$.

4. Explain the concept of signed numbers in your own words.

5. What is a negative number? Can you think of some other examples in which negative numbers are used?

1 Inequalities

We have been using the symbol $<$ for less than and the symbol $>$ for greater than since Chapter 1. These two symbols are used for **inequalities**, expressions in which the left side does not equal the right side. This section shows some additional symbols used for inequalities:

symbol	*read as*
$\neq$	is not equal to
$<$	less than
$\leq$	less than or equal to
$>$	greater than
$\geq$	greater than or equal to

An easy way to remember which inequality is which is to keep in mind that the arrow portion of the inequality always points to the smaller of the two numbers.

points towards smaller number

$6 < 9$	6 is less than 9
$5 > 0$	5 is greater than 0
$-5 \leq 9$	-5 is less than or equal to 9

The symbol $\leq$ means that the number on the left side of the inequality is either less than *or* equal to the number on the right side. The inequality $6 \leq 9$ is true since 6 is less than 9. The inequality $9 \leq 9$ is also true since 9 is equal to 9. However, the inequality $9 < 9$ is *false* since 9 is *not* less than 9.

EXAMPLE 1

State which are true and which are false:

(a) $25 < 31$ **(b)** $0 \geq -3$ **(c)** $7 < 7$
(d) $-4 \leq 15$ **(e)** $-5.5 \leq -5.5$ **(f)** $-3 \geq -1$

SOLUTION

(a) true **(b)** true **(c)** false
(d) true **(e)** true **(f)** false ■

◇ *Do Practice Exercise 1.*

A slash through an inequality symbol means that the inequality is not true. For example:

$6 \not> 9$	6 is *not* greater than 9
$-5 \not\geq 0$	-5 is *not* greater than or equal to 0
$-1 \not< -2$	-1 is *not* less than -2
$-2 \not\leq -4$	-2 is *not* less than or equal to -4

OBJECTIVES

1 Use inequalities.
2 Calculate absolute values.

1. State which are true and which are false:

(a) $-1 < 5$

(b) $3 \geq 8$

(c) $-5 \leq -5$

(d) $4.5 \geq 4.5$

(e) $-8 > -\frac{1}{2}$

(f) $-7.8 \leq -7.2$

ANSWERS
1. (a) true (b) false (c) true (d) true (e) false (f) true

2. State which are true and which are false:

(a) $4 \not> 3$

(b) $-5 \neq 7$

(c) $-13 \leq -5$

(d) $-1 \neq +1$

(e) $-2 \not< 7$

(f) $-3 \nleq -3.1$

3. Evaluate each:

(a) $|-9|$

(b) $|+6.2|$

(c) $\left|-6\frac{3}{4}\right|$

(d) $|-3.5|$

(e) $|5.7|$

(f) $|-5.2|$

(g) $|0|$

(h) $\left|+\frac{5}{6}\right|$

ANSWERS

2. (a) false (b) true (c) true (d) true (e) false (f) true

3. (a) 9 (b) 6.2 (c) $6\frac{3}{4}$ (d) 3.5 (e) 5.7 (f) 5.2 (g) 0 (h) $\frac{5}{6}$

EXAMPLE 2

State which are true and which are false:

(a) $16 \nleq 16$ (b) $12 \neq 12$ (c) $7.5 = 7.5$

(d) $-5 \geq -1$ (e) $-16 \neq +16$ (f) $4\frac{1}{2} \neq 4.5$

SOLUTION

(a) false (b) false (c) true

(d) false (e) true (f) false ■

◇ *Do Practice Exercise 2.*

2 Absolute Values

The **absolute value** of a number is its distance from the number 0 on a number line. The integer -3 is 3 units from 0, therefore the absolute value of -3 is the number 3.

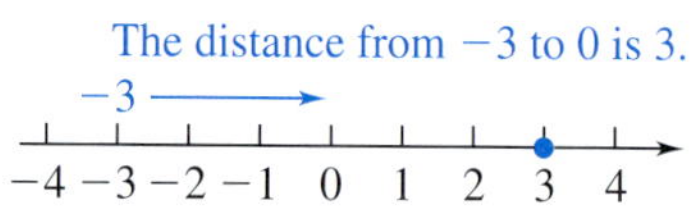

The integer $+3$ is also 3 units from 0 on the number line so that its absolute value is also 3. Both -3 and $+3$ have the absolute value 3 since both are three units from 0.

The symbol for absolute value is two vertical lines written as follows:

$|-3| = 3$ The absolute value of -3 is 3.

$|+7.5| = 7.5$ The absolute value of $+7.5$ is 7.5.

Some other examples of absolute values are:

$$|-12| = 12 \qquad |0| = 0 \qquad \left|-\frac{2}{3}\right| = \frac{2}{3} \qquad |6.4| = 6.4$$

EXAMPLE 3

Evaluate each:

(a) $|-5|$ (b) $|+2.2|$ (c) $\left|-3\frac{1}{2}\right|$

(d) $|0|$ (e) $\left|4\frac{1}{4}\right|$ (f) $|-5.2|$

SOLUTION

(a) $|-5| = 5$ (b) $|+2.2| = 2.2$ (c) $\left|-3\frac{1}{2}\right| = 3\frac{1}{2}$

(d) $|0| = 0$ (e) $\left|4\frac{1}{4}\right| = 4\frac{1}{4}$ (f) $|-5.2| = 5.2$ ■

◇ *Do Practice Exercise 3.*

name date hour

Place either the $<$ or $>$ symbol between the numbers to make the inequality true.

1. $6 \quad 8$ **2.** $5 \quad 0$ **3.** $-1.9 \quad 4.2$ **4.** $-5 \quad -4$

5. $\frac{1}{2} \quad -\frac{1}{4}$ **6.** $-1.4 \quad -1.5$ **7.** $4 \quad -1$ **8.** $-7 \quad 12$

9. $3.5 \quad -6.2$ **10.** $2.1 \quad -3.2$ **11.** $5 \quad -\frac{1}{2}$ **12.** $-\frac{4}{5} \quad -3\frac{1}{2}$

Place either a $\leq$ or a $\geq$ symbol in between the numbers to make the inequality true.

13. $12 \quad 14$ **14.** $6 \quad -4$ **15.** $43 \quad -43$

16. $-\frac{1}{4} \quad \frac{1}{4}$ **17.** $3.5 \quad 3.8$ **18.** $-1\frac{1}{2} \quad -1\frac{1}{4}$

19. $-5 \quad 3.5$ **20.** $+2.7 \quad -1$ **21.** $1\frac{3}{4} \quad -\frac{1}{2}$

22. $52\frac{2}{3} \quad 4\frac{1}{2}$ **23.** $-50 \quad -5.1$ **24.** $-3\frac{1}{4} \quad -3$

Place either an $=$ or $\neq$ symbol in between the numbers to make the statement true.

25. $6 \quad 4$ **26.** $5 \quad -5$ **27.** $-3 \quad -3$

28. $\frac{4}{10} \quad \frac{2}{5}$ **29.** $-4 \quad -\frac{40}{10}$ **30.** $4 \quad -7$

31. $-11 \quad -11$ **32.** $13 \quad -12$ **33.** $+10 \quad -10$

34. $\frac{3}{4} \quad 0.75$ **35.** $-\frac{7}{10} \quad -0.7$ **36.** $7.9 \quad -2$

Evaluate each.

37. $|12|$ **38.** $|+7|$ **39.** $\left|-\frac{3}{4}\right|$ **40.** $\left|-4\frac{4}{5}\right|$

41. $|-7|$ **42.** $|0|$ **43.** $|-2.4|$ **44.** $|3.8|$

45. $|+100|$ **46.** $|-0.004|$ **47.** $|6|$ **48.** $|-5.555|$

49. $|-3.4|$ **50.** $\left|2\frac{1}{2}\right|$ **51.** $\left|-4\frac{1}{3}\right|$ **52.** $\left|-5\frac{1}{4}\right|$

53. $|3.25|$ **54.** $|-2.75|$ **55.** $|-4.5|$ **56.** $\left|-\frac{1}{3}\right|$

57. $|-7|$ **58.** $|3.84|$ **59.** $|-0.009|$ **60.** $|-0.1|$

State which are true *and which are* false.

61. $-6 > +2$ **62.** $3 \neq -7$ **63.** $4 < -5.2$ **64.** $3.4 \geq 1$

65. $-1 > -2$ **66.** $-1 \geq 1$ **67.** $+3 < -5$ **68.** $-6 \leq 2\frac{1}{2}$

69. $0 \leq 0$ **70.** $-8 \geq +8$ **71.** $-40 < -4$ **72.** $7.5 < 7.9$

73. $-3.1 > -3.0$ **74.** $7\frac{1}{4} \neq 7.25$ **75.** $-10 < -10$ **76.** $-8\frac{1}{3} > -7\frac{3}{4}$

JOURNAL EXERCISES

Follow the directions given for each.

1. Explain each of the symbols for inequalities, including the meaning of a / through an inequality.
2. In your own words, explain absolute values. At this time, do you know of a use for absolute values?
3. Is it necessary to use either a positive sign or a negative sign with the number 0? Why, or why not?
4. Is $-6\frac{2}{3} \ngeq -6.5$ true or false? Make up five similar problems of your own and bring them to class to exchange with another student.

1 Add Numbers with Like Signs

The addition of positive signed integers is simply the addition of whole numbers:

$$(+6) + (+3) = 6 + 3 = 9$$

This addition problem can be shown on the number line. Begin at the first number of the sum $(+6)$ and add the second number $(+3)$ by moving to the right 3 units as shown.

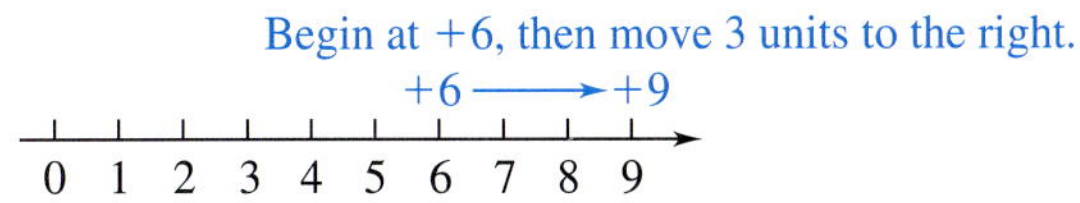

Addition with signed numbers looks somewhat different if one of the numbers is negative. For example, look at the following problem:

$$(+5) + (-2)$$

The answer cannot be the same as $(+5) + (+2)$ since the second number is now negative. Again a number line can help in understanding the process. Begin with $+5$ on the number line and move two units to the *left* since -2 is negative.

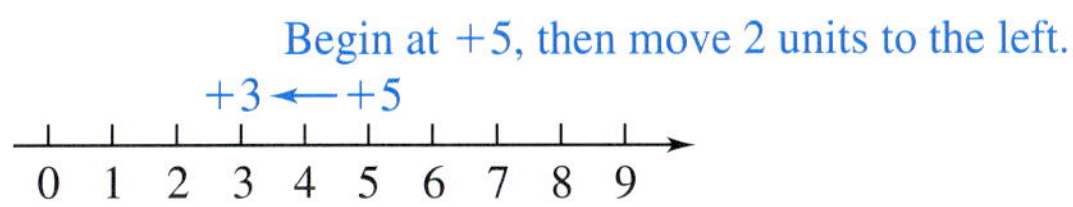

Thus $(+5) + (-2)$ is equal to $+3$.

EXAMPLE 1

Add $(+3) + (+4)$ using a number line.

SOLUTION

Start at $+3$, then move 4 to the right since you are adding a positive number.

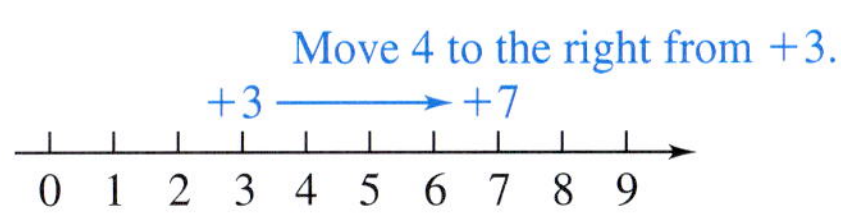

$$(+3) + (+4) = +7$$ ■

EXAMPLE 2

Add $(-4) + (-5)$ using a number line.

SOLUTION

Start at -4, then move 5 units to the *left* since you are adding a negative number.

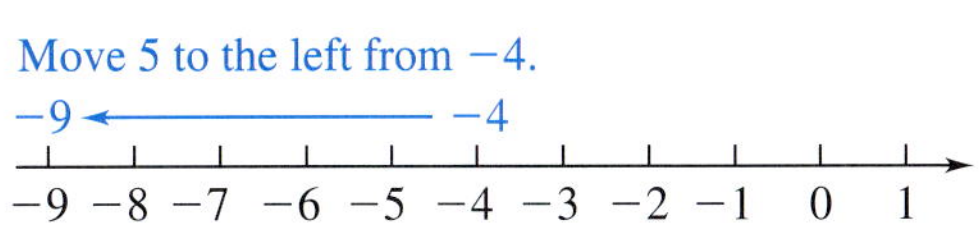

$$(-4) + (-5) = -9$$ ■

OBJECTIVES

1 Add numbers with like signs.

2 Add numbers with unlike signs.

1. Add $(+1) + (-3)$.

2. Add $(-4) + (-2)$.

3. Do $-3\frac{1}{2}$ and -2 have like signs?

4. Do $+5.2$ and -7 have like signs?

5. Do 3 and 2.7 have like signs?

6. Add $(-15) + (-10)$.

7. Add positive twelve plus a positive four.

You may have noticed from the examples above that the sum of two positive numbers is a positive number. Also, the sum of two negative numbers is a negative number.

◇ *Do Practice Exercises 1 and 2.*

Two numbers have **like signs** if they both have the same sign. Thus, $+5$ and $+8.2$ have like signs since both have the $+$ sign. The signed numbers -4 and -6 also have like signs since both have the $-$ sign. Two numbers are said to have **unlike signs** if the signs of the numbers differ. Therefore, -6 and $+2.3$ have unlike signs, as do $4\frac{1}{2}$ and -0.55. Although no sign is in front of $4\frac{1}{2}$, we understand that it is a positive number, or $4\frac{1}{2} = +4\frac{1}{2}$.

◇ *Do Practice Exercises 3 through 5.*

The rule that follows gives the steps to add two numbers with like signs. Step 3 of the rule can be remembered by recalling that the sum of two positive numbers is positive and that the sum of two negative numbers is negative.

Rule: To Add Two Numbers with Like Signs

Step 1 Find the absolute value of each number.

Step 2 Add the absolute values.

Step 3 The sum receives the sign of the original numbers.

EXAMPLE 3

Add $+6$ and $+9$.

SOLUTION

First note that $+6$ and $+9$ have like signs so that the rule above applies.

Step 1 $|+6| = 6$ $\quad |+9| = 9$ — absolute values

Step 2 $6 + 9 = 15$ — Add absolute values.

Step 3 $+15$ — give sign of original numbers ■

EXAMPLE 4

Evaluate $(-5) + (-12)$.

SOLUTION

(-5) and (-12) have like signs.

Step 1 $|-5| = 5$ $\quad |-12| = 12$ — absolute values

Step 2 $5 + 12 = 17$ — Add absolute values.

Step 3 -17 — give sign of original numbers ■

◇ *Do Practice Exercises 6 and 7.*

ANSWERS
1. -2
2. -6
3. yes
4. no
5. yes
6. -25
7. $+16$

2 Add Numbers with Unlike Signs

The sum of two numbers with unlike signs has the sign of that number which is further from 0 on a number line. For example, the sum of $(-4) + (+3)$ is negative since -4 is further from 0 than is $+3$. This is the idea behind Step 4 of the rule that follows.

Rule: To Add Two Numbers with Unlike Signs

Step 1 Find the absolute value of each number.

Step 2 Find the greater absolute value.

Step 3 *Subtract* the smaller absolute value from the greater absolute value.

Step 4 Give the result the sign of the number with the greater absolute value.

8. Add $(-18) + (+14)$.

9. Find the sum of -20 and $+70$.

EXAMPLE 5

Add $(+4) + (-8)$.

SOLUTION

First, notice that $+4$ and -8 have unlike signs.

Step 1	$\lvert +4 \rvert = 4$	$\lvert -8 \rvert = 8$	absolute values
Step 2	8		greater absolute value
Step 3	$8 - 4 = 4$		Subtract.
Step 4	-4		give sign of -8

The number -8 has a greater absolute value than does $+4$, so the final result has the same sign as -8. ■

EXAMPLE 6

Add $(-5.5) + (+3.5)$.

SOLUTION

Note that -5.5 and $+3.5$ have unlike signs.

Step 1	$\lvert -5.5 \rvert = 5.5$	$\lvert +3.5 \rvert = 3.5$	absolute values
Step 2	5.5		greater absolute value
Step 3	$5.5 - 3.5 = 2$		Subtract.
Step 4	-2		give sign of -5.5 ■

◇ *Do Practice Exercises 8 and 9.*

HELPFUL HINT

The first question to answer when adding signed numbers is whether they have like signs or unlike signs.

ANSWERS
8. -4
9. 50

10. Add $(-4.5) + (+4)$.

11. Add negative three-eighths plus negative two and one tenth.

ANSWERS

10. -0.5

11. $-2\frac{19}{40}$ or -2.475

EXAMPLE 7

Add $(-30) + (+5)$.

SOLUTION

(-30) and $(+5)$ have unlike signs, so absolute values must be subtracted.

$\lvert -30 \rvert = 30 \qquad \lvert +5 \rvert = 5$	absolute values
$30 - 5 = 25$	Subtract.
-25	Result. ■

Fractions and decimal numbers can also be positive or negative. The process of adding fractions or decimals with signs follows the same rules.

EXAMPLE 8

Add $\left(-\frac{1}{2}\right) + (-0.75)$.

SOLUTION

First note these two values have like signs, thus absolute values must be added.

$\left(-\frac{1}{2}\right) + (-0.75)$	
$(-0.5) + (-0.75)$	fraction to decimal
$\lvert -0.5 \rvert = 0.5 \qquad \lvert -0.75 \rvert = 0.75$	absolute values
$0.5 + 0.75$	Add.
-1.25	Result. ■

◇ *Do Practice Exercises 10 and 11.*

Draw a number line, graph the first number in the addition problem, then add using the number line.

1. $(+6) + (+4)$

2. $0 + (+5)$

3. $(-3) + (+2)$

4. $(+7) + (+3)$

5. $(+4) + (-2)$

6. $(-10) + (-3)$

7. $(-4) + (-3)$

8. $(-5) + (+4)$

9. $(+8) + (-6)$

10. $(-3) + (-5)$

11. $(-2) + (+4)$

12. $(+6) + (-5)$

Evaluate each. (Review)

13. $|14|$

14. $|+3|$

15. $|-6|$

16. $|+5|$

17. $|0|$

18. $\left|-\frac{1}{2}\right|$

19. $\left|+\frac{3}{4}\right|$

20. $|-6.25|$

Add each.

21. $(+6) + (+3)$ **22.** $(+8) + (-3)$ **23.** $(+4) + (+3)$ **24.** $(-9) + (+3)$

25. $(-4) + (-2)$ **26.** $-6 + 5$ **27.** $(-4) + (-5)$ **28.** $(-6.1) + (+3)$

29. $(+5) + (-2.5)$ **30.** $(-1) + (-4)$ **31.** $(-6) + (+4)$ **32.** $(+5) + (-8)$

33. $(+3) + (+9)$ **34.** $(-8) + (-7)$ **35.** $(-6) + (+12)$ **36.** $(-8) + (-14)$

37. $(+3) + (-8)$ **38.** $(-7) + (-12)$ **39.** $(+3) + (+4)$ **40.** $(-7) + (+15)$

41. $(-60) + (-12)$ **42.** $(-3) + (-7)$ **43.** $\left(\frac{1}{2}\right) + (-2.4)$ **44.** $\left(\frac{3}{4}\right) + \left(-4\frac{1}{8}\right)$

45. $3.7 + (-2.6)$ **46.** $(+2.7) + (-2)$ **47.** $(-4) + \left(3\frac{3}{4}\right)$ **48.** $(-3.5) + (-8.1)$

49. $\left(-\frac{1}{2}\right) + \left(-\frac{1}{4}\right)$ **50.** $\left(-\frac{3}{4}\right) + 5.6$

Set each of the following up as a signed number problem using addition, then add.

51. Terri has a checking account balance of \$43.06 and writes a check for \$57.02. What would be the new balance if the check were cashed?

52. Tom has a net worth of \$16,000 but loses \$24,000 on a real estate investment. What is his new net worth?

53. During a scientific investigation in northern Greenland, the temperature was $-12°C$ before increasing by 9°C. What is the new temperature?

54. At age 40, a woman was 65 inches tall. However, she was two inches shorter at age 80 due to curvature of the spine. What is her new height?

55. A scuba diver was 8 meters below the surface before diving 1.5 meters deeper. Write the new depth as a negative number.

56. A supply store sold 40 lawn mowers although they only had 12 on hand at the store. What is their current inventory of lawn mowers? (*Hint:* It is a negative number.)

57. An automobile dealer sold a fleet of 50 sedans to a police department, but only had 8 of them on hand on the car lot. What is their current inventory of sedans? (*Hint:* It is a negative number.)

JOURNAL EXERCISES

1. Explain the differences between adding two signed numbers with like signs and adding two signed numbers with unlike signs. Why are the two processes different?

2. Is the concept of absolute value useful in adding signed numbers? Explain.

3. Do you think the following statement is true or false? There are more positive whole numbers than negative whole numbers. Explain your answer.

1 Subtract Signed Numbers

The subtraction of signed numbers can be shown on the number line. For example, $(+5) - (+2) = 5 - 2 = 3$ can be subtracted by starting at $+5$ and then moving two to the left.

Begin at +5, then move to the left 2 units.

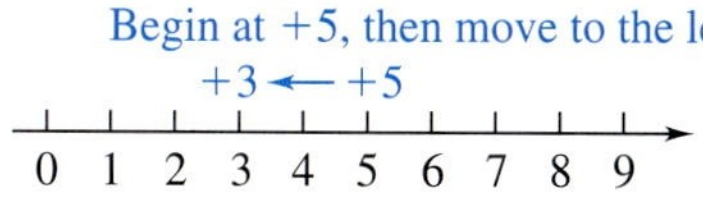

Therefore, $(+5) - (+2) = +3$.

The subtraction of a positive number results in movement to the left on the number line. Therefore, subtraction of a negative number results in movement in the opposite direction, or to the right on a number line. Therefore,

$(+1) - (-4)$ positive one minus a negative four

can be solved by beginning at $+1$ on the number line and moving 4 units to the right.

Begin at +1, then move to the right 4 units.

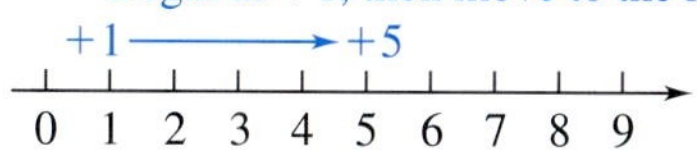

Therefore, $(+1) - (-4) = +5$.

EXAMPLE 1

Subtract $(-3) - (+4)$ using a number line.

SOLUTION

Begin at -3, then move to the *left* four units since a positive number is being subtracted.

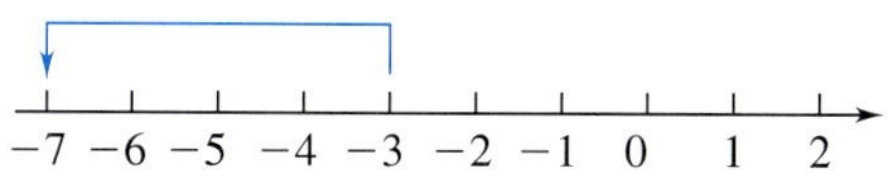

$$(-3) - (+4) = -7$$ ■

EXAMPLE 2

Subtract $(-3) - (-2)$ using a number line.

◇ *Do Practice Exercises 1 and 2.*

OBJECTIVE

1 Subtract signed numbers.

1. Subtract $(-5) - (-8)$.

2. Subtract $(-5) - (+3)$.

ANSWERS
1. $+3$
2. -8

3. Evaluate each:

(a) $(+12) - (-7)$

(b) $(-21) - (+8)$

(c) $(+20) - (-1)$

(d) $(+7) - (+2)$

(e) $(+2) - (-9)$

(f) $(-4) - (+9)$

ANSWERS
3. (a) $+19$ (b) -29 (c) $+21$
(d) $+5$ (e) $+11$ (f) -13

SOLUTION

Start at -3, then move to the *right* two units since a negative number is being subtracted.

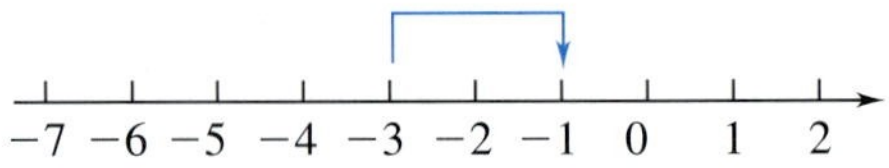

$$(-3) - -2) = -1 \quad ■$$

The following numerical examples help you understand the rule for subtracting signed numbers. First, recall that a negative sign in front of a number means ''the opposite of.'' For example:

$$-(-8) = \text{the opposite of } -8 = +8 = +(+8)$$
$$-(+9) = \text{the opposite of } +9 = -9 = +(-9)$$

The facts that $-(-8) = +(+8)$ and $-(+9) = +(-9)$ are used below to change each subtraction problem to an addition problem:

$(+7) - (-8)$		$(-12) - (+9)$
$(+7) + (+8)$	⟵ change to addition ⟶	$(-12) + (-9)$

This is the idea behind Step 1 of the rule that follows.

Rule: To Subtract Signed Numbers

Step 1 Change the subtraction sign to addition and change the sign of the number that follows.

Step 2 Solve the problem using the rules for adding signed numbers.

EXAMPLE 3

Find $(+8) - (-11)$.

SOLUTION

	$(+8) - (-11)$	
Step 1	$(+8) + (+11)$	Replace $-(-11)$ with $+(+11)$.
Step 2	$8 + 11$	Drop parentheses.
	19	Result. ■

EXAMPLE 4

Find $(-14) - (+5)$.

SOLUTION

$(-14) - (+5)$	
$(-14) + (-5)$	Change signs.
$\lvert -14 \rvert = 14 \qquad \lvert -5 \rvert = 5$	absolute values
$14 + 5 = 19$	Add absolute values.
-19	Result. ■

◇ *Do Practice Exercise 3.*

Subtraction problems involving signed numbers are first changed to an addition problem, then solved using the rules for *addition* of signed numbers from the prior section. Some additional examples follow.

EXAMPLE 5

Find $(+3) - (+20)$.

SOLUTION

$(+3) - (+20)$	
$(+3) + (-20)$	Change signs.
$\lvert +3 \rvert = 3 \qquad \lvert -20 \rvert = 20$	absolute values
$20 - 3 = 17$	Subtract absolute values.
-17	Result. ■

EXAMPLE 6

Subtract negative four from negative two and five tenths.

SOLUTION

$(-2.5) - (-4)$	-4 subtracted from -2.5.
$(-2.5) + (+4)$	Change signs.
$\lvert -2.5 \rvert = 2.5 \qquad \lvert +4 \rvert = 4$	absolute values
$4 - 2.5 = 1.5$	Subtract absolute values.
$+1.5$	Result. ■

◇ *Do Practice Exercises 4 and 5.*

4. Subtract $(+9) - (+12)$.

5. Subtract $(+7) - (-4)$.

The next two examples are of a type that some students have difficulty with. Look at them carefully. They show the importance of keeping the signs on any signed numbers separate from the math operator such as addition or subtraction.

EXAMPLE 7

Evaluate $-8 - 19$.

SOLUTION

It is important to first separate the signed numbers from the math operator such as addition or subtraction.

$-8 - 19$	negative 8 minus 19
	→ math operation (subtraction)
$(-8) - (19)$	
	→ signed numbers
$(-8) - (+19)$	$19 = +19$
$(-8) + (-19)$	Change signs.
-27	Result. ■

ANSWERS
4. -3
5. $+11$

6. Find $(-7) - 12$.

7. Subtract $4 - (-3.2)$.

ANSWERS
6. -19
7. 7.2

EXAMPLE 8

Evaluate $10 - (-4)$.

SOLUTION

$10 - (-4)$	10 minus negative 4
$(+10) - (-4)$	Identify signed numbers.
$(+10) + (+4)$	Change signs.
$+14$	Result. ■

◇ *Do Practice Exercises 6 and 7.*

Draw a number line, graph the first number in the subtraction problem, then subtract using the number line.

1. $(+5) - (+2)$

2. $(+3) - (-5)$

3. $(-5) - (+3)$

4. $0 - (-4)$

5. $(+5) - (+6)$

6. $(+8) - (-2)$

7. $(-4) - (+5)$

8. $(-3) - (-7)$

Evaluate each. (Review)

9. $|5|$

10. $|-4|$

11. $|-6.4|$

12. $|+8|$

Add. (Review)

13. $(+6) + (-5)$

14. $(-4) + (-3)$

15. $8 + (-2)$

16. $(-5) + (0)$

17. $(-20) + (-40)$

18. $(7.5) + \left(-8\frac{1}{2}\right)$

19. $(+4) + (-6)$

20. $(-5) + (-3)$

21. $(-8) + (+4.5)$

22. $(+9) + (-45)$

Subtract.

23. $(-4) - (-2)$

24. $(+11) - (-5)$

25. $8 - (-4)$

26. $(-14) - (-6)$

27. $12 - (+7)$

28. $(-13) - 4$

29. $17 - (-5.5)$

30. $(-7) - (-13)$

31. $(+8) - (-2)$

32. $(-7) - (+8)$

33. $(+3) - (-12)$

34. $(-7) - (-40)$

35. $(+7.5) - (+41)$

36. $(-36) - (-8)$

37. $(-17) - (+19)$

38. $(+16) - (+2)$

39. $\left(-14\frac{1}{2}\right) - (+8)$

40. $(+13) - \left(-4\frac{1}{3}\right)$

41. $(-5) - 7$

42. $8 - (-3)$

43. $(+6) - (-4)$

44. $9 - (-3)$

45. $12 - 18$

46. $-3 - 3.5$

47. $4 - (-2.7)$

48. $+0.3 - 6.8$

49. $-2\frac{1}{2} - \left(+\frac{2}{3}\right)$

50. $(-7) - (3.5)$

Add or subtract as indicated.

51. $(-6) + (-3)$

52. $4 - (-2)$

53. $(+8) - (+4)$

54. $(-12) + (-11)$

55. $(+3) - (-9)$

56. $6 - 9$

57. $(+4) - (-2)$

58. $3 - (+2)$

59. $(-4) + (-7)$

60. $(-7) - (+13)$

61. $(+51) + (-9)$

62. $3 - (-6)$

63. $(+95) - 12$

64. $4 - 21$

65. $9 - (-13)$

66. $-24 - 9$

67. $-3 - 2.5$

68. $+8 - 4.2$

Work each problem as a signed number problem.

69. The temperature is thirteen degrees above zero before falling fifteen degrees. What is the final temperature?

70. The lowest temperature recorded in Fairbanks, Alaska, is -40°F. The highest temperature recorded in Cairo, Egypt, is 117°F. What is the difference between the two temperatures?

The following table is to be used for the two problems that follow.

Wind Chill Table

		Temperature (°F)				
		20	10	0	−10	−10
	5	16	7	−5	−15	−26
Wind	10	3	−9	−22	−34	−46
Speed	15	−5	−18	−31	−45	−58
(mph)	20	−10	−24	−39	−53	−67

71. What is the effective wind chill temperature of 20°F in the presence of a wind of 15 mph? Could this temperature be dangerous for humans or animals?

72. At 6:30 A.M. the temperature was 0°F with a wind speed of 5 mph. By 7:30 the wind had changed directions and was averaging 20 mph although the temperature remained the same. What is the difference in the wind chill? Does the direction of the wind affect wind chill?

JOURNAL EXERCISES

Follow the direction given for each.

1. Explain how the subtraction of signed numbers differs from the addition of signed numbers.

2. Write and solve three word problems of your own that require addition or subtraction of signed numbers.

OBJECTIVES

1 Multiply using signed numbers.

2 Divide using signed numbers.

1 Multiplication

The multiplication of positive numbers is used for repeated addition of the same number. For example:

$$4 \cdot 5 = 5 + 5 + 5 + 5 = 20$$ Four 5s are added together.

Similarly the multiplication of a positive number times a negative number can also be shown using addition.

$$(+4) \cdot (-5) = (-5) + (-5) + (-5) + (-5) = -20$$ Four -5s are added together.

The multiplication of a positive number times a negative number can always be written as the sum of negative numbers. Further, the addition of negative numbers always results in a negative number. Therefore, a positive number times a negative number will always be a negative number. For example:

$$(+5) \cdot (-3) = \underbrace{(-3) + (-3) + (-3) + (-3) + (-3)}_{\text{add five } -3\text{s}} = -15$$

$$(-2) \cdot (+3) = (+3) \cdot (-2) = \underbrace{(-2) + (-2) + (-2)}_{\text{add three } -2\text{s}} = -6$$

EXAMPLE 1

Multiply $(+4) \cdot (-2)$ using repeated multiplication.

SOLUTION

$$(+4) \cdot (-2) = (-2) + (-2) + (-2) + (-2) = -8$$ ■

◇ *Do Practice Exercise 1.*

1. Multiply each using repeated addition:

(a) $(+4) \cdot (-8)$

(b) $(-8) \cdot (+9)$

(c) $(+4) \cdot (-2.5)$

The multiplication of two negative numbers can be seen indirectly as the continuation of a pattern.

$$(-5)(+3) = -15$$
$$(-5)(+2) = -10$$
$$(-5)(+1) = -5$$
$$(-5)(0) = 0$$
$$(-5)(-1) = +5$$
$$(-5)(-2) = +10$$

increasing by 5s

Notice that the result of the multiplication increases by 5 each time the second factor decreases by 1. The pattern continues by multiplying -5 times 0, then by -1, then by -2 an so on. It is a natural continuation of the pattern that the product of two negative numbers is a positive number.

The patterns identified above show that the product of two signed numbers with unlike signs is negative an that the product of two negative signed numbers is positive. This information is used in Step 3 of the following rule.

ANSWERS

1. (a) -32 (b) -72 (c) -10

2. Multiply $(-7) \cdot (+8)$.

3. Find the product of -9 and -12.

4. Multiply eight times negative four and one-half.

ANSWERS
2. -56
3. $+108$
4. -36

Rule: To Multiply Signed Numbers

Step 1 Find the absolute value of each number.

Step 2 Multiply the absolute values.

Step 3 Find the correct sign:
- $+$ if numbers have like signs
- $-$ if numbers have unlike signs

EXAMPLE 2

Multiply (-3) times $(+4)$.

SOLUTION

First, notice they have unlike signs so that the result will be negative.

Step 1	$\lvert -3 \rvert = 3$	$\lvert +4 \rvert = 4$	absolute values
Step 2	$3 \cdot 4 = 12$		Multiply absolute values.
Step 3	-12		Result. ■

EXAMPLE 3

Find $\left(-5\frac{1}{2}\right) \cdot \left(-2\frac{1}{4}\right)$.

SOLUTION

The product of two factors with like signs is positive.

$$\left|-5\frac{1}{2}\right| = 5\frac{1}{2} \qquad \left|+2\frac{1}{4}\right| = 2\frac{1}{4} \qquad \text{absolute values}$$

$$5\frac{1}{2} \cdot 2\frac{1}{4} \qquad \text{Multiply.}$$

$$\frac{11}{2} \cdot \frac{9}{4} \qquad \text{Change to fractions.}$$

$$\frac{99}{8} \qquad \text{Multiply fractions.}$$

$$+12\frac{3}{8} \qquad \text{Result.}$$ ■

◇ *Do Practice Exercises 2 through 4.*

The multiplication rule for signed numbers can be shortened to the following.

HELPFUL HINT TO MULTIPLY TWO NUMBERS

1. Multiply the two absolute values.
2. The result is $+$ if the signed numbers have like signs.
3. The result is $-$ if the signed numbers have unlike signs.

EXAMPLE 4

Find the product of $+5$ with -2.1.

SOLUTION

$\lvert +5 \rvert = 5$ and $\lvert -2.1 \rvert = 2.1$	absolute values
$5 \cdot 2.1 = 10.5$	Multiply.
-10.5	Give $-$ sign. ■

2 Division

The rule for division of signed numbers is motivated by multiplication, as shown below:

Division		*Multiplication*
$\frac{42}{+7} = +6$	since	$(+7) \cdot (+6) = +42$
$\frac{-42}{+7} = -6$	since	$(+7) \cdot (-6) = -42$
$\frac{+42}{-7} = -6$	since	$(-7) \cdot (-6) = +42$
$\frac{-42}{-7} = +6$	since	$(-7) \cdot (+6) = -42$

The quotient of two numbers with like signs is positive. The quotient of two numbers with unlike signs is negative. This idea is used in Step 3 of the following rule.

Rule: To Divide Signed Numbers

Step 1 Find the absolute value of each number.

Step 2 Divide the absolute values.

Step 3 Find the correct sign:
- $+$ if numbers have like signs
- $-$ if numbers have unlike signs

EXAMPLE 5

Find $(+600) \div (-5)$.

SOLUTION

Note that they have unlike signs so that the result will be negative.

Step 1	$\lvert +600 \rvert = 600$ $\lvert -5 \rvert = 5$	absolute values
Step 2	$600 \div 5 = 120$	Divide.
Step 3	-120	Result. ■

The rules for dividing signed numbers can be simplified as follows:

5. Divide -200 by -10.

6. Divide $+\frac{1}{3}$ by $-\frac{4}{5}$.

7. Divide negative four tenths by two tenths.

HELPFUL HINT TO DIVIDE TWO NUMBERS

1. Divide the two absolute values.
2. The result is + if the signed numbers have like signs.
3. The result is − if the signed numbers have unlike signs.

As you can see, the rule for dividing signed numbers is very similar to the rule for multiplying signed numbers. In both cases the process of finding the sign is the same. The sign of the result is + if the signs are like, and − if the signs are unlike.

EXAMPLE 6

Find $-17 \div -2$.

SOLUTION

They have like signs, so the result will be positive.

$$(-17) \div (-2)$$

$$\frac{-17}{-2}$$

$$+8.5 \quad \text{or} \quad 8\frac{1}{2}$$ Result. ■

The process is the same even if dividing decimals, fractions, or mixed numbers. Look at the next example.

EXAMPLE 7

Divide $4\frac{1}{4}$ by $\left(-\frac{1}{5}\right)$.

SOLUTION

First, note that the two numbers have unlike signs so that the result of the division will be negative.

$$+4\frac{1}{4} \div \left(-\frac{1}{5}\right)$$

$$\frac{17}{4} \div \left(-\frac{1}{5}\right)$$ Change to a fraction.

$$\frac{17}{4} \cdot \left(-\frac{5}{1}\right)$$ ÷ to ·, invert second fraction

$$\frac{17 \cdot -5}{4 \cdot 1}$$ Multiply.

$$\frac{-85}{4}$$

$$-21\frac{1}{4} \quad \text{or} \quad -21.25$$ Result. ■

◇ *Do Practice Exercises 5 through 7.*

ANSWERS
5. $+20$
6. $-\frac{5}{12}$
7. -2

HELPFUL HINT

Division by zero is undefined. Therefore, both 25 ÷ 0 and (−7.3) ÷ 0 are undefined.

EXAMPLE 8

The temperature at Point Barrow, Alaska, was 8°F and fell to −19°F in three days. What was the average drop in temperature per day?

SOLUTION

The average drop per day is the total temperature drop divided by the number of days over which the temperature dropped.

$$\frac{\text{temp. drop}}{\text{number days}} = \frac{8 - (-19)}{3 \text{ days}} = \frac{8 + 19}{3} = \frac{27}{3} = 9°\text{F/day}$$

(8 − (−19) → change in temperature; 3 days → number of days)

name date hour

Multiply.

1. $(+5)(+7)$

2. $(+3)(-8)$

3. $(-4)(-5)$

4. $(-6)\cdot(+9)$

5. $6\cdot(-4)$

6. $(-12)\times(-11)$

7. $(-8)(-7)$

8. $7\cdot 7$

9. $4\cdot(-9)$

10. $(-1)(-1)$

11. $(+1)(-1)$

12. $(-3)(+2)$

13. $(+3)(-8)$

14. $(+7)(-9)$

15. $(-8)(+8)$

16. $(+12)\cdot(-3)$

17. $(-19)\cdot(-4)$

18. $(+67)(-3)$

Divide.

19. $12 \div (-4)$

20. $(+35) \div 7$

21. $(-64) \div (-8)$

22. $(-600) \div (-6)$

23. $-7 \div (+2)$

24. $48 \div (-3)$

25. $(-250) \div (-50)$

26. $(-425) \div (-25)$

27. $(-63) \div (+9)$

28. $-81 \div (+3)$

29. $-110 \div (10)$

30. $65 \div (-5)$

31. $\left(\frac{1}{4}\right) \div \left(-\frac{1}{2}\right)$

32. $\left(-\frac{3}{4}\right) \div \left(-\frac{3}{4}\right)$

33. $-\frac{7}{9} \div \frac{9}{7}$

34. $4\frac{1}{2} \div \left(-\frac{2}{3}\right)$

35. $(-16.8) \div (-0.2)$

36. $(-21.5) \div (-5)$

Evaluate.

37. $(-6) + (-5)$

38. $16 \div (-4)$

39. $(+7)(-8)$

40. $4 \cdot (+5)$

41. $(17)(-5)$

42. $(-40) \div (-8)$

43. $-62 \div (-4)$

44. $(-18) + (-20)$

45. $100 \div (-25)$

46. $14 + (-16.5)$

47. $-12 - \left(-12\frac{1}{2}\right)$

48. $0 \cdot \left(-\frac{5}{8}\right)$

49. $-75 \cdot (+11)$

50. $(-9)(+6)$

51. $0 \div (-3)$

52. $-16 \div 4$

53. $(-200) \div 2$

54. $(6.5)(-5.3)$

55. $4.7 + (-5.91)$

56. $-3 - (-4)$

57. $+14\frac{1}{4} + (-80)$

58. $(-9) \times (-8)$

59. $(-43.2) \div 6$

60. $12 - (-99.9)$

61. $(-8) + 3.4$

62. $\left(-\frac{1}{4}\right) \cdot \frac{1}{3}$

63. $(+18) - (-4)$

64. $7.2 - (-1.1)$

65. $6.4 \cdot (-8.6)$

66. $3.5 + (-8.2)$

67. $(-6) \div 3\frac{1}{2}$

68. $6 \div \left(-\frac{3}{5}\right)$

69. $(-7) \cdot (-3.2)$

70. $\left(-\frac{2}{3}\right) \cdot \left(+\frac{3}{4}\right)$

71. $\frac{6}{7} \div \left(-1\frac{3}{4}\right)$

72. $(+9.6) \cdot (-4.01)$

Work this problem.

73. List the prime numbers that occur between 0 and 10. Recall that 1 is not a prime number. Place a negative sign in front of each, then multiply them together.

JOURNAL EXERCISES

1. Explain the differences between multiplication and division of signed numbers.

2. Give examples of each: whole numbers, proper fractions, improper fractions, mixed numbers, decimal numbers, and signed numbers. Do you know of any number that is not a signed number? Explain.

3. Do sales clerks normally work with negative numbers? Can you think of anyone who works with signed numbers on a daily basis? Why are signed numbers necessary?

1 Work Problems Containing Different Notations

We have previously worked many order of operations problems. The challenge of the problems in this section lie in the wide range in types of numbers we can now work with: integers, fractions, mixed numbers, decimals, and signed numbers. All of these number types can now be combined into a single problem.

The order of operations for signed numbers follows the same rules as those for integers, fractions, mixed numbers, and decimals. The rules are repeated next.

Rule: To Evaluate an Expression

Step 1 Evaluate the contents of the parentheses and above and below the fraction bar.

Step 2 Evaluate exponents.

Step 3 Evaluate multiplication and division in the order in which they occur while moving from the left to the right across the expression.

Step 4 Evaluate addition and subtraction in the order in which they occur while moving from the left to the right across the expression.

"Parentheses" as used in Step 1 of the rule above refers to mathematical operations within the parentheses, not to parentheses simply holding a signed number. Look carefully at the following examples.

EXAMPLE 1

Evaluate $(-1)(+5) + (-6)$.

SOLUTION

$(-1)(+5) + (-6)$

$(-5) + (-6)$ Multiply first.

-11 Add with like signs. ■

EXAMPLE 2

Evaluate $(-5)^2(-2) \div (+25)$.

SOLUTION

$(-5)^2(-2) \div (+25)$

$(+25)(-2) \div (+25)$ Exponent: $(-5)^2 = (-5)(-5) = (+25)$

$(-50) \div (+25)$ Multiply.

-2 Divide to get result. ■

◇ *Do Practice Exercises 1 through 3.*

OBJECTIVE

1 Work with problems containing fractions, decimals, mixed numbers, and signed numbers.

1. Evaluate $(-7) + (-3)(-4)$.

2. Find $(-4)^2 \div (+2)$.

3. Evaluate $(-3)(-2) \div (0)$.

ANSWERS

1. $+5$

2. $+8$

3. Division by zero is undefined.

4. Evaluate
$$(-3) + \left(+2\frac{1}{2}\right)(-4).$$

5. Evaluate $(-1)^2 + (-4)$.

ANSWERS
4. −13
5. −3

EXAMPLE 3

Evaluate $\left(-3\frac{1}{4}\right) - \left(6\frac{1}{2}\right)\left(4\frac{2}{3}\right)$.

SOLUTION

$$\left(-3\frac{1}{4}\right) - \left(6\frac{1}{2}\right)\left(4\frac{2}{3}\right)$$

$$\left(-\frac{13}{4}\right) - \left(\frac{13}{2}\right)\left(\frac{14}{3}\right) \quad \text{Change to fractions.}$$

$$\left(-\frac{13}{4}\right) - \left(\frac{13}{1}\right)\left(\frac{7}{3}\right) \quad \text{Cross reduce.}$$

$$\left(-\frac{13}{4}\right) - \left(\frac{91}{3}\right) \quad \text{Multiply.}$$

$$\left(-\frac{39}{12}\right) - \left(\frac{364}{12}\right) \quad \text{Use common denominator.}$$

$$\frac{(-39) - 364}{12} \quad \text{Subtract.}$$

$$\frac{(-39) + (-364)}{12} \quad \text{Change signs.}$$

$$\frac{-403}{12} \quad \text{Add numerators.}$$

$$-33\frac{7}{12} \quad \text{Result.}$$

■

The next example uses parentheses within brackets for signed numbers. The first step in evaluating the expression is to add the two signed numbers within the brackets.

EXAMPLE 4

Evaluate $(-6)^2 - [(-3) + (-6)]^2$.

SOLUTION

$$(-6)^2 - [(-3) + (-6)]^2$$

$$(-6)^2 - [-9]^2 \quad \text{Add within brackets.}$$

$$(+36) - (+81) \quad \text{Evaluate exponents.}$$

$$(+36) + (-81) \quad \text{Change signs.}$$

$$-45 \quad \text{Result.}$$

■

◇ *Do Practice Exercises 4 and 5.*

name date hour

Evaluate. (Review)

1. $6(-8)$

2. $(+6) - (+15)$

3. $(-16) + (-4)$

4. $(-100) \div (-2)$

5. $(-40) + 120$

6. $(-3)(-7)$

7. $72 \div (-8)$

8. $(-8) - (-22)$

9. $(-3) - (+4)$

10. $7 + (-12)$

11. $(-25) \div (-5)$

12. $(-9.1) \cdot (+6)$

13. $(-8)(+4)$

14. $360 \div (-60)$

15. $(-5.2) \div (+2)$

16. $\left(+\frac{3}{4}\right) \div (-1.2)$

17. $6 - 8\frac{1}{2}$

18. $\left(-\frac{7}{3}\right) \cdot 4\frac{1}{2}$

19. $(-16) - (-4.2)$

20. $(8.5) + (-2.1)$

Evaluate.

21. $(-6)(-8) + (+4)$

22. $(-7)(+2) + (-3)$

23. $-7 + (-20) \div (+5)$

24. $(-1)(-1)(-3)$

25. $(-1) \div (-1) \cdot (-1)$

26. $0 \cdot (-16) + 41$

27. $(-80) \div 4 + 2$

28. $(-16) - (-14) + (+8)$

29. $(-3)(-21) + (-10) \div (-2)$

30. $(-5) - (-3) \cdot 4$

31. $(-28) \div (-2) \cdot (-50)$

32. $[37 + (-22)] \div (-5)$

33. $6 - 8 \cdot (-4)$

34. $0 \cdot (-6) \div (+3)$

35. $18 - (34)(-2)$

36. $[76 \div (-2)] \cdot (-1)$

37. $[(-6) + (+8)]^2$

38. $(-16)^2 + (-2.1)$

39. $(20) - (-5) \cdot (8)$

40. $(+3) - (-4)(+2)^2$

41. $6 \div (-3) - (+4)(-8)$

42. $4 \cdot (-9) + (-1.6)$

43. $(+14) - [(2) \div (-2) + (+1)]$

44. $[(-6) + (-3)] \cdot [(-8) + 2(4)]$

45. $4\frac{1}{2} - \left(-12\frac{1}{4}\right)$

46. $\left(-\frac{1}{2}\right)\left(\frac{3}{4}\right) - \frac{4}{5}$

47. $(2.1)^2 + \left(1\frac{1}{3}\right)$

48. $[(-16.2) \div (0.02)] + 7\frac{1}{2}$

49. $\left(-2\frac{1}{4}\right)^2 + \left(1\frac{1}{3}\right)$

50. $[(-12)\left(-4\frac{1}{2}\right) + (-2.5)]^2$

51. $\left(-5\frac{1}{2}\right) - (-7.5)\left(-3\frac{3}{5}\right)$

52. $(-12)^2 \div (-1.44)$

53. $(-7{,}800{,}000) \div (-0.002)$

54. $(-3{,}000)(-10)^2 + \left(-2\frac{2}{5}\right)$

55. $(-7) \div (-7) \div (-7) + 1$

56. $(-4.5) - \left(3\frac{1}{2}\right)\left(-\frac{1}{4}\right)$

First write as an expression, then evaluate.

57. Find the sum of negative seven with the product of negative two and a positive eight.

58. Find the product of seventeen with the sum of negative four and negative nine.

59. Find the product of negative ten with the quotient of seventy divided by negative two.

60. Find the product of -70 with the quotient of -1.1 divided by 2.

61. Find the average of -6 and -4.4.

62. Find the average of 8.8 and -4.8.

63. Square negative five and add the result to three-fourths.

64. Add the following three numbers: negative sixteen hundred, negative seven million seven, and four hundred thousand sixty. Then multiply the sum times negative four and three tenths.

JOURNAL EXERCISES

1. In some sense, this section is a summary of arithmetic. Explain.

2. Go back through your journal and list subjects that were particularly difficult for you. Are you able to work those problems now?

3. Are you pleased with the progress you have made so far in the course? If not, what would have helped you learn the material better?

CHAPTER 9 REVIEW

SECTION	TOPIC OR TERM	EXAMPLE OR DEFINITION
9.1		
Introduction to Negative Numbers	Positive Number	A positive number is a number greater than 0.
	Negative Number	A negative number is a number less than 0.
	Opposite	-8 is opposite $+8$, $+3.2$ is opposite -3.2
	Signed Numbers	Signed numbers are numbers with a + or − sign in front of them: $+4, \ -5.2, \ +\frac{1}{4}$
	Integers	$\ldots -4, -3, -2, -1, 0, 1, 2, 3, 4, \ldots$
9.2		
Basics	Inequality	An inequality is an expression in which the left side does not equal the right side: $-6 < -3, \quad -5 \geq -10, \quad 7 \neq 11$
	Absolute Value	The absolute value of a number is its distance on the number line from 0: $\lvert -6 \rvert = 6, \quad \lvert -2.5 \rvert = 2.5, \quad \lvert 4 \rvert = 4$
9.3		
Addition	Like Signs	Two numbers have like signs if they have the same signs: -3 and -1 have like signs
	Unlike Signs	Two numbers have unlike signs if they have different signs: $+4$ and -2 have unlike signs
	Add with Like Signs	Add absolute values and give the result the sign of the original numbers. $(+6) + (+4) = +10$, $(-3) + (-2) = -5$
	Add with Unlike Signs	Subtract absolute values and give the result the sign of the number with the greater absolute value. $(-3) + (+5) = +2$, $(+8) + (-60) = -52$
9.4		
Subtraction	Subtract Signed Numbers	Change the sign of the second number and use the rules for addition of signed numbers. $(-6) - (-4) = (-6) + (+4) = -2$ $(-5) - (+8) = (-5) + (-8) = -13$ $(+14) - (-12) = (+14) + (+12) = +26$

SECTION	TOPIC OR TERM	EXAMPLE OR DEFINITION
9.5		
Multiplication and Division	Multiply and Divide Signed Numbers	The result when multiplying or dividing numbers with like signs is positive. The result when multiplying or dividing numbers with unlike signs is negative. $(-6)(-4) = +24, \quad (-8) \cdot (+7) = -56$ $(-24) \div (-12) = +2, \quad 36 \div (-5) = -7\frac{1}{5}$
9.6		
Order of Operations	Order of Operations	$(-12) - 3(-8)$ $(-12) - (+3)(-8)$ Identify signs. $(-12) - (-24)$ Multiply. $(-12) + (+24)$ Change signs. $+12$ Result.

THE BUDDY SYSTEM IN SCUBA DIVING

One rule in scuba diving is to never dive alone. In other words, a diver should always have a buddy with him or her. The following example shows you why.

Tom and Laurie were scuba diving at a depth of 80 feet when Tom's ear drum broke.

1. Write the depth as a negative number.

In pain, Tom panicked and became disoriented or confused. He swam for what he thought was the surface. Actually, he was swimming towards the depths of the ocean. Laurie saw Tom panic and saw that he was mistakenly going deeper. She raced to catch up with him and did so at an additional 12 feet of depth.

2. Beginning with the depth from question 1 above, write an expression for the new depth and evaluate it.

Laurie calmed Tom and took him half way to the surface where they needed to wait for a few minutes. They needed to wait in order to allow their bodies time enough to remove the nitrogen bubbles that had formed within their bodies. Laurie waited with Tom, then took him to the surface.

3. Write the depth at which they waited as a negative number.

The buddy system saved Tom's life.

CHAPTER 9 REVIEW PROBLEMS

Place either $<$ or $>$ between the two numbers in order to make the statement true.

1. $6 \quad 9$

2. $3 \quad -4$

3. $-5 \quad -6$

4. $2 \quad -7$

5. $-4 \quad -3\frac{1}{2}$

6. $2.5 \quad 2.75$

7. $6 \quad -8$

8. $-3 \quad \frac{1}{4}$

9. $-9 \quad 21.7$

10. $12 \quad 4.2$

Place either $\leq$ or $\geq$ between the two numbers in order to make the statement true.

11. $6 \quad 4$

12. $-5 \quad -20$

13. $7 \quad -8$

14. $-4 \quad -3$

15. $-2.7 \quad -2.9$

16. $-6\frac{1}{4} \quad 6\frac{1}{4}$

17. $5.2 \quad (-3)$

18. $-6.5 \quad -6.0$

19. $-2.5 \quad -2\frac{1}{4}$

20. $-6.8 \quad -2.5$

Plot each number on a number line, then find the absolute value.

21. -6

22. $-5, 7.5$

23. $-2, -3$

24. $-1\frac{1}{4}$

25. $2\frac{1}{2}$

26. -5.7

27. 7.5, 0

28. -2, -3.4

29. 5, $-4\frac{1}{2}$

30. 17, -8.5

Evaluate.

31. $(-6) \div (+2)$

32. $(+9)(-7)$

33. $(-28) \div (-4)$

34. $(+6) - (-2)$

35. $(-8) + (-3)$

36. $(+4) - (+5)$

37. $(-6) + (-7)$

38. $(24)(-2)$

39. $(-18) \div (-3)$

40. $(-5) - (-2)$

41. $(-27) \div (+3)$

42. $(-16) + (+7)$

43. $(-8)(+4)$

44. $(+60) + (-8)$

45. $(5) \cdot (-8)$

46. $(+81) \div (+9)$

47. $(-3) - (6)$

48. $(+16) - (-4)$

49. $(-21) + (+3)$

50. $(-13) \cdot (-6)$

51. $(+5) - (-10)$

52. $2 - (-9)$

53. $(+17) + (-21)$

54. $(-42) \div (-6)$

55. $(-1) \cdot (+1)$

56. $(-2) - (-3)$

57. $(-3) + (-4)$

58. $(9)(-8)$

59. $(-54) \div (+3)^2$

60. $(+48) \div (+6) + (-4)$

61. $-4 - 9$

62. $3 - (-25)$

63. $-450 \div -12$

64. $(-4.5) \cdot (-9)$

Evaluate.

65. $3 - 2 \cdot 5$

66. $(+4) - (-2)(+1)$

67. $(+8) \div (-2) \cdot (+3)$

68. $6 - 4(-5)$

69. $(-5) \cdot (-3) \cdot (-8)$

70. $(-75) \cdot (-5) + (-1)$

71. $\frac{1}{2} \div \left(-\frac{3}{4}\right)$

72. $\left(-2\frac{1}{2}\right) \cdot 4\frac{3}{5}$

73. $(-2.75) \div 0.25$

74. $16 \div (-4) + (-5)(-3)$

Write each as an expression using signed numbers, then evaluate.

75. The temperature is $-6°F$ before rising $8°F$. What is the final temperature?

76. The temperature is $-2°F$ and falls at an average rate of $3°F$ per day for five days in a row. What is the final temperature?

77. Multiply negative seven times the sum of negative four and positive nine.

78. Find the quotient of negative seventeen million ten divided by positive three.

CHAPTER 9 TEST

State which are true *and which are* false.

1. $65 < -9$

2. $-3 \leq -30$

3. $7.5 \neq 7\frac{1}{2}$

Plot each on a number line.

4. -7.5

5. $5\frac{3}{4}$

6. $8, -4$

Evaluate each.

7. $|-12.3|$

8. $|-4|$

9. $\left|+6\frac{4}{5}\right|$

Evaluate.

10. $(-6)(+8)$

11. $(-42) \div (-6)$

12. $(-6) - (-8)$

13. $12 + (-90)$

14. $-19 - 5$

15. $(-7) \cdot (-12)$

16. $(+60) \div (-6)$

17. $(-14) + (-8)$

18. $(-9) - (+27)$

Evaluate each expression.

19. $(-6) - (+3)(-4)$

20. $5 - 21(-5)$

21. $(-82) \div 41 \cdot (+3)$

Work each problem.

22. Does $|-12| = |12|$?

23. Subtract $|-41.5| - 1.8$.

24. Is $0 \leq 0$ a true statement?

25. Find the product of negative six with twelve thousand thirty.

26. Divide the product of -12 and -8 by 16.

27. Find the product of negative four with negative two and three tenths.

28. The temperature is 8°F. It falls at the rate of 4°F each day for three days in a row. What is the final temperature?

29. The temperature fell from 4°C to -6°C. By how many degrees did it fall?

30. Jane subtracted the checks she had written and found that the remaining balance was $-\$38.64$. What does that mean?

An Introduction to Algebra

10

APPLICATION *A company estimates that monthly sales are given by the equation:*

$$s = 22.2 \cdot x + \$12{,}000$$

where: s = *monthly sales and*
x = *monthly advertising expenses*

What monthly sales can they expect with an advertising budget of $1,000 per month?

SOLUTION *Total sales are found by substituting $1,000 in for the variable* x, *then solving the equation.*

$s = 22.2 \cdot x + 12{,}000$	
$s = 22.2 \cdot 1{,}000 + 12{,}000$	Replace x with 1,000.
$s = 22{,}200 + 12{,}000$	Multiply.
$s = 34{,}200$	Add.

The expected monthly sales are $34,200.

10.1 VARIABLES AND EXPRESSIONS

OBJECTIVES

1. Identify constants, variables, terms, and expressions.
2. Identify like terms.
3. Add and subtract like terms.

Arithmetic and algebra are similar in many ways. For example, the same types of numbers are used in algebra as are used in arithmetic: integers, fractions, mixed numbers, decimals, and signed numbers. The mathematical operations of addition, subtraction, multiplication, division, and exponents are used in both algebra and arithmetic. Furthermore, the commutative, associative, and distributive properties of mathematics apply to both arithmetic and algebra. In this chapter, we will look at some properties of algebra.

1 Constants, Variables, Terms, and Expressions

The basic building blocks of algebra are constants and variables. **Constants** are numbers such as 4, -10.5, $2\frac{1}{4}$, or $\frac{13}{3}$. They are called constant because their values never change.

Variables are letters or symbols that represent a quantity that is unknown or may vary or change. Variables may be represented by any letter such as a, B, C, m, n, x, y, or z.

EXAMPLE 1

Identify which are variables and which are constants:

(a) 7.5 **(b)** x **(c)** $-3\frac{1}{2}$

(d) A **(e)** g **(f)** $\frac{4}{5}$

SOLUTION

(a) constant **(b)** variable **(c)** constant
(d) variable **(e)** variable **(f)** constant ■

An **expression** is made up of one or more quantities, called **terms**, that are being added. For example:

$$3xy + 6x$$

($6x$ → term; $3xy$ → term)

The expression $3xy + 6x$ has two terms, $3xy$ and $6x$. The expression $9y - 4ab + 3c$ has three terms: $9y$, $-4ab$, and $3c$.

The term $6x$ is written without a multiplication sign even though the constant 6 is being multiplied times the variable x:

$$6 \cdot x = 6x$$

($6x$ → preferred way to write a term)

The constant being multiplied times the variable or variables in a term is called the **numerical coefficient**. Several terms and their numerical coefficients are shown below:

Term	*Numerical Coefficient*
$12y$	12
$-3r$	-3
$5.5A$	5.5
$\frac{1}{2}z$	$\frac{1}{2}$
$81ab$	81

The numerical coefficient of the term y is 1 since 1 times y is y:

$$y = 1 \cdot y$$

(1 → numerical coefficient)

The numerical coefficient of $-5x$ is -5 since $-5x = (-5) \cdot x$. The numerical coefficient of $-x$ is -1 since $-x = (-1) \cdot x$.

EXAMPLE 2

Find the numerical coefficient of each:

(a) $5x$ **(b)** $-9y$ **(c)** $21.3z$ **(d)** $-3\frac{1}{2}a$

(e) x **(f)** $5ab$ **(g)** $-4.5xyz$ **(h)** $-r$

SOLUTION

(a) 5 **(b)** -9 **(c)** 21.3 **(d)** $-3\frac{1}{2}$

(e) 1 **(f)** 5 **(g)** -4.5 **(h)** -1 ■

◇ *Do Practice Exercise 1.*

Some examples of the building blocks of algebra follow:

Building Block	*Examples*
Constants	$3.7, -4, \frac{1}{2}, 5\frac{2}{7}, 8$
Variables	a, b, c, A, m, p, x, y
Terms	$5y, 2x, 5xy, -2y, 2.7z$
Expressions	$3y + 2z, \frac{1}{2} - 5x$

2 Like Terms

Two terms are said to be **like terms** if they both have the exact same variables raised to the same exponent. For example, $5y$ and $8y$ are like terms since both contain the variable y. The terms $5x$ and $-7x$ are also like terms since they both contain the variable x.

Unlike terms are terms that do not contain the same variables or they contain the same variables raised to different exponents. Thus $5y$ and $8.5r$ are unlike terms since they contain different variables. The terms $3x$ and $7xy$ are also unlike terms since one has the variable x and the other has the variables xy.

Some examples of like terms and unlike terms are:

Like Terms	*Unlike Terms*
$3x, -\frac{2}{5}x$	$3x, 2y$
$7xy, 2.3xy$	$7x, 50$
$ab, -ab$	$-9y, 3y^2$

EXAMPLE 3

Which of the following are like terms?

$$5ab, \quad 6x, \quad 9y, \quad -2x \quad 4.5z, \quad 2\frac{1}{2}y, \quad -8.2ab$$

1. Find the numerical coefficient of each:

(a) $7y$

(b) $\frac{1}{3}r$

(c) A

(d) $-2z$

(e) $-\frac{1}{2}r$

(f) $-s$

(g) $4\frac{1}{3}ab$

(h) $-5\frac{1}{2}b$

ANSWERS

1. (a) 7 (b) $\frac{1}{3}$ (c) 1 (d) -2 (e) $-\frac{1}{2}$ (f) -1 (g) $4\frac{1}{3}$ (h) $-5\frac{1}{2}$

2. Are $5y$ and $4xy$ like terms?

3. Are $3XYZ$ and $\frac{1}{2}XYZ$ like terms?

4. Are $3y$, $5xy$, and $-2.7x$ like terms?

SOLUTION

Like terms: $6x, \quad -2x$

Like terms: $9y, \quad 2\frac{1}{2}y$

Like terms: $5ab, \quad -8.2ab$ ■

◇ ***Do Practice Exercises 2 through 4.***

3 Add and Subtract Like Terms

Like terms can be added (or subtracted) even though the value for the variables are unknown. The idea behind the adding of like terms is given by the following example:

□□ □□□□ □□□□□□

2 squares + 4 squares = 6 squares

The same basic concept can be applied when adding $2x + 4x$:

$$\underbrace{x + x}_{2x} + \underbrace{x + x + x + x}_{4x} = \underbrace{x + x + x + x + x + x}_{6x}$$

Or when adding $2ab + 4ab$ as follows:

$$\underbrace{ab + ab}_{2ab} + \underbrace{ab + ab + ab + ab}_{4ab} = \underbrace{ab + ab + ab + ab + ab + ab}_{6ab}$$

The terms $2x$ and $4x$ or the terms $2ab$ and $4ab$ can be added because they are *like terms*. They have the same variables raised to the same power. Unlike terms cannot be added together. For example, the following cannot be combined:

2 squares + 9 circles

Similarly, the following are unlike terms and *cannot* be combined into one term:

$$2x + 9y$$

Like terms are added using the distributive property of mathematics, which is:

$$\text{Distributive Property:} \qquad ax + bx = (a + b)x$$

This is just a variation of the distributive property studied in Chapter 1. An example follows:

$$3 \cdot 5 + 3 \cdot 2 = 3 \cdot (5 + 2)$$

Several examples of the distributive property are shown next.

ANSWERS
2. no, unlike terms
3. yes, like terms
4. no, unlike terms

EXAMPLE 4

Simplify $10z + 27z$ using the distributive property.

SOLUTION

$ax + bx$	$10z + 27z$	
$(a + b)x$	$(10 + 27)z$	Distributive property
	$37z$	Result. ■

◇ *Do Practice Exercises 5 and 6.*

Adding or subtracting algebraic terms involves signed number operations as shown in the next two examples.

EXAMPLE 5

Simplify $20y - 47y$

SOLUTION

$20y - 47y$	$20y$ and $47y$ are like terms.
$(20 - 47)y$	Distributive property
$-27y$	Result. ■

EXAMPLE 6

Add $-20xy + (-10xy)$.

SOLUTION

$-20xy + (-10xy)$	$-20xy$ and $-10xy$ are like terms.
$[(-20) + (-10)]xy$	Distributive property
$-30xy$	Result. ■

EXAMPLE 7

Subtract $25A - (-10)A$.

SOLUTION

First, note that $25A$ and $-10A$ are like terms so that they can be combined into one term.

$25A - (-10)A$	
$(+25)A - (-10)A$	Identify signs.
$[(+25) - (-10)]A$	Distributive property
$[(+25) + (+10)]A$	Change $-(-10)$ to $+(+10)$.
$(+35)A$	Add.
$35A$	Result. ■

◇ *Do Practice Exercise 7.*

5. Simplify $18y + 21y$.

6. Can $31xy + 12.5y$ be simplified?

7. Simplify each:

(a) $22x - 35x$

(b) $(-8)y + (-6)y$

(c) $-10a - (-15)a$

ANSWERS
5. $39y$
6. No simplification is possible.
7. (a) $-13x$ (b) $-14y$ (c) $5a$

8. Simplify each:

(a) $5.2y - 0.3y$

(b) $\left(-6\frac{1}{2}\right)x + (-2)x$

(c) $3z - \left(-\frac{1}{2}\right)z$

9. Simplify each:

(a) $2y + 5y + 9xz$

(b) $3ab - 21ab + 7a$

(c) $-5R - 8A + 2A$

ANSWERS
8. (a) $4.9y$
(b) $-8.5x$ or $-8\frac{1}{2}x$
(c) $3.5z$ or $3\frac{1}{2}z$
9. (a) $7y + 9xz$ (b) $-18ab + 7a$
(c) $-5R - 6A$

The process is the same when adding like terms that contain fractions, decimals, or mixed numbers. Look at the next example.

EXAMPLE 8

Add $4.5r + \left(-2\frac{3}{4}\right)r$

SOLUTION

$$4.5r + \left(-2\frac{3}{4}\right)r$$

$$\left[(+4.5) + \left(-2\frac{3}{4}\right)\right]r \qquad \text{Distributive property}$$

$$[(+4.5) + (-2.75)]r \qquad -2\frac{3}{4} = -2.75$$

$$1.75r \qquad \text{Result.}$$

■

◇ *Do Practice Exercise 8.*

Remember, only like terms can be combined. Unlike terms must be left separate as shown in the next example.

EXAMPLE 9

Simplify $6xy + 3x + 2x$.

SOLUTION

First note that $3x$ and $2x$ are like terms, but $6xy$ is unlike $3x$ or $2x$ and cannot be combined with them.

$$6xy + 3x + 2x$$

$$6xy + (3 + 2)x \qquad \text{Distributive property}$$

$$6xy + 5x \qquad \text{Result.}$$

$6xy$ and $5x$ *cannot* be combined since they are not like terms. ■

◇ *Do Practice Exercise 9.*

name date hour

Decide whether each is a constant or a variable.

1. 25 **2.** x **3.** C **4.** -6.5

5. z **6.** $+\frac{1}{2}$ **7.** s **8.** 80

Find the numerical coefficient of each term.

9. $3x$ **10.** $2xy$ **11.** $-5z$ **12.** $7.2a$

13. $-3.1x$ **14.** $6\frac{1}{2}ab$ **15.** $+3\frac{1}{4}c$ **16.** $\frac{1}{2}y$

17. $\frac{2}{5}z$ **18.** $-\frac{1}{3}r$ **19.** x **20.** $-y$

21. $-2\frac{1}{2}z$ **22.** $-100y$ **23.** $+3.4z$ **24.** $-9.8r$

25. $-A$ **26.** B **27.** $-9C$ **28.** $3.275y$

Simplify, if possible.

29. $3x + 2x$ **30.** $4z + 5z$ **31.** $7a + 81a$

32. $4y + 3y$ **33.** $5b - 2b$ **34.** $16y + 9a$

35. $16x - 24c$ **36.** $7y + 11.5y$ **37.** $17.5x - 3.5x$

38. $3y + 8y$ **39.** $4z - 3z$ **40.** $(-8)a + (+2)a$

41. $(-4)b + (-5)b$

42. $(+3)y - (-2)y$

43. $(-8)z - (+4)z$

44. $-4a + 2a$

45. $-3b - 9m$

46. $6y - 8y$

47. $-7.5a + 2a$

48. $9y - 19y$

49. $\frac{1}{2}b - \frac{3}{4}b$

50. $-16 + 21y$

51. $97r - \frac{1}{2}r$

52. $-4y - 3$

53. $1.5z - 27z$

54. $12r - 200r$

55. $a - \frac{1}{3}r$

56. $4x + 2x + 7y$

57. $9z + \frac{1}{2}z$

58. $18y - 2z + 4$

59. $3x - 8x + 2xy$

60. $2y - 4x + 3x$

JOURNAL EXERCISES

Follow the directions given for each.

1. Explain the concepts of constants, variables, terms, and expressions in your own words.

2. Does $x \cdot y = y \cdot x$ for any value of x and y? What if $y = 0$? Explain.

3. Does $\frac{x}{1} = x$? Explain.

4. Is $\frac{x}{1}$ defined? Explain.

5. Define like terms and unlike terms. Why is it that unlike terms cannot be combined into one term?

6. In what ways are arithmetic and algebra similar? In what ways are they different?

7. Were signed numbers used in this section? Where, and why?

1 Basic Properties of Algebra

Recall from Chapter 1 that the basic mathematical properties of arithmetic are the identity property, commutative property, associative property, and distributive property. Examples of each of these are given below:

	Addition	*Multiplication*
Identity	$5 + 0 = 5$	$6 \cdot 1 = 6$
Commutative	$3 + 7 = 7 + 3$	$3 \cdot 5 = 5 \cdot 3$
Associative	$2 + (5 + 4) = (2 + 5) + 4$	$2 \cdot (5 \cdot 4) = (2 \cdot 5) \cdot 4$

The distributive property involves both addition and multiplication.

Distributive $\quad 6(5 + 4) = 6 \cdot 5 + 6 \cdot 4$

Each of these properties also applies in algebra, as shown below:

	Addition	*Multiplication*
Identity	$x + 0 = x$	$x \cdot 1 = x$
Commutative	$x + y = y + x$	$x \cdot y = y \cdot x$
Associative	$10 + (x + z) = (10 + x) + z$	$x \cdot (y \cdot z) = (x \cdot y) \cdot z$

Once again, the distributive property involves both addition and multiplication.

Distributive $\quad 8(x + y) = 8x + 8y$

The value in front of the parentheses is multiplied times *each term* inside the parentheses.

EXAMPLE 1

State which property is used:

(a) $3 + 6 = 6 + 3$ **(b)** $4 \cdot 5 = 5 \cdot 4$
(c) $5(3 + 2) = 5 \cdot 3 + 5 \cdot 2$ **(d)** $1 + (3 + 5) = (1 + 3) + 5$
(e) $x + y = y + x$ **(f)** $8 \cdot b = b \cdot 8$
(g) $7(r + s) = 7 \cdot r + 7 \cdot s$ **(h)** $5 \cdot (2 \cdot a) = (5 \cdot 2) \cdot a$

SOLUTION

(a) Commutative property of addition
(b) Commutative property of multiplication
(c) Distributive property
(d) Associative property of addition
(e) Commutative property of addition
(f) Commutative property of multiplication
(g) Distributive property
(h) Associative property of multiplication ■

◇ ***Do Practice Exercise 1.***

OBJECTIVES

1 Apply the basic properties of mathematics to algebraic expressions.
2 Simplify terms and expressions.

1. State which property is used:

(a) $a + x = x + a$

(b) $3(4 + 5) = 3 \cdot 4 + 3 \cdot 5$

(c) $1 \cdot A = A$

(d) $a + (5y + t) = (a + 5y) + t$

(e) $AB = BA$

(f) $y \cdot 3 = 3 \cdot y$

ANSWERS

1. (a) Commutative property of addition
(b) Distributive property
(c) Identity property of multiplication
(d) Associative property of addition
(e) Commutative property of multiplication
(f) Commutative property of multiplication

2. Simplify each and write in standard form:

(a) $y \cdot 5$

(b) $3 \cdot A \cdot 2$

(c) $a \cdot 3 \cdot b$

(d) $2 \cdot y \cdot x \cdot (-8)$

ANSWERS
2. (a) $5y$ (b) $6A$ (c) $3ab$ (d) $-16xy$

2 Simplify Terms and Expressions

The standard way to write a term containing a constant and one or more variables is to write the constant first. Variables within a term are usually written in alphabetical order. If the variables in a term are not in alphabetical order, they can be rearranged using the commutative property. For example,

$$a \cdot 10 \cdot x = 10 \cdot a \cdot x = 10ax$$

→ preferred

EXAMPLE 2

Write the term $a \cdot y \cdot 18 \cdot b$ in standard form.

SOLUTION

$a \cdot y \cdot 18 \cdot b$

$a \cdot 18 \cdot y \cdot b$ Commutative property

$18 \cdot a \cdot y \cdot b$ Commutative property

$18 \cdot a \cdot b \cdot y$ Commutative property

$18aby$

→ Variables are in alphabetical order.

→ constant first ■

Simplification or **to simplify** is an important idea in algebra. It is the idea of writing a term or an expression in a simpler way. It usually results in writing an equivalent expression with fewer symbols.

EXAMPLE 3

Simplify $6 \cdot y \cdot 3$.

SOLUTION

$6 \cdot y \cdot 3$

$6 \cdot 3 \cdot y$ Commutative property

$18y$ Result. ■

◇ *Do Practice Exercise 2.*

The next few examples show you how to simplify slightly more complicated expressions. Look at them carefully.

EXAMPLE 4

Simplify $6x + 8y + 0.3x$.

SOLUTION

$6x + 8y + 0.3x$	
$6x + 0.3x + 8y$	Commutative property
$(6 + 0.3)x + 8y$	Distributive property
$6.3x + 8y$	Result. ■

EXAMPLE 5

Simplify $3x - 8z + 2x$.

SOLUTION

$3x - 8z + 2x$	
$(+3)x - (+8)z + (+2)x$	Identify signs.
$(+3)x + (-8)z + (+2)x$	$-(+8)z = +(-8)z$
$(+3)x + (+2)x + (-8)z$	Commutative property of addition
$3x + 2x - 8z$	
$5x - 8z$	Result. ■

The last example shows how the commutative property of addition is used to move a term. You can think of the mathematical operation as moving with the term. Look at the next two examples carefully.

EXAMPLE 6

Simplify $2x + 3y - 8x$.

SOLUTION

$2x + 3y - 8x$	
$2x - 8x + 3y$	Commutative property of addition
$-6x + 3y$	Result. ■

Notice that $-8x$ and $+3y$ were exchanged. It may help to think of the term $-8x$ as $+ (-8x)$.

3. Simplify each:

(a) $3x - 9x$

(b) $4y + 2a + 6y$

(c) $5z - 2x + 1.5z$

(d) $5x - 3a + 2x$

(e) $-4r + B - 2r$

(f) $-7a + 3y - 2a$

4. Simplify each:

(a) $3(x + y) + 4x$

(b) $3A + 2(A + B)$

(c) $-2z + 5(z + y)$

ANSWERS
3. (a) $-6x$ (b) $10y + 2a$
(c) $6.5z - 2x$ (d) $7x - 3a$
(e) $-6r + B$ (f) $-9a + 3y$
4. (a) $7x + 3y$ (b) $5A + 2B$
(c) $3z + 5y$

EXAMPLE 7

Simplify $-7ab + 3y - 2.5ab$.

SOLUTION

$-7ab + 3y - 2.5ab$	
$-7ab - 2.5ab + 3y$	Commutative property of addition
$-9.5ab + 3y$	Result. ■

The answer $-9.5ab + 3y$ from the last example is equal to $3y - 9.5ab$ by the commutative property of addition. Thus, either answer is acceptable.

◇ *Do Practice Exercise 3.*

EXAMPLE 8

Simplify $5x + 8(x + y)$.

SOLUTION

$5x + 8(x + y)$	
$5x + 8x + 8y$	Distributive property
$(5 + 8)x + 8y$	Distributive property
$13x + 8y$	Result. ■

◇ *Do Practice Exercise 4.*

Identify the numerical coefficient of each. (Review)

1. $10x$ **2.** $-15xy$ **3.** $7.5z$ **4.** $4\frac{1}{2}t$

State which property is used.

5. $x + y = y + x$

6. $10 \cdot x = x \cdot 10$

7. $1 \cdot x = x$

8. $3z + 1 = 1 + 3z$

9. $x + 0 = x$

10. $r \cdot 1 = r$

11. $12(a + b) = 12a + 12b$

12. $7 \cdot (x + y) = 7 \cdot x + 7 \cdot y$

13. $10(x + b) = 10x + 10b$

14. $x(yz) = (xy)z$

15. $4x - 2y = -2y + 4x$

16. $-3z + 8a = 8a - 3z$

17. $12y - x = -x + 12y$

18. $-3abc + 2b = 2b - 3abc$

19. $50\left(x + \frac{1}{2}\right) = 50x + 25$

20. $9(r + s) = 9 \cdot r + 9 \cdot s$

Simplify if possible and show the reason for each step.

21. $4y + 9y$ **22.** $21z - 11z$ **23.** $3x + 9x$ **24.** $12A - 9A$

25. $3x + 8x$ **26.** $2.4x + 3x$ **27.** $5y + y$ **28.** $7z + z$

29. $12r - 2r$ **30.** $9y - 12y$ **31.** $7.5a - \frac{1}{2}a$ **32.** $-4.5x + \frac{1}{4}x$

33. $-4y - 2y$ **34.** $3x + x$ **35.** $7\frac{1}{2}y - y$ **36.** $2z - 3z - u$

37. $5z + 2(x + z)$ **38.** $z + z + t$ **39.** $3 + 4y + 7$ **40.** $16x + 2y + 5x$

41. $3(x + y) + 5x$ **42.** $xy + xy$ **43.** $a(b + c) + ab$ **44.** $abc + abc + ab$

45. $5(a + y) + 14y$ **46.** $5(x + y) - 8y$ **47.** $3(a + b) - b$ **48.** $-x + (3 + r)x$

Write each of the following as an expression, then simplify if possible.

49. Four b plus three z plus seven z

50. Six y minus four y plus two z

51. Nine r minus three b plus one-half r

52. Add four times the sum of b and c to three c

53. Add seven times the sum of x and y to a negative five x

54. Subtract $3y$ from two times the sum of y and r

JOURNAL EXERCISES

Work each problem.

1. A motor boat is traveling in still water at y miles per hour. Assume the boat is traveling upstream against a current flowing at x miles per hour. Write an expression for the speed of the boat. (*Hint:* Is the boat slower than y or faster than y when traveling upstream against the current?) In which direction is the boat traveling if $x > y$?

2. In your own words, describe the identity, commutative, and associative properties of both addition and multiplication. Are they useful? Explain.

3. In your own words, describe the distributive property. Is it useful? Explain.

4. On your own, create five algebraic expressions that can be simplified. Write down each expression then simplify.

5. On your own, create five algebraic expressions that can be simplified. Bring them to class and exchange them with another student, who will simplify them.

1 Equations that Have a Variable

An equation in arithmetic looks like:

$$3 + 2 = 5$$

The left and right hand side of an equation are separated by an equal sign. The equation is not true if the left hand side and the right hand side are not equal.

Algebraic equations are similar except that variables can now be used. For example:

$$x + 7 = 12$$

In this equation, x is some unknown quantity. We can guess values for x:

Guesses	replace x with the guess	
Is $x = 1$?	$x + 7 = 1 + 7 = 8 \neq 12$	No $x \neq 1$
Is $x = 2$?	$x + 7 = 2 + 7 = 9 \neq 12$	No $x \neq 2$
.		
.		
.		
Is $x = 5$?	$x + 7 = 5 + 7 = 12 = 12$	Yes $x = 5$

In this manner, we can find a value that x can be so that the equation is true.

A value is a **solution to the equation** if it makes the equation true when the value is substituted in place of the variable. Look at the next two examples.

EXAMPLE 1

Is $x = 12$ a solution to $x - 10 = 0$?

SOLUTION

$$x - 10 = 0$$

$$12 - 10 \stackrel{?}{=} 0 \quad \text{Substitute 12 in place of } x.$$

$$2 \neq 0 \quad \text{The equation is not true.}$$

Therefore, $x = 12$ *is not* a solution to the equation. ■

EXAMPLE 2

Is $y = 28$ a solution to $y - 7 = -21$?

SOLUTION

$$y - 7 = -21$$

$$28 - 7 \stackrel{?}{=} -21 \quad \text{Substitute 28 in place of } y.$$

$$+21 \neq -21 \quad \text{They are not equal.}$$

Therefore, $y = 28$ *is not* a solution to the equation. ■

◇ *Do Practice Exercises 1 and 2.*

OBJECTIVES

1 Work with equations that have a variable.

2 Solve equations using the Addition Property.

1. Is $x = 4$ a solution to $12 - x = 8$?

2. Is $y = 7$ a solution to $y - 2 = -3$?

ANSWERS
1. yes
2. no

2 Solve Equations Using the Addition Property

The process of finding a value for a variable that makes the equation true is called **solving** the equation. The value for the variable is called the **solution** to the equation. Therefore, the solution to an equation is a value for the variable that makes the equation true. The analogy that follows is intended to help you understand the process of solving equations.

Assume two people weigh the same and are balanced on a teeter-totter, as follows:

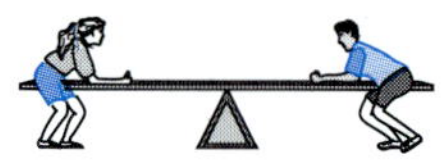

Balanced

What happens if you hand one of them a ten pound weight?

Out of balance

The teeter-totter will no longer be in balance. In order to keep it perfectly balanced, the other person must also have a ten pound weight.

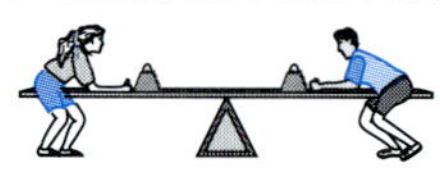

In balance again

Whatever you do to one side of the teeter-totter must also be done to the other side in order to keep it in balance. The same idea applies to equations in arithmetic and algebra. Whatever you do to one side of the equation must also be done to the other side.

We know that an equivalent equation results if the same number is added to both sides of an arithmetic equation. For example:

$$3 + 2 = 5$$

$$3 + 2 \underline{+ 7} = 5 \underline{+ 7} \qquad \text{Add 7 to both sides.}$$

Similarly, the same number can be added to both sides of an algebraic equation. For example:

$$x - 7 = 15$$

$$x - 7 \underline{+ 7} = 15 \underline{+ 7} \qquad \text{Add 7 to both sides.}$$

These equations are **equivalent** to one another. Notice that both sides of the second equation can be simplified by adding like terms:

$$x - 7 \underline{+ 7} = 15 \underline{+ 7}$$

$$x + 0 = 22 \qquad \text{Add like terms.}$$

$$x = 22 \qquad \text{solution to equation}$$

Therefore, x *must* equal 22 in order for $x - 7 = 15$ to be true. We have solved the equation for x.

The answer $x = 22$ can be checked by substituting 22 in place of x. The original equation should be true:

$$x - 7 = 15$$

$$22 - 7 \stackrel{?}{=} 15 \qquad \text{Substitute 22 for } x$$

$$15 = 15 \qquad \text{true}$$

This shows that $x = 22$ is the solution to the equation.

The following property is an important one for you to understand.

ADDITION PROPERTY OF EQUATIONS

The same value can be added to both sides of an equation. In other words, if $a = b$, then $a + c = b + c$ for any value of c.

The idea behind solving equations of this type is to get the variable on one side of the equation, *by itself.* If x is the variable, you should end up with:

$$x = \text{some number}$$

which will be the solution to the equation. Note that x is all alone on one side of the equation.

The first step is to identify the term you are seeking to remove from one side of the equation. Then, add the opposite value to both sides of the equation as shown in the next few examples.

EXAMPLE 3

Solve $y - 4 = 16$ for y.

SOLUTION

Note first that we need to remove $-(+4)$ from the left side of the equation. This can be done by adding 4 to both sides.

$$y - 4 = 16$$

$$y - 4 + 4 = 16 + 4 \quad \text{Add 4 to both sides.}$$

$$y = 20 \quad \text{Solution.}$$

Check the answer:

$$y - 4 = 16$$

$$20 - 4 \stackrel{?}{=} 16 \quad \text{Substitute 20 for } y.$$

$$16 = 16 \quad \text{true}$$

Therefore, $y = 20$ is a solution to $y - 4 = 16$. ■

EXAMPLE 4

Solve $x - 9 = -20$ for x.

SOLUTION

$$x - 9 = -20$$

$$x - 9 + 9 = -20 + 9 \quad \text{Add 9 to both sides.}$$

$$x = -11 \quad \text{Solution.}$$

■

Sometimes you may need to add a negative value in order to remove a term as shown in the next example.

EXAMPLE 5

Solve $b + 3 = 40$ for b.

SOLUTION

First note that 3 must be removed from the left hand side of the equation. This is done by subtracting 3 from both sides, or equivalently, by adding (-3) to both sides.

3. Solve each:

(a) $x - 3 = 18$

(b) $y + 7 = 10$

(c) $-3 = y - 4$

(d) $r + 0.5 = \frac{3}{4}$

(e) $s + 2\frac{1}{2} = 4$

(f) $m - 25 = -52$

ANSWERS

3. (a) $x = 21$ (b) $y = 3$
(c) $y = 1$ (d) $r = 0.25$
(e) $s = 1\frac{1}{2}$ or 1.5
(f) $m = -27$

$$b + 3 = 40$$
$$b + 3 + (-3) = 40 + (-3) \qquad \text{Add } -3 \text{ to both sides.}$$
$$b = 37 \qquad \text{Solution.}$$

It does not matter which side of the equation the variable is on, the process is the same, as shown in the next example.

EXAMPLE 6

Solve $-31 = r + 5$ for r.

SOLUTION

The goal is to remove 5 from the right hand side of the equation. This is done by adding a (-5) to both sides.

$$-31 = r + 5$$
$$-31 + (-5) = r + 5 + (-5) \qquad \text{Add } -5 \text{ to both sides.}$$
$$-36 = r \quad \text{or} \quad r = -36$$

The process is the same when working with decimals, fractions, or mixed numbers.

EXAMPLE 7

Solve $s + 4\frac{1}{2} = \frac{1}{2}$ for s.

SOLUTION

In order for s to be by itself on the left side of the equation, remove $4\frac{1}{2}$ from the left side. The mixed number $4\frac{1}{2}$ is removed from the left side by adding $\left(-4\frac{1}{2}\right)$ to both sides.

$$s + 4\frac{1}{2} = \frac{1}{2}$$
$$s + 4\frac{1}{2} + \left(-4\frac{1}{2}\right) = \frac{1}{2} + \left(-4\frac{1}{2}\right) \qquad \text{Add } -4\frac{1}{2} \text{ to both sides.}$$
$$s = \frac{1}{2} + \left(-4\frac{1}{2}\right)$$
$$s = \left(+\frac{1}{2}\right) + \left(-\frac{9}{2}\right) \qquad \text{Change } -4\frac{1}{2} \text{ to } -\frac{9}{2}.$$
$$s = \frac{(+1) + (-9)}{2} \qquad \text{Add with common denominators.}$$
$$s = \frac{-8}{2}$$
$$s = -4 \qquad \text{Solution.}$$

◇ *Do Practice Exercise 3.*

name date hour

Identify like terms. (Review)

1. $3x, \quad 2y, \quad 5x$

2. $2.7y, \quad 4y, \quad 10xy$

3. $2z, \quad 3y, \quad 5\frac{1}{2}z$

4. $10, \quad 6y, \quad 15$

5. $3z, \quad 2r, \quad 0.5r$

6. $xy, \quad x, \quad y, \quad 2xy$

Simplify. (Review)

7. $3xy + 5xy$

8. $2x - 6x$

9. $-4z - 3z$

10. $3x + 2y - 5x$

11. $y - 8y$

12. $3(r + x) - 2r$

13. $ac + 2ac$

14. $4y + 2x - (-5.5x)$

Solve for the unknown.

15. $x + 9 = 14$

16. $y + 5 = 12$

17. $r + 11 = -2$

18. $y - 8 = 21$

19. $r - 2 = 27$

20. $t - 21 = -5$

21. $x - 7 = -19$

22. $y + 8 = -5$

23. $r + (-4) = 3.1$

24. $5 + y = -30$

25. $r - 3 = 19$

26. $t + 7 = -84.2$

27. $y - 2 = 22.5$

28. $z + 8 = -16$

29. $y - 9.5 = 2$

30. $3 = x - 9$

31. $9 + x = 25$

32. $21 = 4 + r$

33. $t - 0.2 = 10$

34. $y - 8 = -4$

35. $s - 4 = \frac{1}{2}$

36. $r + 20 = -5$

37. $7 + z = 24$

38. $r + 1.2 = -1.4$

39. $s + 6 = 6$

40. $35 = y - 6$

41. $y + 5\frac{1}{2} = 12.4$

42. $x + 11\frac{1}{4} = -4$

43. $p + 2.8 = \frac{3}{4}$

44. $-16 + f = -7$

45. $A + 0 = 21$

46. $B - 0 = -35.2$

47. $-3.5 + x = 7.2$

48. $y = 21 + 3$

49. $x = 19 - 24$

50. $p + 2.5 = -6.6$

Write an equation for each of the following, then solve for the unknown.

51. The sum of x and 8 equals 41.

52. The sum of y and 9 equals -4.

53. s minus 7 equals 3.

54. The sum of 12 and x equals 6.

55. The sum of -15 and y equals -8.

56. 7 equals y minus 10.

57. -9 equals y plus 4.

58. 3.2 plus s equals $-\frac{1}{2}$.

59. An unknown minus forty equals two thousand.

60. Two-fifths plus an unknown equals twenty-one.

61. Seventy-four equals an unknown minus three.

62. Negative seven and one-tenth equals negative four plus an unknown.

JOURNAL EXERCISES

Follow the directions for each problem.

1. Explain the Addition Property of Equations. How is it used in solving equations?

2. If you do something to one side of an equation, why must the same thing also be done to the other side?

3. Can the Addition Property of Equations be used to solve $7x = 91$? Explain.

In the last section we learned how to remove a term from one side of an equation, by adding the same term with an opposite sign to both sides. In this section, we will look at the steps to remove a numerical coefficient from in front of a variable.

OBJECTIVE

1 Solve equations using the multiplication property.

1 Solve Equations Using the Multiplication Property

Both sides of an arithmetic equation can be multiplied times the same value. Similarly, both sides of an algebraic equation can be multiplied times the same value. For example:

$$\frac{x}{5} = 12$$

$$\frac{x}{5} \cdot 5 = 12 \cdot 5 \qquad \text{Multiply both sides by 5.}$$

The result is an equivalent equation, or one that requires the same value for the unknown. The second equation above can be simplified:

$$\frac{x}{5} \cdot 5 = 12 \cdot 5$$

$$\frac{x}{5} \cdot \frac{5}{1} = 12 \cdot 5 \qquad \text{Write 5 as } \frac{5}{1}.$$

$$\frac{x}{\cancel{5}} \cdot \frac{\cancel{5}}{1} = 12 \cdot 5 \qquad \text{Cross reduce.}$$

$$x = 60 \qquad \text{Solution.}$$

Check the answer by substituting 60 in place of x in the original equation:

$$\frac{x}{5} = 12$$

$$\frac{60}{5} \stackrel{?}{=} 12 \qquad \text{Substitute 60 for } x.$$

$$12 = 12 \qquad \text{true}$$

$x = 60$ is a solution to the equation $\frac{x}{5} = 12$.

The idea behind the example above is given in the following important property.

MULTIPLICATION PROPERTY OF EQUATIONS

Both sides of an equation can be multiplied times the same value. In other words, if $a = b$, then $a \cdot c = b \cdot c$ for any value of c.

Again, the idea of solving these types of equations is to get the variable on one side of the equation, *by itself.* First, identify the value you are seeking to remove from one side of the equation. Then multiply both sides of the equation by its multiplicative inverse or reciprocal. Recall from Chapter 3 on fractions that the inverse of 8 is $\frac{1}{8}$ and the inverse of $\frac{2}{5}$ is $\frac{5}{2}$.

EXAMPLE 1

Solve $\frac{x}{9} = 20$ for x.

SOLUTION

First note that we need to remove the 9 in the denominator of the fraction on the left hand side. This is done by multiplying both sides of the equation times 9:

$$\frac{x}{9} = 20$$

$$\frac{x}{9} \cdot 9 = 20 \cdot 9 \quad \text{Multiply both sides by 9.}$$

$$\frac{x}{9} \cdot \frac{9}{1} = 20 \cdot 9 \quad \text{Write 9 as } \frac{9}{1}.$$

$$\frac{x}{\not{9}} \cdot \frac{\not{9}}{1} = 20 \cdot 9 \quad \text{Cross reduce.}$$

$$x = 180 \quad \text{Solution.}$$

Check the answer by substituting 180 in place of x in the original equation:

$$\frac{x}{9} = 20$$

$$\frac{180}{9} \stackrel{?}{=} 20 \quad \text{Substitute 180 in place of } x.$$

$$20 = 20 \quad \text{true}$$

Therefore, $x = 180$ is a solution to $\frac{x}{9} = 20$. ■

EXAMPLE 2

Solve $\frac{A}{4.5} = \frac{1}{2}$ for A.

SOLUTION

$$\frac{A}{4.5} = \frac{1}{2}$$

$$\frac{A}{4.5} \cdot 4.5 = \frac{1}{2} \cdot 4.5 \quad \text{Multiply both sides by 4.5.}$$

$$A = \frac{1}{2} \cdot 4.5 \quad \text{Simplify the left hand side.}$$

$$A = 0.5 \cdot 4.5 \quad \text{Replace } \frac{1}{2} \text{ with 0.5.}$$

$$A = 2.25 \quad \text{Solution.}$$

■

The next example shows how to solve an equation in which an integer is being multiplied times a variable. In this case, both sides will be multiplied by a fraction.

EXAMPLE 3

Solve $6x = 486$ for x.

SOLUTION

The goal is to remove the 6 from in front of $6x$. This can be done by multiplying both sides by the fraction $\frac{1}{6}$, or by dividing both sides by 6.

$$6x = 486$$

$$6x \cdot \frac{1}{6} = 486 \cdot \frac{1}{6} \quad \text{Multiply both sides by } \frac{1}{6}.$$

$$\frac{6x}{1} \cdot \frac{1}{6} = \frac{486}{1} \cdot \frac{1}{6} \quad \text{Write } 6x \text{ as } \frac{6x}{1}.$$

$$\frac{\cancel{6}x}{1} \cdot \frac{1}{\cancel{6}} = \frac{\overset{81}{\cancel{486}}}{1} \cdot \frac{1}{\cancel{6}} \quad \text{Cross reduce.}$$

$$x = 81 \quad \text{Solution.}$$

◇ ***Do Practice Exercise 1.***

The next example shows the steps for working with a numerical coefficient that is a negative number.

EXAMPLE 4

Solve $-8y = 560$ for y.

SOLUTION

Remove -8 from the left side of the equation by multiplying both sides times $(-\frac{1}{8})$:

$$-8y = 560$$

$$-8y \cdot \left(-\frac{1}{8}\right) = 560 \cdot \left(-\frac{1}{8}\right) \quad \text{Multiply both sides by } -\frac{1}{8}.$$

$$\frac{(-8)y}{1} \cdot \frac{1}{-8} = \frac{560}{1} \cdot \frac{-1}{8} \quad \text{Identify signs and fractions. Note that } \left(-\frac{1}{8}\right) = \left(\frac{1}{-8}\right).$$

$$\frac{(-\overset{1}{\cancel{8}})y}{1} \cdot \frac{1}{-\underset{1}{\cancel{8}}} = \frac{\overset{70}{\cancel{560}}}{1} \cdot \frac{-1}{\underset{1}{\cancel{8}}} \quad \text{Cross reduce.}$$

$$y = -70 \quad \text{Solution.}$$

1. Solve each:

(a) $5y = 50$

(b) $75z = 600$

(c) $2.5r = -10$

(d) $9s = -63$

ANSWERS

1. (a) $y = 10$ (b) $z = 8$ (c) $r = -4$ (d) $s = -7$

2. Solve each:

(a) $\frac{x}{12} = 6$

(b) $\frac{y}{3} = -13$

(c) $-5y = 70$

(d) $9r = -270$

(e) $\frac{3}{4}m = \frac{1}{2}$

(f) $7a = 2\frac{1}{4}$

ANSWERS
2. (a) $x = 72$ (b) $y = -39$
(c) $y = -14$ (d) $r = -30$
(e) $m = \frac{2}{3}$ (f) $a = \frac{9}{28}$

EXAMPLE 5

Solve $\frac{3}{2}b = -24$ for b.

SOLUTION

The fraction $\frac{3}{2}$ can be removed from the left hand side by multiplying both sides times the inverse or reciprocal of $\frac{3}{2}$ which is $\frac{2}{3}$.

$$\frac{3}{2}b = -24$$

$$\frac{3}{2}b \cdot \frac{2}{3} = -24 \cdot \frac{2}{3} \quad \text{Multiply both sides by } \frac{2}{3}.$$

$$b \cdot \frac{\cancel{3}}{\cancel{2}} \cdot \frac{\cancel{2}}{\cancel{3}} = -24 \cdot \frac{2}{3} \quad \text{Commutative property}$$

$$b = -24 \cdot \frac{2}{3} \quad \frac{3}{2} \cdot \frac{2}{3} = 1$$

$$b = \frac{-24}{1} \cdot \frac{2}{3} \quad \text{Write } -24 \text{ as } \frac{-24}{1}.$$

$$b = \frac{-48}{3} \quad \text{Multiply.}$$

$$b = -16 \quad \text{Solution.}$$

■

◇ *Do Practice Exercise 2.*

Solve for the unknown.

1. $5y = 45$ **2.** $7z = 63$ **3.** $9r = 99$

4. $12s = 48$ **5.** $2t = -8$ **6.** $7w = -490$

7. $3r = -87$ **8.** $4s = 17$ **9.** $9r = 25$

10. $3t = -34$ **11.** $4x = -50$ **12.** $7t = -65$

13. $8y = -3$ **14.** $7t = -4$ **15.** $-3y = 60$

16. $-4y = 28$ **17.** $-5n = 65$ **18.** $-4v = -48$

19. $\frac{x}{5} = 20$ **20.** $\frac{y}{3} = 7$ **21.** $\frac{u}{2} = 7$

22. $\frac{v}{4} = 21$ **23.** $\frac{t}{6} = 45$ **24.** $\frac{g}{8} = 90$

25. $\frac{x}{21} = -40$ **26.** $\frac{r}{7} = -12$ **27.** $\frac{y}{19} = -67$

28. $\frac{y}{12} = 21$ **29.** $\frac{b}{100} = 100$ **30.** $\frac{x}{97} = -25$

31. $\frac{2y}{3} = 20$ **32.** $\frac{3x}{4} = 160$ **33.** $\frac{5t}{8} = 150$

34. $\frac{7y}{9} = 14$ **35.** $\frac{9b}{11} = 21$ **36.** $\frac{3y}{5} = -2$

37. $\frac{4g}{5} = -7$ **38.** $\frac{3x}{4} = -12$ **39.** $0.2x = 52$

40. $0.5y = 1.5$ **41.** $1.5r = 4.5$ **42.** $2.2x = 9.9$

43. $0.1s = 7$ **44.** $0.2t = -3$ **45.** $\frac{y}{1.4} = 2$

46. $\frac{x}{1.5} = 8$ **47.** $\frac{r}{3} = -2.9$ **48.** $\frac{z}{5} = -6.5$

49. $\frac{t}{3.2} = 1.5$ **50.** $\frac{3y}{2} = 7.5$

Write an equation for each of the following, then solve for the unknown.

51. The product of 6 and y equals 94.

52. 17 equals the product of 4 and x.

53. 53 equals the product of three and t.

54. The product of 7 and an unknown equals 77.

55. The quotient of x divided by 7 equals 40.

56. The quotient of y divided by 9 equals 23.

57. s divided by 3 equals negative 7.

58. y divided by 0.2 equals 4.

59. $3x$ divided by 5 equals -70.

60. $2g$ divided by 7 equals -5.

61. $3z$ equals negative four and two tenths.

62. y divided by -4 equals 1.8.

JOURNAL EXERCISES

Follow the directions for each problem.

1. Explain the Multiplication Property of Equations. When should this property be applied?

2. Explain the difference between the Addition Property of Equations from the last section and the Multiplication Property of Equations introduced in this section. When should each be used?

3. Write ten equations similar to those of this section and the last section. In each case, state whether the Addition Property of Equations or the Multiplication Property of Equations is needed to solve the equation.

1 Solve Equations Requiring More than One Operation

OBJECTIVE

1 Solve equations using both the addition and multiplication properties.

Several steps may be needed to solve an equation. The following list of steps is a guideline for solving equations of this type. *Not all* of these steps will be needed for every problem.

HELPFUL HINT

1. Place all terms with a variable on one side of the equation and those without a variable on the other side.
2. Combine like terms.
3. Remove numerical coefficients by multiplying or dividing.
4. Check your answer.

For each example, the solution can be checked by substituting the value for the variable into the original equation and simplifying both sides.

EXAMPLE 1

Solve $5x + 8 = 43$ for x.

SOLUTION

$$5x + 8 = 43$$

$$5x + 8 + (-8) = 43 + (-8) \qquad \text{Add } -8 \text{ to both sides.}$$

$$5x = 35 \qquad \text{Combine like terms.}$$

$$5x \cdot \frac{1}{5} = 35 \cdot \frac{1}{5} \qquad \text{Multiply both sides by } \frac{1}{5}.$$

$$x = 7 \qquad \text{Solution.}$$

Check the answer:

$$5x + 8 = 43$$

$$5 \cdot 7 + 8 \stackrel{?}{=} 43 \qquad \text{Replace } x \text{ with 7.}$$

$$35 + 8 \stackrel{?}{=} 43$$

$$43 = 43 \qquad x = 7 \text{ is the solution.}$$

1. Solve each:

(a) $6x - 4 = 14$

(b) $2Y + 8 = 20$

(c) $7r + 2 = -20$

(d) $\frac{2}{3} = 2y - \frac{1}{3}$

(e) $4x + 0.5 = 20.5$

(f) $3t + 5 = -2$

ANSWERS

1. (a) $x = 3$ (b) $Y = 6$ (c) $r = -3\frac{1}{7}$
(d) $y = \frac{1}{2}$ (e) $x = 5$ (f) $t = -2\frac{1}{3}$

EXAMPLE 2

Solve $-3 + 2y = 7$ for y.

SOLUTION

$$-3 + 2y = 7$$

$$-3 + 2y + (+3) = 7 + (+3) \qquad \text{Add } +3 \text{ to both sides.}$$

$$2y = 10 \qquad \text{Combine like terms.}$$

$$2y \cdot \frac{1}{2} = 10 \cdot \frac{1}{2} \qquad \text{Multiply both sides by } \frac{1}{2}.$$

$$y = 5 \qquad \text{Solution.}$$

Check the answer:

$$-3 + 2y = 7$$

$$-3 + 2 \cdot 5 \stackrel{?}{=} 7 \qquad \text{Replace } y \text{ with 5.}$$

$$-3 + 10 \stackrel{?}{=} 7$$

$$7 = 7 \qquad y = 5 \text{ is the solution.} \quad ■$$

From now on, check the solutions to each equation on your own.

EXAMPLE 3

Solve $5t - 2.5 = -3$ for t.

SOLUTION

$$5t - 2.5 = -3$$

$$5t - 2.5 + 2.5 = -3 + 2.5 \qquad \text{Add 2.5 to both sides.}$$

$$5t = -0.5 \qquad \text{Combine like terms.}$$

$$5t \cdot \frac{1}{5} = -0.5 \cdot \frac{1}{5} \qquad \text{Multiply by } \frac{1}{5}.$$

$$t = -0.5 \cdot 0.2 \qquad \text{Change } \frac{1}{5} \text{ to 0.2.}$$

$$t = -0.1 \qquad \text{Solution.} \quad ■$$

◇ ***Do Practice Exercise 1.***

Sometimes, like terms containing variables must be added before working the problem. Look at the next few examples.

EXAMPLE 4

Solve $3x + 5x = 24$ for x.

SOLUTION

$$3x + 5x = 24$$

$$(3 + 5)x = 24 \quad \text{Combine like terms.}$$

$$8x = 24$$

$$8x \cdot \frac{1}{8} = 24 \cdot \frac{1}{8} \quad \text{Multiply both sides by } \frac{1}{8}.$$

$$x = 3 \quad \text{Solution.}$$

■

EXAMPLE 5

Solve $5y + 20 - 2y = 0$ for y.

SOLUTION

$$5y + 20 - 2y = 0$$

$$5y - 2y + 20 = 0 \quad \text{Commutative property}$$

$$3y + 20 = 0 \quad \text{Combine like terms.}$$

$$3y + 20 + (-20) = 0 + (-20) \quad \text{Add } -20 \text{ to both sides.}$$

$$3y = -20 \quad \text{Combine like terms.}$$

$$3y \cdot \frac{1}{3} = -20 \cdot \frac{1}{3} \quad \text{Multiply by } \frac{1}{3}.$$

$$y = -\frac{20}{3} = -6\frac{2}{3}$$

■

◇ *Do Practice Exercise 2.*

EXAMPLE 6

Solve $13r - 9 - 2r = -51$ for r.

SOLUTION

$$13r - 9 - 2r = -51$$

$$13r - 2r - 9 = -51 \quad \text{Commutative property}$$

$$11r - 9 = -51 \quad \text{Combine like terms.}$$

$$11r - 9 + 9 = -51 + 9 \quad \text{Add 9 to both sides.}$$

$$11r = -42 \quad \text{Combine like terms.}$$

$$11r \cdot \frac{1}{11} = -42 \cdot \frac{1}{11} \quad \text{Multiply both sides by } \frac{1}{11}.$$

$$r = -\frac{42}{11} = -3\frac{9}{11}$$

■

2. Solve each:

(a) $3r + 8r = -11$

(b) $5z - 2z = 14$

(c) $2y - 4y - 3 = 0$

(d) $5 = 6x - 9x$

ANSWERS

2. (a) $r = -1$ (b) $z = 4\frac{2}{3}$ (c) $y = -1\frac{1}{2}$ (d) $x = -1\frac{2}{3}$

3. Solve each:

(a) $2y + 5y - 2 = 12$

(b) $3x - 0.5x = -10$

(c) $5r = 21 - 3r$

(d) $17 + 2y = 5y$

ANSWERS
3. (a) $y = 2$ (b) $x = -4$
(c) $r = 2\frac{5}{8}$ (d) $y = 5\frac{2}{3}$

Sometimes a term with a variable is found on both sides of the equation. In that event, pick either term and remove it from one side of the equation as shown in the next example.

EXAMPLE 7

Solve $6x = 90 - 3x$ for x.

SOLUTION

$$6x = 90 - 3x$$

$$6x + 3x = 90 - 3x + 3x \qquad \text{Add } 3x \text{ to both sides.}$$

$$9x = 90 \qquad \text{Combine like terms.}$$

$$9x \cdot \frac{1}{9} = 90 \cdot \frac{1}{9} \qquad \text{Multiply both sides by } \frac{1}{9}.$$

$$x = 10$$

◇ *Do Practice Exercise 3.*

name date hour

Solve for the unknown.

1. $x + 9 = 75$
2. $y - 4 = 14$
3. $7t = 56$
4. $\frac{A}{9} = 7$
5. $2x - 2 = 40$
6. $39 - 7 = y$
7. $7x - 4 = 24$
8. $3y + 12 = 15$
9. $9r - 2 = 41$
10. $\frac{1}{8}x - 4 = 20$
11. $91 = 11 + 20p$
12. $9 - 7 = 4 + x$
13. $6 + 2y = 800$
14. $2.4x = 12$
15. $\frac{t}{3} + 5 = 17$
16. $25x + \frac{1}{2} = 76.5$
17. $10y + 8 = -32$
18. $-75 = -7a$
19. $20x - 20 = 3{,}000$
20. $-8y + 3 = +9$
21. $22r - 44 = 0$
22. $6x - 4 = -14$
23. $3r - 2 = \frac{1}{2}$
24. $5y + 2.5 = -20$
25. $6 = 3t - 7$
26. $7 - 21 = 7t$
27. $9y + 6 = -5$
28. $7 + 0.2x = 3$
29. $-5 + 0.3y = -2$
30. $\frac{1}{2}x + 2 = 16$
31. $7p = 4 + 3p$
32. $9 + 2y = 5y$
33. $13r = 24 - 7r$
34. $9 - 7f = 3f$
35. $3m + 9 = 7m$
36. $x + 9 = 2x$
37. $90 = 5(p - 8)$
38. $6(y - 2) = 300$
39. $75 = 7(r - 2)$
40. $3y = 2y - 9$
41. $7(x - 2) = 21$
42. $2(s + 2) = s$

43. $\frac{r}{8} - 1 = 14$

44. $3x - 2 = 2x$

45. $5z - \frac{z}{2} = 8.4$

46. $\frac{2}{3}y - 7 = y$

47. $\frac{1}{3}y = -8$

48. $\frac{3}{8}x = -2$

49. $3z - \frac{1}{8} = 0$

50. $6.7x - 0.4x = 12$

Work this problem.

51. Is $y = 3$ a solution to $5y - 14 = -1$?

JOURNAL EXERCISES

1. Does $9y = 8y$ have a solution? If so, what is it? Explain.

2. Try to solve $\frac{1}{3}x + \frac{1}{4}x + \frac{1}{5}x = \frac{1}{10}$ for x.

3. What is the meaning of "the solution to the equation is"?

4. Can you find an equation that has no solution? (*Hint:* $x = x + 5$) What does it mean if an equation has no solution?

5. List all values for x that satisfy the equation $x = x$. Explain.

1 Algebraic Expressions from Phrases

The first step in solving a word problem is to use the information in the problem to write an equation. This is a skill that takes practice. First, identify the unknown quantity and give it a variable name such as x, y, or z. Then look at the wording of the word problem very carefully as you write out the expressions. Several examples follow.

EXAMPLE 1

Write an expression for the sum of six with an unknown number.

SOLUTION
Let x be the unknown number.

$$\text{The sum of six plus an unknown} = 6 + x$$

Notice that if y were chosen to represent the unknown the expression would be $6 + y$, which has the same meaning. ■

EXAMPLE 2

Write an expression for negative 4 minus an unknown.

SOLUTION
Let $y =$ the unknown.

$$\text{Negative 4 minus an unknown} = (-4) - y$$

The result can also be written as $-4 - y$. ■

◇ *Do Practice Exercise 1.*

EXAMPLE 3

Write an equation for the sum of six and an unknown is equal to 9.

SOLUTION
Let $x =$ the unknown.
6 plus an unknown is equal to 9.

$$6 + x = 9$$ ■

EXAMPLE 4

Write an equation for the sum of eight and an unknown is equal to two times the unknown.

SOLUTION
Let $y =$ the unknown.
8 plus an unknown equals 2 times the unknown.

$$8 + y = 2y$$ ■

◇ *Do Practice Exercise 2.*

OBJECTIVES

1 Write an algebraic expression from words.
2 Solve word problems using equations.

1. Write an expression for each using x for the unknown:

(a) the sum of an unknown plus forty

(b) the sum of negative seven plus an unknown

(c) twenty-seven minus an unknown

2. Write an equation for each using y for the unknown:

(a) Nine minus an unknown equals negative five.

(b) Seven equals the sum of one-half and an unknown.

(c) An unknown is equal to the sum of 4 and negative 7.

ANSWERS
1. (a) $x + 40$ (b) $-7 + x$ (c) $27 - x$
2. (a) $9 - y = -5$ (b) $7 = \frac{1}{2} + y$
(c) $y = 4 + (-7)$

3. Write an equation for each using r for the unknown:

(a) Three times an unknown plus seven equals the unknown.

(b) -4 minus an unknown equals 2 times the unknown.

(c) -5 plus -3 times an unknown equals 408.

ANSWERS
3. (a) $3r + 7 = r$ (b) $-4 - r = 2r$
(c) $(-5) + (-3)r = 408$

EXAMPLE 5

Write an equation for 3 times an unknown equals 71.

SOLUTION
Let r = the unknown.
3 times an unknown equals 71.

$$3r = 71 \quad \blacksquare$$

EXAMPLE 6

Write an equation for -7 plus 2 times an unknown equals 5 times the same unknown.

SOLUTION
Let y = the unknown.
-7 plus 2 times the unknown equals 5 times the unknown.

$$-7 + 2y = 5y \quad \blacksquare$$

◇ *Do Practice Exercise 3.*

2 Word Problems

The following is a list of recommended steps to help you solve word problems that involve variables.

HELPFUL HINT

1. Read the problem carefully; more than once if needed.
2. Look up the meaning of any words you do not understand.
3. Clearly identify the known and unknown quantities.
4. Give the unknown a variable name.
5. Set up the equation.
6. Reread the problem to make sure that the equation represents the problem.
7. Solve the equation.
8. Answer the question with words, and make sure you have answered the correct question.
9. Check your answer to make sure that it is a reasonable answer to the word problem.

EXAMPLE 7

The sum of six with an unknown quantity is equal to forty-two. Find the unknown quantity.

SOLUTION

Let $x =$ the unknown quantity.
6 plus an unknown equals 42.

$$6 + x = 42$$

$$6 + x + (-6) = 42 + (-6) \qquad \text{Add } -6 \text{ to both sides.}$$

$$x = 36 \qquad \text{Result.}$$

Check the answer: the sum of 6 and 36 (the unknown) should equal 42. It does, so that we know that the unknown is 36. ■

EXAMPLE 8

Fourteen subtracted from three times an unknown quantity is equal to twenty-five. Find the unknown quantity.

SOLUTION

Let $x =$ the unknown quantity.
14 subtracted *from* $3x$ equals 25.

$$3x - 14 = 25$$

$$3x - 14 + 14 = 25 + 14 \qquad \text{Add 14 to both sides.}$$

$$3x = 39$$

$$3x \cdot \frac{1}{3} = 39 \cdot \frac{1}{3} \qquad \text{Multiply both sides by } \frac{1}{3}.$$

$$x = 13 \qquad \text{Solution.}$$

Again, you should check to make sure the solution to the equation is the answer to the word problem. ■

◇ *Do Practice Exercises 4 and 5.*

Some word problems refer to consecutive numbers. **Consecutive numbers** are numbers that fall naturally after one another. For example, the consecutive whole numbers are:

$$1, \quad 2, \quad 3, \quad 4, \quad 5, \quad 6, \quad 7, \quad . \quad . \quad .$$

The **consecutive odd numbers** are 1, 3, 5, 7, 9, 11, . . .
The **consecutive even numbers** are 2, 4, 6, 8, 10, 12, . . .

4. The sum of seven with an unknown is equal to twenty-one. Write an equation for the unknown and solve.

5. Seventy equals two times an unknown minus four. Write an equation for the unknown and solve.

ANSWERS
4. $7 + x = 21, \quad x = 14$
5. $70 = 2y - 4, \quad y = 37$

6. The sum of two consecutive odd numbers is eighty-four. Find both numbers.

7. The sum of two consecutive even numbers is 214. Find both numbers.

ANSWERS
6. 41, 43
7. 106, 108

EXAMPLE 9

The sum of two consecutive whole numbers is forty-seven. Find the two numbers.

SOLUTION

Let $x =$ the smaller of the two. Since they are consecutive, the larger number is 1 larger than the smaller and can be written as $x + 1$.

$$\text{smaller number} + \text{larger number} = 47$$
$$x + (x + 1) = 47$$
$$x + x + 1 = 47 \quad \text{Remove parentheses.}$$
$$2x + 1 = 47 \quad \text{Combine like terms.}$$
$$2x + 1 + (-1) = 47 + (-1) \quad \text{Add } -1 \text{ to both sides.}$$
$$2x = 46 \quad \text{Combine like terms.}$$
$$2x \cdot \frac{1}{2} = 46 \cdot \frac{1}{2} \quad \text{Multiply times } \frac{1}{2}.$$
$$x = 23$$
$$(x + 1) = 24$$

The two numbers are 23 and 24. Note that they are consecutive whole numbers and they add to 47. ■

EXAMPLE 10

The sum of two consecutive odd numbers is one hundred sixty-eight. Find the numbers.

SOLUTION

Let $x =$ the smaller odd number. Since they are consecutive odd numbers, the larger number is 2 more than the smaller and can be written as $x + 2$.

$$\text{smaller odd number} + \text{larger odd number} = 168$$
$$x + (x + 2) = 168$$
$$x + (x + 2) = 168$$
$$x + x + 2 = 168 \quad \text{Drop parentheses.}$$
$$2x + 2 = 168 \quad \text{Combine like terms.}$$
$$2x + 2 - 2 = 168 - 2 \quad \text{Subtract 2 from both sides.}$$
$$2x = 166 \quad \text{Combine like terms.}$$
$$2x \cdot \frac{1}{2} = 166 \cdot \frac{1}{2} \quad \text{Multiply both sides by } \frac{1}{2}.$$
$$x = 83$$

The smaller number $= x = 83$

The larger number $= x + 2 = 85$

Notice that both 83 and 85 are odd numbers and that their sum is equal to 168. ■

◇ *Do Practice Exercises 6 and 7.*

EXAMPLE 11

A rectangle has a width of 50 meters and a perimeter of 300 meters. Find the length.

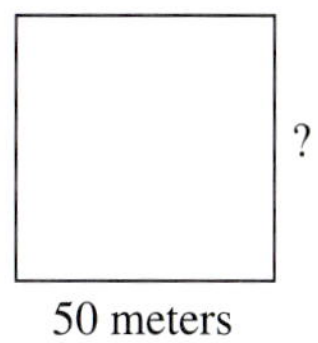

SOLUTION

First, use the formula for the perimeter of a rectangle from Chapter 8.

$$P = 2 \cdot L + 2 \cdot W$$

$$300 = 2 \cdot L + 2 \cdot 50 \quad \text{Substitute values.}$$

$$300 = 2L + 100 \quad \text{Multiply.}$$

$$300 - 100 = 2L + 100 - 100 \quad \text{Subtract 100 from both sides.}$$

$$200 = 2L \quad \text{Combine like terms.}$$

$$200 \cdot \frac{1}{2} = 2L \cdot \frac{1}{2} \quad \text{Multiply both sides by } \frac{1}{2}.$$

$$100 = L \quad \text{or} \quad L = 100 \text{ meters}$$

The length of the rectangle is 100 meters. The perimeter is $2 \cdot 100 + 2 \cdot 50 = 300$ meters, which checks. ■

EXAMPLE 12

A piece of steel is 37 feet long. It is to be cut so that one piece is 9 feet longer than the other. Find the lengths of both pieces.

SOLUTION

Let $x =$ the length of the shorter piece.
Then $(x + 9)$ is the length of the longer piece.

$$\text{short piece} + \text{long piece} = 37$$

$$x + (x + 9) = 37$$

$$2x + 9 = 37$$

$$2x = 28 \quad \text{Add } -9 \text{ to both sides.}$$

$$x = 14 \text{ feet}$$

$$x + 9 = 23 \text{ feet}$$

■

Write an algebraic expression for each.

1. The sum of 6 with an unknown number

2. Four times an unknown quantity

3. Twelve and one-half minus some number

4. An unknown number minus negative seven and four tenths

5. Negative three tenths minus twice some unknown quantity

6. Seven times an unknown number plus four hundred

7. An unknown number plus three times the same unknown

8. The sum of three different unknown numbers

9. Four times an unknown number minus seventy and two tenths

10. Fourteen minus seven times an unknown quantity

Solve for the unknown. (Review)

11. $x + 3 = 10$

12. $y - 5 = -4$

13. $\frac{z}{3} = 8$

14. $6t = 42$

15. $5z + 3 = z$

16. $\frac{1}{2}x = -4$

17. $6x - 8 = 4x$

18. $9z = \frac{1}{3}$

19. $y = 3y - 5$

20. $\frac{z}{5} + 4 = 21.2$

First set up each problem as an algebraic equation, then solve the equation and check your answer.

21. The sum of an unknown with eight is twenty. Find the unknown.

22. The sum of four plus an unknown equals fifty-one. Find the unknown.

23. An unknown minus fourteen is equal to forty. Find the unknown.

24. An unknown minus four equals negative 7. Find the unknown.

25. Six times an unknown is fifty-four. Find the unknown.

26. Seventy-five is five times an unknown. Find the unknown.

27. An unknown times thirty equals one thousand eight hundred. Find the unknown.

28. Negative twenty-four equals three times an unknown. Find the unknown.

29. An unknown divided by eight is four thousand eight hundred. Find the unknown.

30. Negative twenty-seven equals an unknown divided by ten. Find the unknown.

31. Three times an unknown plus five is equal to forty-nine. Find the unknown.

32. Four plus seven times an unknown equals negative one and one-half. Find the unknown.

33. An unknown divided by seven is equal to nine. Find the unknown.

34. Twelve minus two times an unknown is equal to forty and three tenths. Find the unknown.

35. Negative seven is equal to six minus four times an unknown. Find the unknown.

36. The sum of two consecutive integers is equal to ninety-one. Find both integers.

37. The sum of two consecutive odd integers is equal to two hundred and forty. Find both integers.

38. The sum of two consecutive even integers is equal to ninety-four. Find both integers.

39. The sum of two consecutive odd whole numbers is equal to six hundred and twenty-four. Find the two whole numbers.

40. The sum of two consecutive odd integers is equal to three hundred sixty. Find both numbers.

41. A rectangle with a width of twelve feet has an area of one hundred sixty-eight square feet. Find the length of the rectangle.

42. A rectangle has a perimeter of 22 and a length of $3\frac{1}{2}$. Find the width.

43. Darla has saved \$51 more than her brother Freddie. Together, they have saved \$369. How much has Darla saved?

44. Tom has saved \$42 more than Laurie. Together, they have saved \$78. How much has Laurie saved?

45. A pipe is 68 inches long. It is to be cut so that one piece is four inches longer than the other. Find the length of the two pieces.

46. A husband and wife team earn exactly \$2,000 per month. If the wife earns \$350 per month more than the husband, how much does she earn?

47. A company must have sales of \$35,000 per month to pay their bills. One salesman usually sells \$22,500 per month. How much must a second salesman sell if the company is to break even?

JOURNAL EXERCISES

1. Can you find more than one pair of integers that satisfy the following constraint: The difference of two consecutive even integers is two? Explain.

2. Write and solve your own word problem that requires an algebraic equation.

3. Write your own word problem with an unknown and bring it to class where it will be exchanged with a word problem written by another student.

CHAPTER 10 REVIEW

SECTION	TOPIC OR TERM	EXAMPLE OR DEFINITION
10.1		
Variables and Expressions	Constant	A constant is a number such as 3 or -5.7.
	Variable	A variable is a symbol such as x or y for an unknown quantity or a quantity that can vary.
	Term	Terms are quantities that are added or subtracted, such as $3x$ or $5t$.
	Expression	$3x + 5y, \quad 7.5z - 2t + 3.2$
	Numerical Coefficient	The numerical coefficient is the number being multiplied times a variable.
	Like Terms	Like terms are terms with the same variables raised to the same exponents.
	Unlike Terms	Unlike terms are terms with different variables or with the same variables raised to different exponents.
	Add Like Terms	$-9y + 7y = (-9 + 7)y = -2y$
10.2		
Simplify Terms and Expressions	Simplification	Simplification is the process of writing a term or an expression with fewer symbols.
	Simplify Terms	$6 \cdot 3 \cdot x = 18x, \quad 5 \cdot y \cdot 4 = 20y,$ $b \cdot a \cdot 7 = 7ab$
	Simplify Expressions	$4(x + y) + 5y$ $4x + 4y + 5y$ Distributive property $4x + 9y$ Result.
10.3		
Solving Equations Using the Addition Property	Solving Equations	Solving an equation is the process of finding a value for a variable that makes the equation true.
	Addition Property	The addition property states that the same value can be added to both sides of an equation.
	Solve Using Addition Property	$y - 4 = 12$ $y - 4 + 4 = 12 + 4$ Add 4 to both sides. $y = 16$ Solution.

SECTION	TOPIC OR TERM	EXAMPLE OR DEFINITION
10.4		
Solving Equations Using the Multiplication Property	Multiplication Property	The multiplication property states that the same value can be multiplied times both sides of an equation.
	Solve Using Multiplication Property	$\frac{x}{7} = -5$ $\frac{x}{7} \cdot 7 = -5 \cdot 7$ Multiply both sides by 7. $x = -35$ Result.
10.5		
Equations Requiring More than One Operation	Equations that Require More than One Operation	$3y - 5 = -8$ $3y - 5 + 5 = -8 + 5$ Add 5. $3y = -3$ $3y \cdot \frac{1}{3} = -1 \cdot \frac{1}{3}$ Multiply by $\frac{1}{3}$. $y = -1$ Result.
10.6		
Word Problems	Consecutive Whole Numbers	Consecutive whole numbers are numbers that naturally fall after one another: 1, 2, 3, 4, . . .
	Consecutive Odd Numbers	1, 3, 5, 7, 9, . . .
	Consecutive Even Numbers	2, 4, 6, 8, 10, . . .
	Word Problems	The sum of two consecutive whole numbers is 93. Find both numbers. Let x = smaller number, then $x + 1$ = larger number $x + (x + 1) = 93$ $2x + 1 = 93$ $2x + 1 + (-1) = 93 + (-1)$ $2x = 92$ $2x \cdot \frac{1}{2} = 92 \cdot \frac{1}{2}$ $x = 46$ and $x + 1 = 47$

APPLICATION WORD PROBLEM

THE EARTH

The earth is in the shape of a sphere (ball) with a radius of approximately 4,000 miles. The radius of a sphere is the distance from the center of the sphere to the surface.

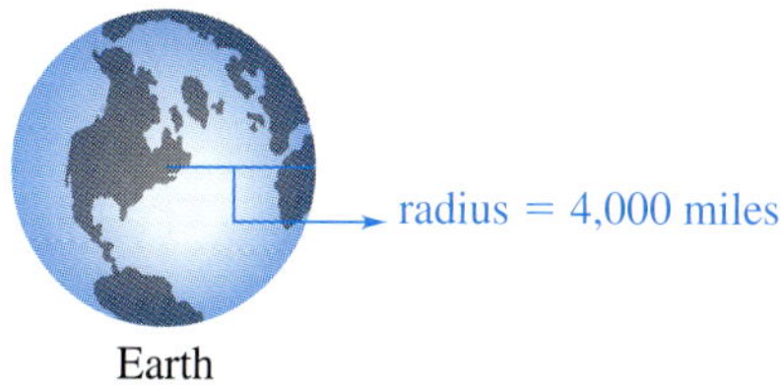

The area of the surface of a sphere is given by the following algebraic formula.

$$A = 4\pi r^2$$

where A = area, π = 3 (approximately), and r = radius.

1. Use the algebraic formula to calculate the surface area of the earth.

The seven continents of the earth and their approximate land areas are:

Continent	*Land Area*
North America	9,400,000 sq miles
South America	6,900,000 sq miles
Europe	3,800,000 sq miles
Asia	17,300,000 sq miles
Africa	11,700,000 sq miles
Australia	3,300,000 sq miles
Antarctica	5,400,000 sq miles

2. What is the land area of all seven continents combined?

The surface of the earth is either land or water. The answer to question 1 is the total surface area of the earth. Assume that the answer to question 2 is the total land area.

3. Find the surface area of the oceans, and round your answer to the nearest million square miles.

4. Complete the following table. Add the values in the second column and use this sum when calculating values in the last column. Round the percents to the nearest percent.

Region	*Surface Area (Square Miles)*	*Surface Area Rounded to Nearest Million*	*Percent of Total*
North America	9,400,000	9,000,000	
South America	6,900,000		
Europe			
Asia			
Africa			
Australia			
Antarctica			
Oceans			

Use data from the next to last column to answer the following problems.

5. Find the ratio of the area of North America to the area of Europe. Which is larger?

6. Find the ratio of the area of North and South America combined to the area of Asia. Which is larger?

7. Find the ratio of the area of Africa to North America. Which is larger?

8. Find the ratio of the area of the oceans to the total surface area of the world. What fraction of the earth is covered with water? Is that more than one-half?

CHAPTER 10 REVIEW PROBLEMS

Identify variables and constants.

1. $6\frac{1}{2}$, x, y

2. 3, 0, A

3. B, -5.2, t

4. 7, x, -2

Identify like terms.

5. x, y, $3x$, $2z$

6. $9y$, $8y$, $3xy$

7. $2t$, $3t$, $4tz$

8. $8y$, $2x$, $9y$, $3x$, $9y^2$

9. $5Y$, $7Y$, $12X$

10. $3r$, $5y$, $2x$, $-3.8y$

Find the numerical coefficient of each.

11. $6xy$

12. $-4z$

13. x

14. $-2\frac{1}{3}y$

15. $7.2y$

16. $-9rt$

17. ab

18. $\frac{3}{2}z$

Simplify, if possible.

19. $3x + 2y$

20. $4y + 5 + 2y$

21. $14x - 7 - 2x$

22. $3x + 2y - 5x$

23. $16 + 4(x + 2)$

24. $-7y + 8 - 9y$

25. $(-6)y + (-4)y$

26. $-3x - 2x$

27. $x + x$

28. $5x + 3 - 2x$

29. $7(x + y) - 4y$

30. $-10.5r + 4(r + y)$

31. $3y - 6 + 5.5y$

32. $2z - z$

Solve for the unknown.

33. $y - 4 = 8$

34. $z + 12 = 65$

35. $r + 2 = -5$

36. $y - 9 = -4$

37. $x + 5 = 9$

38. $x = 8 - 4$

39. $y - 4 = 18$

40. $27 = 3y$

41. $\frac{z}{4} = 24$

42. $6y = -9 - 4$

43. $9y - 2 = 47.5$

44. $5z = 5{,}500$

45. $-8 + 6x = -4$

46. $3y + 2y = 41$

47. $4z = 12 - 5z$

48. $\frac{y}{2} + 5 = 16$

49. $3x = 2x - 2$

50. $8y = -1$

51. $4.5 = z - 9$

52. $4x + x = -2$

53. $3y = 5y - 91$

54. $7z = 4z - 1$

Set each problem up as an equation, then solve.

55. The sum of an unknown and four hundred is equal to one. Find the unknown.

56. Four times an unknown minus sixteen is equal to seven and one-half. Find the unknown.

57. Three times an unknown is equal to the sum of four plus the unknown. Find the unknown.

58. Negative seven plus four times an unknown equals twenty-one. Find the unknown.

59. The sum of two consecutive odd numbers is two hundred forty-four. Find both numbers.

60. Tom throws some coins onto the table. Joe throws twice as much money onto the table. The total amount of money on the table is \$2.61. How much money did Joe place on the table?

CHAPTER 10 TEST

Identify like terms.

1. $3y, \quad 4xy, \quad 5y$

2. $4x, \quad 2y, \quad 3z, \quad 5.1y$

3. $ab, \quad bc, \quad abc, \quad 4.5bc$

4. $6y, \quad -2z, \quad -y$

Simplify.

5. $3x + 5x$

6. $2y - 8y$

7. $7z - 8z$

8. $5(x + 2) + 19$

9. $5r + 2(x + r)$

10. $xy + 2xy + x$

Solve for the unknown.

11. $x - 9 = 4.7$

12. $2y + 3 = -y$

13. $6 + 5z = 34$

14. $6z - z = -500$

15. $3y - 2 = 9$

16. $\frac{y}{5} + 3 = 8$

17. $4z + 2 = -z$

18. $6x = 4x$

Work each problem.

19. What is the numerical coefficient of $-7.5y$?

20. What is the numerical coefficient of $3\frac{1}{2}r$?

Write an algebraic equation for each, then solve for the unknown.

21. The sum of four plus three times an unknown is equal to forty. Find the unknown.

22. Ten y minus forty is equal to sixty. Find the value for y.

23. The sum of two consecutive numbers is 113. Find both numbers.

24. Together, a man and his son weighed 330 pounds. How much did the man weigh if he was 50 pounds heavier than his son?

25. A rental car rents for \$35 per day plus 30 cents per mile. How many miles were driven if the rental fee for two days was \$127?

CUMULATIVE TEST CHAPTERS 9 AND 10

Work each problem.

1. $(-6) + (-4)$

2. $(+3) - (-5)$

3. $6 \cdot (-8)$

4. $(-42) \div (-3)$

5. $-21 - 14$

6. $-6 \cdot 9$

Evaluate each expression.

7. $-5 - (+3)(+2)$

8. $6 \div (-4) + (-1.3)$

9. $(-4)^2 + (-3)$

10. $\left(-\frac{1}{2}\right) - \left(-\frac{1}{4}\right)$

Identify like terms, if any.

11. $3x, \quad -y, \quad 2xy, \quad 4\frac{1}{2}x$

12. $9xyz, \quad 3t, \quad -xyz$

Simplify, if possible.

13. $7x + 12x$

14. $-8y - 15y$

15. $3y + 2x - 5y$

16. $2.7r - 3.2t$

Solve for the unknown.

17. $x + 5 = 9$

18. $y - 3 = 14$

19. $8r = 21$

20. $\frac{y}{9} = 3.2$

21. $6x - 4 = 7$

22. $2y = 5y - 3$

Work each problem.

23. List -4, -3.09, 4.1, and $-3\frac{1}{2}$ from smallest to largest.

24. Write an equation for the following: the sum of an unknown plus forty-one equals fifty-nine. Then solve for the unknown.

25. Write an equation for the following: subtract the product of eight times an unknown from one-half and set this quantity equal to 15. Then solve for the unknown.

11 Statistics

APPLICATION *Data from one insurance company shows that 4.5% of their 16–20 year old drivers will be in an automobile accident within one year. If the company insures 3,000 drivers in this age range, how many accidents should they expect in one year?*

SOLUTION *The answer is found by multiplying 4.5% times 3,000 drivers.*

$$0.045 \cdot 3{,}000 = 135 \text{ car accidents}$$

The company should expect 135 accidents.

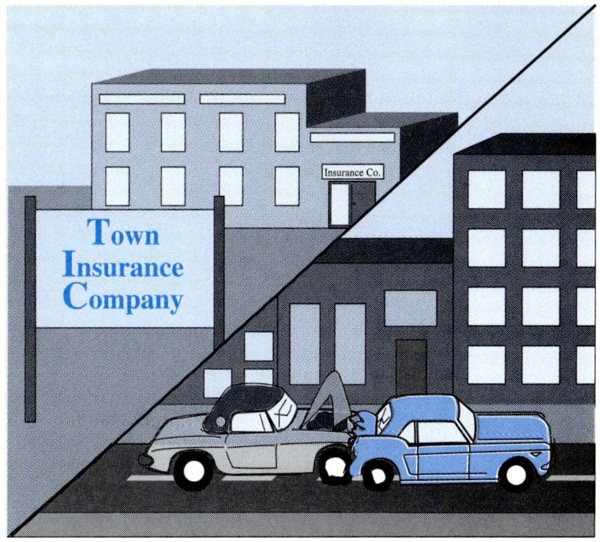

11.1 MEAN, MEDIAN, AND MODE

Statistics is a branch of mathematics that works with data. One of the first things commonly looked for in a **set** of data, or group of numbers, is a single number representing the center or middle. The single number is said to be a measure of the **central tendency**, or middle, of the data. Several measures of central tendency are given in this section.

OBJECTIVES

1 Calculate the mean.
2 Calculate the median.
3 Calculate the mode.

1 Mean

The **mean** is another word for the arithmetic average of a set or group of numbers. This is the average we have calculated throughout the text. It is found by adding the values together and dividing by the number of values.

EXAMPLE 1

Rose's test scores are 83%, 91%, and 74%. Find the mean.

SOLUTION

$$\text{Mean} = \text{average} = \frac{83\% + 91\% + 74\%}{3} = \frac{248\%}{3} = 82\frac{2}{3}\% \quad \blacksquare$$

◇ *Do Practice Exercise 1.*

1. Find the mean of 4.7, 3.5, and 3.5.

ANSWER
1. 3.9

2. What grade would it take to bring the average of 32%, 80%, 75%, and one other test up to 70%?

The following is a problem that may be of particular interest to you near the end of a course.

EXAMPLE 2

A class grade is determined by the average of four tests. Arin's test scores for the first three tests were 75%, 81%, and 79%. What grade must she make on the last test in order to have an average of 80%?

SOLUTION

Arin wants her final average or mean score to be 80%. The mean is the sum of all of the test scores, including the test she has not yet taken, divided by 4:

$$\frac{\begin{matrix}\text{Score} \\ \text{Test 1}\end{matrix} + \begin{matrix}\text{Score} \\ \text{Test 2}\end{matrix} + \begin{matrix}\text{Score} \\ \text{Test 3}\end{matrix} + \begin{matrix}\text{Score} \\ \text{Test 4}\end{matrix}}{4} = 80\% \qquad \text{definition of mean}$$

Let x = the grade on the test she has not yet taken, then:

$$\frac{75\% + 81\% + 79\% + x}{4} = 80\% \qquad \text{Substitute.}$$

$$\frac{235 + x}{4} = 80 \qquad \text{Add percents.}$$

$$\frac{235 + x}{4} \cdot 4 = 80 \cdot 4 \qquad \text{Multiply by 4.}$$

$$\frac{235 + x}{\not{4}} \cdot \not{4} = 80 \cdot 4 \qquad \text{Divide forms of one.}$$

$$235 + x = 320 \qquad \text{Multiply.}$$

$$235 + x + (-235) = 320 + (-235) \qquad \text{Add } -235.$$

$$x = 85\% \qquad \text{Result.}$$

Arin must make an 85% or better on the last test in order to have an average grade of at least 80%. ■

◇ *Do Practice Exercise 2.*

Sometimes, items occur more than once in a list. In this event, each value must first be multiplied times a weight before summing. This method is referred to as **weighted mean**. The next example applies this concept to the calculation of grade point average.

EXAMPLE 3

Hilda's grades for her first semester at college were:

Subject	*Credit Hours*	*Grade*
English	3	C
History	3	C
Math	4	B
Physical Education	1	A
Psychology	3	B

ANSWER
2. 93%

Find her grade point average to the nearest hundredth assuming: A = 4.0, B = 3.0, C = 2.0, D = 1.0, and F = 0.0.

SOLUTION

English is a 3 hour class whereas math is a 4 hour class. Accordingly, the English grade will have less weight in determining the final grade point average than will the math grade. The problem is solved by first weighing or multiplying each grade by the associated number of credit hours. The weighted sum is then divided by the number of credit hours.

Subject	*Credit Hours*	*Points*	*Credit Hours × Points*
English	3	2.0 (a C equals 2.0)	6.0
History	3	2.0	6.0
Math	4	3.0	12.0
Physical Education	1	4.0	4.0
Psychology	3	3.0	9.0
Totals	14		37.0

Grade Point Average = 37 ÷ 14 ≈ 2.64 or a low B average. ■

◇ *Do Practice Exercise 3.*

Calculator Application

Ten years after graduation, a survey was taken of 100 students who graduated from the same high school:

Highest Education Level Completed	*Number of Students*	*Average Annual Income*
High school diploma	55	$10,850
Trade school diploma	18	$16,200
Four year college degree	27	$22,500
Total	100	

Find the weighted average salary of all 100.

Answer: The average salary of all 100 is: $14,958.50

3. Find the grade point average of a student with 3—3 credit hour As, 1—3 credit hour C, and 1—4 credit hour B. Round your answer to the nearest hundredth.

ANSWER
3. 3.38

2 Median

Sometimes, the mean gives a value for the central tendency that does not seem reasonable, as shown in the next example.

EXAMPLE 4

The annual incomes of five different families are: \$20,000, \$21,000, \$36,000, \$28,000, and \$450,000. Find the mean.

SOLUTION

$$\text{Mean} = \frac{\$555{,}000}{5} = \$111{,}000 \quad \blacksquare$$

It does not seem reasonable that the "middle" income of the group in the last example is \$111,000 since four of the five families make less than \$40,000. The one family with the high income of \$450,000 is distorting the value of the mean. We will come back to this example after defining median. The **median** is another measure of the central tendency of a set of data. The steps to calculate the median are given next.

Rule: To Calculate the Median

Step 1 List the numbers from smallest to largest.

Step 2 Find the median:

- **(a)** It is the one in the middle if there are an odd number of numbers.
- **(b)** It is an average of the two in the middle if there are an even number of numbers.

EXAMPLE 5

Find the median of 1, 14, 3, 7, and 11.

SOLUTION
First, list the numbers from smallest to largest. The median is the number in the middle since there are 5, or an odd number, of numbers.

Step 1 1, 3, 7, 11, 14 — Arrange from smallest to largest.

Step 2 7 = median — The median is the one in the middle. ■

EXAMPLE 6

Find the median of 61, 3, 11, and 22.

SOLUTION

Since there are four or an even number of numbers, the median is the average of the two numbers in the middle:

Step 1 3, 11, 22, 61 — from smallest to largest

Step 2 $\frac{11 + 22}{2}$ — Average the two in the middle, since there are 4 (even) numbers.

16.5 ■

◇ *Do Practice Exercises 4 and 5.*

4. Find the median of 3, 15, 11, 21, and 44.

5. Find the median of 9, 7, 14, and 12.

The next example compares the mean to the median for the data on family incomes from Example 4.

EXAMPLE 7

The mean of the family incomes given in Example 4 was $111,000. The median is found by ranking the five annual incomes from smallest to largest and finding the value in the middle.

$20,000 $21,000 $28,000 $36,000 $450,000

median income = $28,000

A value of $28,000 seems more in line as a measure of central tendency than does the mean of $111,000. ■

3 Mode

A third measure of central tendency is the **mode**. The mode is the number that occurs the most often in the data set.

EXAMPLE 8

Find the mean, median, and mode of the elements of the following numbers: 3, 4, 4, 5, 5, 5, 6, 8, 18.5

SOLUTION

$$\text{Mean} = \frac{3 + 4 + 4 + 5 + 5 + 5 + 6 + 8 + 18.5}{9} = 6.5$$

Median = 5 — the number in the middle

Mode = 5 — since 5 occurs more often than any other number in the list ■

ANSWERS
4. 15
5. 10.5

6. Find the mean, median, and mode of 3, 6, 4, 5, 9, and 4.

In the last example, the median and the model were equal. Frequently, the three measures of central tendency will be different even though they are all used to estimate central tendency. Try the practice exercise.

◇ *Do Practice Exercise 6.*

A data set may have more than one mode. For example, a data set with two modes is called **bimodal**. A data set in which no number occurs more than once has no mode.

EXAMPLE 9

Find the mode of the following: 3, 3, 4, 5, 5, 6, and 7.

SOLUTION

The data has two modes, 3 and 5. Thus the data set is called bimodal. ■

EXAMPLE 10

Find the mode of the following: 1, 2, 3, 4, and 7.

SOLUTION

This set has no mode since no number occurs twice. You may wish to check that it does have both a mean and a median. ■

ANSWER

6. mean = 5.2 approximately
median = 4.5
mode = 4

Find the mean, median, and mode of the elements of each. If the mode does not exist, say so.

1. {3, 6}

2. {2, 4, 4, 1}

3. {4, 6, 4, 9, 4}

4. {1, 5, 5}

5. {2, 3, 4}

6. {3, 5, 9}

7. {6, 8, 54}

8. {6, 100, 200}

9. {3.4, 3.5}

10. {10, 15, 12, 19, 20}

11. {6, 8, 2, 6, 4}

12. {4, 4, 4, 4, 4}

13. {1, 2, 1, 2, 1, 2}

14. {1, 100, 2, 4, 4}

15. {3, 6, 2, 17}

16. {3, −6, 5}

17. $\left\{-2.5, 3\frac{1}{4}, -2.5\right\}$

18. {7, −4.5, 0, 0}

19. {5, 5.02, 5.02}

20. {1, 999}

Work each problem.

21. Andrea's four course grades were A, B, B, and C. Find her grade point average if each course is a 3 hour course. (A = 4.0, B = 3.0, C = 2.0, D = 1.0, F = 0.0)

22. The populations of the three largest cities in the world are:

Tokyo, Japan—26,000,000 people
Mexico City, Mexico—18,000,000 people
Sao Paolo, Brazil—16,000,000 people

Find the total population of the three, then the mean population.

23. A salesman flies the following distances: 420 miles, 330 miles, 560 miles, and 1,200 miles. Find the mean and median.

24. At weekly intervals an individual's weight was 185 lbs, 183 lbs, 181 lbs, 186 lbs, and 185 lbs. Find the mean, median, and mode.

25. Which is larger, the mean of {3, 7, 6} or the median of {4, 6}?

26. The daily fluid intakes of a critically ill hospital patient were 2,100 ml, 2,050 ml, 1,925 ml, and 2,075. Find the average.

27. A study was underway to determine the reproductive capacity of a rare bird. A sample of ten nests showed the following number of eggs per nest: 6, 8, 5, 0, 2, 7, 0, 6, 4, and 0. Find the mean, median, and mode.

28. The average salary paid to full-time nurses at one hospital is $37,500. What payroll must the hospital make if they have forty-two full-time nurses?

JOURNAL EXERCISES

1. Explain mean, median, and mode in your own words. Why are different measures of central tendency useful?

2. Look up the definitions of probability and statistics, then give several examples of uses of each.

Collect statistics on each of the following and find the mean, median, and mode if it exists.

3. The height of all of the members in your family

4. The weight of all of the members in your family

5. The amount of money you spend per day for 3 days in a row

6. The amount of television you watch for 3 days in a row

7. The amount of time you spend doing your homework for 3 days in a row

1 Statistical Data

OBJECTIVE

1 Work with statistical data.

In modern society, we are faced with a tremendous amount of statistical data. Statistical data can be shown in the form of tables, bar charts, pie charts, pictographs, and line graphs. Recall that we looked at each of these ways of presenting data in Chapter 2. In this section, additional examples are shown and some conclusions are drawn.

EXAMPLE 1

The Olympics is an international sports competition held only once every four years. One event at the Olympics is the marathon, a race of more than 26 miles. The line graph below shows the winning times, in minutes, of marathons at past Olympics. What conclusions can be drawn?

Marathon Times at Past Olympics

SOLUTION

The fastest marathons today are faster than those of 60 years ago. However, marathon times have not improved much in recent years. From the data given, one cannot say that future marathon times will not improve. In fact, future marathons may be run significantly faster. ■

EXAMPLE 2

A recent survey of 175,000 thirteen year old students showed the following scores on a science test:

Country	*Average Score*
South Korea	78%
Taiwan	76%
Switzerland	74%
Russia	71%
Canada	69%
France	69%
Spain	67%
United States	59%
Jordan	57%

Results on a similar math test showed almost exactly the same rankings by country and an even larger difference between high ranking and low ranking countries. What conclusions can be drawn?

SOLUTION

Students from both Jordan and the United States scored lower than students from South Korea and Taiwan. It is not possible to conclude why this occurred based on the data given. ■

EXAMPLE 3

The number of known deaths of United States citizens is shown for several major wars. What conclusions can be drawn?

War	*Period*	*Deaths*	*Approximate Population of the U.S.*
Civil War	1861–65	498,332	31,500,000
World War I	1917–18	116,516	106,000,000
World War II	1941–46	407,316	132,100,000
Korean War	1950–53	54,246	151,300,000
Vietnam	1960–72	60,151	200,000,000

Data Source: 1992 *World Almanac*

SOLUTION

In terms of number of deaths, the two worst wars fought by the United States were the Civil War and World War II. In comparison, World War I, the Korean War, and the Vietnam War caused fewer deaths. ■

It is interesting to add one column of data to the table from the last example, namely:

War	*Period*	*Deaths as a Percent of Population at that Time*
Civil War	1861–65	1.58%
World War I	1917–18	0.11%
World War II	1941–46	0.31%
Korean War	1950–53	0.04%
Vietnam	1960–72	0.03%

In this regard, the Civil War was clearly the worst, since an average of $1\frac{1}{2}$ out of every 100 citizens died as a result of the war.

Use each table of data to answer the questions that follow.

Company Profits for One Year

Quarter	*Profit*	*Percent of Total Profit*
First	$240,000	
Second	$209,000	
Third	$312,000	
Fourth	$490,000	

1. Find the total profit.

2. Complete the percent column.

3. In which quarter did they make the most profit?

4. What percent of their annual profit occurred in the first two quarters?

5. From the data above, can you conclude that the highest quarterly profits for the company will always occur in the fourth quarter? Explain.

Estimated New Cases of Cancer per Year

Type of Cancer	*Number*	*Percent of Total*
Oral	30,500	
Colon-Rectum	110,000	
Lung	157,000	
Skin	27,600	
Breast	150,900	
Uterus	13,500	

Data Source: *World Almanac*

6. Find the total number of estimated new cases of cancer per year.

7. Complete the percent column.

8. Which two cancers occur most often?

9. How many of those estimated to get lung cancer will die of the cancer if 70% of all lung cancers end in deaths?

10. Can you tell which type of cancer causes the most deaths based on the data given in the table?

Population Estimates of a Rare Species of Bird

Year	*Estimated Population*
1988	1,347
1990	1,100
1992	250
1994	246

11. Use a line graph to plot these. See Chapter 2 if you need to review line graphs.

12. What is the trend (up or down) of the number of birds?

13. Look up the word ''extinction'' in a dictionary and copy down the definition.

14. Would you think that this species might be headed for extinction?

15. Can it be concluded from the data given that there is no way to save this species of bird from extinction?

Some Famous Waterfalls

Name of Falls	*Country*	*Height*
Niagara	Canada—U.S.	182 feet
Glass	Brazil	1,325 feet
Angel	Venezuela	2,648 feet
Wollomombi	Australia	1,100 feet

16. Which waterfall has the greatest height?

17. The height of the Niagara falls is what percent of the height of the Wollomombi Falls of Australia?

18. From the data given, can it be concluded that more water moves through Niagara Falls than through Glass Falls?

JOURNAL EXERCISES

Follow the directions for each problem.

1. Get some data from a magazine or another course, present it in table form, and analyze it.

2. Create your own table of data from something you know about. Bring your data set to class where it will be exchanged with that of another student who will find the mean, median, and mode.

OBJECTIVE

1 Work with histograms.

1 Histograms

A **histogram** is a special bar chart used to represent data. The width of each bar represents a group of numbers called a **class interval**. The height of each bar represents the number of times a value in that class interval occurred. The number of times is referred to as **class frequency**. The steps for creating a histogram will be given by this next example.

A teacher was trying to determine whether or not her students were confused by some recent material she had covered. She gave a quiz over the material with the following results:

50%, 64%, 47%, 63%, 66%, 75%, 58%, 68%, 70%, 72%,
69%, 74%, 70%, 81%, 42%, 57%, 38%, 71%, 73%, 69%

where:

90%–100%	is an A
80%–89%	is a B
70%–79%	is a C
65%–69%	is a D
<65%	is an F

The grades are the class intervals and the number of students making each grade is the class frequency. The teacher can analyze the grades by using a **tally** to find the number of students with each letter grade.

Scale	*Grade*	*Tally of Students*	*Number of Students*
90–100	A		0
80–89	B	|	1
70–79	C	|||||||	7
65–69	D	||||	4
<65	F	||||||||	8

Next, the teacher can draw a histogram:

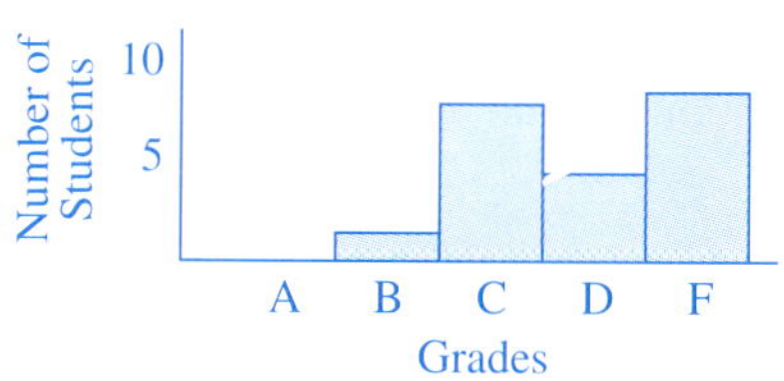

The histogram shows that no students made As, yet there were several Ds and Fs. Based on this information, the teacher decided to review the material with the students once more before giving a major test. Notice that the first step was to divide the grades into smaller groups, namely <65%, 65%–69%, and so on. The number of grades falling into each small group was then counted.

EXAMPLE 1

Jason recorded the distances jumped by his pet flea. In inches, the distances were:

9.2,	8.5,	2.8,	4.0,	0.5,	4.0,	5.0,	7.1,	5.0,	6.4,
4.0,	8.5,	9.0,	10.0,	7.5,	5.0,	4.8,	7.6,	3.4,	1.5,
4.5,	3.8,	6.4,	7.8,	4.2,	3.1,	5.7,	0.7,	1.8,	5.4

Find the mean of the data and plot the data in a histogram.

SOLUTION

$$\text{Mean} = \frac{\text{Sum of distances}}{\text{Number of jumps}} = \frac{157.2}{30} = 5.24 \text{ inches}$$

In order to plot the data in a histogram, the data is first divided up into intervals. The values range from a low of 0.5 inches to a high of 10.0 inches. We arbitrarily divide the numbers from 0 to 10 into five equal size groups: 0.0–1.9, 2.0–3.9, 4.0–5.9, 6.0–7.9, and 8.0–10.0. The number of occurrences in each interval are then counted.

Inches	*Tally*	*Number*
8.0–10.0	|||||	5
6.0–7.9	||||||	6
4.0–5.9	|||||||||||	11
2.0–3.9	||||	4
0.0–1.9	||||	4

Finally, a histogram is drawn.

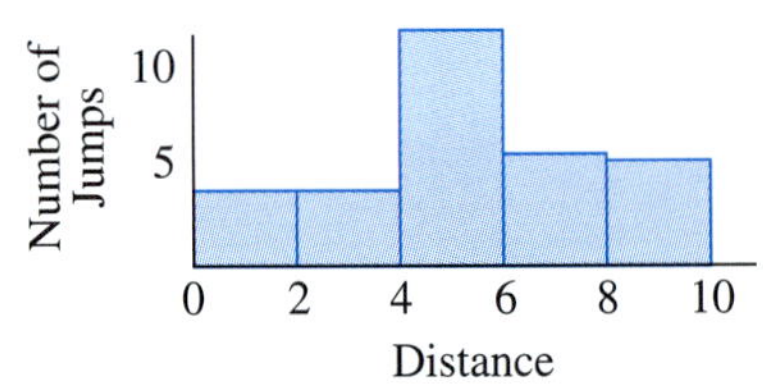

The peak (mode) in the histogram occurs at 4 to 5.9 inches, a number that agrees nicely with the mean of 5.24 inches. ■

Make a histogram from each set of data. Use the intervals given. Put labels on the histogram.

1. Grades on a Test: 37%, 80%, 91%, 85%, 70%, 98%, 72%, 73%, 68%, 89%, 90%, 74%, 79%, 75%, 81%

Intervals: <65%, 65%–69%, 70%–79%, 80%–89%, 90%–100%

2. Sales Tax Rate in Different Cities: 7.5%, 8.2%, 6.6%, 7.2%, 8.5%, 6.8%, 7.8%, 6.4%, 7.1%

Intervals: 6.0%–6.9%, 7.0%–7.9%, 8.0%–8.9%

3. Daily Fluid Intake for a Hospital Patient, in Cubic Centimeters: 2,000, 2,350, 1,900, 2,100, 1,900, 2,300, 2,100, 2,050, 2,000, 1,950, 2,100, 1,800

Intervals: 1,800cc–1,999cc, 2,000cc–2,199cc, 2,200cc–2,399cc

4. Grades on a College Entrance Test: 70%, 62%, 45%, 52%, 63%, 58%, 90%, 42%, 65%, 72%, 48%, 39%, 81%, 62%

Intervals: 21%–40%, 41%–60%, 61%–80%, 81%–100%

5. Weight of Boys Age 12, in Pounds: 140, 125, 130, 112, 127, 135, 138, 121, 124, 118

Intervals: 111–120, 121–130, 131–140

6. Number of Children per Family: 3, 4, 2, 3, 1, 0, 5, 1, 3, 4, 0, 2, 1, 3, 3, 6

Intervals: 0–2, 3–5, 6–8

7. Age of Automobiles, in Years: 3, 1, 10, 7, 5, 3, 2, 4, 1, 5, 3, 6, 3, 2, 6, 8
Intervals: 0–2, 3–5, 6–8, 9–11

8. Weight of Policemen, in Pounds: 240, 210, 190, 175, 168, 183, 195, 200, 175, 190
Intervals: 151–175, 176–200, 201–225, 226–250

9. Dollars in Checking Accounts: \$125, \$185, \$220, \$65, \$800, \$925, \$450, \$1,200, \$600, \$850
Intervals: \$1–\$500, \$501–\$1,000, \$1,001–\$1,500

10. Length of Human Lives, in Years: 75, 73, 81, 78, 45, 83, 70, 66, 67, 49, 58, 76

Intervals: 41–50, 51–60, 61–70, 71–80, 81–90

11. Number of Cavities per Person: 3, 7, 2, 4, 1, 0, 2, 0, 6, 3, 9, 1, 1, 1, 3, 4, 1

Intervals: 0–2, 3–5, 6–8, 9–11

12. Inches of Rain per Year on One Farm: 20, 28, 22, 18, 26, 30, 27, 19, 23, 24, 35

Intervals: 16–20, 21–25, 26–30, 31–35

13. Number of Hours of Television Watched per Day in One Household: $2\frac{1}{2}$, $4\frac{1}{2}$, 2, $3\frac{1}{4}$, 5, $\frac{1}{2}$, $3\frac{1}{2}$, $2\frac{1}{4}$, 2, $1\frac{1}{2}$

Intervals: 0–1.9, 2.0–2.9, >3.0

14. Number of Pets in Different Households: 2, 3, 0, 0, 1, 5, 2, 1, 3, 5, 2, 1, 1, 0, 0, 1

Intervals: 0–1, 2–3, 4–5, 6–7

15. Amount of Memory in Different Personal Computers, in Millions of Bytes: $\frac{1}{2}$, 2, $\frac{1}{2}$, 4, $\frac{1}{2}$, 3, $\frac{1}{2}$, 1, 2, 1, 1, 5, 2, $\frac{1}{2}$

Intervals: 0–1, 2–3, 4–5

16. Length of Movies, in Minutes: 120, 108, 113, 126, 109, 118, 126, 115, 130, 122, 120, 137, 127

Intervals: 101–110, 111–120, 121–130, 131–140

17. Shoe Sizes: 12, $9\frac{1}{2}$, 10, $10\frac{1}{2}$, 7, 9, 11, 8, 8, $7\frac{1}{2}$, 8, 13, $8\frac{1}{2}$, 11, 12

Intervals: $7–8\frac{1}{2}$, $9–10\frac{1}{2}$, $11–12\frac{1}{2}$, $13–14\frac{1}{2}$

JOURNAL EXERCISES

Follow the directions for each problem.

1. Create a data set on your own and use the data to make a histogram.

2. Find a set of data in a magazine or from a different course and use the data to make a histogram.

3. Use your homework grades in the math class you are now enrolled in to create a histogram.

CHAPTER 11 REVIEW

SECTION	TOPIC OR TERM	EXAMPLE OR DEFINITION
11.1		
Mean, Median, and Mode	Central Tendency	Central tendency is a measure of the middle of a set of numbers.
	Mean	The mean of a set of numbers is the average of the numbers: $\text{Mean of } \{3, 4\} = \frac{(3 + 4)}{2} = 3.5$
	Median	The median of a set of numbers is the value in the middle: $\text{Median of } \{1, 7, 8\} = 7$; $\text{Median of } \{4, 6, 7, 12\} = \frac{(6 + 7)}{2} = 6.5$
	Mode	The mode of a set of numbers is the value that occurs most often. The mode does not exist if no number occurs more than once. Mode of $\{3, 3, 5, 7\} = 3$. $\{1, 5, 9\}$ has no mode.
11.2		
Statistical Data	Statistical Data	Statistical data are facts or figures frequently given in the form of tables, graphs, or charts.
11.3		
Histograms	Histograms	A histogram is a bar chart used to show the frequency with which events occur. Make a histogram using the following grades: 69%, 65%, 99%, 95%, 82%, 91%, 75% 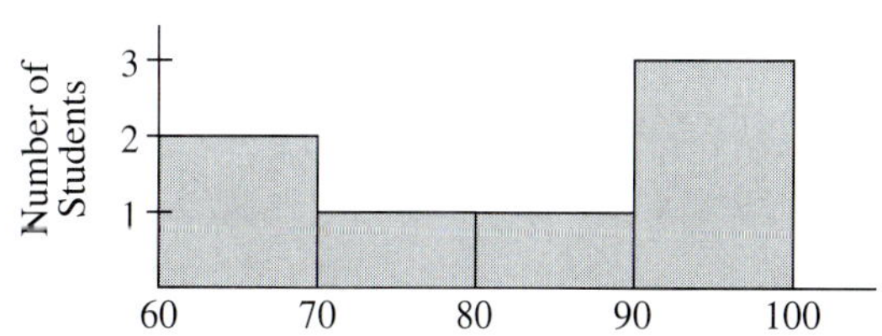 Note that the data is bimodal.

GENETICS

Genetics is the study of how traits such as eye color and height are passed from parents to offspring. A current issue in genetics is **genetic engineering**. The idea is that scientists may be able to change traits such as baldness or perhaps even the likelihood of getting cancer by altering the genetic makeup of an individual. Genetics promises to be a rich and exciting area of study for years to come.

Gregor Mendel (1822–1884) is known as the father of genetics. Mendel was a monk and a teacher. He looked at specific traits of a specific plant, the *garden pea*. Some of these traits are:

1. Vines are either tall or dwarf.
2. Unripe pods are either green or yellow.
3. Flowers are either along the stem or bunched at the top.

Mendel then crossed (fertilized) specific plants with others. He kept careful records of their offspring and compared the results.

In one experiment, Mendel raised and then crossed tall and dwarf varieties of garden peas. All of the offspring in the first generation were tall—the dwarf trait had disappeared entirely. However, when the first generation plants were fertilized from their own pollen, some second generation plants were tall and some were dwarf—the dwarf trait reappeared. The schematic below shows the pattern:

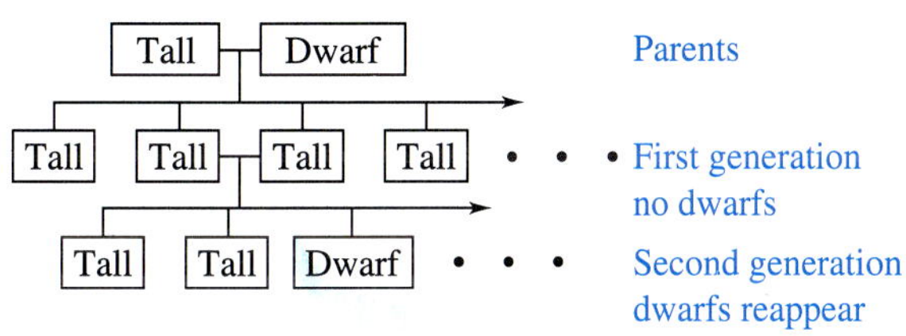

From his experiments, Mendel reported 787 tall and 277 dwarfs in the second generation. Instead, assume that Mendel's numbers were 750 tall and 250 dwarfs in the second generation. Use these two numbers to answer the following questions.

1. Write the number of tall to the number of dwarfs in the second generation as a ratio and reduce to lowest terms.
2. Find the probability, or likelihood, that a member of the second generation is Dwarf as follows:

$$\text{probability of dwarf} = \frac{\#\text{ of dwarfs in second generation}}{\text{total }\#\text{ of plants in second generation}}$$

3. Look up the definitions of statistics and probability in a dictionary and write them down.
4. Explain the following paragraph in your own words:

   Mendel did an experiment from which statistics were gathered. The statistics were used to form probabilities. The probabilities can be used to predict traits of the second generation of a tall and dwarf parent.

5. Mendel's experiment has shown a connection between statistics and probability. Explain.

CHAPTER 11 REVIEW PROBLEMS

Find the mean, median, and mode (if it exists) of each.

1. $\{3, 5, 6, 8\}$

2. $\{4, 4, 1, 5\}$

3. $\{6, 7, 8\}$

4. $\{1, 2, 2, 3, 4\}$

5. $\{1, 2, 6, 8, 22\}$

6. $\{3, 5\}$

7. $\{-8, -8, -6\}$

8. $\{-1, 0, +1\}$

9. $\{-9, +6, +13\}$

10. $\{2, 4, 20\}$

11. $\{3, 7, 8, 25\}$

12. $\{4, 9, 12, 22\}$

13. $\{1.2, 1.5, 2\}$

14. $\{3.6, 8, 5.1\}$

15. $\{7, 9.2, 12\}$

16. $\{4, 7, 52\}$

Follow the directions for each problem.

17. Students test grades were: 71%, 75%, 68%, 73%, 52%, 95%, 90%, 81%, 78%, 77%, 71%, 68%. Find the mean and median.

18. Does the mean of $\{5, 9\}$ equal the median?

19. Does the mean of $\{4, 5, 6\}$ equal the median?

20. Is the mean of $\{3, 4, 6\}$ greater than or less than the median?

21. Is the median of $\{1, 2, 22, 4\}$ equal to the mean?

22. Is the following statement always true: Since the youngest person in a group of three is 42 and the oldest is 50, the median age must be 46.

Use the data from the table to answer the last three questions.

The Four Largest Oceans

Ocean	*Area (Square Miles)*	*Average Depth (Feet)*
Pacific	64,186,000	12,925
Atlantic	33,420,000	11,730
Indian	28,350,000	12,598
Arctic	5,105,700	3,407

23. On average, which ocean is the deepest?

24. Which ocean holds the most water? How can you tell?

25. Using only the data given above, can you tell which ocean has the lowest water temperature?

CHAPTER 11 TEST

Find the mean, median, and mode (if it exists) of each.

1. $\{3, 5, 8, 12\}$

2. $\{1, 4, 4, 2\}$

3. $\{3, 6, 9, 8, 9\}$

4. $\{1, 5, 5\}$

5. $\{5, 9\}$

6. $\{-4, -1, +2\}$

Follow the directions for each problem.

7. Does the mean of $\{12,13\}$ equal the median?

8. Does every set of numbers with more than two elements have a median? A mode? Explain.

9. What does it mean if the mean, median, and mode are all equal in a set with only two numbers?

10. The mean weight of two bass (fish) is 4 pounds 3 ounces. Find the weight of the smaller of the two given that the larger weighs 4 pounds 8 ounces.

11. Draw a histogram using: $\{7, 8, 8, 7, 9, 7, 10, 7, 4, 7, 6, 6, 10, 12\}$
Use the intervals: 4–6, 7–9, and 10–12

Use the data from the table to answer the questions that follow.

Deserts of the World

Desert	*Country*	*Area (Square Miles)*
Chihuahuan	Southern U.S. and Mexico	140,000
Gobi	Mongolia and China	500,000
Great Sandy	Australia	150,000
Kalahari	Southern Africa	225,000
Sahara	Northern Africa	3,500,000

12. What is the combined land area of all five deserts?

13. Which desert is the largest of all?

14. The Chihuahuan Desert of North America is what percent of the size of the Sahara Desert of North Africa?

15. From the information given, can you tell which desert is the driest?

FINAL EXAMINATION

Work each problem.

1. $\begin{array}{r} 2{,}652 \\ 423 \\ +9{,}817 \\ \hline \end{array}$

2. $\begin{array}{r} 76{,}021 \\ -\quad 740 \\ \hline \end{array}$

3. $\begin{array}{r} 654 \\ \times\ 81 \\ \hline \end{array}$

4. $210 \times 10{,}000$

5. $7{,}586 \div 3$

6. 7^2

7. 1^3

8. 0^2

9. $527 \cdot 0$

10. $\frac{1}{12} + \frac{5}{12}$

11. $\frac{1}{6} + \frac{3}{4}$

12. $\frac{31}{51} - \frac{12}{51}$

13. $\frac{17}{20} - \frac{1}{6}$

14. $\frac{2}{3} \cdot \frac{1}{4}$

15. $\frac{4}{5} \div \frac{1}{8}$

16. $\left(\frac{1}{2}\right)^2$

17. $4\frac{1}{4} + 3\frac{1}{4}$

18. $15\frac{1}{5} + 12\frac{2}{5}$

19. $7\frac{1}{8} + 4\frac{1}{3}$

20. $14\frac{1}{3} - 12\frac{1}{3}$

21. $27\frac{3}{5} - 1\frac{1}{8}$

22. $52 - 4\frac{1}{3}$

23. $75 - \frac{3}{8}$

24. $16 + \frac{3}{7}$

25. $\left(2\frac{1}{3}\right)\left(3\frac{1}{4}\right)$

26. $5\frac{1}{5} \div 1\frac{1}{9}$

27. $16 \div 4\frac{4}{5}$

28. $(-1) + (-7)$

29. $(-8) - (+4)$

30. $-12 - 15$

31. $(-6)(-4)$

32. $(+8) \cdot (-3)$

33. $75 \div (-25)$

34. $(-20) \div (-4)$

35. $\frac{1}{2} \cdot (-8)$

36. $\frac{3}{4} \div 0.25$

37. $4\frac{2}{3} \div \frac{1}{5}$

38. $(-4.2) \cdot 3\frac{1}{2}$

39. $\frac{3}{8} \div 0$

40. $0 \div 7.9$

Evaluate each expression.

41. $16 - 2 \cdot 3$

42. $45 \div 9 - 2$

43. $8 \cdot (450 \div 45)$

44. $9 \cdot 3 - 2 \div 2$

45. $(3 + 7.5) \cdot 9$

46. $9 - (4 + 22)$

47. $\frac{1}{2} - \left(\frac{3}{4} \div \frac{2}{3}\right)$

48. $2\frac{1}{2} \cdot 7 - 6\frac{1}{3}$

49. $(7 + 2.2)^2$

50. $(-3) + (-8) \cdot (+4)$

51. $[(-5) - (+2)] \cdot (-8)$

52. $-(-2.7)$

Work each problem.

53. Find 65% of 21.

54. 270 is 12% of what number?

55. 45 is what percent of 225?

56. A sales tax of 7.5% applies to a purchase of \$250.00. Find the total cost.

57. Round 1,650,835 to the nearest thousand.

58. Write 4,350,275 using words.

59. Write the digits of fourteen billion, sixty thousand.

60. Solve $\frac{7}{40} = \frac{x}{480}$ for x.

61. Write a ratio for \$25 in taxes for every \$100 in earnings.

62. Toni bicycled 4 miles in 35 minutes. At that same rate, how far will he be able to bicycle in 60 minutes?

63. Find the perimeter and area of:

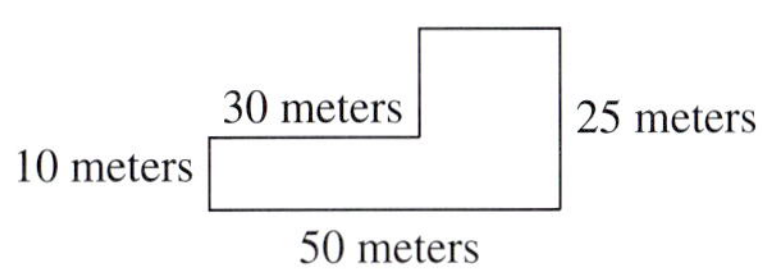

64. Find the circumference and area of a circle with radius seven feet. Use $A = \pi r^2$, $C = 2\pi r$, and $\pi \approx 3.14$.

65. An individual weighs 160 pounds. Find the weight to the nearest tenth of a kilogram if 1 kilogram = 2.2 pounds.

66. Two angles of a triangle measure 40° and 55° respectively. Find the measure of the third angle.

67. Find the perimeter and area of the right triangle with hypotenuse of 13″ and one leg of 12″.

Use $P = a + b + c$ and $A = \frac{1}{2}bh$.

68. Simplify $3x + 21x$.

69. Simplify $14y - 21y$.

70. Simplify $6.5x + 22y - 0.20x$.

71. Solve $x - 3 = 21$ for x.

72. Solve $\frac{4}{3}y + 2 = -0.5$ for y.

73. Find the mean, median, and mode of $\{6, 8, 10, 10, 12\}$.

74. Find the prime factorization of 60.

75. Use the following data on test grades and construct a histogram.

81%, 92%, 40%, 75%, 80%, 92%, 77%, 68%, 78%, 75%, 60%, 99%, 72%, 79%, 84%, 68%, 85%, 89%, 94%, 66%

Group Intervals: 90%–100%
80%–89%
70%–79%
60%–69%
<60%

APPENDIX A

APPENDIX B

ANSWERS TO SELECTED EXERCISES

GLOSSARY

INDEX

APPENDIX A
GROUP APPLICATION WORD PROBLEMS

All individuals in the group should contribute to the solution. Each Group Application Word Problem requires a report, preferably typed.

CHAPTERS 1–2

GAME WITH THREE DISCS

A game is played with three pegs and three discs. The pegs are identical but the discs are of different sizes. The diagram shows the beginning position of the pegs and discs:

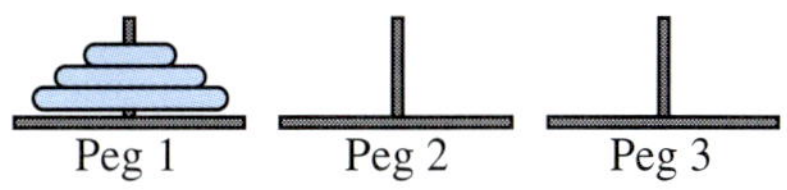

The rules of the game are:

1. Only one disc can be moved at a time.

2. Discs can only go on one of the three pegs.

3. A larger disc *cannot be* placed on top of a smaller one.

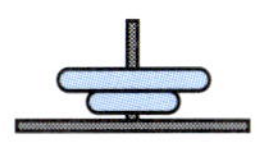

4. A smaller disc *can be* placed on top of a larger disc, or on an empty peg.

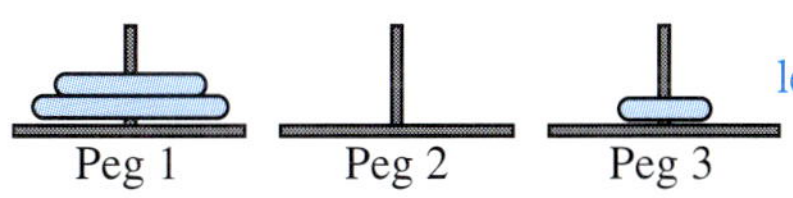

What is the minimum number of moves needed to move the entire stack from Peg 1 to Peg 3? Is it better to begin the game by moving the smallest disc to Peg 2, or to Peg 3?

Extend the game to four discs. What should the first move be?

INFORMATION FROM CHARTS AND GRAPHS

Pick a broad topic such as the economy, business, or science. Each person in the group is to find a chart or graph in a magazine or newspaper that is related to the topic and write three questions that refer to the information in the graph. Bring the graph and related questions to the group to be exchanged. Answer each other's questions and discuss.

BACTERIA IN THE KITCHEN

Initially, a single bacteria is lurking in a dark corner of your kitchen. Assume that the number of bacteria doubles every day. How many bacteria do you have in your kitchen after 30 days? after 45 days? Are the numbers larger than you thought they would be before you began working on the problem?

CHAPTERS 3–4

STOCKS

Stock prices are usually quoted in your local newspaper. Find the prices for IBM stock and General Motors stock by looking in your newspaper under New York Stock Exchange. Look under **IBM** for IBM and **GnMotr** for General Motors.

Ignoring any commissions, what would be your cost to purchase 100 shares of each? Exactly a week later, find the prices of both stocks. What would the price be at this time for 100 shares of each?

INVESTMENT GROUP

Assume you are in an investment group that has $10,000 to invest. As a group, choose three companies listed on the New York Stock exchange and decide how many shares of each to buy. One month later, find the stock prices of these three companies and find the amount by which your investment has increased or decreased in value.

FOREIGN CURRENCY

The foreign currency exchange rate can be found in most newspapers in either the business section or the travel section. A bank is another source for this information. Find the U.S. exchange rate for Canadian, Mexican, Italian, and French money. Suppose a restaurant bill is $25 in U.S. currency. Use ratios and proportions to find the cost of the bill in the other four currencies.

CHAPTERS 5–6

MINERALS IN YOUR WATER

The following figures represent the amount of various minerals in the water system of one major city. Find out the amounts of each of the listed minerals in the water in your area. This information can be found by calling your local water department. Find the difference between the given amounts and the amounts in your area.

Mineral	*Amount*
Calcium	80 mg/1
Magnesium	20 mg/1
Sodium	10 mg/1
Iron	0.02 mg/1

WHICH JOB SHOULD I TAKE?

A person has a job offer in each of two different states. Each job is offering $30,000 per year although the tax rates in the two states differ:

Tax	*State 1*	*State 2*
State Income Tax	4.0%	None
Sales tax	2.5%	8.25%

The individual estimates his family will spend $5,000 per year on items with sales taxes. Find the total tax for this family for both states. Which job would result in the least amount of total taxes paid?

Figure out the pay after State Income and Sales taxes for this family if they lived in your area. Use an average State Income tax rate. Make a list of other items that could influence a family decision in regards to which job to take.

TAXES, TAXES, TAXES

Collect information on all of the taxes that you can think of that apply to you. Include information on tax rates. Be sure to include state and federal income taxes, social security taxes, sales taxes, property taxes, and taxes on gasoline.

Try to estimate the total amount of all taxes paid by a typical family of four living in a $50,000 home, making $30,000 per year. Do not be discouraged if you are only able to get approximate figures—this can be a complicated problem for anyone.

CHAPTERS 7–8

DESIGN YOUR OWN BACK YARD

Each person in the group is to individually design the landscape of a backyard for a home. The yard is rectangular with dimensions 50 feet wide and 30 feet deep. Each design must include a patio, a flower garden, and a grassy area.

The following economic factors should be taken into consideration when designing the landscape. Grass sod costs $2.00 per square foot, cement for the patio costs $4.00 per square foot, and a flower garden averages $8.00 per square foot. Your objective is to make the yard attractive without spending more than $6,000.

Bring a sketch of each design and its total cost to the group and compare the desirability of each plan. Work together in the group to come up with a design you all agree on.

COST OF A FURNACE OR AIR CONDITIONER

Measure/estimate the outside dimensions of a house. Sketch the house and show the measurements.

Calculate the number of cubic feet in the house. Contact a retail store that sells furnaces or air conditioners and find the recommended size of unit for the house. Ask for the price of the unit and the cost of installation. Calculate the sales tax on the price of the unit and find the total cost to buy and install a unit.

VOLUME OF A POND

Measure the dimensions of either a swimming pool, a small lake, or a pond. Sketch, showing all measurements, and estimate its volume in cubic feet and in gallons. At what rate (in gallons per day) would water have to flow into the pool (or lake or pond) in order for the total volume to be replaced every 30 days?

CHAPTERS 9–10

DAILY STOCK PRICES

Each person in the group will choose two companies whose stock is sold on the New York Stock Exchange. Find the selling price of each share of stock each day for a week. Use signed numbers to describe the amount each stock rises or falls each day.

At the end of the week compare all of the companies in the group and determine which stock had the greatest overall change in value. Was it an increase or decrease? List factors you think might be important in deciding upon a stock to invest in.

APPLICATIONS OF ALGEBRA

Look through introductory textbooks in biology, chemistry, physics, engineering, electronics, or computer science for examples (formulas) in which algebra is used. Copy the formula and briefly explain each application.

CHAPTER 11

UTILITY BILLS AND AVERAGE TEMPERATURES

Obtain average monthly temperatures for each of the last four months in the city or area in which you live. This data can be obtained from a radio station, a TV station, or from a local newspaper. Then get a copy of a utility bill for a home or apartment for these same months.

Make a graph with utility cost on the vertical axis and average monthly temperature on the horizontal axis. Is there a relationship between the average temperature and the utility bill? Would you expect there to be one? Explain.

TO GO TO SCHOOL, OR NOT TO GO TO SCHOOL

Assume that you are trying to convince a 16 year old to finish high school and to go on to college. What types of data could you use to help convince the teenager? Where do you think you might be able to find that type of data? Collect some of this data and present it in tables and figures. Be prepared to present your ideas to your class.

WHICH MAJOR?

Assume that you are trying to decide what to major in at a college or a technical school. What types of data could help you in your decision making? Where do you think you might be able to find that type of data? Collect some of this data and present it in tables and figures. Be prepared to present your ideas to your class.

APPENDIX B
BASIC OPERATIONS WITH A CALCULATOR

There are many different types of calculators available. In this appendix, the basic functions of one common type of calculator are given. If these directions do not seem to apply to your particular calculator, refer to the manual that came with it. Remember, not all calculators are identical.

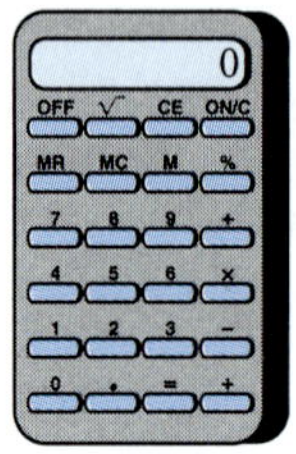

Before working a problem, it is best to clear any numbers remaining from a prior calculation. This is done by pressing [C]. Examples of the basic functions are shown below:

ADDITION:	7 [+] 11 [=]	results in 18
	12.5 [+] 14.6 [=]	results in 27.1
	3 [+] 8.2 [+] 5.9 [=]	results in 17.1
SUBTRACTION:	15 [−] 4 [=]	results in 11
	6.7 [−] 4.5 [=]	results in 2.2
MULTIPLICATION:	50 [×] 87 [=]	results in 4,350
	21.7 [×] 3.05 [=]	results in 66.185
DIVISION:	360 [÷] 3 [=]	results in 120
	49.2 [÷] 0.4 [=]	results in 123

Pressing the [CE] key only erases the number shown on the display. This key allows the user to correct an error without having to start over. The following shows an attempt to add 4 + 6.8. An error was made when 6.4 was mistakenly entered. The error was cleared using [CE] and the problem is then completed.

4 [+] 6.4 [CE] 6.8 [=] results in 10.8

Error corrected.

The percent key is used when finding a percent. First enter the base or whole, followed by a multiplication sign. Then enter the digits of the percent followed by the percent sign.

$150 [×] 6.5 [%] results in 9.75

The square root of a number can be found by pressing the [√] key after entering the number, for example:

1 [√] results in 1

36 [√] results in 6

More complicated expressions can also be solved on the calculator. For example:

Problem: $\frac{25 - 16.4}{5}$

Calculator: 25 [−] 16.4 [÷] 5 [=] results in 1.72

Problem: $22 \cdot 3.1 - 6$

Calculator: 22 [×] 3.1 [−] 6 [=] results in 62.2

However, you must be careful and always use the rules for Order of Operations even when using a calculator. For example, the ÷ must be done before the + in: $21 - 16 \div 4.2$

Therefore, $16 \div 4.2$ must be solved as the first step. This value can then be stored in memory by pressing [M]. It can be recalled from memory at a later time by pressing [MR]. The problem $21 - 16 \div 4.2$ can be solved by:

16 [÷] 4.2 [=] results in 3.809523810 . . .

Store by pressing [M].

21 [−] [MR] [=] results in 17.19047619 . . .

→ Recalls value previously stored.

It is important to clear the memory between problems. This can be done by pressing [MC] or by pressing [C] which clears all displays and all memories. Here are some additional problems you can practice on. The steps to solve each are also given.

EXERCISES

(a) $75 \div 21 \cdot 35.2$

(b) $\frac{168.65 - 76.2}{4.3}$

(c) $\frac{6 \cdot 54 - 5.6}{25}$

(d) $4.5 \cdot 6.9 - 2.1 \times 4.0$

(e) $(\$2{,}165.21 + \$187.00) \times 7\%$

SOLUTIONS

(a) 75 [÷] 21 [×] 35.2 [=] results in 125.7142857 . . .

(b) 168.65 [−] 76.2 [÷] 4.3 [=] results in 21.5

(c) 6 [×] 54 [−] 5.6 [÷] 25 [=] results in 12.736

(d) 2.1 [×] 4.0 [=] results in 8.4

Store value by pressing [M].

4.5 [×] 6.9 [=] results in 31.05

[−] [MR] [=] results in 22.65

→ Recalls value previously stored.

(e) 2,165.21 [+] 187.00 [×] 7 [%] Results in 164.6547 which rounds to $164.65.

ANSWERS TO SELECTED EXERCISES

CHAPTER 1

Section 1.1 (page 7)

1. 3,840; three thousand, eight hundred forty **3.** 4,900; four thousand, nine hundred **5.** 7,250; seven thousand, two hundred fifty **7.** 96,050; ninety-six thousand, fifty **9.** 148,200; one hundred forty-eight thousand, two hundred **11.** 600,221; six hundred thousand, two hundred twenty-one **13.** 16,240; sixteen thousand, two hundred forty **15.** 125,200; one hundred twenty-five thousand, two hundred **17.** 100,000; one hundred thousand **19.** 180,000; one hundred eighty thousand **21.** 4,642,000; four million, six hundred forty-two thousand **23.** 3,600,000; three million, six hundred thousand **25.** 10,000,000; ten million **27.** 6,500,080; six million, five hundred thousand, eighty **29.** 800,000,050; eight hundred million, fifty **31.** 6,400 **33.** 9,321 **35.** 78,000 **37.** 17,300 **39.** 70,001 **41.** 12,020 **43.** 6,004 **45.** 70,450 **47.** 8,002,007 **49.** 9,100,024 **51.** 20,009,000,005 **53.** 99,000,000,000,000 **55.** hundreds **57.** tens **59.** ten thousands **61.** one millions **63.** ten thousands **65.** ten millions **67.** 200 + 40 + 5 **69.** 2,000 + 100 + 30 + 7 **71.** 3,000,000 + 200,000 + 60,000 + 5,000 + 800 + 90 **73.** 20,000,000 + 9,000,000 **75.** 99,999 **77.** 239,000 **79.** sixty-two thousand

Section 1.2 (page 15)

1. 5,871; five thousand, eight hundred seventy-one **3.** 25,000; twenty-five thousand **5.** 32,200,001; thirty-two million, two hundred thousand, one **7.** 700 **9.** 100 **11.** 6,900 **13.** 104,800 **15.** 16,000 **17.** 1,200 **19.** 1,900 **21.** 2,000 **23.** 16,600 **25.** 812,200 **27.** 70,000 **29.** 110,000 **31.** 390,000 **33.** 400,000 **35.** 370,000 **37.** 6,300,000 **39.** 400 **41.** 90,000 **43.** 79,900 **45.** 46,000 **47.** 280,000 **49.** 1,700,000 **51.** 36,000,000 **53.** 10,000,000 **55.** 2,300,000,000 **57.** true **59.** true **61.** true **63.** false **65.** true **67.** $7 > 5$ **69.** $15 > 2$ **71.** $36 < 37$ **73.** $16,000; $20,000 **75.** 700,000 **77.** $35,400,000 < $41,500,000

Section 1.3 (page 25)

1. 18 **3.** 4 **5.** 8 **7.** 5 **9.** 7 **11.** 0 **13.** 3 **15.** 20 **17.** 20 **19.** 19 **21.** 19 **23.** 696 **25.** 0 **27.** 13 **29.** 1,246 **31.** 1,066 **33.** 9,112 **35.** 3,389 **37.** 5,788 **39.** 367 **41.** 688 **43.** 18,321 **45.** 193 **47.** 18,306 **49.** 3,376 **51.** 100 **53.** 4,300 **55.** 82,102 **57.** 16,984 **59.** 89,601 **61.** 1,170,196 **63.** 98,700 **65.** 7,726 **67.** 591 **69.** 3,990,000 **71.** 18,044,000,000 **73.** commutative **75.** associative **77.** identity **79.** commutative **81.** identity **83.** 100 **85.** 39 cases of soda with caffeine; 72 cases of diet soda **87.** Company C; $10,160 **89.** 29

Section 1.4 (page 35)

1. 30 **3.** 49 **5.** 42 **7.** 56 **9.** 32 **11.** 63 **13.** 36 **15.** 25 **17.** 24 **19.** 54 **21.** 27 **23.** 64 **25.** 30 **27.** 4,368 **29.** 8,991 **31.** 19,616 **33.** 35,705 **35.** 15,750 **37.** 65,625 **39.** 57,672 **41.** 3,263 **43.** 55,744 **45.** 476,133 **47.** 10,788 **49.** 142,044 **51.** 416,142 **53.** 94,655,730 **55.** 741,414,339 **57.** 815,100,000 **59.** 23,440 **61.** 34,001,000 **63.** 3,627,000 **65.** commutative **67.** distributive **69.** commutative **71.** identity **73.** 76,000 **75.** 800,000 **77.** 797 **79.** 7,000,012 **81.** 9,801 **83.** 300,000; 302,400 **85.** 665 **87.** 344,400 people

Section 1.5 (page 45)

1. 5 **3.** 7 **5.** 8 **7.** 3 **9.** 7 **11.** 7 **13.** 56 **15.** 21 **17.** 1 r3 **19.** 6 **21.** 5 **23.** 9 **25.** 12 **27.** 50 **29.** 10 **31.** 4 **33.** 20 **35.** 6 **37.** 30 **39.** 217 **41.** 10 **43.** 8 **45.** 16 **47.** 34 **49.** 32 **51.** 400 r7 **53.** 93 r7 **55.** 40 **57.** 89 r14 **59.** 3 r67 **61.** 115 r47 **63.** 29 r1100 **65.** 20 r3 **67.** 900 **69.** 784 **71.** 2,048 **73.** 6,000,000 **75.** 8 **77.** 5 r6 **79.** one hundred twenty-two thousand, eight hundred fifty-seven r5 **81.** 22 points **83.** 9 containers; 30 left over **85.** eighty-four thousand dollars **87.** 68 troops

Section 1.6 (page 53)

1. 11 **3.** 10 **5.** 7 **7.** 17 **9.** 1,280 **11.** 12 **13.** 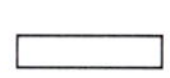 **15.** 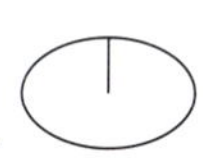 **17.** 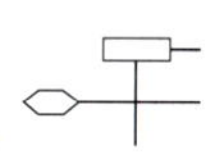 **19.** deductive **21.** inductive **23.** deductive **25.** deductive **27.** deductive **29.** deductive **31.** inductive

Section 1.7 (page 59)

1. 10,000,000,000; both **3.** job #1 = \$10/hour; job #2 = \$9/hour; job #1 **5.** 500 miles **7.** \$13
9. Product 1: \$4 \$3,200 Product 2: \$6 \$3,600 Product 3: \$4 \$4,000 Product 3 gives the largest total profit.
11. \$480/month; \$360/month **13.** 50 capsules

Section 1.8 (page 65)

1. 53 **3.** 22 **5.** 6 **7.** 56 **9.** 13 **11.** 9 **13.** 6 **15.** 0 **17.** 5 **19.** 12 **21.** 2,001 **23.** 2,430
25. 0 **27.** 6,570 **29.** 72 **31.** 6 **33.** $X - 40$ cents **35.** $X - 20$ cents **37.** $2N$ marbles
39. $2{,}800 - W$ caribou **41.** $2{,}800 - W + N$ caribou **43.** $T + J$ books **45.** $T - D$ books **47.** $X + 18$
49. $A + B$ **51.** $C - 1000$

Chapter 1 Review Problems (page 71)

1. 6,243; six thousand, two hundred forty-three **2.** 14,512; fourteen thousand, five hundred twelve **3.** 920,046; nine hundred twenty thousand, forty-six **4.** 302,721; three hundred two thousand, seven hundred twenty-one **5.** 5,123,450; five million, one hundred twenty-three thousand, four hundred fifty **6.** 32,400,521; thirty-two million, four hundred thousand, five hundred twenty-one **7.** 17,407 **8.** 84,010,012 **9.** 168,800 **10.** 9,000,007,620 **11.** 400,956
12. 6,000,000,004,000 **13.** 16,000 **14.** 47,300 **15.** 1,000,000 **16.** 764,320,000 **17.** 16,700,000,000
18. 6,000,000,000 **19.** 56 **20.** 7 **21.** 21 **22.** 42 **23.** 7 **24.** 5 **25.** 192 **26.** 16,650 **27.** 18,668
28. 75 **29.** 101 r5 **30.** 10 r15 **31.** 2,939 **32.** 2,992,079 **33.** 819,404 **34.** 40 **35.** 3,397 **36.** 785,000
37. 1,969 **38.** 7,611 **39.** $3 < 7$ **40.** $4 > 1$ **41.** $12 < 16$ **42.** $75 < 120$ **43.** 15 **44.** 19 **45.** 12
46. 14 **47.** 9 **48.** 61 **49.** Identity property of multiplication **50.** Commutative property of addition
51. Distributive property **52.** Commutative property of multiplication **53.** 16 **54.** $8 + X$ **55.** $N - 75$ dollars
56. 360 notes **57.** 68 **58.** 330 miles **59.** 34 years old **60.** 40,922 people

Chapter 1 Test (page 75)

1. 45,225 **2.** 1,573 **3.** 243 r2 **4.** 85 r5 **5.** 96 **6.** 7,652 **7.** 17,520 **8.** 405 r10 **9.** 1,776
10. 10,412 **11.** 9,202 **12.** 107 r5 **13.** Associative property of multiplication **14.** Distributive property
15. Identity property of multiplication **16.** Commutative property of multiplication **17.** 32 **18.** 961 **19.** 23
20. 21 **21.** 3,000 **22.** 58,000 **23.** 7,400,000 **24.** thirty-seven million, fifty-three **25.** No; 16 is not greater than 24. **26.** 20,700 **27.** 20,000 people **28.** 100 people/square mile **29.** 80 miles **30.** $F + 4$ miles per hour

CHAPTER 2

Section 2.1 (page 81)

1. 5,070 **3.** 178 **5.** 383,040 **7.** three thousand, four hundred **9.** nine million, forty-seven **11.** 25 **13.** 8
15. 81 **17.** 1 **19.** 64 **21.** 10,000 **23.** 256 **25.** undefined **27.** 81 **29.** 8,000 **31.** 75 **33.** 2,401
35. 64 **37.** 1 **39.** 9 **41.** 2^2 **43.** 7^2 **45.** 1^4 **47.** 11^2 **49.** 4^3 **51.** 5^4 **53.** $3^2 4^3$ **55.** $3^2 8^3$ **57.** 18^3
59. $5^2 7^3$ **61.** 2^{12} **63.** 9^2 **65.** 2^8 **67.** 4^{13} **69.** 207^7 **71.** 100^4 **73.** three to the third power
75. six thousand, six hundred **77.** no **79.** 30 **81.** 5,120 rabbits; 5,242,880 rabbits

Section 2.2 (page 87)

1. 13 **3.** 5 **5.** 64 **7.** 34 **9.** 115 **11.** 0 **13.** 406 **15.** 59 **17.** 4 **19.** 53 **21.** 9 **23.** 6
25. 16 **27.** 2 **29.** 5,000 **31.** 25 **33.** 81 **35.** 4 **37.** 729 **39.** 78 **41.** $10 - 7 \cdot 1$; 3
43. $16 + 7 \cdot 3$; 37 **45.** $7 \cdot 8 - 1$; 55 **47.** $256 \div (6 + 2)$; 32 **49.** $(3 + 5) - 8$; 0 **51.** $7 \cdot 8 \div 2$; 28
53. yes **55.** yes, Commutative property **57.** yes

Section 2.3 (page 91)

1. divisible by 2, 4 **3.** divisible by 2, 3, 5, 6, 10 **5.** divisible by 2, 3, 4, 5, 6, 10 **7.** divisible by 2 **9.** divisible by 2, 4, 5, 8, 10 **11.** divisible by 5 **13.** divisible by 2, 3, 4, 5, 6, 8, 9, 10 **15.** divisible by 2, 4, 5, 8, 10 **17.** divisible by 2, 4, 5, 8, 10 **19.** divisible by none **21.** divisible by 2, 5, 10 **23.** divisible by 2, 5, 10 **25.** divisible by none **27.** divisible by 2, 4 **29.** divisible by 2, 3, 4, 6, 8, 9 **31.** divisible by 3, 9 **33.** divisible by none **35.** divisible by 2, 3, 4, 5, 6, 8, 10 **37.** 20 **39.** 1 **41.** 160 **43.** 44 **45.** 36 **47.** 3,000,000,000, for example **49.** 60, for example **51.** 3, 6, or 9 **53.** no; yes

Section 2.4 (page 97)

1. composite **3.** composite **5.** composite **7.** composite **9.** composite **11.** prime **13.** composite **15.** composite **17.** 3^2 **19.** $2^2 \cdot 3$ **21.** 2^5 **23.** $2 \cdot 3 \cdot 11$ **25.** $3 \cdot 5^2$ **27.** $2^2 \cdot 5^2$ **29.** 5^3 **31.** 11^2 **33.** $2^3 \cdot 3 \cdot 11$ **35.** $5 \cdot 41$ **37.** $2 \cdot 3 \cdot 5 \cdot 7$ **39.** $2^2 \cdot 5 \cdot 23$ **41.** $3 \cdot 5^3$ **43.** $2^2 \cdot 5^4$ **45.** 9 **47.** 1 **49.** no remainder **51.** no remainder **53.** no remainder **55.** remainder **57.** 2, 3, 5, 7, 11, 13, 17, 19, 23, 29, 31, 37, 41, 43, 47 **59.** no

Section 2.5 (page 103)

1. 5^2 **3.** $5 \cdot 17$ **5.** $2^3 \cdot 3 \cdot 5$ **7.** $2 \cdot 7^2$ **9.** yes **11.** no **13.** no **15.** yes **17.** 20 **19.** 60 **21.** 24 **23.** 63 **25.** 300 **27.** 108 **29.** 56 **31.** 70 **33.** 144 **35.** 84 **37.** 60 **39.** 504 **41.** 70 **43.** 432 **45.** 420 **47.** 72 **49.** 10 **51.** 18 **53.** 6 **55.** 4 **57.** 5 **59.** 8 **61.** 24 hours; 3 laps **63.** 512 **65.** 4,000; 4,032

Section 2.6 (page 111)

1. College Algebra **3.** 305 students **5.** no **7.** \$29,700 **9.** \$4,956 **11.** \$8,850 **13.** manufacturing **15.** \$12,404,000 **17.** 55 years **19.** decrease **21.** Mt. Pelee; 40,000 deaths **23.** 79 A.D. **25.** 125,000 deaths; smaller number of deaths **27.** 4th quarter **29.** approximately \$80,000 **31.** 4th quarter of 1993 **33.** up

35.

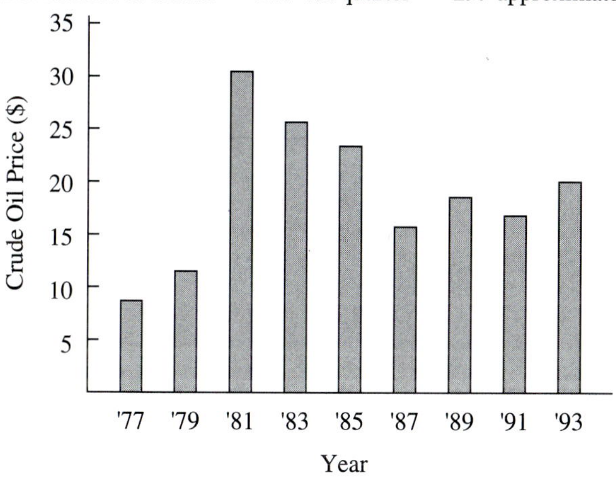

37.

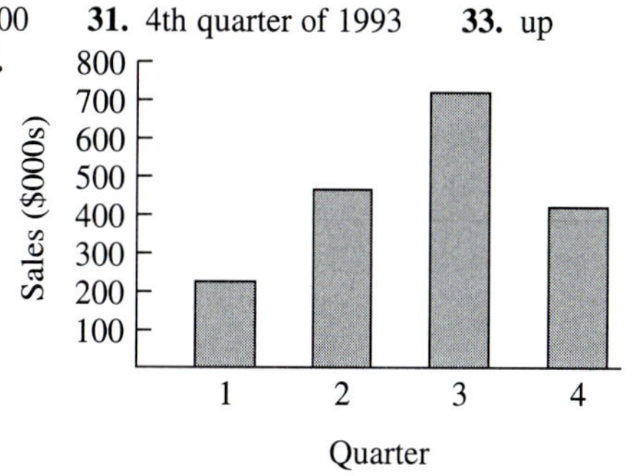

Section 2.7 (page 121)

1.

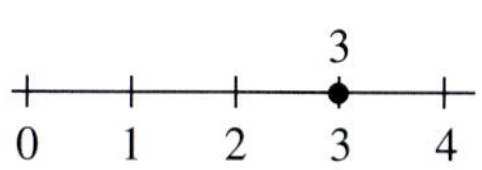

3.

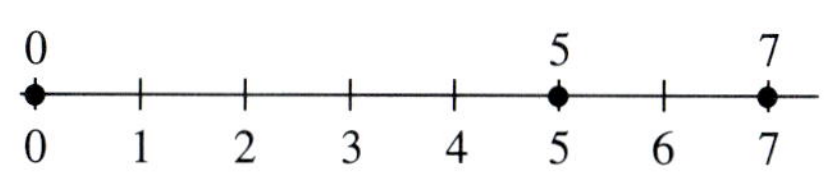

5.

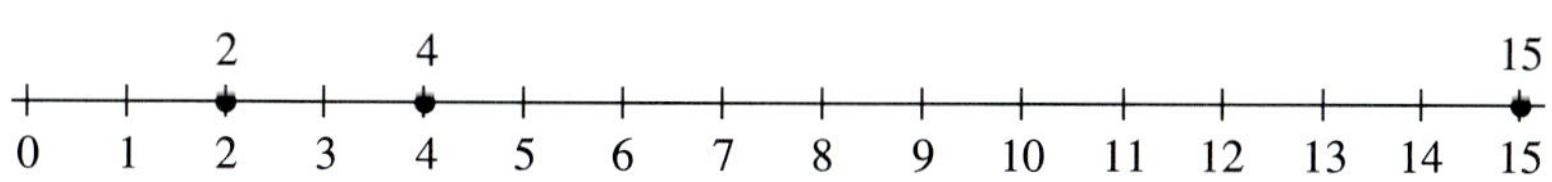

7. A = (0,0); B = (3,1); C = (1,3); D = (2,5) **9.** (1,1); (1,5); (3,1); (3,5)

11.

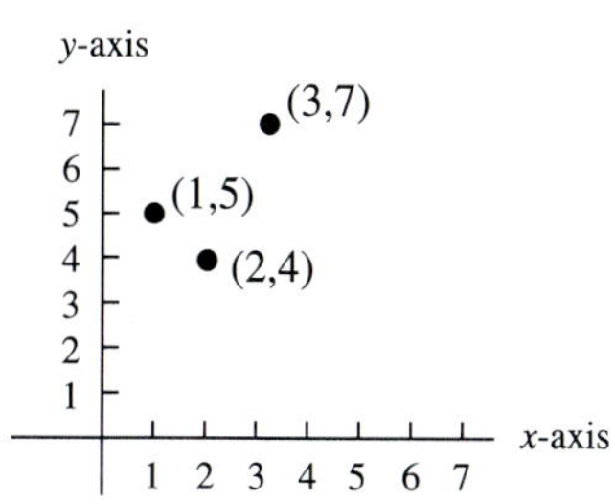

13.

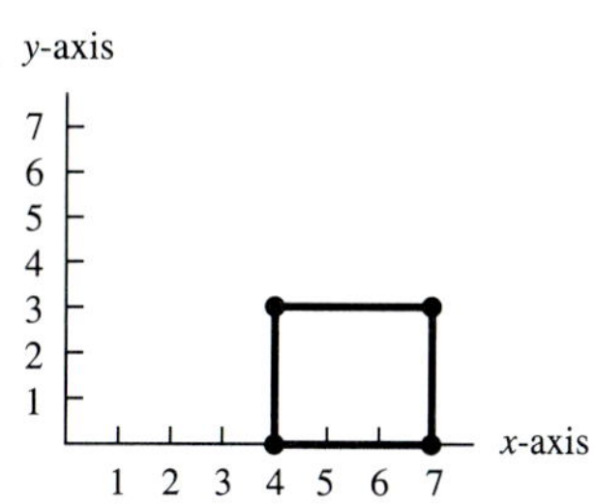

15.

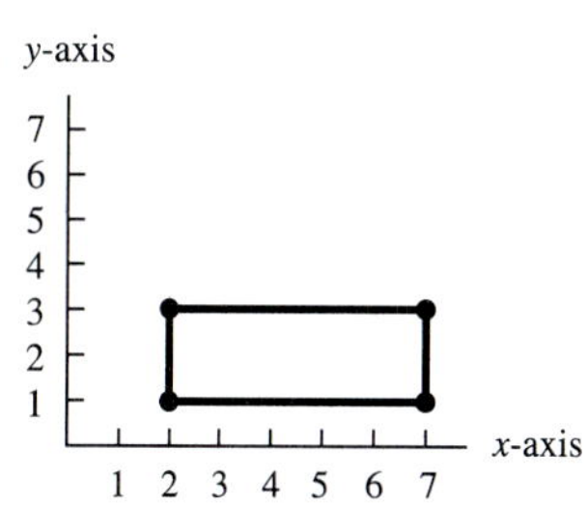

17.

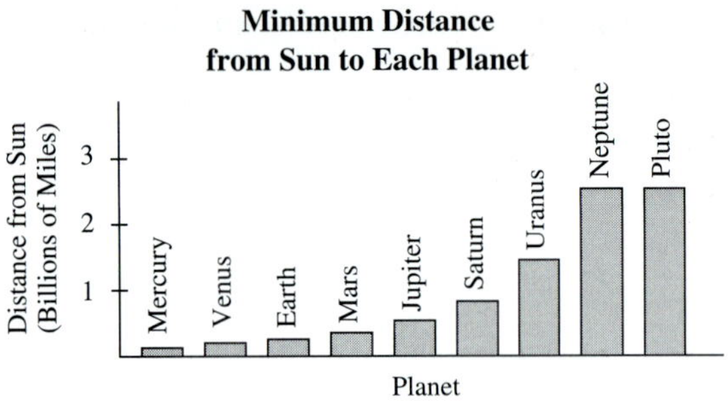

19. 16,000 barrels **21.** 1990 **23.** $2,100,000 **25.** 8–12 years old **27.** age 10

Chapter 2 Review Problems (page 129)

1. 49 **2.** 81 **3.** 144 **4.** 125 **5.** 200 **6.** 70 **7.** 86 **8.** 27 **9.** 4 **10.** 2 **11.** 22 **12.** 63 **13.** 51 **14.** 13 **15.** 378 **16.** 2 **17.** 8,127 **18.** 1 **19.** 5^3 **20.** 7^5 **21.** $3^2 + 5$; 14 **22.** $2^3 + 4^2$; 24 **23.** $8 \cdot (2 + 4)$; 48 **24.** $6 \cdot 64 \div 16$; 24 **25.** $2^2 + 4^3 + 8$; 76 **26.** $2 \cdot 3 \cdot 5$; 30 **27.** divisible by 2, 4, 5, 8, 10 **28.** divisible by 2,4 **29.** divisible by none **30.** divisible by 2, 3, 4, 5, 6, 10 **31.** $2 \cdot 3 \cdot 3$ **32.** $5 \cdot 7$ **33.** $2 \cdot 2 \cdot 3 \cdot 5$ **34.** $3 \cdot 5 \cdot 5$ **35.** 24 **36.** 70 **37.** 60 **38.** 75 **39.** 4 **40.** 6 **41.** yes **42.** 40 **43.** 4,410 **44.** 900 **45.** composite **46.** 100 accidents **47.** 1st quarter **48.** year, location, and so on

49.

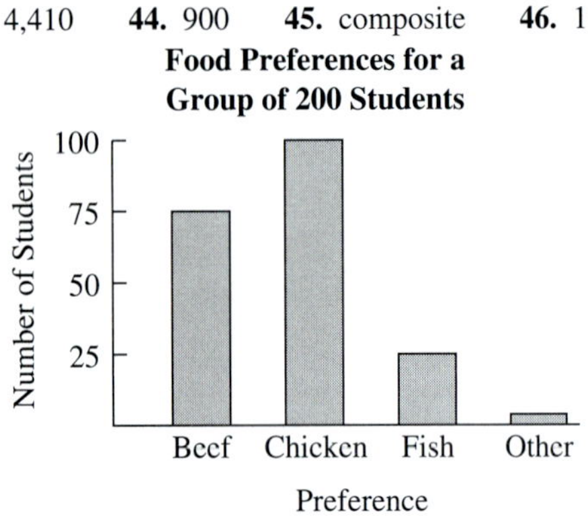

50.

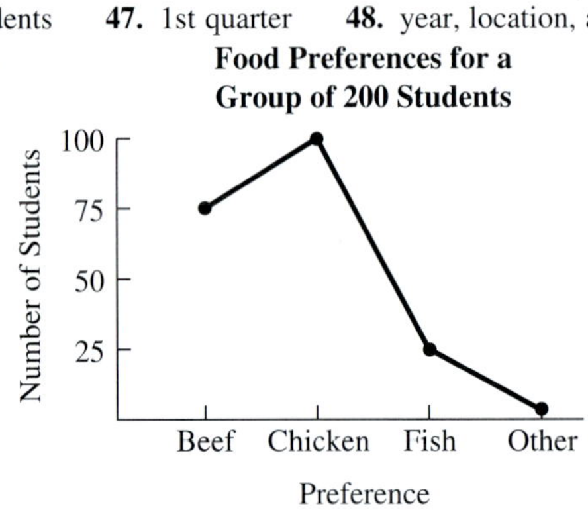

Chapter 2 Test (page 133)

1. 343 **2.** 134 **3.** 1 **4.** 2 **5.** 19 **6.** 59 **7.** 3 **8.** 8 **9.** 4 **10.** 14 **11.** 1 **12.** yes, no **13.** yes, no, yes **14.** $3 \cdot 5 \cdot 5$ **15.** $2 \cdot 2 \cdot 5 \cdot 7$ **16.** 41 **17.** 24 **18.** 112 **19.** no **20.** yes **21.** 75 **22.** 4 **23.** 45 **24.** 23 **25.** 24 **26.** 23, 29 **27.** 1989 **28.** 1990

29.

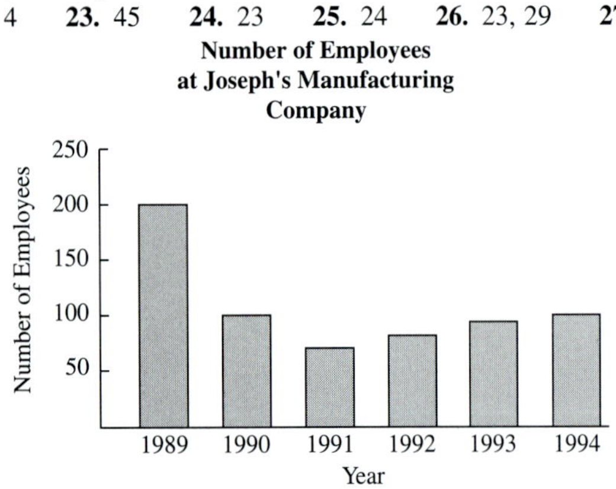

30.

Cumulative Test Chapters 1 and 2 (page 135)

1. three thousand, six hundred fifty-four **2.** forty-one thousand, seven hundred ten **3.** seventy-three thousand, two hundred fifteen **4.** forty thousand, six hundred seventy-nine **5.** two million, one hundred forty-seven thousand **6.** one billion, two hundred sixty-four million, one hundred thousand, one **7.** 54 **8.** 9 **9.** 152 **10.** 65 **11.** 680

12. 367 r6 **13.** 1,933 **14.** 1,273 **15.** 55,335 **16.** 3,305 **17.** 1,958,800 **18.** 141 r16 **19.** 9 **20.** 64 **21.** 32 **22.** 3,136 **23.** 84 **24.** 11 **25.** 125 **26.** 17 **27.** 21 **28.** 41 **29.** 0 **30.** 1,152 **31.** $2 \cdot 2 \cdot 3 \cdot 5$ **32.** $3 \cdot 3 \cdot 5$ **33.** 72 **34.** 60 **35.** 45,000 deaths **36.** 265,000 deaths **37.** 3 times as many **38.** 4; 68 No, they are not equal. **39.** 160 staples **40.** inductive reasoning

CHAPTER 3

Section 3.1 (page 145)

1. $\frac{2}{3}$ **3.** $\frac{3}{4}$ **5.** $\frac{3}{4}$ **7.** $\frac{1}{2}$ **9.** $\frac{7}{12}$ **11.** $\frac{4}{9}$ **13.** $\frac{5}{6}$ **15.** proper, <1; proper, <1; proper, <1 **17.** improper; improper; proper, <1; proper, <1 **19.** $\frac{10}{0}$ **21.** all are defined **23.** $\frac{8}{8}$ **25.** $\frac{9}{9}, \frac{1}{1}$ **27.** 36 **29.** 56 **31.** 54 **33.** 1 **35.** $\frac{1}{4}$ **37.** $\frac{1}{5}$ **39.** $\frac{1}{18}$ **41.** $\frac{4}{5}$ **43.** $\frac{3}{4}$ **45.** $\frac{2}{3}$ **47.** $\frac{7}{8}$ **49.** $\frac{21}{40}$ **51.** $\frac{6}{11}$ **53.** $\frac{3}{8}$ **55.** $\frac{1}{3}$ **57.** $\frac{2}{7}$ **59.** $\frac{9}{10}$ **61.** $\frac{17}{18}$ **63.** $\frac{21}{62}$ **65.** $\frac{7}{15}$ **67.** $\frac{5}{11}$ **69.** $\frac{1}{4}$ **71.** $\frac{1}{4}$ **73.** $\frac{19}{41}$ **75.** $\frac{3}{8}$ **77.** $\frac{1}{2}$ **79.** $\frac{1}{2}$ **81.** $\frac{1}{13}; \frac{1}{13}$ **83.** $\frac{1}{365}; \frac{1}{4}$

Section 3.2 (page 153)

1. yes **3.** yes **5.** yes **7.** no **9.** yes **11.** no **13.** no **15.** $\frac{4}{5}$ **17.** $\frac{1}{11}$ **19.** $\frac{9}{10}$ **21.** $\frac{1}{15}$ **23.** $\frac{81}{425}$ **25.** $\frac{1}{3}$ **27.** 15 **29.** 75 **31.** 390 **33.** 60 **35.** 40 **37.** 16 **39.** 25 **41.** 63 **43.** 9 **45.** 20 **47.** 90 **49.** 32 **51.** 900 **53.** 132 **55.** $\frac{49}{63}$ **57.** $\frac{25}{35}, \frac{50}{70}$ **59.** Both ate the same amount of pie. **61.** no **63.** Both bikers are going the same speed.

Section 3.3 (page 159)

1. $\frac{3}{20}$ **3.** $\frac{3}{56}$ **5.** $\frac{27}{56}$ **7.** $\frac{13}{224}$ **9.** $\frac{93}{35}$ **11.** $\frac{36}{35}$ **13.** $\frac{5}{11}$ **15.** $\frac{6}{35}$ **17.** $\frac{4}{35}$ **19.** $\frac{15}{56}$ **21.** $\frac{3}{4}$ **23.** $\frac{5}{32}$ **25.** $\frac{13}{14}$ **27.** $\frac{3}{28}$ **29.** $\frac{4}{35}$ **31.** $\frac{3}{5}$ **33.** $\frac{3}{5}$ **35.** $\frac{5}{9}$ **37.** $\frac{3}{4}$ **39.** $\frac{2}{9}$ **41.** $\frac{1}{18}$ **43.** $\frac{13}{28}$ **45.** $\frac{4}{5}$ **47.** $\frac{1}{5}$ **49.** $\frac{4}{15}$ **51.** $\frac{1}{2}$ **53.** $\frac{1}{9}$ **55.** $\frac{9}{32}$ **57.** $\frac{3}{2}$ **59.** $\frac{6}{1}$ **61.** $\frac{250}{7}$ **63.** $\frac{15}{4}$ **65.** $\frac{14}{3}$ **67.** $\frac{3}{28}$ **69.** $\frac{5}{18}$ **71.** $\frac{1}{27}$ **73.** 0 **75.** 34 failed **77.** \$400 per month **79.** 800 acres **81.** $\frac{1}{500{,}000}$ of the mass of the solar system

Section 3.4 (page 167)

1. $\frac{3}{2}$ **3.** $\frac{2}{5}$ **5.** $\frac{3}{7}$ **7.** $\frac{11}{1}$ **9.** $\frac{3}{1}$ **11.** $\frac{3}{40}$ **13.** $\frac{3}{32}$ **15.** $\frac{3}{55}$ **17.** $\frac{1}{8}$ **19.** $\frac{1}{11}$ **21.** $\frac{9}{20}$ **23.** 1 **25.** 1 **27.** 1 **29.** $\frac{56}{5}$ **31.** 1 **33.** $\frac{27}{20}$ **35.** $\frac{24}{35}$ **37.** $\frac{3}{5}$ **39.** $\frac{24}{5}$ **41.** $\frac{56}{15}$ **43.** $\frac{75}{4}$ **45.** $\frac{18}{7}$ **47.** $\frac{9}{5}$ **49.** $\frac{28}{75}$ **51.** $\frac{2}{5}$ **53.** $\frac{8}{3}$ **55.** $\frac{7}{9}$ **57.** 3 **59.** 0 **61.** $\frac{2}{3}$ **63.** $\frac{21}{16}$ **65.** 600 head, 200 head, $\frac{1}{6}$ **67.** 30,000 business starts

Section 3.5 (page 175)

1. $\frac{3}{4}, \frac{1}{4}$ **3.** $\frac{7}{8}, \frac{3}{8}, \frac{1}{8}$ **5.** $\frac{3}{127}, \frac{5}{127}, \frac{2{,}004}{127}$ **7.** $\frac{2}{21}, \frac{3}{21}, \frac{17}{21}$ **9.** 16 **11.** 12 **13.** 16 **15.** 187 **17.** 180 **19.** $\frac{8}{10}$ **21.** $\frac{35}{49}$ **23.** $\frac{32}{40}$ **25.** $\frac{35}{42}$ **27.** $\frac{36}{39}$ **29.** $\frac{1}{3}, \frac{3}{8}$ **31.** $\frac{3}{15}, \frac{4}{10}$ **33.** $\frac{7}{11}, \frac{6}{7}$ **35.** $\frac{4}{9}, \frac{1}{2}$ **37.** $\frac{2}{9}, \frac{3}{7}$ **39.** $\frac{7}{18}, \frac{25}{27}$ **41.** $\frac{4}{45}, \frac{3}{32}$ **43.** $\frac{7}{8}, \frac{8}{9}$ **45.** $\frac{5}{26}, \frac{4}{13}, \frac{7}{13}$ **47.** $\frac{3}{5}, \frac{8}{15}$ **49.** $\frac{3}{8}, \frac{7}{22}$ **51.** $\frac{1}{7}, \frac{0}{7}$ **53.** $\frac{12}{25}, \frac{10}{21}$ **55.** $\frac{9}{4}, \frac{25}{12}$ **57.** $\frac{1}{2}, \frac{4}{11}$

59. $\frac{6}{7}, \frac{3}{8}$ **61.** $\frac{5}{7}, \frac{4}{9}$ **63.** $\frac{1}{2}, \frac{3}{8}$ **65.** $\frac{2}{5}, \frac{1}{3}, \frac{5}{18}$ **67.** less time **69.** no **71.** First is faster.

Section 3.6 (page 183)

1. $\frac{3}{4}$ **3.** $\frac{4}{5}$ **5.** $\frac{8}{9}$ **7.** $\frac{11}{15}$ **9.** $\frac{6}{7}$ **11.** $\frac{13}{20}$ **13.** $\frac{4}{5}$ **15.** $\frac{8}{9}$ **17.** $\frac{9}{13}$ **19.** $\frac{1}{2}$ **21.** $\frac{1}{2}$ **23.** $\frac{1}{2}$ **25.** $\frac{1}{5}$ **27.** $\frac{1}{4}$ **29.** $\frac{2}{3}$ **31.** $\frac{6}{11}$ **33.** 0 **35.** 2 **37.** $\frac{1}{5}$ **39.** $\frac{3}{7}$ **41.** $\frac{1}{15}$ **43.** $\frac{1}{5}$ **45.** $\frac{9}{35}$ **47.** $\frac{3}{50}$ **49.** $\frac{1}{32}$ **51.** 0 **53.** $\frac{3}{10}$ **55.** $\frac{4}{5}$ **57.** $\frac{3}{25}$ **59.** $\frac{2}{3}$ **61.** $\frac{15}{2}$ **63.** $\frac{1}{2}$ **65.** $\frac{1}{5}$ **67.** $\frac{20}{3}$ **69.** 99 **71.** $\frac{8}{9}$ **73.** $\frac{7}{2}$ **75.** $\frac{1}{6}$ **77.** No, sale price is $\frac{2}{9}$ below the regular price.

Section 3.7 (page 193)

1. 24 **3.** 12 **5.** 12 **7.** 15 **9.** 42 **11.** 15 **13.** 12 **15.** 12 **17.** 60 **19.** $\frac{6}{15}, \frac{5}{15}$ **21.** $\frac{3}{24}, \frac{20}{24}$ **23.** $\frac{10}{25}, \frac{1}{25}$ **25.** $\frac{13}{12}$ **27.** $\frac{5}{6}$ **29.** $\frac{24}{35}$ **31.** $\frac{1}{30}$ **33.** $\frac{1}{12}$ **35.** $\frac{13}{56}$ **37.** $\frac{7}{9}$ **39.** $\frac{1}{2}$ **41.** $\frac{31}{72}$ **43.** $\frac{19}{90}$ **45.** $\frac{13}{12}$ **47.** $\frac{1}{34}$ **49.** $\frac{23}{36}$ **51.** $\frac{97}{702}$ **53.** $\frac{61}{150}$ **55.** $\frac{3}{5}$ **57.** $\frac{3}{8}$ **59.** 1 **61.** $\frac{1}{3}$ **63.** $\frac{7}{20}$ **65.** $\frac{19}{3}$ **67.** $\frac{13}{24}$ **69.** $\frac{11}{25}$ **71.** 2 **73.** $\frac{23}{20}$ **75.** $\frac{1}{12}$ **77.** $\frac{4}{5}$ **79.** $\frac{48}{11}$ **81.** 1,300 survive **83.** $\frac{4}{7}$ were built by General Motors.

Section 3.8 (page 203)

1. $\frac{5}{9}$ **3.** $\frac{5}{8}$ **5.** $\frac{22}{15}$ **7.** $\frac{27}{64}$ **9.** $\frac{64}{125}$ **11.** $\frac{64}{169}$ **13.** $\frac{13}{8}$ **15.** 0 **17.** $\frac{125}{216}$ **19.** $\frac{25}{81}$ **21.** $\frac{8}{27}$ **23.** $\frac{23}{30}$ **25.** 1 **27.** $\frac{9}{400}$ **29.** $\frac{8}{21}$ **31.** 0 **33.** $\frac{12}{7}$ **35.** $\frac{1}{63}$ **37.** 1 **39.** $\frac{13}{16}$ **41.** $\frac{9}{8}$ **43.** $\frac{32}{7}$ **45.** $\frac{27}{5}$ **47.** $\frac{1}{15}$ **49.** $\frac{20}{3}$ **51.** 10 **53.** $\frac{3}{8}$ **55.** $\frac{5}{3}$ **57.** $\frac{12}{25}$ **59.** $\frac{2}{7}$ **61.** $\frac{8}{3}$ **63.** $\frac{144}{25}$

Section 3.9 (page 211)

1. $\frac{13}{8}$ **3.** $\frac{49}{12}$ **5.** $\frac{8}{15}$ **7.** $\frac{5}{28}$ **9.** $\frac{13}{12}$ **11.** $\frac{5}{36}$ **13.** $\frac{1}{36}$ **15.** 72 **17.** $\frac{7}{10}$ **19.** $\frac{35}{12}$ **21.** $\frac{105}{8}$ **23.** $\frac{1}{10}$ **25.** $\frac{7}{18}$ **27.** $\frac{11}{72}$ **29.** $\frac{23}{14}$ **31.** $\frac{3}{100}$ **33.** $\frac{12}{35}$ **35.** $\frac{15}{32}$ **37.** $\frac{13}{56}$ **39.** $\frac{1}{6}$ **41.** $\frac{21}{4}$ **43.** $\frac{9}{8}$ **45.** $\frac{1}{10}$ **47.** $\frac{2}{75}$

Chapter 3 Review Problems (page 221)

1. $\frac{1}{6}$ **2.** $\frac{6}{7}$ **3.** $\frac{1}{8}$ **4.** $\frac{9}{4}$ **5.** $\frac{7}{40}$ **6.** $\frac{8}{3}$ **7.** $\frac{5}{48}$ **8.** $\frac{13}{12}$ **9.** $\frac{11}{40}$ **10.** $\frac{35}{32}$ **11.** $\frac{13}{18}$ **12.** $\frac{7}{15}$ **13.** $\frac{8}{7}$ **14.** $\frac{17}{140}$ **15.** 0 **16.** $\frac{2}{441}$ **17.** $\frac{16}{45}$ **18.** $\frac{9}{10}$ **19.** $\frac{7}{20}$ **20.** $\frac{27}{10{,}000}$ **21.** $\frac{1}{3}$ **22.** $\frac{3}{4}$ **23.** $\frac{7}{5}$ **24.** $\frac{12}{17}$ **25.** $\frac{4}{5}$ **26.** $\frac{1}{8}$ **27.** $\frac{3}{4}$ **28.** $\frac{14}{19}$ **29.** $\frac{5}{21}$ **30.** $\frac{65}{82}$ **31.** $\frac{2}{9}$ **32.** $\frac{1}{10}$ **33.** 2 **34.** $\frac{57}{50}$ **35.** $\frac{7}{2}$ **36.** $\frac{29}{24}$ **37.** $\frac{14}{3}$ **38.** $\frac{1}{6}$ **39.** $\frac{41}{294}$ **40.** $\frac{47}{72}$ **41.** $\frac{2}{3}, \frac{5}{6}$ **42.** $\frac{3}{10}, \frac{3}{5}, \frac{7}{8}$ **43.** $\frac{1}{4}, \frac{1}{3}, \frac{1}{2}$ **44.** $\frac{3}{6}, \frac{4}{6}, \frac{5}{6}$ **45.** $\frac{5}{8}, \frac{2}{3}, \frac{3}{4}$ **46.** $\frac{1}{3}, \frac{3}{8}, \frac{2}{5}$ **47.** $\frac{0}{6}, \frac{7}{9}, \frac{6}{7}$ **48.** $\frac{11}{40}, \frac{6}{5}, \frac{7}{4}$ **49.** $\frac{13}{8}$ **50.** $\frac{7}{24}$ **51.** $\frac{9}{20}$ **52.** $\frac{35}{72}$ **53.** $\frac{80}{21}$ **54.** $\frac{29}{196}$ **55.** $\frac{1}{3}$ **56.** $\frac{9}{64}$ **57.** $\frac{31}{72}$ **58.** $\frac{55}{144}$ **59.** $\frac{9}{64}$ **60.** 2 **61.** $\frac{23}{40}$ **62.** $\frac{24}{35}$ **63.** $\frac{20}{27}$ **64.** $7,500 **65.** Joan ran faster.

Chapter 3 Test (page 223)

1. $\frac{5}{12}$ **2.** $\frac{4}{5}$ **3.** $\frac{9}{10}$ **4.** $\frac{2}{19}$ **5.** $\frac{1}{18}$ **6.** $\frac{1}{42}$ **7.** $\frac{17}{14}$ **8.** $\frac{1}{5}$ **9.** 1 **10.** $\frac{24}{5}$ **11.** $\frac{8}{5}$ **12.** $\frac{23}{20}$ **13.** 0 **14.** $\frac{21}{10}$ **15.** $\frac{91}{135}$ **16.** $\frac{2}{3}$ **17.** $\frac{13}{147}$ **18.** $\frac{103}{200}$ **19.** $\frac{25}{6}$ **20.** 0 **21.** 200 **22.** 3 **23.** $\frac{39}{40}$ **24.** $\frac{5}{8}$ **25.** $\frac{1}{21}$ **26.** $\frac{2}{27}$ **27.** $\frac{3}{4}$ **28.** $\frac{3}{4}$ **29.** $105 **30.** 300

CHAPTER 4

Section 4.1 (page 229)

1. proper **3.** proper **5.** improper **7.** $5 + \frac{7}{8}$ **9.** $41 + \frac{2}{3}$ **11.** $5\frac{1}{2}$ **13.** $1\frac{7}{8}$ **15.** 3 and 4 **17.** 18 and 19

19.

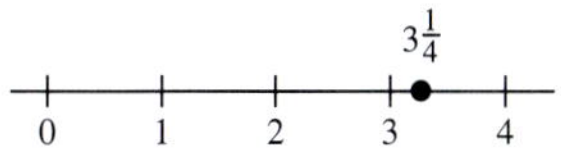

21.

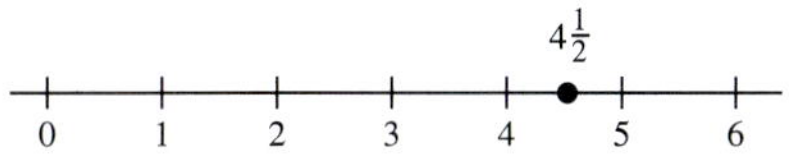

23. $4\frac{1}{3}$ **25.** $\frac{17}{5}$ **27.** $\frac{31}{7}$ **29.** $\frac{65}{7}$ **31.** $\frac{71}{8}$ **33.** $\frac{127}{10}$ **35.** $\frac{13}{3}$ **37.** $\frac{53}{8}$ **39.** $\frac{47}{5}$ **41.** $\frac{149}{12}$ **43.** $\frac{103}{4}$ **45.** $1\frac{1}{2}$ **47.** $1\frac{1}{9}$ **49.** $1\frac{4}{11}$ **51.** $5\frac{1}{8}$ **53.** $2\frac{3}{20}$ **55.** $26\frac{1}{2}$ **57.** 8 **59.** $112\frac{4}{9}$ **61.** $4\frac{4}{7}$ **63.** $4\frac{6}{19}$ **65.** $2 + \frac{7}{8}$ **67.** $10\frac{1}{2}$ **69.** $2 + \frac{3}{4}$ **71.** $7 + \frac{1}{8}$ **73.** one million, eighty-five thousand, twenty-five and two-thirds **75.** $2\frac{2}{3}$ **77.**

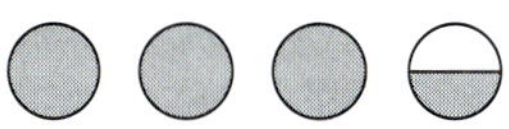

Section 4.2 (page 237)

1.

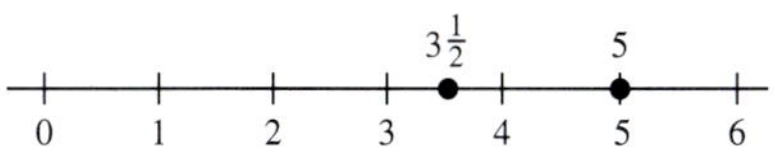

3.

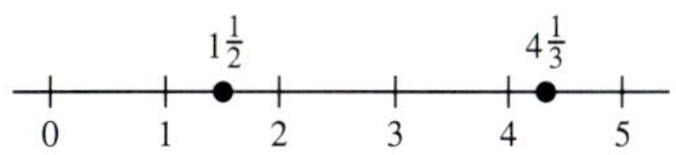

5.

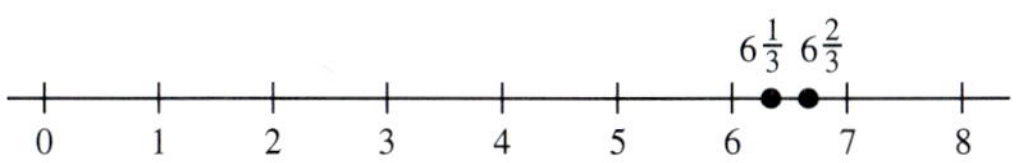

7.

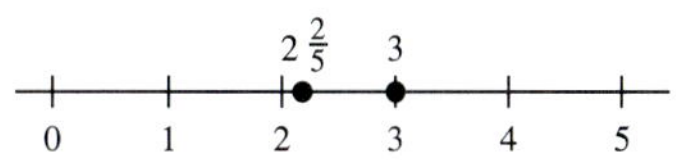

9. $4\frac{3}{4}$ **11.** $4\frac{1}{5}$ **13.** $3\frac{2}{3}$ **15.** $19\frac{1}{3}$ **17.** $11\frac{3}{5}$ **19.** $4\frac{1}{2}, 5\frac{2}{3}$ **21.** $4\frac{1}{2}, 8$ **23.** $6\frac{1}{2}, 6\frac{3}{5}$ **25.** $2\frac{1}{2}, 2\frac{4}{5}$ **27.** $2\frac{1}{5}, 3\frac{4}{5}$ **29.** $2\frac{2}{3}, 5\frac{1}{2}$ **31.** $1\frac{3}{8}, \frac{10}{7}, 1\frac{1}{2}$ **33.** $2\frac{1}{2}, \frac{7}{3}$ **35.** $6\frac{2}{3}, 6\frac{3}{5}$ **37.** $5, \frac{25}{6}$ **39.** $2\frac{3}{8}, 2$ **41.** $7\frac{2}{5}, 6\frac{2}{9}$ **43.** $7\frac{1}{2}, 6\frac{3}{4}, 6$ **45.** $7\frac{1}{2}, \frac{18}{5}, 2\frac{1}{4}$ **47.** $4\frac{9}{10}, 4\frac{7}{8}, 4\frac{1}{3}$ **49.** 7 **51.** more flour **53.** $25\frac{1}{2}, 38\frac{3}{4}, 40\frac{2}{3}, 42\frac{1}{2}$

Section 4.3 (page 247)

1. $9\frac{3}{8}$ **3.** $2\frac{1}{4}$ **5.** $1\frac{7}{8}$ **7.** $8\frac{1}{20}$ **9.** $1\frac{11}{15}$ **11.** $4\frac{13}{20}$ **13.** 32 **15.** $9\frac{1}{12}$ **17.** $2\frac{14}{45}$ **19.** $1\frac{1}{15}$ **21.** $8\frac{2}{5}$ **23.** $3\frac{3}{64}$ **25.** $9\frac{11}{12}$ **27.** 2 **29.** $4\frac{5}{6}$ **31.** $5\frac{4}{21}$ **33.** $8\frac{4}{5}$ **35.** $1\frac{47}{65}$ **37.** $6\frac{5}{8}$ **39.** $8\frac{13}{30}$ **41.** $3\frac{103}{180}$ **43.** $890 **45.** $70\frac{262}{425}$ **47.** $76\frac{3}{4}$ pounds **49.** $20\frac{3}{4}$ yards **51.** $137.50 **53.** $\frac{7}{48}$ liters per hour

Section 4.4 (page 255)

1. $\frac{6 \text{ men}}{5 \text{ boys}}$ **3.** $\frac{3}{100}$ **5.** $\frac{25 \text{ students}}{1 \text{ classroom}}$ **7.** $\frac{2}{5}$ **9.** $\frac{16 \text{ miles}}{1 \text{ gallon}}$ **11.** $\frac{1}{10}$ **13.** $\frac{1 \text{ mile}}{3 \text{ laps}}$ **15.** $\frac{10 \text{ students}}{1 \text{ teacher}}$ **17.** $\frac{500 \text{ cattle}}{1 \text{ ranch}}$ **19.** $\frac{3}{8}$ **21.** $\frac{3 \text{ books}}{1 \text{ class}}$ **23.** $\frac{\$12}{1 \text{ ticket}}$ **25.** $\frac{3 \text{ cups milk}}{5 \text{ cups flour}}$ **27.** $\frac{3 \text{ wrecks}}{100 \text{ drivers}}$ **29.** $\frac{\$925 \text{ tuition}}{1 \text{ semester}}$ **31.** $\frac{10 \text{ females}}{9 \text{ males}}; \frac{9 \text{ males}}{10 \text{ females}}$ **33.** 1:81 **35.** 3 parts of sand plus 1 part of concrete = 4 parts cement, 3 parts of sand in 4 parts of concrete **37.** 3:1; Europe is smaller. **39.** 225:2 **41.** 6:1

Section 4.5 (page 263)

1. extremes: 3, 100 means: 4, 75 **3.** extremes: 5, 35 means: 7, 25 **5.** extremes: 4, 1,000 means: 5, 800 **7.** false **9.** true **11.** false **13.** true **15.** true **17.** true **19.** false **21.** false **23.** false **25.** false **27.** false **29.** true **31.** Company A: $\frac{6 \text{ females}}{5 \text{ males}}$ Company B: $\frac{6 \text{ females}}{5 \text{ males}}$ The ratios are equal. **33.** 1992: $\frac{\$1 \text{ profit}}{\$10 \text{ sales}}$ 1993: $\frac{\$3 \text{ profit}}{\$40 \text{ sales}}$ The ratios are not equal. 1992 has highest profit to sales ratio.

Section 4.6 (page 269)

1. true **3.** false **5.** true **7.** true **9.** $x = 4$ **11.** $a = 3$ **13.** $z = 100$ **15.** $y = 9\frac{3}{5}$ **17.** $b = 2$ **19.** $y = 30$ **21.** $d = 3$ **23.** $y = 98$ **25.** $y = 7$ **27.** $a = 5$ **29.** $x = 11$ **31.** $a = 25$

Section 4.7 (page 275)

1. \$70 **3.** 16 small jobs **5.** 10 more hours **7.** 95 days **9.** 4,000 trees **11.** 240 cups of flour **13.** 32 kilograms of strontium **15.** 15 gallons of alcohol **17.** 80,000 square feet; about 2 acres **19.** \$720 for the year; $\frac{1}{30}$

Chapter 4 Review Problems (page 281)

1. $3\frac{1}{3}$ **2.** $2\frac{2}{5}$ **3.** $7\frac{1}{2}$ **4.** $2\frac{2}{7}$ **5.** $1\frac{11}{13}$ **6.** $5\frac{5}{6}$ **7.** $\frac{9}{2}$ **8.** $\frac{11}{2}$ **9.** $\frac{23}{4}$ **10.** $\frac{23}{8}$ **11.** $\frac{43}{5}$ **12.** $\frac{38}{9}$

13.

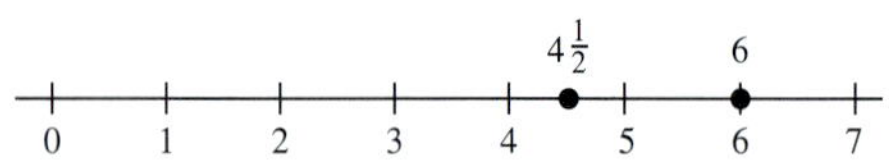

14.

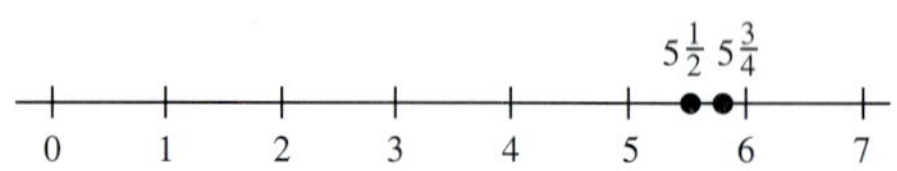

15.

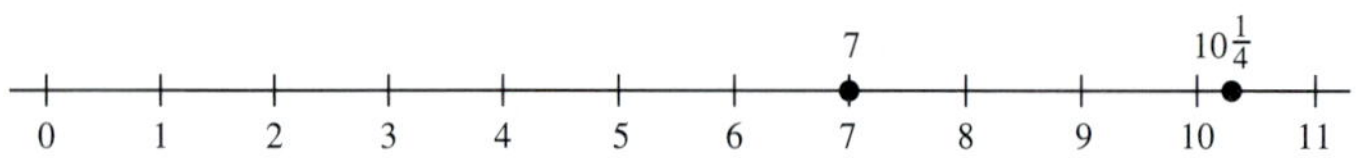

16.

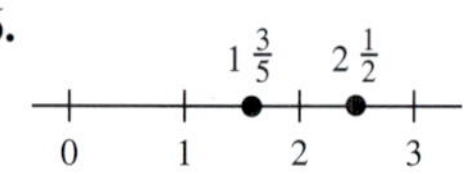

17. $7\frac{1}{6}$ **18.** $9\frac{39}{40}$ **19.** $1\frac{39}{176}$ **20.** $3\frac{3}{8}$ **21.** $1\frac{1}{10}$ **22.** $5\frac{317}{399}$ **23.** $16\frac{3}{4}$ **24.** $5\frac{24}{25}$ **25.** $\frac{1}{75}$ **26.** $\frac{2 \text{ cars}}{1 \text{ family}}$ **27.** $\frac{30 \text{ students}}{1 \text{ faculty}}$ **28.** $\frac{6{,}000{,}000 \text{ red blood cells}}{1 \text{ cubic milliliter}}$ **29.** true **30.** true **31.** false **32.** false **33.** true **34.** true **35.** false **36.** false **37.** $x = 21$ **38.** $y = 8$ **39.** $b = 2$ **40.** $y = 84$ **41.** $r = 10$ **42.** $x = 96$ **43.** $y = 5$ **44.** $x = 40$ **45.** $y = 12$ **46.** $y = 5$ **47.** $a = 1$ **48.** $x = 6$ **49.** $4\frac{7}{36}$ **50.** 1:4 **51.** 1:2 **52.** 27 milliliters **53.** 792 chocolate chips **54.** 48 whales; whales are equally distributed throughout the area. **55.** \$26,250 **56.** 75 gallons of solution A **57.** \$9,100 **58.** 4 more days **59.** $\frac{2}{5}$ **60.** 700 tons of rock

Chapter 4 Test (page 285)

1. $\frac{24}{5}$ **2.** $\frac{31}{4}$ **3.** $\frac{2{,}003}{3}$ **4.** $\frac{108}{7}$ **5.** $4\frac{1}{6}$ **6.** $8\frac{1}{2}$ **7.** $58\frac{1}{3}$ **8.** $28\frac{14}{17}$ **9.** 3:100 **10.** 1:4 **11.** $\frac{8 \text{ goats}}{25 \text{ acres}}$

12. 1:20 **13.** $9\frac{11}{15}$ **14.** $30\frac{9}{32}$ **15.** $4\frac{29}{40}$ **16.** $\frac{19}{22}$ **17.** $4\frac{7}{15}$ **18.** 57 **19.** $2\frac{9}{43}$ **20.** 9 **21.** false
22. true **23.** $x = 55$ **24.** $y = 2$ **25.** $a = 480$ **26.** $b = 11\frac{9}{70}$ **27.** 3:5 **28.** 216,000 revolutions
29. $3\frac{49}{60}$ miles **30.** 525 defectives

Cumulative Test Chapters 3 and 4 (page 287)

1. $\frac{3}{4}$ **2.** $\frac{1}{3}, \frac{12}{23}$ **3.** $\frac{5}{0}$ is undefined **4.** $\frac{9}{0}$ is undefined **5.** $\frac{2}{5}$ **6.** $6\frac{1}{4}$ **7.** $5\frac{5}{6}$ **8.** $2\frac{11}{15}$ **9.** $\frac{4}{7}$ **10.** $\frac{115}{9}$
11. $\frac{361}{12}$ **12.** $\frac{601}{8}$ **13.** $1\frac{1}{4}$ **14.** $5\frac{44}{49}$ **15.** $\frac{14}{27}$ **16.** $\frac{22}{29}$ **17.** 2,666 **18.** 145 **19.** $1\frac{7}{15}$ **20.** $\frac{6}{41}$
21. $x = 2$ **22.** $y = 60$ **23.** $t = 6$ **24.** $y = 10$ **25.** $s = 2$ **26.** $a = 24$ **27.** 1:24 **28.** 31:1
29. 1,092 gallons **30.** 100,000,000,000 cells **31.** $34\frac{1}{6}$ **32.** $4\frac{53}{90}$ **33.** $6\frac{1}{12}$ **34.** \$6,840 **35.** 480 seconds
36. 6,280,000; 6,000,000 **37.** 2,400 **38.** 720 feet **39.** $1\frac{19}{24}$ pounds per day **40.** inductive

CHAPTER 5

Section 5.1 (page 297)

1. six thousand, two hundred twenty-five **3.** twelve million, two hundred, ninety-seven thousand **5.** seven million, four thousand, one hundred twenty-five

7. 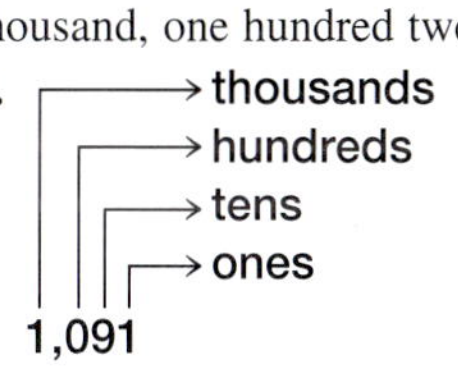

9. 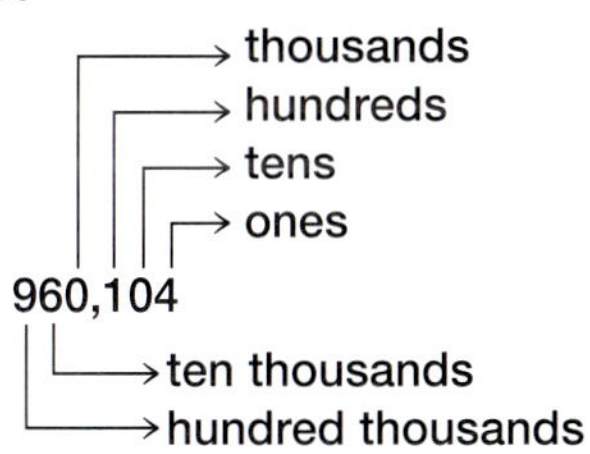

11. 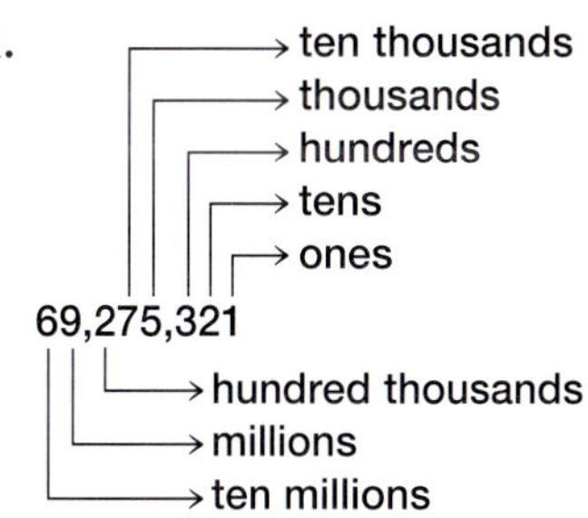

13. 2,075. **15.** 8,000,074. **17.** 2 **19.** 7 **21.** 4 **23.** 9 **25.** 3 **27.** 2 **29.** 3 **31.** 1 **33.** 0 **35.** 0
37. 1 **39.** 7 **41.** 6 **43.** 5 **45.** 6 **47.** 3 **49.** 7 **51.** 1 **53.** $\frac{4}{10} + \frac{5}{100}$; $0.4 + 0.05$ **55.** $\frac{4}{10}$; 0.4
57. $\frac{1}{10} + \frac{6}{100} + \frac{9}{1000} + \frac{1}{10{,}000}$; $0.1 + 0.06 + 0.009 + 0.0001$ **59.** $\frac{6}{10} + \frac{1}{100} + \frac{4}{10{,}000}$; $0.6 + 0.01 + 0.0004$
61. 40 **63.** 0.7 **65.** 10 **67.** 9.967 **69.** $\frac{7}{10}, \frac{3}{100}$; 0.67 **71.** yes **73.** yes

Section 5.2 (page 305)

1. >1 **3.** >1 **5.** >1 **7.** >1 **9.** >1

11. 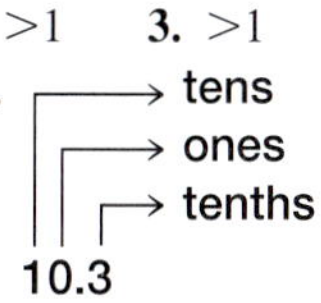

13. 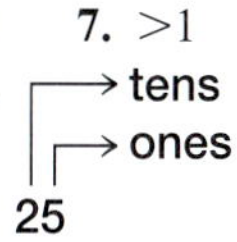

15. 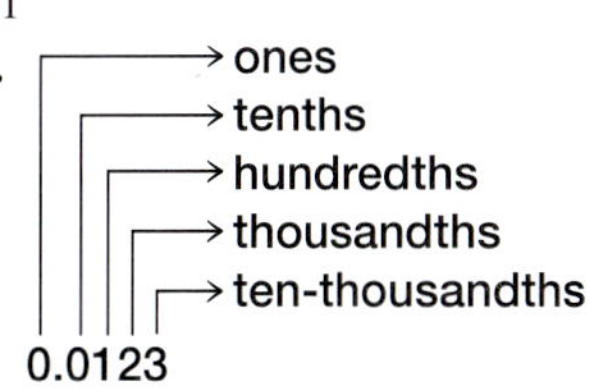

17. five and twenty-nine hundredths **19.** forty-five hundredths **21.** seventeen and fifty-two hundredths **23.** seven and twenty-five hundredths **25.** seventy-five **27.** twelve and four tenths **29.** twenty-one and five thousandths
31. one hundred and one hundredth **33.** three hundred sixty-two and eighty-five hundredths **35.** one thousand, forty and two thousandths **37.** six hundred fifty and three thousandths **39.** eight million, two hundred twenty-one thousand, three

hundred twenty-four and thirty-five hundredths **41.** 0.18 **43.** 12.04 **45.** 4.71 **47.** 0.0068 **49.** 412.003 **51.** 600.004 **53.** 750.012 **55.** 92,400 **57.** 0.004 **59.** 400,000.09 **61.** 0.000001 **63.** 60,000,000,400.06 **65.** twenty-seven dollars and five cents **67.** one thousand, three hundred forty dollars **69.** \$1,010.03 **71.** thirty-eight hundredths **73.** twenty-five thousandths

Section 5.3 (page 315)

1. six and two hundredths **3.** seven thousand, four hundred twenty and two hundredths **5.** six thousand, two hundred twenty-one and ninety-nine hundredths **7.** 7.4 **9.** 65 **11.** 1,051.0035 **13.** 1,340.5 **15.** 57 **17.** 4,000 **19.** 50,000 **21.** 416 **23.** 4,250.1 **25.** 3,000 **27.** 9,681.01 **29.** 25,000.1 **31.** 3.9 **33.** 7.31 **35.** 9.6 **37.** 1.07 **39.** 903.07 **41.** 8.28 **43.** 12.51 **45.** 8.94 **47.** 4.46 **49.** 21.9 **51.** 48.93 **53.** 22.403 **55.** 165.8 **57.** 31.1 **59.** 16,968.143 **61.** 10; 9.7 **63.** 10; 9.75 **65.** 7,033; 7,032.2 **67.** 8; 7.61 **69.** 21; 21.25 **71.** \$13.50 **73.** \$13.49 **75.** 7.09

Section 5.4 (page 325)

1. 7.98 **3.** 22.6 **5.** 24.14 **7.** 56.26 **9.** 24.64 **11.** 0 **13.** 48.8 **15.** 59.16 **17.** 13.08 **19.** 163.06 **21.** 0.065 **23.** 21.6 **25.** 632 **27.** 12.75 **29.** 6.4072 **31.** 51.181 **33.** 19.84 **35.** 5.9 **37.** 47.45 **39.** 3,024.02 **41.** 10.22 **43.** 0.018 **45.** 114.144 **47.** 0.093 **49.** 10.75 **51.** 9.0601 **53.** 3.11 **55.** 7.2 **57.** 11.7195 **59.** 1.84 **61.** 724.35 **63.** \$4,512.50 **65.** 2,431,050 **67.** 16,005 **69.** six hundred forty-five and twelve hundredths **71.** one thousand, nine hundred twelve and three hundred seventy-five ten-thousandths **73.** 0.07 **75.** 12.047 **77.** 0.153 **79.** 0.0144 **81.** 102.204 **83.** \$12.96 **85.** \$441 **87.** \$61.50

Section 5.5 (page 335)

1. 2.3 **3.** 2 **5.** 3 **7.** 46.9 **9.** 23 **11.** 840 **13.** 0.15 **15.** 72 **17.** 4 **19.** 2.1 **21.** 6.9 **23.** 2,010 **25.** 3.8 **27.** 0.8 **29.** 260.0 **31.** 6.9 **33.** 7.2 **35.** 0.03 **37.** 24.03 **39.** 2.04 **41.** 22.30 **43.** 1.86 **45.** 1.65 **47.** 473 **49.** 0.00295 **51.** 160.8 **53.** 15.6 **55.** 3.39 **57.** 2.9 **59.** 41.3 **61.** 10.1 **63.** 129.6 **65.** 168.24 **67.** 33.5 **69.** 0.25 **71.** 7.7 **73.** \$525.40 per month **75.** 127 square yards

Section 5.6 (page 345)

1. $7\frac{1}{3}$ **3.** $2\frac{2}{5}$ **5.** $\frac{21}{4}$ **7.** $\frac{97}{12}$ **9.** 0.25 **11.** 0.5 **13.** 0.625 **15.** $0.33\overline{3}$ **17.** 0.0375 **19.** 0.37 **21.** 1.45 **23.** 1.6 **25.** $\frac{1}{2}$ **27.** $\frac{1}{5}$ **29.** $\frac{17}{20}$ **31.** $\frac{1}{4}$ **33.** $\frac{7}{8}$ **35.** $\frac{7}{50}$ **37.** $\frac{9}{20}$ **39.** $\frac{81}{200}$ **41.** 4.5 **43.** 3.75 **45.** 12.13 **47.** 3.35 **49.** 21.8 **51.** 47.0125 **53.** 17.6 **55.** 99.01 **57.** $6\frac{1}{4}$ **59.** $8\frac{11}{20}$ **61.** $3\frac{7}{20}$ **63.** $20\frac{2}{25}$ **65.** $14\frac{3}{8}$ **67.** $10\frac{1}{8}$ **69.** $13\frac{1}{10}$ **71.** $37\frac{21}{50}$ **73.** 7.6 **75.** $5\frac{3}{25}$ **77.** 15.25 **79.** $35\frac{1}{5}$

Section 5.7 (page 351)

1. 18.703 **3.** 8.4 **5.** 175.05 **7.** 420 **9.** 25.85 **11.** 8.56 **13.** 21.34 **15.** 1.95 **17.** 77.44 **19.** 759.5 **21.** 148 **23.** 1.925 **25.** 4.81 **27.** 0.575 **29.** 10.8 **31.** $6 \cdot \$6.49 = \$38.94$ **33.** $[2{,}400 \div (270 \div 4.5)] \cdot 1.20 = \$48$ **35.** $\frac{4}{7} = \frac{x}{100}; x = 57\frac{1}{7}$ pounds **37.** $36 \cdot 27.75 = \$999; 999 - 800 = \$199$ **39.** $2.5 \cdot 30 = 75$ pounds of raw meat

Chapter 5 Review Problems (page 357)

1. sixteen and seventy-five hundredths **2.** two hundred ten and three hundred seventy-five thousandths **3.** one thousand, four hundred five and eight hundredths **4.** six thousand, two hundred forty-one and fifty-seven hundredths **5.** 7.004 **6.** 50.21 **7.** 2,007.7 **8.** 0.0075 **9.** 7.5 **10.** 12.41 **11.** 18 **12.** 751.520 **13.** 10.05 **14.** 38.4 **15.** 7.8 **16.** 24.15 **17.** 0.344 **18.** 8.67 **19.** 15.47 **20.** 5.67 **21.** 15.2 **22.** 4.67 **23.** 9.93 **24.** 25.74 **25.** 24.3 **26.** 9.03 **27.** 26.418 **28.** 0.1225 **29.** 19.697 **30.** 0.225 **31.** $\frac{7}{10}$ **32.** $21\frac{1}{2}$ **33.** $\frac{7}{200}$ **34.** $\frac{1}{100}$ **35.** $14\frac{3}{25}$ **36.** $1{,}050\frac{13}{20}$ **37.** 3.8 **38.** 0.6 **39.** 12.875 **40.** 75.75 **41.** 27.375 **42.** 18.75 **43.** 22.495 **44.** 5.18 **45.** 73.1 **46.** 9.8 **47.** 4.6 **48.** 27.44 **49.** $142.3\overline{3}$ **50.** 9.1 **51.** 4.507

52. 0.0001 **53.** 5.98 **54.** 365 **55.** 0.0065 **56.** 0.1 **57.** \$10,572.96 **58.** 22.6 gallons **59.** Joan **60.** $\frac{11}{4}$; $2\frac{3}{4}$; 2.75

Chapter 5 Chapter Test (page 361)

1. six hundred five and twenty-seven hundredths **2.** five hundred twelve and eight tenths **3.** seven thousand, two hundred forty-three and five hundredths **4.** one million and eight hundredths **5.** 6.014 **6.** 0.75 **7.** 2,001.4 **8.** 80,001.05 **9.** 18.3 **10.** 2.35 **11.** 14.965 **12.** 6.96 **13.** 0.0037 **14.** 5.824 **15.** 18.3 **16.** 14.59 **17.** 8.96 **18.** 3.928 **19.** 3.63 **20.** 8.9 **21.** 0.75 **22.** 0.625 **23.** $\frac{2}{5}$ **24.** $\frac{7}{50}$ **25.** 7.98 **26.** \$7.59 **27.** 1.65 **28.** $4\frac{1}{4}$; 4.25 **29.** 0.499, $4\frac{1}{5}$, 4.7, $\frac{20}{3}$ **30.** 3,000 business cards

CHAPTER 6

Section 6.1 (page 369)

1. 8% **3.** 90% **5.** 50% **7.** 30% **9.** 70% **11.** 0% **13.** 50% **15.** 60 **17.** 130 **19.** 0 **21.** 7 **23.** 25 **25.** 12 **27.** 20 **29.** 50 **31.** 100 **33.** 50 **35.** $\frac{15}{100}$; 0.15 **37.** $\frac{21}{100}$; 0.21 **39.** $\frac{42}{100}$; 0.42 **41.** $\frac{79}{100}$; 0.79 **43.** $\frac{125}{100}$; 1.25 **45.** $\frac{3}{100}$; 0.03 **47.** 24 **49.** 2 **51.** 30 **53.** 104 **55.** 0 **57.** 200 **59.** 1 **61.** 40,000 **63.** 3.6 **65.** $\frac{55}{100}$; 55% **67.** $\frac{0}{100}$; 0% **69.** $\frac{80}{100}$; 80% **71.** $\frac{10}{100}$; 10% **73.** $\frac{60}{100}$; 60% **75.** $\frac{90}{100}$; 90% **77.** 30% are males; 70% are females. **79.** 90% of the students passed; 10% did not. **81.** $\frac{1}{2}$ of the birds ate crickets; 50% ate crickets.

Section 6.2 (page 379)

1. 0.35 **3.** 0.02 **5.** 0.9 **7.** 0.99 **9.** 0.13 **11.** 0.21 **13.** 0.0 **15.** 0.01 **17.** 1.4 **19.** 0.8 **21.** 0.085 **23.** 0.374 **25.** 0.798 **27.** 0.2525 **29.** 0.125 **31.** 1.02125 **33.** 51% **35.** 25% **37.** 99% **39.** 0% **41.** 125% **43.** 75% **45.** 3% **47.** 300% **49.** 120% **51.** 55% **53.** 81.5% **55.** 60% **57.** 75.2%; $75\frac{1}{5}$% **59.** 99.9%; $99\frac{9}{10}$% **61.** 125.5%; $125\frac{1}{2}$% **63.** 3.25%; $3\frac{1}{4}$% **65.** 0.75; 75% **67.** 0.7; 70% **69.** 0.8; 80% **71.** 0.9; 90% **73.** 0.8; 80% **75.** 0.1; 10% **77.** six-tenths is either a decimal or a fraction; 60% of a mile **79.** 14 quarters **81.** 0.0925 **83.** 14 free throws

Section 6.3 (page 387)

1. $\frac{3}{10}$ **3.** $\frac{1}{20}$ **5.** 0 **7.** 1 **9.** $\frac{37}{100}$ **11.** $\frac{37}{50}$ **13.** $\frac{41}{20}$ **15.** $\frac{16}{5}$ **17.** $\frac{83}{500}$ **19.** $\frac{239}{500}$ **21.** $\frac{37}{400}$ **23.** $\frac{3}{400}$ **25.** 25% **27.** 75% **29.** 10% **31.** 80% **33.** 37% **35.** 45% **37.** 160% **39.** 30% **41.** 150% **43.** 95% **45.** 162.5% **47.** $177\frac{7}{9}$% **49.** 0.7 **51.** 20% **53.** 62.5% **55.** 1.4 **57.** $1\frac{2}{5}$ **59.** 1 **61.** 110% **63.** $1\frac{1}{10}$ **65.** 3.5 **67.** 650% **69.** $6\frac{1}{2}$ **71.** $266.\overline{6}$ **73.** 75% passed; 25% failed **75.** $\frac{2}{5}$ do not own their own home. **77.** 55% of the seats are empty.

Section 6.4 (page 395)

1. 7% **3.** 90% **5.** 112% **7.** 70% **9.** 0.16 **11.** 0.27 **13.** 0.03 **15.** 4.2 **17.** 14 **19.** 1,980 **21.** 4.2 **23.** 100 **25.** 189 **27.** 20 **29.** 10% **31.** 200% **33.** 5 **35.** $16\frac{2}{3}$% **37.** 800 **39.** 7,500 **41.** 200 **43.** 400 **45.** 200 **47.** 25% **49.** 112 **51.** 1,300 **53.** 7% **55.** 375 **57.** 54 **59.** 500%

61. 500% **63.** 190 **65.** 2.5% **67.** 0.05 **69.** 0.2 **71.** 0.1 **73.** $\frac{1}{5}$, 25%, 0.27 **75.** $4\frac{1}{4}$, 450%, 6.75
77. $1\frac{3}{5}$, 165%, 1.75 **79.** $12,960 per year **81.** Jan ate 20%.

Section 6.5 (page 403)

1. $900 interest; $1,200 interest **3.** 12.5% interest rate **5.** 7.5% tax rate **7.** $7,600 per year **9.** 0.8% tax rate **11.** 78.5% of income **13.** 25% of total **15.** 0.5% tax rate **17.** $450,000 profit **19.** 8.5% interest **21.** $1,500,000 **23.** She receives $262.50 per week. **25.** $960 interest

Section 6.6 (page 409)

1. $12,000 commission **3.** $1,800 commission **5.** $280,000 total sales **7.** 2% commission

9.

Group	*Number*	*Percent*
Very young	415	61.9%
Mature	220	32.8%
Old	35	5.2%
	670	99.9%

11. $160; $64; $88 **13.** $60,000 in sales **15.** $950; $1,100; $1,250

17. $3\frac{1}{3}$% commission rate **19.** $40,000 **21.** $96,000; $14,400; $9,600

23.

Age	*Number*	*Percent*
0–14	5,677,000	21.4%
15–59	16,871,000	63.6%
>59	3,979,000	15.0%
	26,527,000	100%

25.

Quarter	*Sales*	*Percent*
1st	$ 450,000	24.6%
2nd	426,000	23.3%
3rd	375,000	20.5%
4th	580,000	31.7%
	1,831,000	100.1%

Section 6.7 (page 417)

1. $15,725 **3.** $54 **5.** $713,000 **7.** 30% markdown **9.** $2,720 **11.** coyote #1 60%; coyote #2 30%; coyote #3 10% **13.** $50; 25% **15.** $12,000 **17.** 2.5 **19.** $1,408 increase in take-home pay

Chapter 6 Review Problems (page 427)

1. 37.5% **2.** $41\frac{2}{3}$% **3.** 12% **4.** 25% **5.** 75% **6.** 100% **7.** 65% **8.** 5% **9.** 12% **10.** 1% **11.** 125% **12.** 0.1% **13.** 58.1% **14.** 150% **15.** 75% **16.** 50% **17.** 87.5% **18.** 80% **19.** 30% **20.** 15% **21.** 20% **22.** 300% **23.** 0.14 **24.** 0.7 **25.** 2.2 **26.** 0.33 **27.** 0.028 **28.** 0.99 **29.** 1 **30.** 10 **31.** $\frac{2}{5}$ **32.** $\frac{3}{20}$ **33.** $\frac{19}{20}$ **34.** $\frac{21}{100}$ **35.** 1 **36.** $\frac{9}{5}$ **37.** $\frac{9}{200}$ **38.** $\frac{43}{500}$ **39.** $\frac{1}{8}$ **40.** $\frac{3}{20}$ **41.** 0.15 **42.** 0.875 **43.** 12.5% **44.** 87.5% **45.** 8 **46.** 300 **47.** 25 **48.** 170 **49.** 400 **50.** 25 **51.** 800 **52.** 7.52 **53.** 750 **54.** 1 **55.** 35%, $\frac{1}{2}$, $\frac{2}{3}$ **56.** 4%, $\frac{1}{4}$, 0.45 **57.** 12%, $\frac{1}{8}$, 0.13 **58.** 75%, 2, 5.14 **59.** 260%, $\frac{12}{4}$, 22 **60.** 12%, 2.14, $3\frac{1}{4}$ **61.** 60 **62.** 300 **63.** 30% **64.** 80 **65.** 70% **66.** 1.68 **67.** $79.50 **68.** $140 per year **69.** $9.54 per hour **70.** $120 **71.** 6% discount **72.** 2% improvement **73.** $2,000,000 **74.** 3 deliveries **75.** average = 68.3%, for a D

Chapter 6 Test (page 431)

1. 65% **2.** 1% **3.** 125% **4.** 37.5% **5.** 0.16 **6.** 2.15 **7.** 1 **8.** 0.155 **9.** $\frac{1}{5}$ **10.** $\frac{3}{4}$ **11.** $\frac{5}{8}$ **12.** $\frac{7}{200}$ **13.** $\frac{11}{50}$ **14.** $\frac{1}{20}$ **15.** 0.80 **16.** 0.05 **17.** 80% **18.** 22% **19.** 375 **20.** 800 **21.** 5% **22.** $3,200 per year **23.** $11.52 **24.** $\frac{3}{4}$ of 80 is greater than 50% of 70 **25.** 75%

Cumulative Test Chapter 5 and 6 (page 433)

1. 9.25 **2.** 16.254 **3.** 2.7 **4.** 10.6 **5.** $1\frac{2}{15}$ or $1.13\overline{3}$ **6.** 0.64 **7.** 0.275, $\frac{3}{8}$, 40% **8.** 300%, 3.1, $\frac{16}{3}$

9. 20.8 **10.** 140 **11.** 1,200 **12.** 10 **13.** 80 **14.** 5 **15.** 28%; $\frac{7}{25}$ **16.** 225%

17.

Grades	*Percent*
A	10%
B	15%
C	45%
Lower than C	30%
	100%

18. 30.17 **19.** 240 **20.** $1,360.06; round to the nearest penny, since payments will be made to the nearest cent **21.** 0.65 of a foot **22.** 6.25% of his original weight **23.** $3,720 tax **24.** $3.85 each **25.** $9.60

CHAPTER 7

Section 7.1 (page 441)

1. 36 **3.** 1,760 **5.** 8 **7.** $5\frac{1}{4}$ **9.** 8 **11.** 104 **13.** $16\frac{2}{3}$ **15.** 7 **17.** 18,480 **19.** 6 **21.** 10 ft 8 in **23.** 8 gal 2 qt **25.** 12 yd 2 ft **27.** 5 wk 4 da **29.** 7 da 3 hr **31.** 12 ft 5 in **33.** 8 gal 1 qt **35.** 5 cups 6 oz **37.** 2 wk 4 da **39.** 17 gal 1 qt **41.** 8 wk 1 da **43.** 5 gal 2 qt **45.** 3 hr 12 min **47.** 34,320 feet **49.** 120,000,000 pounds **51.** 1 hour 40 minutes **53.** $173\frac{1}{3}$ yards **55.** 675 calories **57.** $33\frac{1}{3}$ feet per second **59.** 6 hours 40 minutes **61.** 13 hours **63.** $\frac{22}{25}$; 0.88

Section 7.2 (page 451)

1. kilo **3.** hecto **5.** deci **7.** 1,000 **9.** 100 **11.** 0.55 **13.** 6,000 **15.** 0.0025 **17.** 0.2 **19.** 200,000 **21.** 5.2 **23.** 400 **25.** 6 **27.** 743,000 **29.** 0.48 **31.** 0.5 **33.** 70 **35.** 0.0004 **37.** 1.5 **39.** 2,900 **41.** 1.29 grams of sodium **43.** 12 glasses **45.** 4,500 milliliters **47.** 2 tablets **49.** 38 grams **51.** 800 centigrams

Section 7.3 (page 461)

1. 2.2 **3.** 0.6 **5.** 0.4 **7.** 15.2 **9.** 1.5 **11.** 9.5 **13.** 18.9 **15.** 9,070 **17.** 39.4 **19.** 2.7 **21.** 3.2 **23.** 25.4 **25.** 907,000,000 **27.** 6 **29.** 240 **31.** 483 **33.** 0.1 **35.** 0.1 **37.** 19.7 **39.** 11,352 **41.** 64.4 **43.** 29.4 **45.** 18.6 kilograms **47.** 13.1 feet **49.** 45.7 meters **51.** 0.4 kilometers **53.** 11.2 gallons; $13.13 **55.** No, the one-half inch wrench is too large. **57.** 60 inches; 1.5 meters **59.** the one with 400 grams; 59.8 grams **61.** 1.9 miles **63.** 140°F **65.** 4.5 kilograms

Chapter 7 Review Problems (page 469)

1. 100 **2.** 12 **3.** 10 **4.** 0.45 **5.** 36 **6.** 1,000 **7.** 1,000 **8.** 1.61 **9.** 60 **10.** 100 **11.** 10 **12.** 2,000 **13.** 2.54 **14.** 0.95 **15.** 5,280 **16.** 4 **17.** 0.91 **18.** 2 **19.** 104 **20.** 3.33 **21.** 240 **22.** 20.3 **23.** 12.9 **24.** 8 **25.** 22.9 **26.** 364 **27.** 6.6 **28.** 81.8 **29.** 7,256 **30.** 1,200 **31.** 2 days 12 hours 18 minutes **32.** 87 pounds 8 ounces **33.** 6 feet 3 inches **34.** 9 hours 15 minutes **35.** 5 gallons 2 quarts **36.** 5,110 feet **37.** 6 feet 11 inches **38.** 2 quarts 1 pint **39.** 129.5 centimeters **40.** 1,611.8 meters **41.** 8.5 liters **42.** 3,238.2 grams **43.** 193.0 centimeters; 109.1 kilograms **44.** 213.4 centimeters; 127.3 kilograms **45.** 36,280 kilograms **46.** 4,400.4 pounds **47.** 6.6 pounds **48.** 96 **49.** 4 miles 640 yards **50.** 2,508 pounds; 1,139.9 kilograms **51.** 6 feet 6 inches **52.** 4,669 kilometers **53.** 56 minutes **54.** 62.1 miles/hour **55.** 44 pounds of gold **56.** 6,600.7 pounds of garbage

Chapter Test (page 473)

1. 15.2 **2.** 336 **3.** 3,000 **4.** 48.4 **5.** 400 **6.** 3,600 **7.** 6,000 **8.** 4,545 **9.** 11.8 **10.** 84 **11.** 10,800 **12.** 252 **13.** 23.9 **14.** 3.2 **15.** 153 pounds **16.** 3 feet 3 inches **17.** 65 minutes 5 seconds **18.** 3 quarts **19.** 3,960.4 pounds **20.** 6 feet 2 inches **21.** 0.007 ounces **22.** 133.5 pounds; 60.7 kilograms **23.** 19.8 gallons **24.** 128.8 kilometers per hour **25.** 36.4 kilograms

CHAPTER 8

Section 8.1 (page 481)

1. 

3. 

5. 

7. 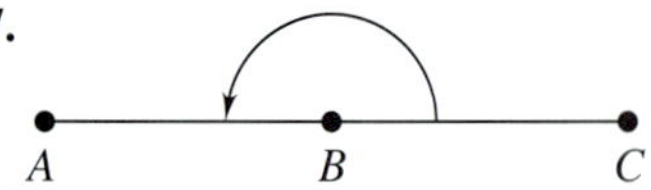

9. 270° 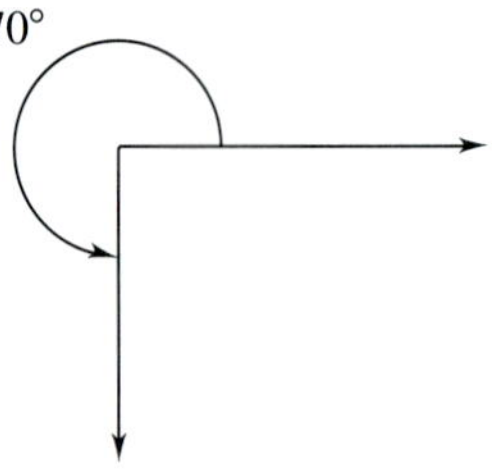

11. 90° **13.** 72° **15.** 270° **17.** $\overline{AC}$ **19.** $\overline{GH}$ **21.** 1° **23.** 105° **25.** 56° **27.** 89° **29.** 25°; 115° **31.** true **33.** false **35.** false **37.** true **39.** false **41.** false **43.** 32° **45.** 75° **47.** 115° **49.** 90° **51.** 270° **53.** 78°

Section 8.2 (page 491)

1. 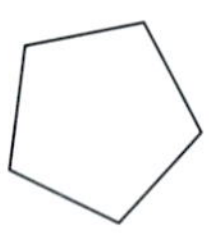

3. 

5. 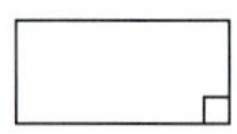

7. 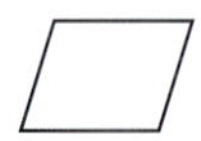

9. 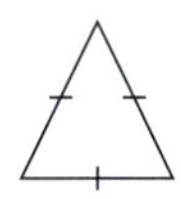

11. 60° **13.** 50° **15.** 60° **17.** 360° **19.** 71 cm **21.** true **23.** true **25.** true **27.** false **29.** false **31.** false **33.** false **35.** 36° **37.** 110° **39.** 18°

41.

rectangle	*length*	*width*
HIJG	15	10
GJEF	15	15
HIEF	15	25
GCDF	40	15
JEDC	25	15
ABCK	20	20

43. 83°

Section 8.3 (page 501)

1. rectangle; 16″ **3.** triangle; 26 in **5.** circle; 31.4 ft **7.** 6.28 in **9.** 17.27 cm **11.** 12 ft **13.** 9.2 cm **15.** 68 cm **17.** 28 m **19.** 9.7 cm **21.** 110 mm **23.** 57 in **25.** 30.5 cm **27.** 25.7 mm **29.** 50 in **31.** 20 ft **33.** 22 in **35.** 22.5 in **37.** 51.4 cm **39.** 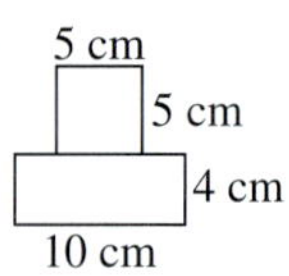

; 38 cm **41.** 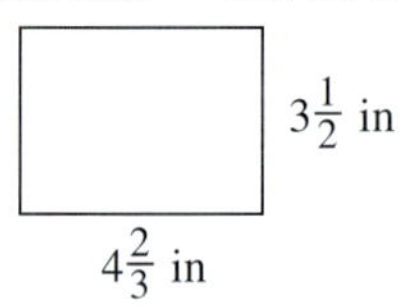

; $16\frac{1}{3}$ in

43. 100 ft **45.** 7,140 m **47.** 1,920 mi **49.** 2,700,400 mi; 25,120 mi; 282,600 mi; 6,280 mi

Section 8.4 (page 511)

1. 50 sq in **3.** 6 cm^2 **5.** 39.25 cm^2 **7.** 20 sq in **9.** 48 cm^2 **11.** 51.63 sq in **13.** f **15.** a **17.** b **19.** 25 sq in **21.** 200.96 sq cm **23.** 7 sq in **25.** 0.785 sq yd **27.** 11.52 cm^2 **29.** 117 sq in **31.** 252 cm^2 **33.** 75.36 cm^2 **35.** 48 sq in **37.** 100.75 cm^2 **39.** 28.28 sq ft **41.** \$42.86; \$41.67; the \$75,000 home costs less per square foot **43.** 150 sq ft; 150 sq ft; 160 sq ft; either plan **45.** .20; 20% **47.** the rectangular piece

Section 8.5 (page 525)

1. 28 in; 48 sq in **3.** 31.4 ft; 78.5 sq ft **5.** 34 cm; 76 cm^2 **7.** 6,280 mm^3 **9.** 0.14 in^3 **11.** 50 cm^3
13. 1,808.64 cm^3 **15.** 4,710 cm^3 **17.** 856.67 ft^3 **19.** 2,093.33 mm^3 **21.**

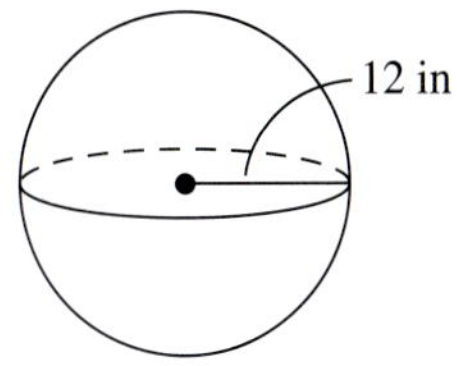

23.

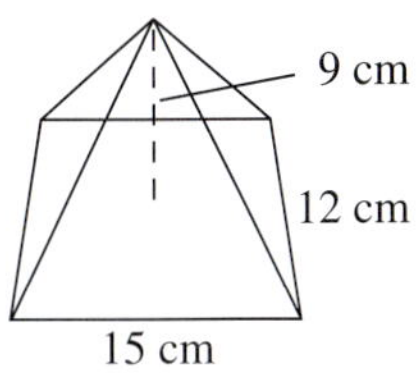

25.

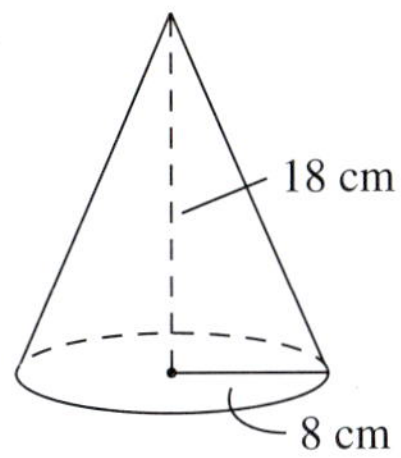

27.

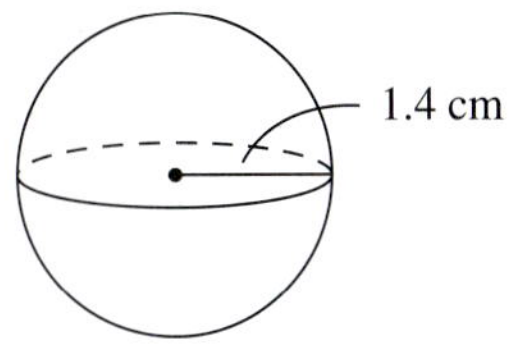

29.

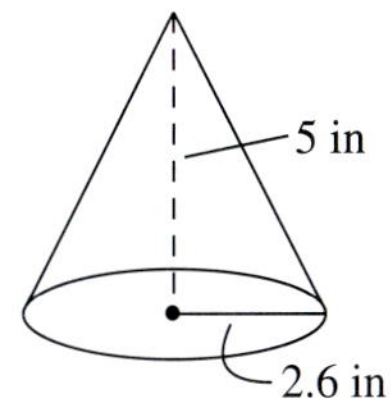

31. f **33.** c **35.** b **37.** 94,067,987 cu ft **39.** 36,000 cu cm **41.** 5,184 cu in, or 3 cu ft
43. volume is 4 times; no **45.** 600 cu ft **47.** 496 sq in **49.** 326.56 sq ft

Section 8.6 (page 537)

1. $\overline{RQ} \leftrightarrow \overline{YX}$; $\overline{PR} \leftrightarrow \overline{ZY}$; $\overline{PQ} \leftrightarrow \overline{ZX}$ **3.** $\overline{AE} \leftrightarrow \overline{BD}$; $\overline{EC} \leftrightarrow \overline{DC}$; $\overline{AC} \leftrightarrow \overline{BC}$ **5.** $\overline{AB} \leftrightarrow \overline{AD}$; $\overline{AC} \leftrightarrow \overline{AE}$; $\overline{BC} \leftrightarrow \overline{DE}$
7. $\frac{8\text{ m}}{12\text{ m}} = \frac{10\text{ m}}{15\text{ m}} = \frac{2}{3}$ **9.** $\frac{8'}{24'} = \frac{10'}{30'} = \frac{1}{3}$ **11.** 120 ft **13.** 1.5 in **15.** 8 in **17.** 100 in **19.** 16 ft
21. 1,020 m **23.** 15 ft

Section 8.7 (page 547)

1. yes **3.** no **5.** no **7.** 5 cm **9.** 8 m **11.** 13′ **13.** $\sqrt{89}$ yd **15.** $\sqrt{1.75}$ in **17.** $\sqrt{3}$ cm **19.** $\sqrt{1.6}$
21. 4 **23.** 9 **25.** 13 m **27.** 44.24 sq in **29.** 6 ft more **31.** $\sqrt{10{,}900}$ kph; 104 kph

Chapter 8 Review Problems (page 557)

1.

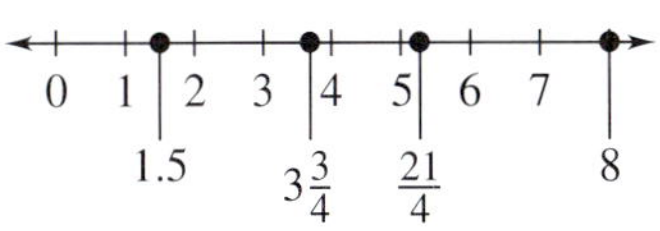

2.

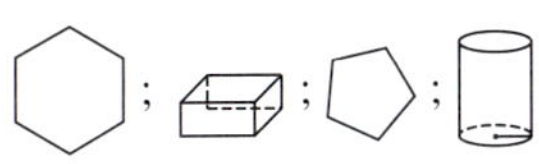

3. $\overline{AH}$, $\overline{CD}$ **4.** 240° **5.** 90° **6.** 35°, 110°

7. 70° **8.** 270° **9.** 60° **10.** 40° **11.** 82° **12.** 25.8 m **13.** 28 ft **14.** 28 cm **15.** 27 m **16.** 69.08 in
17. 28 mm **18.** 41.12 cm **19.** 16 cm **20.** 16 in **21.** 123.48 sq yd **22.** 180 sq in **23.** 97.12 sq ft
24. 42 sq m **25.** 200.96 cm^2 **26.** 72.88 ft^2 **27.** 7.56 in^2 **28.** 102 sq in **29.** 96 cm^2 **30.** 69.08 cu in
31. 150 cu in **32.** 339.12 cm^3 **33.** 162 in^3 **34.** 15.07 cu ft **35.** 904.32 cu ft **36.** 12.83 cu in **37.** 198 sq in
38. 122° **39.** 16° **40.** 165° **41.** 27° **42.** 30° **43.** 145° **44.** 40° **45.** 75° **46.** 180° **47.** $\frac{6}{15} = \frac{8}{20} = \frac{2}{5}$
48. $\frac{15}{30} = \frac{9}{18} = \frac{1}{2}$ **49.** 6 ft **50.** 3 in **51.** 15 in **52.** $\sqrt{29}$ m **53.** 5 ft **54.** 8.89 cm **55.** 10 gal
56. 23.86 cm **57.** 102 in **58.** 1,210 ft **59.** 48

Chapter 8 Test (page 563)

1.

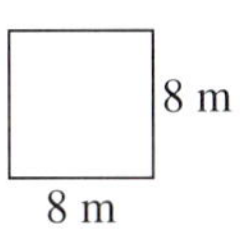

; 32 m

2.

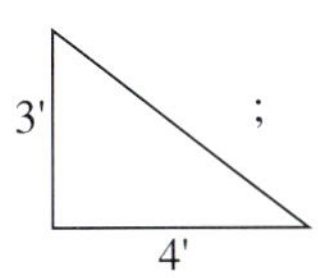

; 12 ft

3.

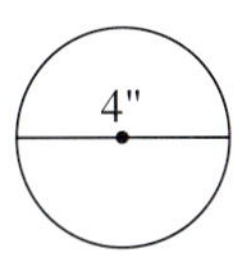

; 12.6 in

4.

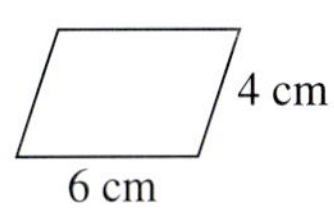

; 20 cm

5.

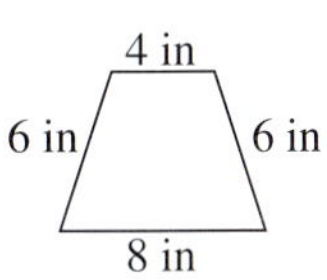

; 24 in

6.

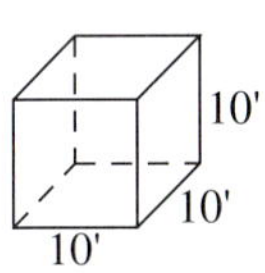

; 1,000 cu ft

7.

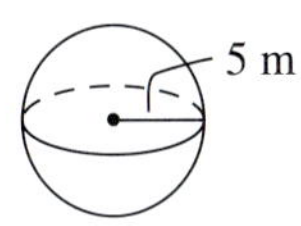

; 523.3 m^3

8. 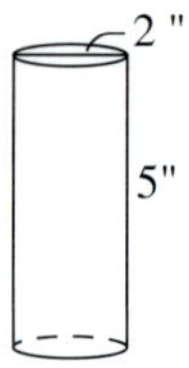

; 15.7 in³

9. 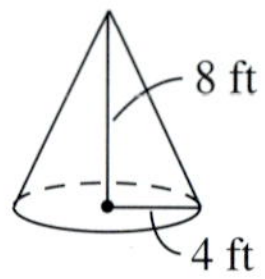

; 134 cu ft

10. 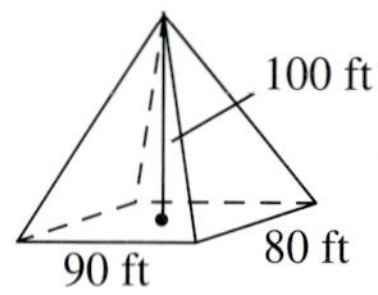

; 240,000 cu ft

11. 79° **12.** 66° each **13.** 288° **14.** 3 sq ft **15.** 40.69 in³ **16.** 3,300 sq ft **17.** 24 cm **18.** 120° **19.** $\angle A = 75°$ **20.** yes; all three angles are equal

Cumulative Test Chapters 7 and 8 (page 565)

1. 3,000 **2.** 5 **3.** 80 **4.** 78 **5.** 480 **6.** 0.75 **7.** 400 **8.** 8 **9.** 3.636 **10.** 45.72 **11.** 46.934 **12.** 322 **13.** 21 pounds 4 ounces **14.** 2 hours 25 minutes **15.** $\angle A = 12°$ **16.** 35°, 110° **17.** 18.8 inches **18.** perimeter = 36 feet; area = 54 square feet **19.** 44.0 cm **20.** 199.25 sq cm **21.** 904.32 sq feet **22.** sphere **23.** $x = 9$ yd **24.** 13 **25.** 2

CHAPTER 9

Section 9.1 (page 573)

1. 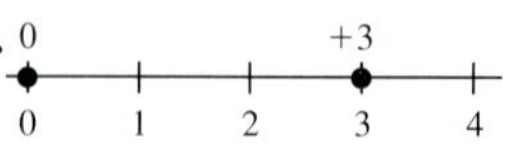

3. 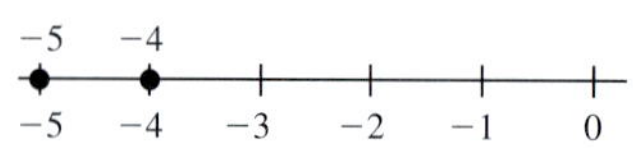

5. 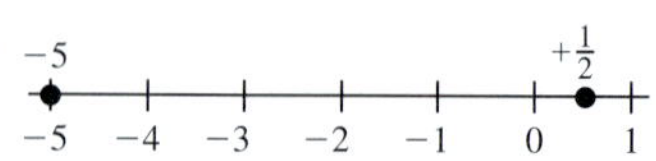

7. 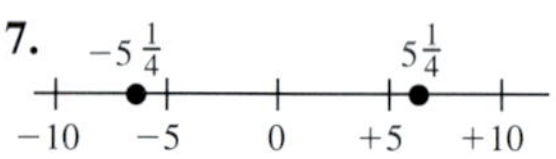

9. 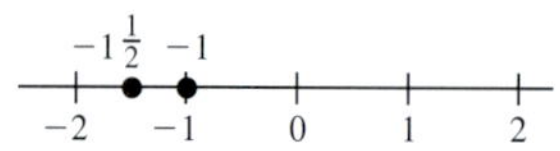

11. 

13. 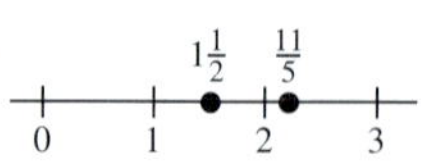

15. −7 **17.** $-\frac{21}{2}$ **19.** −2.7 **21.** −5,000 **23.** −7.5 **25.** 6 **27.** 4

29. −7.0 **31.** 3 **33.** 5.5 **35.** true **37.** true **39.** false **41.** true **43.** true **45.** false **47.** false **49.** false **51.** true **53.** true **55.** true **57.** true **59.** true **61.** false **63.** 12 degrees below zero **65.** −8°F

Section 9.2 (page 577)

1. $<$ **3.** $<$ **5.** $>$ **7.** $>$ **9.** $>$ **11.** $>$ **13.** $\leq$ **15.** $\geq$ **17.** $\leq$ **19.** $\leq$ **21.** $\geq$ **23.** $\leq$ **25.** $\neq$

27. = **29.** = **31.** = **33.** $\neq$ **35.** = **37.** 12 **39.** $\frac{3}{4}$ **41.** 7 **43.** 2.4 **45.** 100 **47.** 6 **49.** 3.4

51. $4\frac{1}{3}$ **53.** 3.25 **55.** 4.5 **57.** 7 **59.** 0.009 **61.** false **63.** false **65.** true **67.** false **69.** true

71. true **73.** false **75.** false

Section 9.3 (page 583)

1. 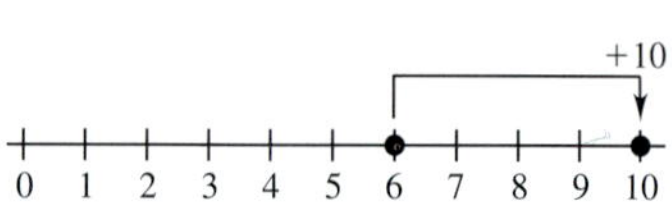

3. 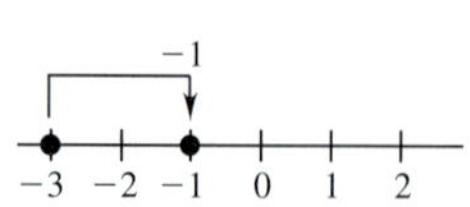

5. 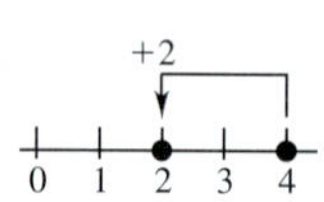

7. 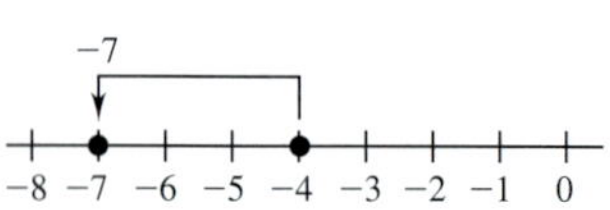

9. 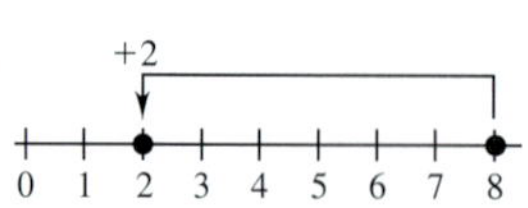

11. 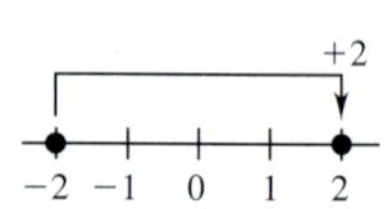

13. 14 **15.** 6 **17.** 0 **19.** $\frac{3}{4}$ **21.** +9 **23.** +7 **25.** −6 **27.** −9 **29.** 2.5 **31.** −2 **33.** +12

35. +6 **37.** −5 **39.** +7 **41.** −72 **43.** −1.9 **45.** 1.1 **47.** −0.25 **49.** $-\frac{3}{4}$

51. \$43.06 − 57; 02; −\$13.96 **53.** −12°C + 9°C; −3°C **55.** −8 − 1.5; −9.5 meters **57.** 8 − 50; −42 sedans

Section 9.4 (page 591)

1. 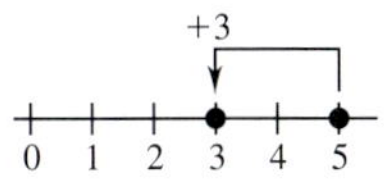

3. 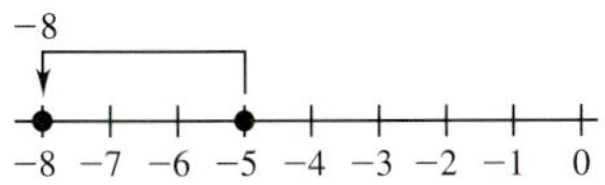

5. 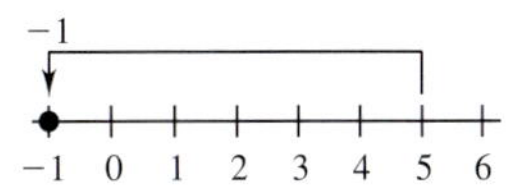

7. 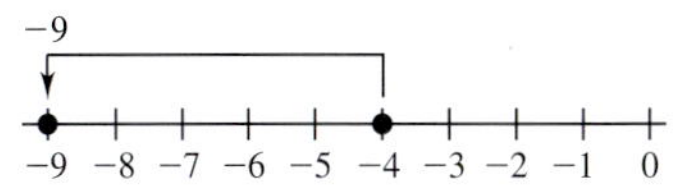

9. 5 **11.** 6.4 **13.** +1 **15.** +6 **17.** −60 **19.** −2 **21.** −3.5 **23.** −2 **25.** +12 **27.** +5 **29.** 22.5 **31.** +10 **33.** +15 **35.** −33.5 **37.** −36 **39.** $-22\frac{1}{2}$ **41.** −12 **43.** +10 **45.** −6 **47.** +6.7 **49.** $-3\frac{1}{6}$ **51.** −9 **53.** +4 **55.** +12 **57.** +6 **59.** −11 **61.** +42 **63.** +83 **65.** +22 **67.** −5.5 **69.** $+13° - 15° = -2°$ **71.** −5°F; yes, it could be very dangerous

Section 9.5 (page 601)

1. +35 **3.** +20 **5.** −24 **7.** +56 **9.** −36 **11.** −1 **13.** −24 **15.** −64 **17.** +76 **19.** −3 **21.** +8 **23.** $-3\frac{1}{2}$ **25.** +5 **27.** −7 **29.** −11 **31.** $-\frac{1}{2}$ **33.** $-\frac{49}{81}$ **35.** +84 **37.** −11 **39.** −56 **41.** −85 **43.** +15.5 **45.** −4 **47.** $+\frac{1}{2}$ **49.** −825 **51.** 0 **53.** −100 **55.** −1.21 **57.** $-65\frac{3}{4}$ **59.** −7.2 **61.** −4.6 **63.** +22 **65.** −55.04 **67.** $-1\frac{5}{7}$ **69.** +22.4 **71.** $-\frac{24}{49}$ **73.** 2, 3, 5, and 7; $(-2)(-3)(-5)(-7) = +210$

Section 9.6 (page 607)

1. −48 **3.** −20 **5.** +80 **7.** −9 **9.** −7 **11.** +5 **13.** −32 **15.** −2.6 **17.** $-2\frac{1}{2}$ **19.** −11.8 **21.** +52 **23.** −11 **25.** −1 **27.** −18 **29.** +68 **31.** −700 **33.** +38 **35.** +86 **37.** +4 **39.** +60 **41.** +30 **43.** +14 **45.** $+16\frac{3}{4}$ **47.** $+5.74\overline{3}$ **49.** $+6\frac{19}{48}$ **51.** −32.5 **53.** +3,900,000,000 **55.** $\frac{6}{7}$ **57.** $-7 + (-2)(+8) = -23$ **59.** $-10 \cdot (70 \div -2) = +350$ **61.** $\frac{-6 + (-4.4)}{2}$; −5.2 **63.** $(-5)^2 + \frac{3}{4} = 25\frac{3}{4}$

Chapter 9 Review Problems (page 615)

1. < **2.** > **3.** > **4.** > **5.** < **6.** < **7.** > **8.** < **9.** < **10.** > **11.** ≥ **12.** ≥ **13.** ≥ **14.** ≤ **15.** ≥ **16.** ≤ **17.** ≥ **18.** ≤ **19.** ≤ **20.** ≤

21. 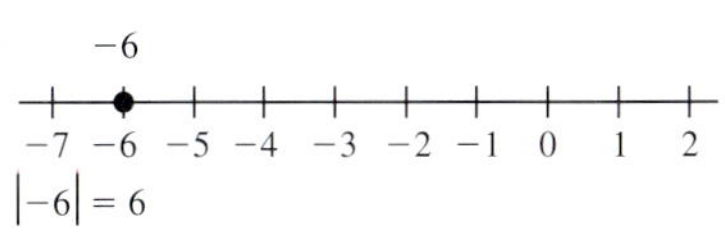

22. 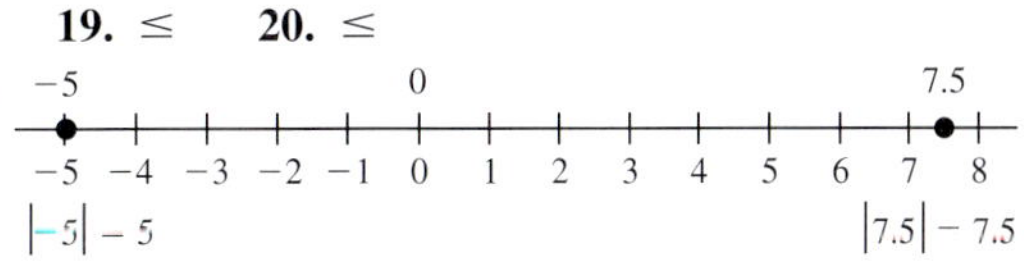

23. 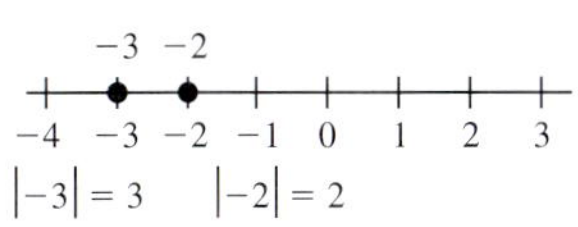

24. 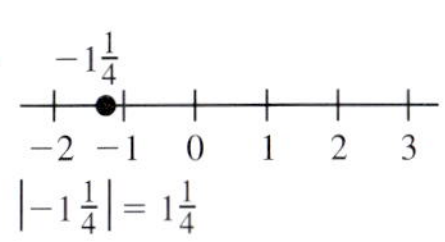

25. 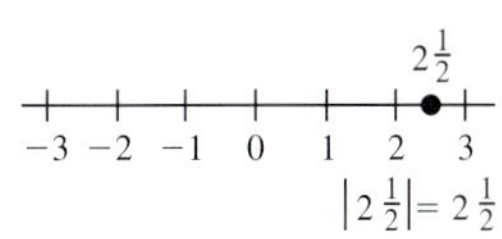

26. 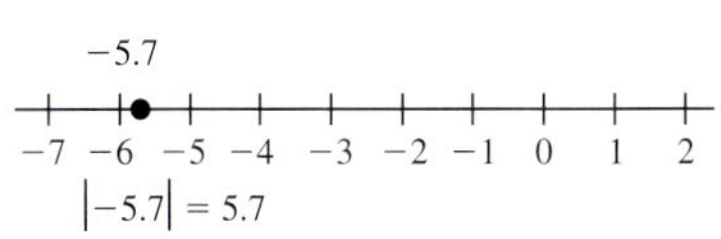

27. 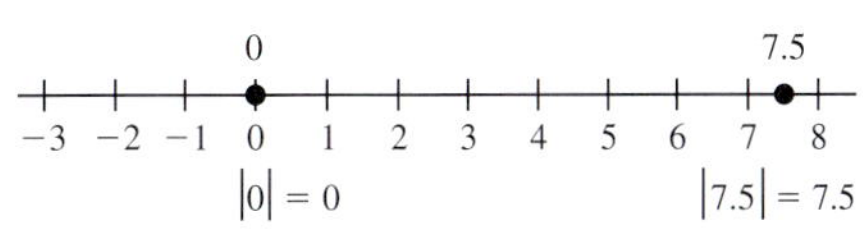

28. 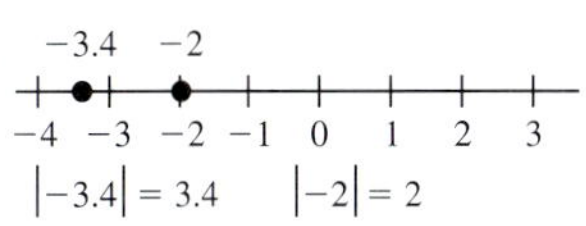

29. 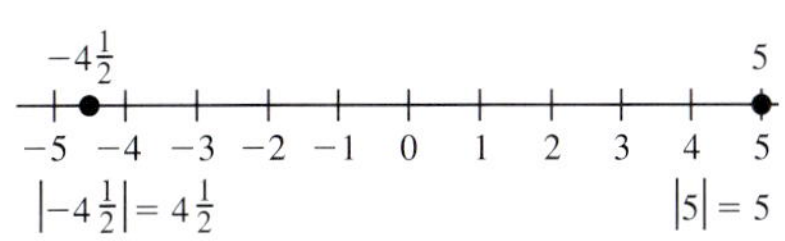

30. 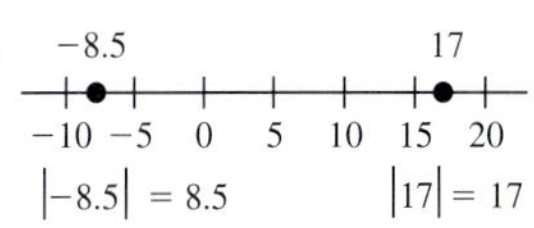

31. -3 **32.** -63 **33.** $+7$ **34.** $+8$ **35.** -11 **36.** -1 **37.** -13 **38.** -48 **39.** $+6$ **40.** -3 **41.** -9 **42.** -9 **43.** -32 **44.** $+52$ **45.** -40 **46.** $+9$ **47.** -9 **48.** $+20$ **49.** -18 **50.** $+78$ **51.** $+15$ **52.** $+11$ **53.** -4 **54.** $+7$ **55.** -1 **56.** $+1$ **57.** -7 **58.** -72 **59.** -6 **60.** $+4$ **61.** -13 **62.** $+28$ **63.** $+37\frac{1}{2}$ **64.** 40.5 **65.** -7 **66.** $+6$ **67.** -12 **68.** $+26$ **69.** -120 **70.** $+374$ **71.** $-\frac{2}{3}$ **72.** $-11\frac{1}{2}$ **73.** -11 **74.** $+11$ **75.** $-6°F + 8°F = +2°F$ **76.** $-2°F - 3°F \cdot 5 = -17°F$ **77.** $-7 \cdot [(-4) + (+9)] = -35$ **78.** $-17{,}000{,}010 \div (+3) = -5{,}666{,}670$

Chapter 9 Test (page 619)

1. false **2.** false **3.** false **4.**

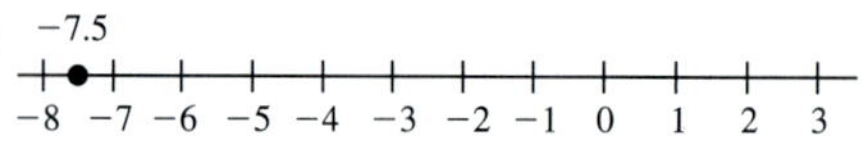

5.

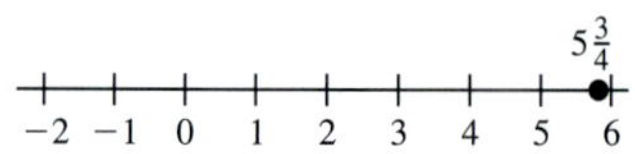

6.

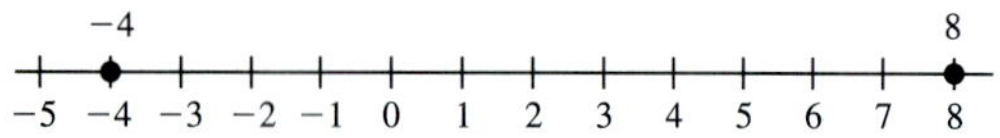

7. 12.3 **8.** 4 **9.** $6\frac{4}{5}$ **10.** -48 **11.** $+7$ **12.** $+2$ **13.** -78 **14.** -24 **15.** $+84$ **16.** -10 **17.** -22 **18.** -36 **19.** $+6$ **20.** $+110$ **21.** -6 **22.** yes, $|-12| = |12|$ **23.** $+39.7$ **24.** yes **25.** $-72{,}180$ **26.** $+6$ **27.** $+9.2$ **28.** $-4°F$ **29.** 10°C **30.** The checks exceed her balance by $38.64.

CHAPTER 10

Section 10.1 (page 627)

1. constant **3.** variable **5.** variable **7.** variable **9.** $+3$ **11.** -5 **13.** -3.1 **15.** $+3\frac{1}{4}$ **17.** $\frac{2}{5}$ **19.** $+1$ **21.** $-2\frac{1}{2}$ **23.** $+3.4$ **25.** -1 **27.** -9 **29.** $5x$ **31.** $88a$ **33.** $3b$ **35.** $16x - 24c$ **37.** $14x$ **39.** z **41.** $-9b$ **43.** $-12z$ **45.** $-3b - 9m$ **47.** $-5.5a$ **49.** $-\frac{1}{4}b$ **51.** $96\frac{1}{2}r$ **53.** $-25.5z$ **55.** $a - \frac{1}{3}r$ **57.** $9\frac{1}{2}z$ **59.** $-5x + 2xy$

Section 10.2 (page 633)

1. 10 **3.** 7.5 **5.** Commutative property of addition **7.** Identity property of multiplication **9.** Identity property of addition **11.** Distributive property **13.** Distributive property **15.** Commutative property of addition **17.** Commutative property of addition **19.** Distributive property **21.** $13y$; Distributive property **23.** $12x$; Distributive property **25.** $11x$; Distributive property **27.** $6y$; Distributive property **29.** $10r$; Distributive property **31.** $7a$; Distributive property **33.** $-6y$; Distributive property **35.** $6\frac{1}{2}y$; Distributive property **37.** $7z + 2x$; Distributive property **39.** $4y + 10$; Distributive property **41.** $8x + 3y$; Distributive property **43.** $2ab + ac$; Distributive property **45.** $5a + 19y$; Distributive property **47.** $3a + 2b$; Distributive property **49.** $4b + 3z + 7z$; $4b + 10z$ **51.** $9r - 3b + \frac{1}{2}r$; $9\frac{1}{2}r - 3b$ **53.** $7(x + y) + (-5x)$; $2x + 7y$

Section 10.3 (page 639)

1. $3x$, $5x$ **3.** $2z$, $5\frac{1}{2}z$ **5.** $2r$, $0.5r$ **7.** $8xy$ **9.** $-7z$ **11.** $-7y$ **13.** $3ac$ **15.** $x = 5$ **17.** $r = -13$ **19.** $r = 29$ **21.** $x = -12$ **23.** $r = 7.1$ **25.** $r = 22$ **27.** $y = 24.5$ **29.** $y = 11.5$ **31.** $x = 16$ **33.** $t = 10.2$ **35.** $s = 4\frac{1}{2}$ **37.** $z = 17$ **39.** $s = 0$ **41.** $y = 6.9$ **43.** $p = -2.05$ **45.** $A = 21$ **47.** $x = 10.7$ **49.** $x = -5$ **51.** $x + 8 = 41$; $x = 33$ **53.** $s - 7 = 3$; $s = 10$ **55.** $-15 + y = -8$; $y = 7$ **57.** $-9 = y + 4$; $y = -13$ **59.** $x - 40 = 2{,}000$; $x = 2{,}040$ **61.** $74 = x - 3$; $x = 77$

Section 10.4 (page 645)

1. $y = 9$ **3.** $r = 11$ **5.** $t = -4$ **7.** $r = -29$ **9.** $r = 2\frac{7}{9}$ **11.** $x = -12\frac{1}{2}$ **13.** $y = -\frac{3}{8}$ **15.** $y = -20$ **17.** $n = -13$ **19.** $x = 100$ **21.** $u = 14$ **23.** $t = 270$ **25.** $x = -840$ **27.** $y = -1{,}273$ **29.** $b = 10{,}000$ **31.** $y = 30$ **33.** $t = 240$ **35.** $b = 25\frac{2}{3}$ **37.** $g = -8\frac{3}{4}$ **39.** $x = 260$ **41.** $r = 3$ **43.** $s = 70$ **45.** $y = 2.8$ **47.** $r = -8.7$ **49.** $t = 4.8$ **51.** $6y = 94; y = 15\frac{2}{3}$ **53.** $53 = 3t; t = 17\frac{2}{3}$ **55.** $x \div 7 = 40; x = 280$ **57.** $s \div 3 = -7; s = -21$ **59.** $3x \div 5 = -70; x = -116\frac{2}{3}$ **61.** $3z = -4.2; z = -1.4$

Section 10.5 (page 651)

1. $x = 66$ **3.** $t = 8$ **5.** $x = 21$ **7.** $x = 4$ **9.** $r = 4\frac{7}{9}$ **11.** $p = 4$ **13.** $y = 397$ **15.** $t = 36$ **17.** $y = -4$ **19.** $x = 151$ **21.** $r = 2$ **23.** $r = \frac{5}{6}$ **25.** $t = 4\frac{1}{3}$ **27.** $y = -1\frac{2}{9}$ **29.** $y = 10$ **31.** $p = 1$ **33.** $r = 1\frac{1}{5}$ **35.** $m = 2\frac{1}{4}$ **37.** $p = 26$ **39.** $r = 12\frac{5}{7}$ **41.** $x = 5$ **43.** $r = 120$ **45.** $z = 1\frac{13}{15}$ **47.** $y = -24$ **49.** $z = \frac{1}{24}$ **51.** no

Section 10.6 (page 659)

1. $6 + x$ **3.** $12\frac{1}{2} - x$ **5.** $-0.3 - 2x$ **7.** $x + 3x$ **9.** $4x - 70.2$ **11.** $x = 7$ **13.** $z = 24$ **15.** $z = -\frac{3}{4}$ **17.** $x = 4$ **19.** $y = 2.5$ **21.** $x + 8 = 20; x = 12$ **23.** $x - 14 = 40; x = 54$ **25.** $6x = 54; x = 9$ **27.** $30x = 1{,}800; x = 60$ **29.** $x \div 8 = 4{,}800; x = 38{,}400$ **31.** $3x + 5 = 49; x = 14\frac{2}{3}$ **33.** $x \div 7 = 9; x = 63$ **35.** $-7 = 6 - 4x; x = 3.25$ **37.** $x + (x + 2) = 240;\ x = 119, (x + 2) = 121$ **39.** $x + (x + 2) = 624;\ x = 311, (x + 2) = 313$ **41.** $L = 168 \div 12; L = 14$ feet **43.** $(x + 51) + x = 369;$ $x = 159; (x + 51) = \$210$ Darla has saved \$210. **45.** $x + (x + 4) = 68$; 32 inches, 36 inches **47.** $35{,}000 = x + 22{,}500$; \$12,500 per month

Chapter 10 Review Problems (page 667)

1. constant; $6\frac{1}{2}$; variables: x, y **2.** constants: 3, 0; variable: A **3.** constant: -5.2; variables: B, t **4.** constants: 7, -2; variable: x **5.** $x, 3x$ **6.** $9y, 8y$ **7.** $2t, 3t$ **8.** $8y, 9y;\ 2x, 3x$ **9.** $5Y, 7Y$ **10.** $5y, -3.8y$ **11.** 6 **12.** -4 **13.** 1 **14.** $-2\frac{1}{3}$ **15.** 7.2 **16.** -9 **17.** 1 **18.** $\frac{3}{2}$ **19.** $3x + 2y$ **20.** $6y + 5$ **21.** $12x - 7$ **22.** $2y - 2x$ **23.** $4x + 24$ **24.** $-16y + 8$ **25.** $-10y$ **26.** $-5x$ **27.** $2x$ **28.** $3x + 3$ **29.** $7x + 3y$ **30.** $-6.5r + 4y$ **31.** $8.5y - 6$ **32.** z **33.** $y = 12$ **34.** $z = 53$ **35.** $r = -7$ **36.** $y = 5$ **37.** $x = 4$ **38.** $x = 4$ **39.** $y = 22$ **40.** $y = 9$ **41.** $z = 96$ **42.** $y = -2\frac{1}{6}$ **43.** $y = 5.5$ **44.** $z = 1{,}100$ **45.** $x = \frac{2}{3}$ **46.** $y = 8\frac{1}{5}$ **47.** $z = 1\frac{1}{3}$ **48.** $y = 22$ **49.** $x = -2$ **50.** $y = -\frac{1}{8}$ **51.** $z = 13.5$ **52.** $x = -\frac{2}{5}$ **53.** $y = -45.5$ **54.** $z = -\frac{1}{3}$ **55.** $x + 400 = 1; x = -399$ **56.** $4x - 16 = 7\frac{1}{2}; x = 5.875$ **57.** $3x = 4 + x; x = 2$ **58.** $-7 + 4x = 21; x = 7$ **59.** $x + (x + 2) = 244; x = 121, (x + 2) = 123$ **60.** $x + 2x = \$2.61;\ x = \$0.87$; Joe threw \$1.74 onto the table.

Chapter 10 Test (page 669)

1. $3y, 5y$ **2.** $2y, 5.1y$ **3.** $bc, 4.5bc$ **4.** $6y, -y$ **5.** $8x$ **6.** $-6y$ **7.** $-z$ **8.** $5x + 29$ **9.** $7r + 2x$ **10.** $3xy + x$ **11.** $x = 13.7$ **12.** $y = -1$ **13.** $z = 5.6$ **14.** $z = -100$ **15.** $y = 3\frac{2}{3}$ **16.** $y = 25$ **17.** $z = -\frac{2}{5}$ **18.** $x = 0$ **19.** -7.5 **20.** $3\frac{1}{2}$ **21.** $4 + 3x = 40; x = 12$ **22.** $10y - 40 = 60; y = 10$ **23.** $x + (x + 1) = 113; x = 56, (x + 1) = 57$ **24.** $x + (x + 50) = 330$ pounds; $x = 140, (x + 50) = 190$ pounds **25.** $2 \cdot 35 + 0.30x = \$127; x = 190$ miles

Cumulative Test Chapters 9 and 10 (page 671)

1. -10 **2.** $+8$ **3.** -48 **4.** $+14$ **5.** -35 **6.** -54 **7.** -11 **8.** -2.8 **9.** $+13$ **10.** $-\frac{1}{4}$ **11.** $3x, 4\frac{1}{2}x$ **12.** $9xyz, -xyz$ **13.** $19x$ **14.** $-23y$ **15.** $2x - 2y$ **16.** $2.7r - 3.2t$ **17.** $x = 4$ **18.** $y = 17$ **19.** $r = 2\frac{5}{8}$ **20.** $y = 28.8$ **21.** $x = 1\frac{5}{6}$ **22.** $y = 1$ **23.** $-4, -3\frac{1}{2}, -3.09, 4.1$ **24.** $x + 41 = 59; x = 18$ **25.** $\frac{1}{2} - 8x = 15; x = -1.8125$

CHAPTER 11

Section 11.1 (page 679)

1. mean = 4.5, median = 4.5, mode does not exist **3.** mean = 5.4, median = 4, mode = 4 **5.** mean = 3, median = 3, mode does not exist **7.** mean = $22\frac{2}{3}$, median = 8, mode does not exist **9.** mean = 3.45, median = 3.45, mode does not exist **11.** mean = 5.2, median = 6, mode = 6 **13.** mean = 1.5, median = 1.5, mode = 1, 2 (bimodal) **15.** mean = 7, median = 4.5, mode does not exist **17.** mean = 0.583, median = -2.5, mode = -2.5 **19.** mean = 5.0133, median = 5.02, mode = 5.02 **21.** grade point average = 3.00 **23.** mean = 627.5 miles, median = 490 miles **25.** The mean of {3, 7, 6} is greater. **27.** mean = 3.8, median = 4.5, mode = 0

Section 11.2 (page 683)

1. $1,251,000 **3.** Fourth quarter **5.** No, future years may show different patterns. **7.** 6.2%; 22.5%; 32.1%; 5.6%; 30.8%; 2.8%; 100.0% **9.** 109,900 deaths **11.**

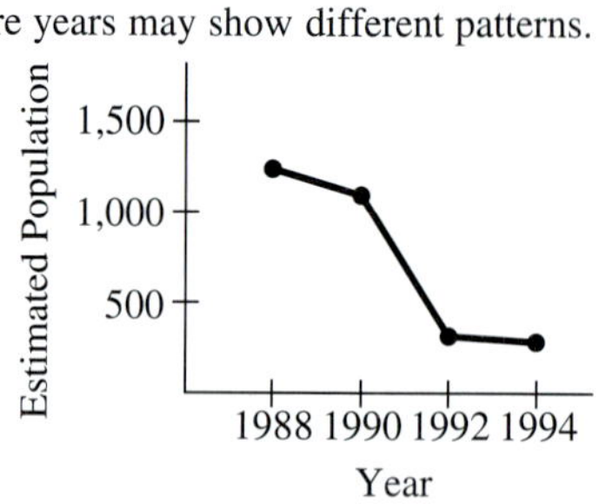

13. extinction—the state or process of something being no longer in existence **15.** That conclusion cannot be drawn from the data given. **17.** 16.5%

Section 11.3 (page 687)

1.

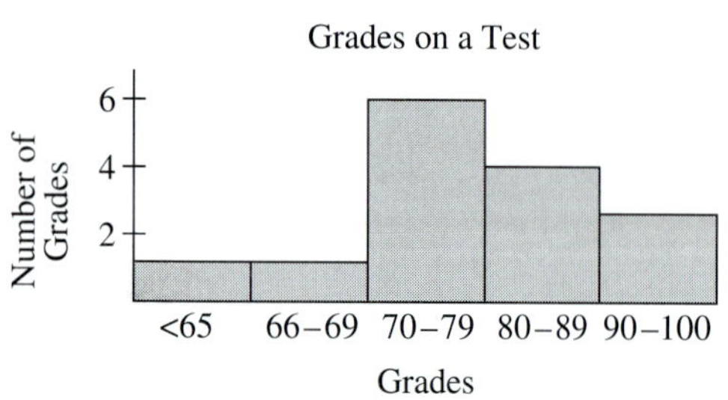

3.

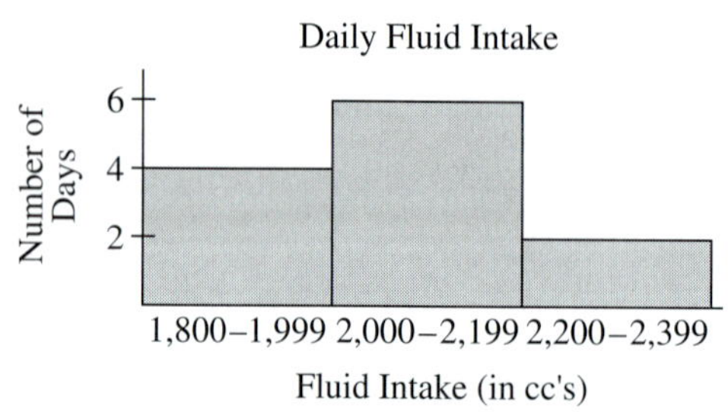

5.

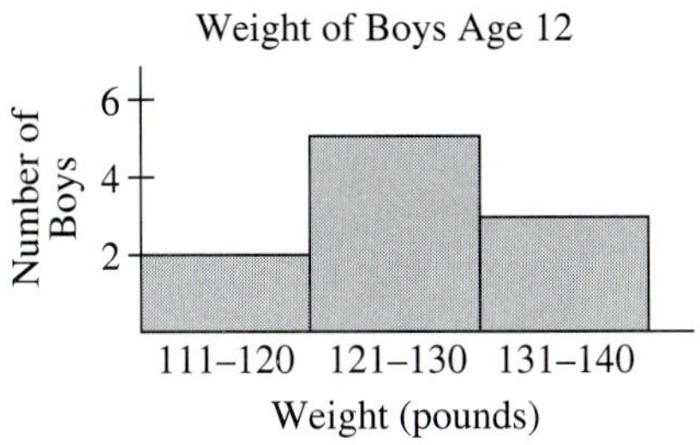

7.

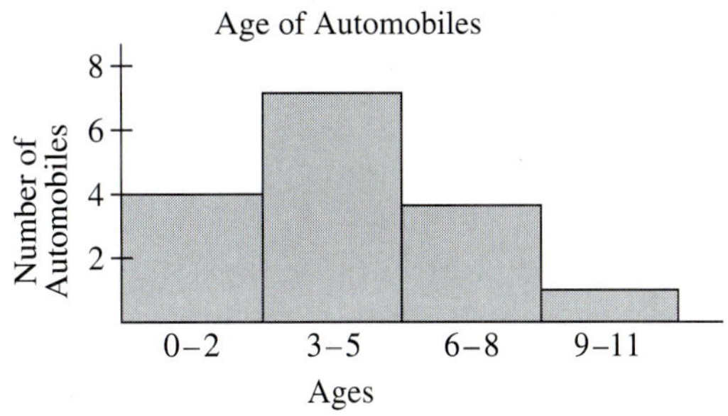

9.

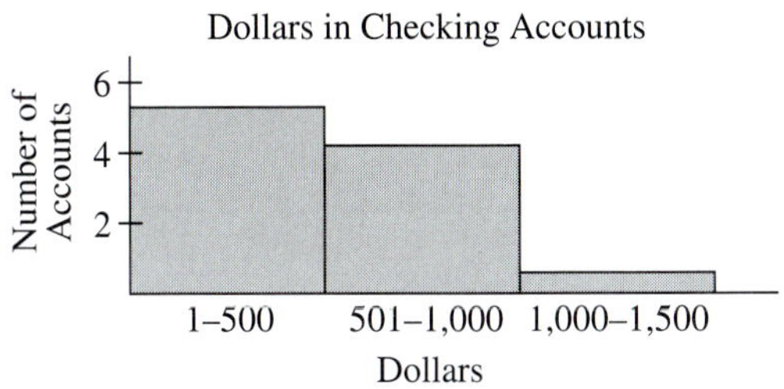

11.

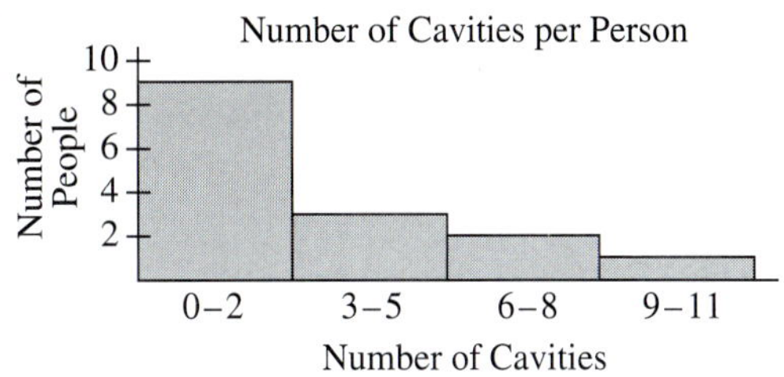

13.

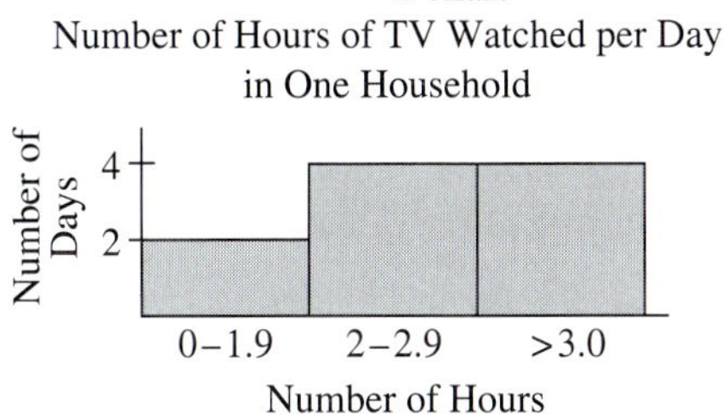

15.

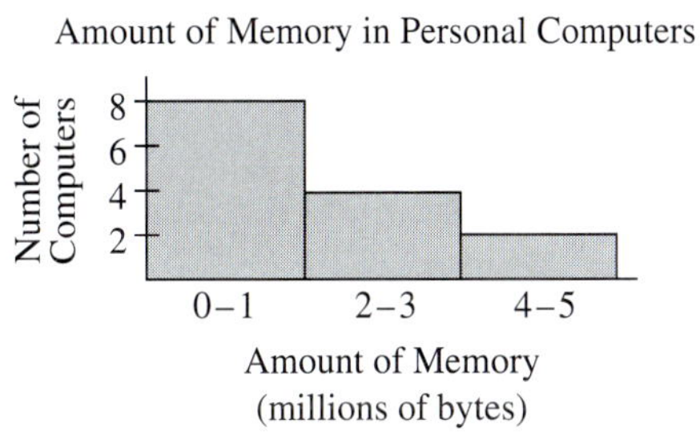

17.

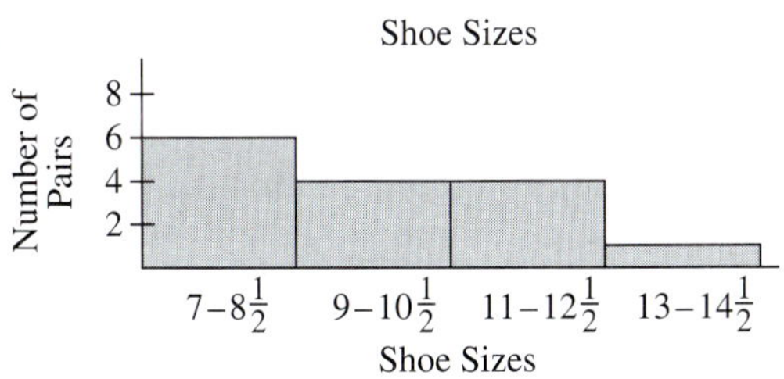

Chapter 11 Review Problems (page 697)

1. mean = 5.5; median = 5.5; mode does not exist **2.** mean = 3.5; median = 4; mode = 4 **3.** mean = 7.0; median = 7.0; mode does not exist **4.** mean = 2.4; median = 2; mode = 2 **5.** mean = 7.8; median = 6; mode does not exist **6.** mean = 4.0; median = 4.0; mode does not exist **7.** mean ≅ −7.3; median = −8.0; mode = −8.0 **8.** mean = 0.0; median = 0.0; mode does not exist **9.** mean ≅ 3.3; median = 6.0; mode does not exist **10.** mean ≅ 8.7; median = 4.0; mode does not exist **11.** mean ≅ 10.8; median = 7.5; mode does not exist **12.** mean ≅ 11.8; median = 10.5; mode does not exist **13.** mean ≅ 1.6; median = 1.5; mode does not exist **14.** mean ≅ 5.6; median = 5.1; mode does not exist **15.** mean = 9.4; median = 9.2; mode does not exist **16.** mean = 21.0; median = 7.0; mode does not exist **17.** mean = 74.9%; median = 74% **18.** yes **19.** yes **20.** greater than **21.** no **22.** no **23.** Pacific Ocean **24.** Pacific Ocean—It is the deepest and has the largest surface area. **25.** no

Chapter 11 Test (page 699)

1. mean = 7; median = 6.5; mode does not exist **2.** mean = 2.75; median = 3; mode = 4 **3.** mean = 7; median = 8; mode = 9 **4.** mean = $3\frac{2}{3}$; median = 5; mode = 5 **5.** mean = 7; median = 7; mode does not exist **6.** mean = −1; median = −1; mode does not exist **7.** The mean of 12 and 13 does equal its median, namely 12.5.

8. yes; not necessarily a mode; A mode requires two or more equal values. **9.** The two numbers are identical. **10.** 3 pounds 14 ounces **11.**

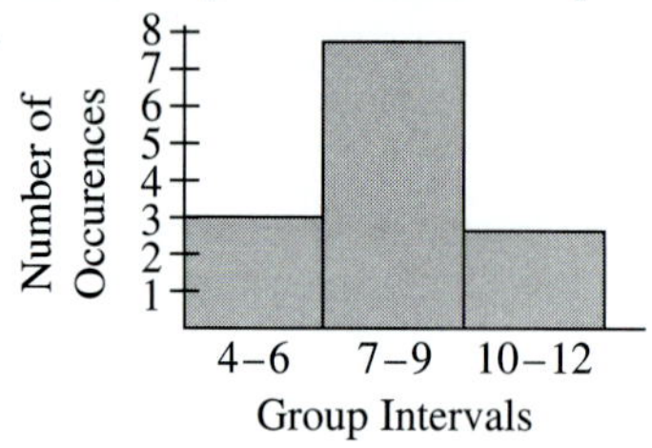

12. 4,515,000 square miles **13.** Sahara **14.** 4% **15.** no

FINAL EXAM (page 701)

1. 12,892 **2.** 75,281 **3.** 52,974 **4.** 2,100,000 **5.** $2528\frac{2}{3}$ **6.** 49 **7.** 1 **8.** 0 **9.** 0 **10.** $\frac{1}{2}$ **11.** $\frac{11}{12}$ **12.** $\frac{19}{51}$ **13.** $\frac{41}{60}$ **14.** $\frac{1}{6}$ **15.** $6\frac{2}{5}$ **16.** $\frac{1}{4}$ **17.** $7\frac{1}{2}$ **18.** $27\frac{3}{5}$ **19.** $11\frac{11}{24}$ **20.** 2 **21.** $26\frac{19}{40}$ **22.** $47\frac{2}{3}$ **23.** $74\frac{5}{8}$ **24.** $16\frac{3}{7}$ **25.** $7\frac{7}{12}$ **26.** $4\frac{17}{25}$ **27.** $3\frac{1}{3}$ **28.** -8 **29.** -12 **30.** -27 **31.** $+24$ **32.** -24 **33.** -3 **34.** $+5$ **35.** -4 **36.** 3 **37.** $23\frac{1}{3}$ **38.** -14.7 **39.** undefined **40.** 0 **41.** 10 **42.** 3 **43.** 80 **44.** 26 **45.** 94.5 **46.** -17 **47.** $-\frac{5}{8}$ **48.** $11\frac{1}{6}$ **49.** 84.64 **50.** -35 **51.** $+56$ **52.** $+2.7$ **53.** 13.65 **54.** 2,250 **55.** 20% **56.** \$268.75 **57.** 1,651,000 **58.** four million, three hundred fifty thousand, two hundred seventy-five **59.** 14,000,060,000 **60.** $x = 84$ **61.** $\frac{\$1 \text{ tax}}{\$4 \text{ earnings}}$ **62.** $6\frac{6}{7}$ miles **63.** perimeter = 150 meters; area = 800 square meters **64.** circumference = 43.96 feet; area = 153.86 square feet **65.** 72.7 pounds **66.** 85° **67.** perimeter = 30 inches; area = 30 square inches **68.** $24x$ **69.** $-7y$ **70.** $6.3x + 22y$ **71.** $x = 24$ **72.** $y = -1.875$ **73.** mean = 9.2; median = 10; mode = 10 **74.** $2 \cdot 2 \cdot 3 \cdot 5$ **75.**

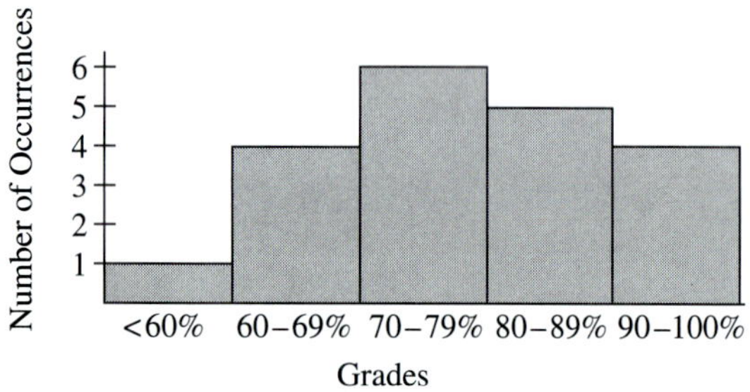

GLOSSARY

Either an example or a definition is given for each key term or topic used in the text. The section in which the topic is introduced is also given.

A

absolute value The absolute value of a number is its distance from 0 on a number line. **[9.2]**

addition property of equations The addition property of equations is the idea that the same value can be added to both sides of an equation. The result is an equivalent equation. **[10.3]**

angle An angle is formed when two lines or rays meet at a point. **[8.1]**

angle-angle rule Two triangles are similar if two angles of one triangle equal two angles of the other. **[8.6]**

associative property The associative property applies to both addition and multiplication, for example: $3 + (7 + 8) = (3 + 7) + 8$ and $4 \cdot (9 \cdot 3) = (4 \cdot 9) \cdot 3$. **[1.3]**

average The average of a group of numbers is the sum of the numbers divided by the amount of numbers. **[1.5]**

B

base The base in the percent formula is the total, or whole. **[6.4]**

C

Celsius Celsius is the temperature scale used in most countries other than the United States **[7.3]**

circle A circle is all points equally distant from a center point. **[8.3]**

circumference The circumference of a circle is the distance around the edge of the circle. **[8.3]**

commission Commission is income calculated as a percent of sales. **[6.6]**

common denominator A common denominator is a denominator common to two or more fractions. **[3.5]**

commutative property The commutative property applies to both addition and multiplication, for example: $2 + 8 = 8 + 2$ and $7 \cdot 9 = 9 \cdot 7$. **[1.3]**

complementary angle Complementary angles are two angles that add to 90°. **[8.1]**

complex fraction A complex fraction is a fraction in which the numerator, the denominator, or both are fractions. **[3.6]**

composite number A composite number is a whole number greater than one that has two or more factors, each greater than one. **[2.4]**

congruent angles Congruent angles are angles of equal measure. **[8.1]**

congruent line segments Congruent line segments are line segments of equal length. **[8.1]**

counting numbers The counting numbers are: 1, 2, 3, 4, . . . **[1.1]**

critical thinking Critical thinking is the process of understanding and finding a reasonable solution to a problem. **[1.5]**

cross multiply $\frac{5}{8} = \frac{35}{56}$

$5 \cdot 56 = 8 \cdot 35$ Cross multiply. **[3.3]**

cubed Cubed means raised to the third power. **[2.1]**

D

deductive reasoning Facts are used to work towards a solution when using deductive reasoning. **[1.6]**

degree Degree is the unit of measure used for angles. There are 360 degrees in a complete circle. **[8.1]**

denominate number A denominate number is a number followed by a unit of measurement, such as 6 feet 4 inches. **[7.1]**

denominator The denominator of a fraction is the number on the bottom. **[3.1]**

diameter The diameter of a circle is twice the radius. **[8.3]**

difference The difference is the result of subtracting two numbers. **[1.3]**

distributive property $6 \cdot (3 + 9) = 6 \cdot 3 + 6 \cdot 9$ **[1.4]**

divisible One number is divisible by a second if the first divides into the second number without remainder. **[2.3]**

E

equilateral triangle An equilateral triangle is one in which all three sides have the same length. **[8.2]**

estimate An estimate is an approximate answer. **[1.4]**

even numbers The even numbers are: 2, 4, 6, 8, 10, 12, . . . **[1.1]**

expanded notation Expanded notation is the process of writing a number as the sum of its place value pieces. **[1.1]**

exponent $5^2 = 5 \cdot 5$ (2 → exponent) **[2.1]**

expression $3y + 5x$ **[1.8]**

F

factor A factor is a value that is being multiplied times another value. **[1.3]**

factor tree

```
  18
  / \
 3 · 6
 /  / \
3 · 2 · 3   prime factorization
```

[2.4]

Fahrenheit Fahrenheit is the temperature scale used in the United States. **[7.3]**

forms of one Forms of one are fractions in which the numerator equals the denominator. **[3.1]**

G

greatest common divisor The greatest common divisor of two numbers is the largest whole number that divides both without remainder. **[2.5]**

H

hexagon A hexagon is a six sided figure. **[8.2]**

histogram A histogram is a bar chart used to show the frequency with which events occur. **[11.3]**

hypotenuse The hypotenuse of a right triangle is the side opposite the 90° angle, which happens to be the longest side. **[8.7]**

I

identity property The identity property applies to both addition and multiplication, for example: $6 + 0 = 6$ and $7 \cdot 1 = 7$. **[1.3]**

improper fraction An improper fraction is one in which the numerator is greater than or equal to the denominator. **[3.1]**

inductive reasoning Patterns are used to work towards the solution when using inductive reasoning. **[1.6]**

inequalities $6 < 8$, $7 \geq -5$, or $7 \neq 3$ **[9.2]**

integers The integers are: . . . -3, -2, -1, 0, $+1$, $+2$, $+3$, . . . **[9.1]**

isosceles triangle An isosceles triangle is one in which two sides have equal length. **[8.2]**

K

keyword A keyword (or phrase) is a word (phrase) in a word problem that helps identify which math operation to use. **[1.2]**

L

least common multiple The least common multiple of two numbers is the smallest whole number that is evenly divisible by both. **[2.5]**

like signs Two numbers have like signs if both are positive numbers, or both are negative numbers. **[9.3]**

like terms Two terms are like terms if they have the same variables raised to the same powers. **[10.1]**

line A line is a one-dimensional figure of infinite length. **[8.1]**

line segment A line segment is a portion of a line that ends with two endpoints. **[8.1]**

M

markdown An item has been marked down if its price has been reduced. For example, an item was marked down from \$80 to \$65. **[6.7]**

markup Markup refers to the increase in price between the cost and sale price of an item. For example, an item costing \$35 was marked up to \$69. **[6.7]**

mean The mean is another word for average. It is found by adding the numbers together and dividing by the number of numbers. **[11.1]**

median The median of an odd number of numbers is the number in the middle. The median of an even number of numbers is the average of the two numbers in the middle. **[11.1]**

metric system The metric system is a system of measurement based on meters, liters, and grams. **[7.2]**

mixed number A mixed number is a whole number plus a fraction, for example $5\frac{1}{4}$. **[4.1]**

mode The mode of a group of numbers is the number that occurs more often than any other number. A group of numbers may have more than one mode. A group of numbers is said to have no mode if no number occurs more than once. **[11.1]**

multiplication property of equations The multiplication property of equations is the idea that the same value can be multiplied times both sides of an equation. The result is an equivalent equation. **[10.4]**

N

number line 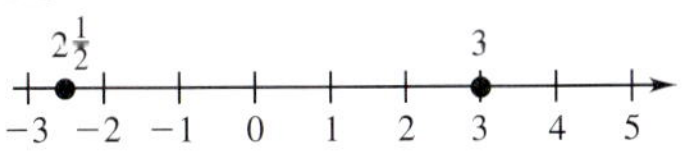

[1.2]

numerator The numerator of a fraction is the number on the top. **[3.1]**

numerical coefficient $\underline{7.6}xy$ **[10.1]**
→ numerical coefficient

O

odd numbers The odd numbers are: 1, 3, 5, 7, 9, 11, . . . **[1.1]**

order Order is the idea that a number is either less than, equal to, or greater than another number. **[1.2]**

order of operations The order of operations are the rules used to evaluate expressions. **[2.2]**

P

parallel Two lines that are always the same distance apart are said to be parallel. **[8.2]**

parallelogram A parallelogram is a quadrilateral in which opposite sides are parallel. **[8.2]**

pentagon A pentagon is a five sided figure. **[8.2]**

percent Percent is the number out of one hundred. **[6.1]**

percent formula

Percent of Whole		=	Amount
Rate ·	Base	=	Amount
R ·	B	=	A

[6.4]

perfect square A perfect square is a number whose square root is a whole number. **[8.7]**

perimeter The perimeter of a figure is the distance around the outer edge of the figure. **[8.3]**

period 26,478,265 **[1.1]**
→ millions period (26)
→ thousands period (478)

perpendicular Two lines that meet at a right angle are perpendicular. **[8.2]**

pictograph A pictograph is a chart showing information with pictures. **[2.6]**

pie chart A pie chart is a chart shaped like a pie. **[2.6]**

place value 4,563
→ ones place value
→ tens place value
→ hundreds place value
→ thousands place value **[1.1]**

point A point is a location or position. **[8.1]**

polygon A polygon is a two-dimensional figure with three or more sides. **[8.2]**

power of ten The first few powers of ten are: 10, 100, 1,000, and 10,000. **[1.4]**

prime factorization The prime factorization of a number is the unique group of primes whose product equals that number. **[2.4]**

prime number A prime number is a whole number greater than one that has exactly two factors: itself and the number 1. **[2.4]**

product The product is the result of multiplying two numbers. **[1.4]**

proper fraction A proper fraction is one in which the numerator is less than the denominator. **[3.1]**

proportion A proportion is an equation in which two ratios are equal to one another. **[4.5]**

Pythagorean Theorem In a right triangle, the length of the hypotenuse squared equals the sum of the squared lengths of the other two sides: $c^2 = a^2 + b^2$ **[8.7]**

Q

quadrilateral A quadrilateral is a four sided figure. **[8.2]**

quotient The quotient is the result of dividing one number by another. **[1.5]**

R

radius The radius of a circle is the distance from the center to the circle. **[8.3]**

rate The rate in the percent formula is the percent. **[6.4]**

ratio A ratio is a comparison of two numbers. **[4.4]**

ray A ray is a line with one endpoint. **[8.1]**

reciprocals The reciprocal of $\frac{3}{4}$ is $\frac{4}{3}$. **[3.4]**

rectangle A rectangle is a parallelogram with four right angles. **[8.2]**

rectangular pyramid A rectangular pyramid is a three-dimensional figure shaped like a pyramid. **[8.5]**

rectangular solid A rectangular solid is a three-dimensional figure shaped like a box. **[8.5]**

reduce $\frac{6}{8} = \frac{3}{4}$ **[3.2]**

remainder The remainder is the amount left over after division. **[1.5]**

rhombus A rhombus is a parallelogram with equal sides. **[8.2]**

right angle A right angle is one that measures 90°. **[8.1]**

right circular cylinder A right circular cylinder is a three-dimensional figure shaped like a soda can. **[8.5]**

right triangle A right triangle is a triangle with one right angle. **[8.7]**

rounding 26,821 rounded to the nearest thousand is 27,000 **[1.2]**

S

signed numbers Signed numbers are numbers with either a + or a − sign, such as +9.5 or $-3\frac{1}{2}$. **[9.1]**

similar figures Two geometric figures are similar if one is an enlargement of the other. **[8.6]**

simplify To simplify is to write a quantity in a simpler way, usually with fewer symbols. **[4.2]**

square A square is a rectangle with sides of equal length. **[8.2]**

squared Squared means raised to the second power. **[2.1]**

square root $\sqrt{64} = 8$, $\sqrt{121} = 11$ **[8.7]**

statistics Statistics is the study of data. **[11.1]**

straight angle A straight angle is an angle measuring 180°. **[8.1]**

sum The sum is the result of adding two or more numbers. **[1.3]**

supplementary angle Supplementary angles are two angles that add to 180°. **[8.1]**

T

term $3x + 5$ **[1.8]** ($3x$ → term; 5 → term)

trailing zeros 6.52000 **[5.2]** (000 → trailing zeros)

trapezoid A trapezoid is a quadrilateral with two parallel sides. **[8.2]**

U

unit factor A unit factor is a fraction equal to 1 in which the numerator is equal to the denominator. **[7.1]**

unlike signs Two numbers have unlike signs if one is positive and the other is negative. **[9.3]**

unlike terms Two terms are unlike terms if they have different variables or if the variables are raised to different powers. **[10.1]**

U.S. System The U.S. System is a system of measurement used in the United States. **[7.1]**

V

variable A variable is a letter that represents a number. **[1.8]**

W

whole numbers The whole numbers are: 0, 1, 2, 3, 4, 5, . . . **[1.1]**

INDEX

E

F

G

H

P

Q

R

S

T

U

V

W

X

Y

Z